TOXIC PLANTS AND
OTHER NATURAL TOXICANTS

Toxic Plants and Other Natural Toxicants

Edited by

Tam Garland and A. Catherine Barr

Texas A&M University
College of Veterinary Medicine
and
Texas Veterinary Medical Diagnostic Laboratory
College Station, Texas 77843, USA

Associate Editors
Joseph M. Betz
US Food and Drug Administration
Center for Food Safety and Applied Nutrition
Washington, DC 20204, USA

John C. Reagor
Texas Veterinary Medical Diagnostic Laboratory
College Station, Texas 77843, USA

Assistant Editor
E. Murl Bailey, Jr.
Texas A&M University
College of Veterinary Medicine
College Station, Texas 77843, USA

CAB INTERNATIONAL

CAB INTERNATIONAL
Wallingford
Oxon OX10 8DE
UK

Tel: +44 (0)1491 832111
Fax: +44 (0)1491 833508
E-mail: cabi@cabi.org

CAB INTERNATIONAL
198 Madison Avenue
New York, NY 10016-4314
USA

Tel: +1 212 726 6490
Fax: +1 212 686 7993
E-mail: cabi-nao@cabi.org

A catalogue record for this book is available from the British Library,
London, UK.
A catalogue record for this book is available from the Library of Congress,
Washington DC, USA.

ISBN 0 85199 263 3

Printed in the UK by Biddles Ltd, Guildford and King's Lynn from copy
supplied by the editors.

Contents

Preface x
Acknowledgments xi
Abbreviations xii
Dedication to Dr Alan Seawright xiii

Regional and Epidemiological Issues
 1 Recently Encountered Poisonous Plants of Rio Grande do Sul
    and Uruguay 1
 2 Tribute to Dr Wayne Binns (1911-1994) 6
 3 Genesis to Genesis: A Historic Perspective of Plant Toxicology 9
 4 Poisonous Plant Names 13
 5 A Tribute to Dr James W. Dollahite 17
 6 Locoweed Research to Reduce Poisoning on Western US Rangelands . 19
 7 Use of a Native Insect to Deter Grazing and Prevent
    Poisoning in Livestock 23
 8 Estimating Lethal Dose of Larkspur for Cattle 29
 9 Control Technology Selection for Poisonous Plants with
    the EXSEL Expert System 35
10 Management and Edaphic Factors Related with the Incidence of
    Marsh Ragwort 40
11 Management, Environmental and Livestock Interactions Impact on
    Perennial Ryegrass/*Neotyphodium*/Livestock Associations 45
12 Development of an Immunoassay for Corynetoxins 49
13 [^{14}C]Senecionine Production in *Senecio vulgaris* Root Culture 55

Cardiopulmonary System
14 Hematological Changes in Sheep with Experimentally
    Produced Acute and Chronic Lupinosis 62
15 Atypical Pneumonia Associated with Ryegrass Staggers in Calves 69
16 Dissimilatory Metabolism by Ruminal Microbes:
    Impact on Ruminant Toxicoses 73
17 Yew (*Taxus* spp.) Poisoning in Domestic Animals 78
18 Pasture Mediated Acute Bovine Pulmonary Emphysema
    and Edema in British Columbia 81
19 A Study of Persin, the Mammary Cell Necrosis Agent from Avocado
    (*Persea americana*), and its Furan Derivative in Lactating Mice 86

20 Bioassay-directed Isolation of Active Principle from
   Lavendula stoechas 91
21 Analysis of Gossypol in Whole Cottonseed and Rumen Fluid by HPLC . 97
22 The Excretion of Minerals from Broilers Fed Tannic Acid,
   Polyethylene Glycol and Protein 101
23 Influence of Tannic Acid on Amino Acid Digestibility in Broilers 106
24 Effect of Polyethylene Glycol on the Chelation of
   Trace Elements by Tannins 111
25 The Use of Copper Boli for Cattle Grazing High-molybdenum Forage . 115
26 A Newly Discovered Toxic Plant, *Stemodia kingii*, in Western Australia 120
27 Site and Sequence Selectivity of DNA Binding by Divalent Cadmium:
   Evidence for Inhibition of BPDE-induced DNA Lesions in the Major
   Groove but Not in the Biologically Relevant Minor Groove 125
28 Monitoring of Physiological and Pathological Changes in Turkey
   Poults Fed Leaves of Potentially Cardiomyotoxic
   Nerium oleander and *Persea americana* 131
29 Enzootic Geophagia and Hepatitis of Calves and the Possible
   Role of Manganese Poisoning 137
30 The Feeding of Lupin Grain Can Cause Rumen Acidosis and Rumenitis 143
31 Effects of Dietary Cottonseed Meal and Cottonseed Oil on
   Nutrient Digestibility and Rumen Protozoa 149
32 Detoxification Potential of a New Species of Ruminal Bacteria that
   Metabolize Nitrate and Naturally Occurring Nitrotoxins 154

Diagnostic/Treatment Issues
33 Vitamin K1 Therapy of *Ferula communis* variety *brevifolia*
   Toxicosis in Sheep 159
34 Development of a Vaccine Against Annual Ryegrass Toxicity 165
35 Conditioned Food Aversions to Oxalic Acid in the Food Plants
   of Sheep and Goats 169
36 Fungi Microbiota and Oxalate Formation on *Brachiaria brizantha*
   and their Possible Toxic Effects on Calves Kept at Pasture
   without Supplementation of Calcium and Magnesium 174
37 Protective Effects of Cyclodextrins on Tunicaminyluracil Toxicity . . . 179
38 Pyrrolizidine Alkaloid Detoxification by an Ovine Ruminal
   Consortium and its Use as a Ruminal Supplement in Cattle 185
39 A Comparison of a Nursling Rat Bioassay and an ELISA to
   Determine the Amount of Phomopsins in Lupin Stubbles 191
40 Towards a Commercial Vaccine Against Lupinosis 196
41 Long-term Effects of Acute or Chronic Poisoning by
   Tunicaminyluracil Toxins 201
42 Experimental Modification of Larkspur (*Delphinium* spp.) Toxicity . . . 205
43 Treatment and Prevention of Livestock Poisoning:
   Where to from Here? 211

44 Toxicity and Diagnosis of Oleander (*Nerium oleander*)
    Poisoning in Livestock 215
45 Are there Toxic and Non-toxic Varieties of *Eupatorium
    rugosum* Houttuyn? 220
46 The Chemical Identification of Plant Toxins in Ingesta and Forage ... 223
47 Conditioned Food Aversion: A Management Tool to
    Prevent Livestock Poisoning 227
48 Development and Validation of a Multiresidue Alkaloid Screen 233
49 Tolerance in Cattle to Timber Milkvetch (*Astragalus miser* var.
    serotinus) due to Changes in Rumen Microbial Populations 239
50 Fourier Transform Infrared Spectroscopy for Rapid Analysis of
    Toxic Compounds in Larkspurs and Milkvetches 243

Miscellaneous Topics

51 DNA Alkylation Properties of Dehydromonocrotaline: Sequence
    Selective N7 Guanine Alkylation and Heat and Alkali-resistant
    Multiple Crosslinks 249
52 Acute and Chronic Toxicity Induced by Activated Ptaquiloside in
    Rats: Comparison of Pathogenesis due to Oral and Intravenous
    Administrations 255
53 Immunotoxic Effects of Selenium in Mammals 260
54 *In vivo* and *in vitro* Effects of L-Canavanine and L-Canaline 267
55 The Isolation of a Hepatotoxic Compound from *Eupatorium
    adenophorum* (Spreng) and Some Preliminary Studies on its
    Mode of Action 271
56 Glycosidase Inhibitors in British Plants as Causes of Livestock
    Disorders 276
57 The Immunologic and Toxic Effects of Chronic Locoweed
    (*Astragalus lentiginosus*) Intoxication in Cattle 285
58 Sawfly (*Perreyia flavipes*) Larval Poisoning of Cattle, Sheep
    and Swine 291

The Reproductive System and the Embryo

59 Comparison of the Reproductive Effects of *Baptisia australis,
    Iva annua* and *Sophora nuttalliana* in Rats 297
60 Livestock Poisoning by Teratogenic and Hepatotoxic Range Plants ... 303
61 Pine Needle (*Pinus ponderosa*) and Broom Snakeweed
    (*Gutierrezia* spp.) Abortion in Livestock 307
62 Comparative Study of Prenatal and Postnatal Monocrotaline
    Effects in Rats 312
63 Embryotoxic Effect of *Plumeria rubra* 317
64 Evaluation of the Toxicity of *Solanum malacoxylon*
    During the Perinatal Period 323

65 Induction of Mammary Gland Carcinoma, Monocytosis and
   Type II Pneumonocyte Proliferation by Activated Ptaquiloside 329
66 Ingestion of Nitrate-containing Plants as a Possible Risk Factor
   for Congenital Hypothyroidism in Foals 334
67 Diterpene Acid Chemistry of Ponderosa Pine and Implications
   for Late-term Induced Abortions in Cattle 339
68 Toxic and Teratogenic Piperidine Alkaloids from *Lupinus,*
   Conium and *Nicotiana* Species 345
69 Toxic Amines and Alkaloids from Texas Acacias 351
70 A Urushiol Component Isolated from the Australian Native
   Cashew (*Semecarpus australiensis*) 356

Public Health and Herbal Medicine Concerns

71 Plant Toxicology and Public Health: Critical Data Needs and
   Perspectives on Botanical (Herbal) Medicines 359
72 Public Health and Risk Assessment 363
73 Perspectives on Plant Toxicology and Public Health 367
74 Evaluation of the Occurrence of Algae of the Genus *Prototheca*
   in Cheese and Milk from Brazilian Dairy Herds 373
75 Shamans and Other Toxicologists: Current Approaches to the
   Discovery of New Pharmaceuticals in Traditional Herbal Medicines .. 377

The Musculoskeletal and the Neurological System

76 Morphologic Studies of Selenosis in Herbivores 380
77 Chronic Selenosis in Ruminants 389
78 Effects of Paxilline, Lolitrem B and Penitrem on Skeletal and
   Smooth Muscle Electromyographic Activity in Sheep 397
79 Tall Larkspur Poisoning in Cattle Grazing Mountain Rangeland 402
80 Krimpsiekte, a Paretic Condition of Small Stock Poisoned by
   Bufadienolide-containing Plants of the Crassulaceae in South Africa .. 407
81 A Toxicity Assessment of *Phalaris coerulescens*: Isolation of a
   New Oxindole 413
82 Two New Alkaloids of *Conium maculatum*, and Evidence for a
   Tautomeric Form for "γ"-Coniceine 419
83 Probable Interaction between *Solanum eleagnifolium* and
   Ivermectin in Horses 423
84 *Ipomoea carnea*: The Cause of a Lysosomal Storage Disease in
   Goats in Mozambique 428
85 Investigation of the Neurotoxic Compounds in *Asclepias*
   subverticillata, Western-whorled Milkweed 435
86 Isolation of Karwinol A from Coyotillo (*Karwinskia humboldtiana*)
   Fruits 440
87 The Toxicity of the Australian Cycad *Bowenia serrulata*
   to Cattle 447

88 Evaluation of the Toxic Effects of the Legumes of Mimosa
   (*Albizia julibrissin*) and Identification of the Toxicant 453

Mycotoxins and Mycotoxicoses
89 The Development of Lupinosis in Weaner Sheep Grazed on
   Sandplain Lupins 459
90 Mycotoxin Contamination of Australian Pastures and Feedstuffs 464
91 Control of the Mycotoxic Hepatogenous Photosensitization,
   Facial Eczema, in New Zealand 469
92 Occurrence of *Fusarium moniliforme* and Fumonisins in
   Australian Maize in Relation to Animal Disease 474
93 Equine Leukoencephalomalacia in Brazil 479
94 Isolation of an Extract of *Corallocytostroma ornicopreoides* sp. nov.,
   which Causes the Rumenitis Associated with Black Soil Blindness . . . 483
95 *Corallocytostroma ornicopreoides*: A New Fungus Causing a
   Mycotoxicosis of Cattle Grazing Mitchell Grass Pastures in Australia . 487
96 Ergotism and Feed Aversion in Poultry 493
97 Experimental Black Soil Blindness in Sheep 497

Detoxification, Digestion and Elimination
98 Transient Hepatotoxicity in Sheep Grazing *Kochia scoparia* 504
99 *Calliandra calothyrsus* Leaf Extracts Effects on Microbial
   Growth and Enzyme Activities 509
100 Chinaberry (*Melia azedarach*) Poisoning in Animals 514
101 The Purification and Isolation of Two Hepatotoxic Compounds
    from the Uruguay Sawfly *Perreyia flavipes* 517
102 Toxicity of Monocrotaline (Pyrrolizidine Alkaloid) to the
    Liver of Chicken Embryos 522
103 Experimental Poisoning in Rabbits Fed *Senna occidentalis* Seeds 527
104 Toxicity and Molecular Shape of Pyrrolizidine Alkaloids 531
105 Molecular Interactions of Pyrrolizidine Alkaloids with Critical
    Cellular Targets 537
106 Factors Influencing Urinary Excretion of Immunoreactive
    Sporidesmin Metabolites in Sheep Dosed with Sporidesmin 543
107 Disease in Cattle Dosed Orally with Oak or Tannic Acid 549
108 *Jatropha curcas* Toxicity: Identification of Toxic Principle(s) 554
109 Species Differences in Bioactivation and Detoxification of
    Pyrrolizidine Alkaloids 559
110 Bog Asphodel (*Narthecium ossifragum*) Poisoning in Cattle 564
111 Livestock Poisoning by *Baccharis coridifolia* 569
112 *Narthecium ossifragum* Associated Nephrotoxicity in Ruminants 573

Index 577

Preface

This book embodies the fully edited and refereed proceedings of the 5th International Symposium on Poisonous Plants (ISOPP®5), which assembled in San Angelo, Texas, USA, 18 May-23 May, 1997.

The International Symposium on Poisonous Plants owes its existence primarily to two visionary, scientific gentlemen who realized the value of sharing knowledge and interest in many aspects of toxic plants. These leaders, Alan Seawright of the Queensland Poisonous Plant Committee and Lynn F. James of the US Poisonous Plant Research Laboratory, began collaborating in 1974 to form the US/Australian Poisonous Plant Symposium. The founding fathers decided that this exchange of knowledge would have some rules to guide future collaborations. They made it clear that groups that hosted future meeting must adhere to the guidelines established in 1974. Those six guidelines are: 1) the meetings are dedicated to the exchange of information on effects of poisonous plants on livestock; 2) scientific contributions should be relevant to poisonous plants and livestock; 3) poisonous plant research should be greatly advanced, by providing scientists with an opportunity to exchange information about the plant intoxications of their respective regions; 4) location must be one in which the relationship between poisonous plants and the local range conditions can be demonstrated to the attendees; 5) a priority is to keep the cost to the attendees at a reasonable minimum, including housing, meals and registration; and 6) the organizing committee will be responsible for the quality and content of the meeting, and its proceedings, and must adhere to the stated goals.

These founding fathers watched their fledgling collaboration establish itself in the scientific community. In 1989, it was proposed that the US/Australian alliance become an international meeting. In 1993, the term for the International Symposium on Poisonous Plants was coined, ISOPP®. The founders have attended these meetings and smiled at the advancements in poisonous plant research and awareness, the result of their alliance.

This symposium brought together veterinary and human health professions, agricultural scientists, botanists, chemists, biochemists, regulatory agency representatives, and extension personnel from around the world to discuss the latest developments in this area of research and regulatory surveillance. The scientific research is both fundamental and applied, acknowledging and examining all aspects of plant-associated toxins with relevance to animal health and productivity, safe food production - especially as it affects human health - and improved productivity and commercial returns for agriculture. Poisonous plants represent a tremendous economic impact with an estimated annual impact in the millions of dollars for both Australia and the US and surely for other countries.

The chapters in this book cover aspects of plant poisoning and mycotoxicoses such as 1) the detection, isolation and identification of chemical compounds, biosynthesized by plants and microbes, that are injurious to mammalian health; 2) investigation of the biochemistry of plant-associated toxins to elucidate mechanisms of action; 3) the investigation of toxin-induced adverse effects, which may include fatality, reduced productivity, fetotoxic effects such as spontaneous abortions, deformities and organ-specific toxicity; 4) development of management protocols to reduce the social and economic impact of toxic plants, microbes and other natural toxicants. These protocols might include appropriate eradication methods, immunization programs, controlled animal access to toxin sources, suitable withholding periods to allow for metabolic detoxification, and regulatory advice concerning toxin-residue contamination of agricultural produce.

Every contribution to this book has been fully, mercilessly and ruthlessly edited to conform to consistency of style, presentation and scientific integrity of this work. The strictest of editing was necessary to keep this book within acceptable size limits. Some details may have been omitted to accomplish this, but the emphasis is on awareness of the problems, treatments and solution more than experimental methodology. Authors' addresses and appropriate references are included to facilitate additional inquiries.

The Editors

Tam Garland and A. Catherine Barr
Texas A&M University System
College Station, Texas 77843

Acknowledgments

The Editors thank their Associate Editors for their assistance and tireless hours in refereeing the chapters of this book. All the editors and associate editors thank our spouses who have endured this effort.

The 5th International Symposium on Poisonous Plants was sponsored though a grant from the Texas Agricultural Experiment Service. Thank you E. Murl Bailey, Jr for your diligent efforts in securing funding on behalf of ISOPP®5. Thanks to Merck and Company, Inc. and to Waltham for their financial support. Appreciation is expressed to Dr A. Konrad Eugester of the Texas Veterinary Medical Diagnostic Laboratory for providing resources for the preparation of this book. Further, appreciation is expressed to Dr Samuel W. Page of the US Food and Drug Administration, Joint Institute of Food Safety and Applied Nutrition for providing resources necessary for the completion of this publication.

Abbreviations

A	adenosine	HCl	hydrochloric acid
ACN	acetonitrile	HCT	hematocrit
AIDS	acquired immune deficiency syndrome	H_2O	water
		HOAc	acetic acid
$AlCl_3$	aluminum chloride	HPLC	high performance liquid chromatography
Al_2O_3	aluminum oxide		
ARS	Agricultural Research Service	H_2SO_4	sulfuric acid
AST	aspartate amino transferase	KOH	potassium hydroxide
C	cytosine	LDH	lactic dehydrogenase
CBC	complete blood count	MAFF	Ministries of Agriculture, Fisheries and Food
$CHCl_3$	chloroform		
CNS	central nervous system	MCHC	mean corpuscular hemoglobin concentration
CO_2	carbon dioxide		
CPK	creatine phosphokinase	MCV	mean corpuscular volume
CSF	cerebral spinal fluid	MeOH	methanol
DCM	dichloromethane	$MgSO_4$	magnesium sulfate
DM	dry matter	MS	mass spectrometry
DNA	deoxyribonucleic acid	N_2	nitrogen
ECG	electrocardiogram	NaOH	sodium hydroxide
EDTA	ethylene diamine tetraacetic acid	Na_2SO_4	sodium sulfate
		NH_4OH	ammonium hydroxide
EtOAc	ethyl acetate	NMR	nuclear magnetic resonance
EtOH	ethanol		
Et_2O	diethyl ether	RBC	red blood cells
FDA	Food and Drug Administration	RNA	ribonucleic acid
G	guanine	T	thymidine
GABA	gamma-aminobutyric acid	TCA	trichloroacetic acid
GC	gas chromatography	TLC	thin layer chromatography
GC/MS	gas chromatography/mass spectrometry	USDA	United States Department of Agriculture
GGT	gamma-glutamyl transferase	UV	ultraviolet
GI	gastrointestinal	WBC	white blood cells

Dedication to Dr Alan Seawright

A tribute to Alan Seawright - veterinarian, pathologist, chemist, research scientist, teacher, administrator, scholar, and more for his contributions to science and, particularly on this occasion, his research on the effects of poisonous plants on livestock. In addition we acknowledge his contribution in the initiation of ISOPP and his continued support of it. His contributions have been crucial to its success.

In seeking information about Dr Alan Seawright, I received a letter from Australia conveying some historical data. The letter in part said, "This is not as complete as I would like, but his (Alan's) mobility in recent years has made it difficult to obtain exact detail, he tends to regard this kind of information as private," but then that is expected as Alan Seawright is a modest person. In preparing this tribute to Dr Alan Seawright, I have respected this feeling. I have a résumé of Alan's and let me say that it is impressive. I will not attempt to list these accomplishments but I would like to talk about Alan Seawright the man. To that end I have taken an article written by Don Ornduff for the *Hereford Journal* (pp. 332-336, July, 1983) as a format to describe a complete and outstanding scientist and then show how Dr Seawright fits into that description. Alan, I hope you will like it. I believe you will.

"As most experienced purebred breeders know, a sire is best measured by his get. His own appearance - his individuality - tells us what he seems to be... [His] pedigree tells us what he ought to be. But only his performance tells us what he is. Unfortunately, as a leading animal scientist long ago pointed out, these three aspects - individuality, pedigree and performance - do not always coincide." In Alan's case individuality, degrees (or pedigree) and performance do coincide. Success in any area calls for traits such as integrity, good character, etc.; he has prepared himself well to succeed as a teacher and research scientist by long years of study in veterinary medicine, pathology, chemistry, and toxicology. Further, he understands the need to use these tools for the betterment of society. Alan is a man of integrity, indeed, and he has rendered great service to the society in which he lives."

In outlining the problem that faces anyone who attempts to improve a herd or breed of livestock, Sanders wrote: (Ornduff, 1983) "An artist, in modeling in plastic clay or conjuring with marble, brings forth a conception that the world acclaims as a triumph. He deals, however, with his materials direct, and they respond instantly to his slightest touch, as he toils toward a preconceived ideal. There is no resistance to his manipulations.

"What, then, should be our estimate of the work of one who has first to conceive the figure in his brain; whose only tools are the laws of heredity, selection, inbreeding, outcrossing, and alimentation; whose only materials are flesh and blood, unapproachable except by indirection; who battles ever against the stubborn forces of atavism or reversion to ancestral forms; who seeks, and succeeds in producing, a creature pulsating with life, exquisitely fashioned, down to the minutest detail, not only a thing of beauty in itself - which artists try, sometimes with ill success, to reproduce on canvas or in bronze - but a creation that serves as well the highest utilitarian purpose. The breeder of animals directs the spark of life itself. The possibilities of his art are almost infinite."

As Sanders points out, "Those who would stand out from the crowd must begin by conceiving in mind the image of the animals which they want. Truly, the men of history who have stood out as leaders have been men of vision. Their imagination created in their respective fields images the goals which they sought."

Alan Seawright is truly a man of vision. He has been able to visualize the goals which he has set thus allowing him to go straight to them. He knows where he is going and how to get there.

The importance of an ideal and a goal is well recognized. "This is the first requisite for outstanding success in any field, be it work or play. As time passes, other ideals and refinements come to the fore... This implies that goals and ideals change. They do, indeed, as man directs.

"No ideal, however, no matter how perfect, will ever realize itself. It must be powered by that priceless ingredient - enthusiasm. It has been said, and truly, that the principal difference between successful men and mediocre can be measured to a large extent in terms of enthusiasm. In everyday terms this means nothing more or less than hard work, done ungrudgingly.

"It is a recognized fact that most men succeed in whatever line of work undertaken, not because they are particularly brilliant, but rather because they are willing to pay the price in order to reach a goal they deem worthy. A dull person of no imagination is unlikely to accomplish anything, no matter how much labor he puts into it; but is just as unlikely that a brilliant person could do any better if he approaches his task without enthusiasm. Real success... as in any worthwhile goal, requires both imagination and enthusiasm as prime ingredients.

"The famous American poet, Emerson, once wrote that 'Nothing great was ever achieved without enthusiasm.' Enthusiasm is more than "Lawrence Welk Show bubbles" - Alan can show that kind of enthusiasm, but he also understands the need for hard work, and work is no stranger to him.

The outstanding person has three outstanding characteristics. "First, he will have an active and creative imagination or vision, together with the necessary enthusiasm to put it to work. Second, he will be guided by an insatiable curiosity... and, third, he will possess the courage to chart and the tenacity to hold true to his course... without yielding to frustration and the enticement of nostrums and short cuts" (Ornduff, 1983).

In a tribute to Alan, Roger Kelly said, "He has enormous energy and enthusiasm. I was impressed by his energy and the high expectations he had of himself and others." He also pointed out that, "Alan has, as much as any person, retained the urgent sense of curiosity and excitement about science..." One only needs to be around Alan for a short period of time to feel the energy, enthusiasm, and curiosity that has carried him so far in his research and the other activities in which he has been involved.

I have often said that one of the greatest burdens that can be placed on a person is to give him or her an academic degree - PhD, DVM, MD, etc. Alan has been awarded these kinds of degrees and carries them with humility, grace, and confidence. He understands fully the implications of service to his fellow beings that accompanies their bestowal and has rendered these services in full measure.

I have another veterinary friend who had a little saying that is appropriate here. It was reserved for special friends and Alan is special in more ways than his excellence as a scientist, teacher, and administrator. When he said it, he put his hand on your shoulder and with lower jaw slightly protruded would say, "I love you like a brother." Alan, I and others here today say to you, "We love you like a brother." Love is best defined as sincere concern and respect for another person. Alan, we feel that deep sense of respect for you.

And last I read in the letter from Australia, a "vision of Alan in his 90s, in a wheel chair, maybe, but with a cage of mice at his feet and a microscope beside him." Now my vision of the future. I have a sheepman friend in his early 90s who, when asked how he was doing, responded, "Good, I herded my own sheep this summer. When I started, I could only walk a little way then had to rest, but before long I could walk all day. You know, when you quit, you die." Alan, die we will, but until then, quit we won't.

We have examined many of the features that characterize successful people. Alan fits them all. Those of us who have had the opportunity to meet and associate with him are better because he has passed our way. Alan, we appreciate what you stand for, the high ideals you represent, what you have accomplished, and, above all, your friendship, for which we thank you.

Lynn F. James

P.S. I still have a fond and warm feeling for the welcome given me by you and other members of the Queensland Poisonous Plant Committee when I first visited Australia in August 1974. I appreciate the opportunity to have worked with you on the Poisonous Plant Symposium and getting to know you. The cooperation on these endeavors has been so easy and natural. We wish you the very best in your continued efforts to advance the cause of science and also in your personal endeavors.

Chapter 1

Recently Encountered Poisonous Plants of Rio Grande do Sul and Uruguay

F. Riet-Correa[1], R. Rivero [2], F. Dutra[3], C.D. Timm[1] and M.C. Méndez[1]

[1]Laboratório Regional de Diagnóstico, Facultade de Veterinária, Pelotas University, 96010-900, Pelotas RS, Brazil; [2] DILAVE Miguel C. Rubino, Casilla de Correo 57037, CP 60.000, Paysandú, Uruguay; [3] DILAVE Miguel C. Rubino, Rincón 203, Treinta y Tres, Uruguay

Introduction

Toxic plants affecting livestock in the Brazilian state of Rio Grande do Sul and in Uruguay have been reviewed (Riet-Correa *et al.*, 1993; Riet-Correa *et al.*, 1994). This chapter reports other plant intoxications observed between 1992 and 1996.

Halimium brasiliense in Sheep

Poisoning by *Halimium brasiliense* in sheep is characterized by transient seizures with muscular tremors, ventroflexion of the neck, opisthotonos, nystagmus, tetanic spasms and limb paddling movements. The intoxication has been observed in the municipality of Rio Grande in Rio Grande do Sul, Brazil, and in the departments of Lavalleja, Maldonado, Cerro Largo, Durazno and Treinta y Tres, in Uruguay. The illness is seasonal with most cases occurring from August to November. Morbidity varies (1-15%) but some farmers report up to 50% in years when drought conditions prevailed. When affected sheep are removed from the paddocks after the observation of signs, mortality is 1-5%. In drought conditions, on some farms without this practice, it is as high as 35% (Riet-Correa *et al.*, 1995).

Macroscopic lesions are not significant. The main histologic lesion is vacuoles, sometimes containing macrophages or axonal residues, in the white matter of the brain and spinal cord. Characteristic electron-microscopic lesions are axonal degeneration followed by ballooned myelin sheaths with the disappearance of the axoplasm. A pigment identified as ceroid-lipofuscin is present in neurons, astrocytes, Kupffer cells, and macrophages of the spleen and lymph nodes. Feeding trials in sheep illustrated that the disease is caused by the ingestion of *H. brasiliense* in amounts from 2,100 to 3,000g/kg (Riet-Correa *et al.*, 1995).

Aeschynomene spp. in Swine

Intoxication by rice contaminated with 13% seeds of *Aeschynomene* spp. was seen in swine. Clinical signs included uncoordinated gait, falling, and difficulty rising followed by lateral recumbency. Reflexes, sensibility and limb movements were conserved, as was appetite. The animals recovered 2-14d later. Five pigs were fed a commercial ration of rice containing 13% *Aeschynomene* spp. seeds and showed clinical signs similar to those observed in the spontaneous cases, 24hrs-6d after ingestion began. Two pigs recovered 6d after removal of contaminated feed. The others were euthanized. Histologic lesions were characterized by symmetrical focal degeneration in the cerebellar nuclei, the vestibular nuclei and, to a lesser extent, in white matter. The initial lesion, observed 24hrs after ingestion, was a vacuolation of the neuropil, which progressed with time to a loss of parenchyma, vascular reaction, a few small spheroids, astrocytosis and filling with gitter cells. In some lesions, a peripheral rim of viable neurons from the cerebellar or vestibular nuclei was seen at the margin of the lesion or even just within it. Astrocytosis was more prominent in one pig euthanized 15d after apparent recovery. Large vacuoles, sometimes containing macrophages or axonal residues were often observed and blood vessels were surrounded by numerous gitter cells. Pigs fed food containing 3% and 6% seeds for 25d did not develop clinical signs (Timm, 1996).

Nierembergia hippomanica in Cattle

Intoxication by *Nierembergia hippomanica* has been frequently diagnosed in cattle in North-west Uruguay. Morbidity is 10-80% and deaths do not occur. Most outbreaks are observed in milking cows or in 3-4yr old steers. The intoxication occurs any time of the year from January to November. All outbreaks tend to occur in cultivated pastures or in wheat or barley stubble fields. Clinical signs are salivation, diarrhea, restlessness, abdominal pain and periodic motion of the head and limbs. Cows have a decreased milk production. Affected animals recovered within 1wk after removal from infested pastures (Odini *et al.*, 1995).

The lowest toxic dose of green plant administered to cattle and sheep was 10-15g/kg. Clinical signs were similar to those observed in field cases. Most animals recovered in 1-8d, except one calf that died after the ingestion of 50g/kg. The main lesions were focal hemorrhages in the large intestine and enteritis in the small intestine. The dried plant was not toxic to cattle and sheep. One steer that received ten daily doses of 5g/kg showed clinical signs after the last dose, demonstrating a cumulative effect of the plant (Odini *et al.*, 1995).

Two sheep that received 20g/kg of the plant developed anorexia, diarrhea, abdominal pain, restlessness and excessive salivation (Odini *et al.*, 1995). A pyrrole-3-carbamidine has been identified as the toxic principle of *N. hippomanica* (Buschi and Pomilio, 1987).

Anagallis arvensis in Cattle and Sheep

Four cases of *Anagallis arvensis* poisoning were diagnosed in the Department of Paysandú, Uruguay during December, 1994, and January, 1995, in barley and wheat stubble fields. Cattle of different ages were affected. Morbidity was 7-30% and case fatality was 50-86%. In two cases the animals had been introduced in the stubble fields 7-15d before the observation of clinical cases. In two others the animals were in the fields 30-45d before developing clinical signs. Eight of 289 ewes died after grazing in the same field that had affected cattle. Another case was observed in sheep and cattle with calves with no clinical signs in the calves.

Clinical signs included anorexia, restlessness, weight loss, hemorrhagic diarrhea, muscular tremors and convulsions. Serum urea, creatinine and magnesium were increased. Clinical manifestation period was 2-15d.

Gross lesions were characterized by petechial hemorrhages and edema of the mesentery, presence of clear yellowish fluid in the cavities, erosive and ulcerative lesions in the esophagus, hemorrhagic abomasitis and enteritis, perirenal edema and yellowish discoloration of the kidneys. The main histologic lesion was a severe tubular nephrosis.

The disease was reproduced in two sheep with *A. arvensis*. One died after being dosed with leaves at the rate of 40g/kg daily for 4d. The other died after the ingestion of 175g/kg divided into six daily doses of 27g/kg and one of 13.5g/kg. Clinical and pathologic findings were similar to those in the spontaneous cases.

Xanthium strumarium in Cattle

Xanthium strumarium poisoning has been observed in Rio Grande do Sul and Uruguay during spring (September and October). It occurs in sandy soils after floods at the borders of rivers or creeks. One or 2wks following the return of normal water levels there is a massive germination of the plant and the animals eat enough of the newly germinated seedlings in their cotyledonary stage to became intoxicated. Mortality varies between 3-82% (Méndez *et al.*, 1994).

Clinical signs, observed a few hours after ingestion of the plant, include opisthotonos, depression, muscle fasciculation, increased respiratory and cardiac rates, sternal or lateral recumbency and terminal paddling movements 12-24hrs following the onset of clinical signs. At necropsy, the livers were swollen and dark red, and the gall bladder walls were edematous. The pleural and peritoneal cavities contained a yellowish fluid. Petechiae and ecchymoses were seen on serous membranes. Dry feces with blood or mucus were often observed in the rectum. The livers had a hemorrhagic centrilobular necrosis that occasionally extended to the periportal hepatocytes (Méndez *et al.*, 1994). Experimentally, intoxication was reproduced in calves dosed with 7.5-10g/kg of cotyledons (Méndez *et al.*, 1994).

Cycas revoluta in Cattle

An outbreak of acute intoxication by *Cycas revoluta* was observed in Uruguay in September, 1995. Two bulls exhibited signs of aggressiveness, incoordination and diarrhea, 7-10d after being introduced into an area where *C. revoluta* had been cultivated. The livers were swollen, dark red and mottled. The gall bladder wall, the mesentery, and the abomasum wall were edematous. Hemorrhages were observed in the digestive tract. Microscopically there was centrilobular liver necrosis. Hepatocytes of the midzonal and periportal regions were vacuolated. The disease was reproduced in a calf given 51g/kg of *C. revoluta*.

Lantana camara in Cattle and Sheep

Two cases of *Lantana camara* poisoning in sheep and one in cattle were observed in North-west Uruguay. Mortalities of 87% and 33% were observed in two flocks. Both outbreaks occurred after the flocks and a few cows were placed in parks where *L. camara* had been cultivated. Sheep stayed in the paddocks for 24hrs.

Clinical signs in sheep and cows were evidenced by severe photodermatitis, anorexia, restlessness, jaundice, brown urine, weight loss, ruminal stasis, excessive salivation, lacrimation, and occasionally keratitis. Serum GGT and AST were increased. Some sheep died 24-48hrs after the onset of signs, but in most animals the clinical manifestation period can vary from 5-20d.

Jaundice, subcutaneous yellow edema and swollen, ochre-colored livers with distended and edematous gall bladders were observed at necropsy. Microscopically the liver had severe vacuolization of periportal hepatocytes and mild proliferation of bile duct cells. A mild tubular nephrosis was also observed.

Erechtites hieracifolia in Cattle

Intoxication caused by *Erechtites hieracifolia* was observed in East Uruguay in March, 1993, in a herd of 120 yearling cattle. Twelve affected animals died. Clinical signs were characterized by progressive weight loss, abdominal straining, protracted scouring and prolapse of the rectum. At necropsy there was excessive abdominal fluid, edema of the mesentery and the wall of the abomasum, and a pale, hard liver with an enlarged and edematous gall bladder. Microscopic lesions of the liver were characterized by diffuse fibrosis, megalocytosis and proliferation of bile duct cells. The plant contained 0.2% pyrrolizidine alkaloids.

Nerium oleander in Cattle

Nerium oleander is an ornamental plant found commonly in Uruguay and southern Brazil. An outbreak of the intoxication was observed in North-west Uruguay in a paddock where an oleander plant was trimmed. Eighty 2yr-old heifers were introduced in the area and five of them died 24-48hrs after being placed in the paddock. Some animals were found dead. Others had clinical signs characterized by depression, weakness, anorexia, ataxia and diarrhea. No significant lesions were observed at necropsies. Oleander leaves were found in the rumen.

References

Buschi, C.A. and Pomilio, A.B. (1987) Pyrrole-3-carbamidine: a lethal principle from *Nierembergia hippomanica*. *Phytochemistry* 26, 863-865.

Méndez, M.C., Santos, R.C. and Riet-Correa, F. (1994) Intoxicação por *Xanthium* sp. (carrapicho) em bovinos. *Boletim do Laboratório Regional de Diagnóstico* 14, 27-30.

Odini, A., Rivero, R., Riet-Correa, F., Mendez, M.C. and Giannechinni, E. (1995) Intoxicación por *Nierembergia hippomanica* en bovinos y ovinos. *Veterinária Uruguay* 30, 3-12.

Riet-Correa, F., Méndez, M.C. and Schild, A.L. (1993) *Intoxicações por plantas e micotoxicoses em animais domésticos*. Editorial Hemisferio Sur, Montevideo.

Riet-Correa, F., Méndez, M.C., Barros, C.S.L. and Gava, A. (1994) Poisonous plants of Rio Grande do Sul. In: Colegate, S.M. and Dorling, P.R. (eds), *Plant-Associated Toxins: Agricultural, Phytochemical and Ecological Aspects*. CAB International, Wallingford, Oxon, pp. 13-18.

Riet-Correa, F., Méndez, M.C., Pereira Neto, O., Soares, M.P., Vieira, M.A., Silva, E.A., and Soares, M.P. (1995) Intoxicação por *Halimium brasiliense* em ovinos. *Boletím do Laboratório Regional de Diagnóstico* 15, 32-37.

Timm, C.D. (1996) Intoxicação por *Aeschynomene* species em Suinos. MSc Thesis, Faculdade de Veterinária, Universidade Federal de Pelotas, Brazil.

Chapter 2

Tribute to Dr Wayne Binns (1911-1994)

D.J. Wagstaff[1] and L. James[2]

[1]*FDA, Center for Food Safety and Applied Nutrition, Washington, DC 20204, USA;* [2]*USDA-ARS, Poisonous Plant Research Laboratory, Logan, Utah 84321, USA*

Tribute is paid to Dr Wayne Binns (1911-1994) and to those things for which he stood. He was born on 20 July, 1911, to John Binns and Ellen Thornton Binns at their home in the small rural Utah town of American Fork. At birth, his heart was flawed by a patent foramen ovale. Although the lesions changed over time, his heart was his Achilles' heel. His father was away much of the time tending their sheep either on the summer mountain range or on the winter desert range. When Wayne was nine years old he saw his father suffer a fatal heart attack and fall from the wagon where he had been unloading bundles of grain.

His widowed mother lost the debt-ridden sheep business but kept the family home, where she reared her seven children by hard work and high principles. Times were tough but a large garden, chickens, and a few livestock supplied much of their food. The boys delivered newspapers both before and after school.

Young Wayne loved school and always would. The school song of the red brick high school on a hill overlooking town spoke of "a beacon through the night." He carried the light as student body president.

Veterinary medicine was one consideration as a career path but he was undecided as he pursued general studies during his freshman year at the University of Utah. Still undecided the following year, he enrolled at Utah State Agricultural College in the School of Forestry. Indecision led to discouragement; he decided to quit school. When he told his landlord that he was leaving, they had a long chat. As a result he talked to the head of the Veterinary Science Department and his career path became clear. He traveled far but his roots remained in the West.

In 1934, he was accepted into the veterinary school at Iowa State College. Lacking funds for a bus or railroad ticket he rode there in a sheep train. During the first two years he lived and worked at the college poultry farm. He held a second job washing dishes. In the last two years he lived at the veterinary clinic, where he also worked, and continued to wash dishes. To earn extra money, he enrolled in Reserve Officer Training Corps for 25 cents a day. It was a thrill for his mother to see him graduate in 1938. He typified the "Gentle Doctor" as portrayed by the Christian Peterson statue. He was to follow in the footsteps of L.H. Pammel of Iowa State who wrote the first American book about poisonous plants. After graduation he was employed for two years in a mixed practice in Illinois.

In 1940, he returned to the West, and on May 15, married Lucille Sparrow of Ogden, Utah. Their honeymoon was spent going to St Anthony, Idaho, where he was briefly employed by a farmers' cooperative, but could not resist a job in the Veterinary Science Department at Utah State. Within two years he became head of the department. That same year, 1942, his first child, Ralph, was born.

He was inducted into the Army, inspecting meat and other foods in Salt Lake City. A transfer to the Northwest Service Command Laboratory at Edmonton, Alberta, Canada, was next. He learned, like other soldiers, the frustration of "hurry up and wait." His third post was in Vancouver, Washington. From there he went to Camp Crowder, Missouri, prior to shipping out to Tinian in the Marianas command hospital where he worked, prepared to receive the massive casualties from the expected invasion of Japan. Only later did he realize that history was unfolding before his eyes. That B29 on the runway that was suddenly roped off carried the bomb that ended the war.

By the time he was discharged in 1946, he was a father for the second time: a girl named Ann. In the postwar decade he did what he liked best: teach and advise students at Utah State University. He counseled preveterinary students about the need for them to take rigorous classes in science as the foundation to develop their ability to diagnose disease. Diagnosis was for him the highest application of the veterinary art. He cared for his students, became their friend, and maintained contact with many for the rest of his life. He wanted them to be good citizens.

His love of learning carried him "Far above Cayuga's waters." Although his work for a masters degree at Cornell was mainly clinical in nature, he was exposed to the spirit of Liberty Hyde Bailey, Walter Muenscher, and John Kingsbury, whose works with horticultural and poisonous plants are monumental.

Not long after returning to Utah State, he suffered a heart attack. But through the love of his family, his faith in God and the skill of his physicians, the flame brightened again. Although his health forced him to leave the faculty, another path opened up. He became Head of the USDA Poisonous Plant Research Laboratory.

The legacy he inherited was rich and bright. C. Dwight Marsh started a field station at Salina Canyon, Utah to research sheep spewing sickness caused by *Helenium hoopesii* and other plant problems. The researchers were headquartered in Washington, DC, which was a long commute to their remote summer station. Over time the bright vision of Chesnut and Marsh was dimmed.

When Dr Binns took over the laboratory, he fanned the spark; the work was expanded and based on solid science. The laboratory was moved to Logan, Utah, near USU. He assembled a team of scientists of complementary backgrounds. The laboratory and each of the scientists became world renowned. Field observations, experimental exposures and laboratory study were combined to address livestock plant poisoning problems. Collaborative work was done in other laboratories and universities.

Numerous plant problems such as those caused by *Lupinus* spp., *Halogeton* spp., and *Delphinium* spp., were investigated. One of the main animal disease problems the laboratory addressed on its broadened horizons was a congenital sheep defect in certain flocks in Idaho. Microbial and genetic causes were eliminated and theories shifted to environmental factors. In the summer of 1958, efforts centered on an experimental sheep herd he established at Summit Flat in the Boise National Forest. None of the plants fed that summer produced the disease, but the diligence paid off. A casual remark made by a Basque sheepherder to a summer student employee about the toxicity of *Veratrum californicum* led to observations that, contrary to published statements, sheep on the local range ate this plant. *Veratrum* collected that fall as they left Summit Flat produced the condition in sheep fed at the laboratory in Logan. This was a scientific breakthrough. Almost all plant poisonings reported before that were acute. In contrast, the team proved that a factor that acted at a distant time and place caused the deformity. He was given the Superior Service Award by the Secretary of Agriculture in 1962 for his work.

Because of his work in veterinary toxicology he was one of three veterinarians chosen by the American Veterinary Medical Association to form the American Board of Veterinary Toxicology in 1967. They administered the first examination and certified the successful applicants.

Once again his heart weakened; after cardiac bypass surgery at age 62, he was advised to retire. This was painful, but after a period of adjustment he found gratifying service in his church and in his community. He and his wife Lucille filled a mission for the Church of Jesus Christ of Latter-day Saints in New York City. This was expensive, but they felt it was worthwhile. They later spent several years as officiators in the Logan LDS Temple.

He authored or coauthored 82 publications. He was honored as an invited speaker at several meetings including some in Europe. But of all the honors he received, the only one that appears in his autobiography was presented by little people who knew him not as Doctor or Professor or Elder. The poster made by kindergarten children he tutored said simply, "Thanks, Grandpa Binns."

The greatest tribute to a person is to strengthen the legacy received from them. The Poisonous Plant Research Laboratory in Logan is one of the premier institutions of its type in the world. The board he helped found has thrived. The light carried by Dr Wayne Binns still shines. And now as we bid farewell to our friend and colleague, the words come to mind from the school song of his Cache Valley Alma Mater, "Peace rests over all."

Chapter 3

Genesis to Genesis: A Historic Perspective of Plant Toxicology

D.J. Wagstaff

FDA, Center for Food Safety and Applied Nutrition, Washington, DC 20204, USA

Introduction

Plant toxicology has a rich history, a subset of the human experience. History study helps us avoid mistakes, prevent duplication of research, gain new insights, and improve cooperation by sharing a common legacy. Toxicology is one of the oldest sciences. Aboriginals have long used toxic plants to stun fish and to poison arrows.

The First Genesis

Present knowledge rests upon a foundation of ancient writings from the Levant (Eastern Mediterranean) and successive European cultures. The Eden account in Genesis shows that plants are both good and bad. All except one plant in the garden were good. Consumption of the forbidden fruit of the Tree of Knowledge of Good and Evil brought death into the world (Genesis, Holy Bible). After entering into this terrestrial state humans were prevented from eating of the fruit of another tree special to God, the Tree of Life. Writings as distant as Mesoamerica and the Indian Ayruvedic speak of a tree of life. God has resided on the holy mountain of Sinai, and other gods have had their mountains, Olympus and beyond, from whence they descend to commune with mortals in sacred groves and gardens. Each of the 12 gods of Olympus has a special tree; for Zeus it is the oak. The good/evil duality of plants is seen around the world. Norse gods have a tree of destiny which grows towards heaven but whose roots are eaten by the devil.

As God and other gods have their special trees, so too humans have special plants that permit them to escape the mundane for recreational or spiritual purposes. The effect may be mild euphoria or violent convulsions, heightened awareness or deep trance. According to Homer, the sorceress Circe turned Ulysses' companions into swine by giving them "stupefying juices." Solanaceous plants, one of the groups of psychoactive plants, have been used this way worldwide. People intoxicated by Solanaceae are vulnerable to suggestion and can be made to believe that they are animals. This may be the basis for the Homeric account. In the werwolf tradition a person who has used a particular solanaceous salve acts like a ferocious wolf. The aphrodisiac use of mandrake, another solanaceous plant, may owe its effect to increased susceptibility to suggestion. Chewing henbane root helped oracles speak the will of gods. Peruvian Indians endured violent convulsions brought on by a drink made of *Datura sanguinea* fruit so they could "speak to the mighty gods confidentially." Various solanaceous plants were used to increase the intoxicating effect of alcoholic drinks, which led Germany to ban use of henbane seeds in beer (Fuhner, 1926).

Power and learning were early centered in Egypt and the fertile crescent. Surrounded by desert, the people realized the importance of plants and had knowledge of their toxicity. Pharaoh's people saw the lotus as the source of life and viewed trees as sacred. Writings on their buildings indicate knowledge of different types of plant poisoning, e.g. phototoxicity due to *Ammi majus*.

From Moses' account of toxic quail alighting in the Israelite camp we learn that poisons (possibly *Conium maculatum*) can be passed on through meat and that the severity of the poisoning is related to dose (Exodus, Holy Bible). The concept of a dose-response relationship was known long before Paracelsus was given credit for enunciating it in the Middle Ages.

As Egypt and Babylon declined, power and learning shifted eastward and northward to Medea, Persia and Asia Minor and then westward to Greece and Rome. Despite the accumulated wisdom of the ages, people sometimes failed to learn from history, and so were doomed to suffer anew. Xenophon returning to Greece in 401 BC from a campaign against the Persians at Babylon passed near Trebizond, Asia Minor. His troops ate honey made toxic by nectar from the Pontic azalea (*Rhododendron luteum*) and were incapacitated for a period. Fortunately the harassing Colchian army did not attack, and they escaped. Despite Xenophon's description of the dose-response relationship among his poisoned soldiers, in 67 BC history was repeated. During a campaign by Pompey against King Mithridates of Pontus near the same Trebizond, Roman soldiers who were intoxicated by the local honey were attacked and slaughtered (Baumannn, 1993).

The ancients not only recorded effects of the environment on themselves but also noted their impact on the environment. Homer saw devastating forest fires and pasturing by goats that prevented regrowth. Plato mentioned deforestation to supply lumber for buildings. Warfare consumed enormous quantities of wood. Xerxes used 4,207 ships in his campaign against the Greeks. After shipwreck prevented a marine invasion, he built a wooden bridge across the Dardanelles. It is no wonder that much of the Mediterranean region is now treeless. Bracken and asphodel have replaced oak and yew in that area. Cropping, grazing, logging and urban expansion are just a few of the ways by which deforestation continues. In Britain, bracken invades lands that were logged off centuries ago.

The classical period of plant knowledge occurred during the Age of Reason, which was ushered in by the Greek philosophers supported by the Hellenic Empire of Alexander the Great. Some physicians of the day, notably Hippocrates, insisted that herbal products and other drugs should only be used to treat diagnosed medical conditions. He understood the principle of dose response, for as he said regarding poppy sap (*Papaver somniferum*), a small amount relieves suffering, a larger amount brings drowsiness, and a still larger dose kills.

The best plant scholar of the time was Theophrastus, a disciple of Aristotle, who rejected the miracles of nature and instead used observation and reason to place botany on a scientific footing in his report *Enquiry into Plants*. The most prolific writer about toxic plants was the little-known Nicander. Unfortunately their work was not continued and little progress was made in the next 20 centuries. Roman scholars such as Pliny the Elder, Dioscorides, and Galen advanced the science a little, and then progress halted.

With the disintegration of Rome, power was fragmented among the tribes of Central Europe. Even though these cultures lacked a written language and did not contribute significantly to our scientific knowledge, they nevertheless affected what was to come after them. Their migrations through conquest or colonization greatly affected the distribution of plants. Useful plants and seeds were deliberately transported while weeds were accidentally redistributed. Their experience and attitudes governed which plants they used for food, for medicines and for affecting the psyche. This in turn determined their toxicology experiences. Plants were used as religious symbols by the clergy, as effectors of good and evil by witches, as arbitrary drugs by physicians and as weapons by murderers. Fact and fancy are difficult to disentangle in the records remaining from that period.

Southern Europe was assaulted by Moslems. During the seventh century, Islam spread out of Arabia westward across North Africa to the Atlantic and east across southern Asia to the Islands of the Southwest Pacific. The Moslem Wall broadened in Southern Europe to cover nearly all of the Iberian Peninsula and spread through the Balkans to knock at the gates of Vienna before receding.

Northern Europe was invaded by Scandinavians who, in the latter part of the first millennium expanded in three major sweeps: west across the North Atlantic, southeast through Russia and southwest along the coasts of Europe. They were eventually absorbed but their enduring influence is exemplified in the Norman (Northmen) conquest of Britain.

In eastern Europe the threat during the early part of the present millennium was from Mongolian tribesmen known as the Golden Horde. Though they attacked both Fortress Europe and the Moslem Wall with swift brutal efficiency, neither toppled.

Finally the emerging European nations went on the offensive. They suppressed their pettiness to launch crusades to retake the Holy Land. The crusaders failed to free Palestine, but acquired a taste for Moslem-controlled oriental spices. Pursuit of spices has long been a prime factor in global exploration and international relations. They were used in pre-Roman times for worship, anointing of kings and preparation of bodies. Gradually the religious aura of spices diminished and they were then employed by both priest and peasant as food flavorings. Since the Moslem Wall could not be breached, circumventing it by exploring new sea routes became the obsession of the Portuguese and later the Spanish. They became world powers as their growing navies brought them wealth from the Orient and from their colonies in the newly discovered Western Hemisphere. When these two Iberian empires weakened, power shifted north to western Europe.

Gradually during the Middle Ages, plant books called herbals were written, which contained mixtures of the plausible and the fantastic, but they were the start of a return to the principles of observation and reason. With the way paved, the study of plants was again put on a scientific footing by the Swedish physician Carolus Linnaeus. Plants were organized genealogically using floral structure as the guide without regard to medicinal or mystical uses.

Discovery of the Western Hemisphere was followed by Europeanization of the Americas. But there was a difference in the United States, where the common man was a landowner and a voter. When he demanded help from the government to stop plant poisoning of his livestock, it responded in 1894, with the employment of Victor King Chesnut by the US Department of Agriculture to study plant poisoning. With this, the botanical triumvirate of Theophrastus, Linnaeus and Chesnut was complete and plant toxicology became a recognized field. Different scientific fields cooperated to solve plant poisoning problems. Governments of many nations supported study. The amount of research grew until it plateaued in the 1970s.

The Second Genesis

The worldwide movement of plants and their problems is epitomized by the modern Genesis, a British rock music group, in the song *The Return of the Giant Hogweed*, about the importation of the giant hogweed (*Heracleum mante-gazianum*) from southern Russia into England and subsequent severe dermatologic poisonings. The light of science has flared and flickered over the ages. Its value to us in showing the path ahead is assured only to the extent that we cooperate to enhance the methods and body of knowledge that is plant toxicology.

References

Baumann, H. (1993) *The Greek World in Myth, Art and Literature.* Timber Press, Portland.
Bible, King James Version, Cambridge University Press, Cambridge.
Fuhner, H. (1926) Solanaceae as drugs: An historic-ethnological study. Naunyn-Schmiedenberg (ed), *Archives of Experimental Pathology and Pharmacology*, 111, 281-294.

Chapter 4

Poisonous Plant Names

D.J. Wagstaff[1] and J.H. Wiersema[2]

[1]*FDA, Center for Food Safety and Applied Nutrition, Washington, DC 20204, USA*; [2]*USDA-ARS, Beltsville, Maryland 20705, USA*

Introduction

Plant toxicology is changing; knowledge of poisonous plant names is important to ensure that the same name is universally applied. Both common names and scientific names are needed for communication with the public and for scientific exchange. An International Checklist of Poisonous Plants is being compiled to define the set of vascular plants reported to be toxic for humans or animals.

Checklists of poisonous plants of the world have been compiled four times in the twentieth century. The first list, *Poisonous Plants of All Countries*, was printed in 1905 by Bernhard Smith. The second checklist was published in 1911 as a section of the book *A Manual of Poisonous Plants* by the American botanist Pammel. This is the only list with fully qualified botanical names. The third effort, printed in 1923, is a revision by Bernhard Smith. The fourth list, *Poisonous Plants of the World*, was distributed in mimeograph form in 1949 (3rd revision in 1951) by Moldenke. Other sources of names of toxic plants include monographs, journal articles and books of Kingsbury (1964), Everist (1981), Cooper and Johnson (1984), Lampe and McCann (1985), and Kellerman *et al.* (1988).

From these sources, the checklist at present includes 4,672 species in 1,646 genera and 255 families. There are 11,614 common names. The botanical names are in Botanical Latin and most of the common names are in English. This list of plants reported to be toxic represents about 1% of the known vascular plants of the world.

Poisonous Plant Names

In accordance with the provisions of the International Code of Botanical Nomenclature, a botanical name has two parts: a genus and species binomial and the example, *Abrus precatorius* L. is the name published by Linnaeus for the precatory bean. A description in Botanical Latin is published for each species. The usual standard for identification of each plant species is a dried, or otherwise preserved, specimen of the plant called the "type," which is filed in an herbarium. In some cases an illustration is acceptable. Usage of a name is based upon reference to this "type" and comparison of a given plant with the type specimen. Some species have been given more than one botanical name, thus rules of nomenclature are applied to determine the name with precedence. Plant taxonomy is an evolving field, and differences in classification can result in changes of scientific names.

Whereas botanical names generally convey meaningful information, sometimes they are descriptively inaccurate or may even be whimsical. Examples of names of poisonous plants and their derivations are given in Table 4.1. Geographic place names are easily recognized for a specific epithet like *"californicum,"* but more difficult when the Latinized form is different from that of daily usage. Place names are sometimes used in a broad sense; *"canadense"* was applied to plants from the northeastern United States as well as plants from Canada. Habitat names such as *"montana"* (mountain), *"arvense"* (field) and *"convallaria"* (valley) are straightforward.

Some plants are named for animals. The similarity of bracken fern (*Pteridium aquilinum*) fronds to eagle wings is obvious as is the similarity of the tripartite fiddlehead to the claws. People have wondered how *Delphinium* looked like a dolphin. Other names commemorate people or mythologic characters. Descriptive names indicate such characteristics as the number of leaves, presence of spots and color of fruit. Some root words have special meanings, such as the "pseudo" in *pseudoacacia*. Black locust is no less legitimate in genealogy or nomenclature than *Acacia*. The term "pseudo" is just an odd way of saying that there is a resemblance.

In contrast to botanical names that are often arbitrary, common names when first coined are meaningful. Names are not standardized between groups of people and, as languages shift, these names change also. The common name "dogwood" is not derived from a canine but from the Old English word "dagge," meaning a sharp object, usually a meat skewer. Thus another name is "skewer wood." Names can convey attitudes. A plant name containing "maiden" or "saint" is viewed positively in contrast to the negative view of plants named after "dog" or "devil." Understanding a few old names aids understanding common names. Wort means plant, glob is a round flower, gowan is a yellow or white flower, and bunk is a word derived from Arabic meaning an aromatic root.

An example of name derivation is the term figwort for Scrophulariaceae. Both the common and botanical names refer to tubercles on the roots of some species. The family name comes from scrofula, a tuberculous swelling on the neck of animals. A tubercle is like a pile such as a pile of sticks. If the pile contains blood it describes the human disease known as piles or hemorrhoids. An old term for the lesion is figwart because it is shaped like a fig. In accordance with the now largely discredited Doctrine of Signatures, the figwort is used to treat figwart.

Table 4.1. Names of selected poisonous plants.

Name	Meaning	Example	Comments
canadense	of Canada	*Menispermum canadense* L.	NE US, Canada
Colchicum	of Colchis	*Colchicum autumnale* L.	Old country on Black Sea
halepense	of Aleppo, Syria	*Sorghum halepense* L.	Latinized place name
montana	mountain	*Arnica montana* L.	
arvense	field	*Equisetum arvense* L.	
Ranunculus	little frog	*Ranunculus abortivus* L.	marsh where frogs live
aquilinum	eagle	*Pteridium aquilinum* (L.) Kuhn	
Delphinium	dolphin	*Delphinium barbeyi* Huth	bud resembles a dolphin
Hyoscyamus	pig bean	*Hyoscyamus niger* L.	
Helenium	Helen of Troy	*Helenium hoopseii* L.	
Nicotiana	Jean Nicot	*Nicotiana tabacum* L.	
Mercurialis	mercury	*Mercurialis annua* L.	
triphyllum	3 leaved	*Arisaema triphyllum* (L.)	
maculatum	spotted	*Conium maculatum* L.	spots on the stem
rubra	red	*Actaee rubra* (Iton)	color of the fruit
Avena	ancient oats	*Avena sativa* L.	
Festuca	ancient fescue	*Festuca arundinacea* Schreber	
Linum	ancient flax	*Linum usitatissimum* L.	

Databases

The checklist of poisonous plant names is organized in a DataPerfect database. It contains species binomial and author, synonyms, common names, and references of toxicity in the major plant toxicology texts. In the future journal citations will also be cross-referenced to the plant names. Part of the files from the database are available in the World Wide Web (WWW) at the address HTTP://VM.CFSAN. FDA.GOV/~DJW/README.HTML.

Validity of plant names in the project is based on the GRIN database (Germplasm Resources Information Network) developed by USDA to support the archiving of germplasm materials. This database includes names of economically important and poisonous plants. It contains fully qualified botanical names, synonyms, some common names, a unique identifying code number for each taxon, reference to the Latin description or diagnosis for the plant, and pertinent taxonomic literature. The Internet address is HTTP://WWW.ARS-GRIN.GOV/NPGS/TAX.

References

Bernhard Smith, A. (1905) *Poisonous Plants of All Countries.* J. Wright and Company, London, UK.

Bernhard Smith, A. (1923) *Poisonous Plants of All Countries*, 2nd ed. Bailliere Tindall Cox, London, UK.

Cooper, M.R. and Johnson, A.W. (1984) *Poisonous Plants in Britain and their Effects on Animals and Man.* Ministry of Agriculture Fish Food Reference Book 161.

Everist, S.L. (1981) *Poisonous Plants of Australia*, revised ed. Angus and Robertson, Sydney.

Kellerman, T.S., Coetzer, J.A.W. and Naudé, T.W. (1988) *Plant Poisonings and Mycotoxicoses of Livestock in Southern Africa.* Oxford University Press, Capetown, Republic of South Afica.

Kingsbury, J.M. (1964) *Poisonous plants of the United States and Canada.* Prentice-Hall, Englewood Cliff, NJ.

Lampe, K.F. and McCann, M.A. (1985) *AMA Handbook of Poisonous and Injurious Plants.* Chicago Review Press, Chicago, IL.

Moldenke, H.N. (1951) *Poisonous Plants of the World*, 3rd ed. Mimeographed copy, Yonkers, NY.

Pammel, L.H. (1911) *A Manual of Poisonous Plants.* Torch Press, Cedar Rapids, IA.

Chapter 5

A Tribute to Dr James W. Dollahite

E.M. Bailey, Jr

Department of Veterinary Physiology and Pharmacology, College of Veterinary Medicine, Texas A&M University, College Station, Texas 77843, USA

Dr James W. (Dolly) Dollahite was born on May 1, 1911, and grew up in West-Central Texas near Johnson City, Texas. He received his Doctor of Veterinary Medicine degree from The Agricultural and Mechanical College of Texas in 1933. He worked for the US Government and practiced until World War II. He served as a US Army Veterinarian during the war and later retired as a Lieutenant Colonel in the US Air Force Reserve. Following the war, he went back into veterinary practice in Marfa, Texas, but developed an interest in toxicology. Dr Dollahite combined his practice and a part-time position with the Texas Agricultural Experiment Station's Alpine Station to further his interests in plant toxicology. He also held a research position with the USDA at Beltsville, Maryland. In 1956, he started a full-time Texas Agricultural Experiment Station position and was responsible for moving the Alpine Research Station to become the Marfa Toxic Plant Research Station, during which time he drove many miles over West Texas and Southern New Mexico investigating toxic plant problems and developing his toxic plant research. He closed the Marfa Station and moved his research endeavors to College Station in 1958, where he was in what was then Veterinary Research Section of the College of Veterinary Medicine. He received his MS Degree in Veterinary Physiology (there was no formal Toxicology Program at the time) in 1961. J.W. became an Associate Professor of Pathology in 1962 and Professor in 1965. He transferred to the Department of Veterinary Physiology and Pharmacology in 1968, where he was instrumental in establishing the PhD program in Toxicology in 1969. Dr Dollahite was a Charter and Founding Diplomate of the American Board of Veterinary Toxicology (1966-7). J.W. continued his research on toxic plants until his retirement from Texas A & M in 1975. He carried on his toxic plant research with the USDA-ARS Veterinary Toxicology and Entomology Research Laboratory, College Station, Texas, until his full retirement in 1980. J.W. died July 26, 1984. Dr J.W. Dollahite played a very important role in the development of veterinary toxicology research in Texas, especially toxic plant research, and the development of Veterinary Toxicology as a

Specialty under the American Veterinary Medical Association. Dr Dollahite's greatest attribute was his power of observation, especially for clinical signs in diseased animals. He had over 70 research publications.

Chapter 6

Locoweed Research to Reduce Poisoning on Western US Rangelands

L.F. James[1], M.H. Ralphs[1], K.E. Panter[1], B.L. Stegelmeier[1], J.A. Pfister[1] and R.J. Molyneux[2]

[1] USDA-ARS, Poisonous Plant Research Laboratory, 1150 East 1400 North, Logan, Utah 84321, USA; [2] USDA-ARS, Western Regional Research Center, Albany, California 94710, USA

Introduction

Locoweed is the most widespread group of poisonous plants in the western US. There are over 500 species of *Astragalus* and 22 species of *Oxytropis* in North America, but only 11 species have been verified as causing locoism. Many more species will be found to contain the toxic alkaloid swainsonine. Locoweed populations are cyclic, with outbreaks occurring in wet years, then die-offs in drought (Welsh, 1989). Incidents of poisoning are often catastrophic (Ralphs and Bagley, 1988). Locoweed poisoning is chronic, and signs of poisoning do not become apparent until the animal has grazed the plant for several weeks.

Toxicology

Swainsonine occurs at very low levels in locoweeds (0.01-0.3% dwt) (Molyneux *et al.*, 1989). Swainsonine is a small, sugar-like alkaloid that inhibits the enzymes α-mannosidase and mannosidase II, resulting in accumulation of unmetabolizable glycoproteins in vacuoles, and abnormal glycosylation of hormones, membrane receptors and enzymes (Stegelmeier *et al.*, 1994). Poisoning has been associated with abnormal endocrine, reproductive, immune, and gastrointestinal function. Clinical manifestations include depression, incoordination, emaciation, decreased fertility in both males and females, abortion, birth defects, retarded offspring, and congestive right-heart failure at high elevations (James *et al.*, 1981).

Swainsonine is water soluble and is rapidly absorbed from the GI tract. It circulates through the body systems, and is excreted in the urine, milk and feces. After a short withdrawal time of 5-6d, practically no toxin remains in the serum (Stegelmeier *et al.*, 1995b). Reversal of the effects of intoxication are slower, thus weeks or months may be required for recovery of cell function. Some CNS neurons are lost and cannot be replaced. These irreversible lesions make the value of seriously poisoned animals questionable. Several *in vivo* diagnostic assays of serum α-mannosidase activity and swainsonine are available (Stegelmeier *et al.*, 1995a).

Locoweed is devastating to livestock reproduction. Principal effects on the developing fetus include delayed placentation, decreased vascularization, fetal edema and hemorrhage, and alteration of cotyledon development (James, 1976). Ultrasound imaging has indicated that locoweed reduced fetal heart rate and caused cardiac irregularity (Panter *et al.*, 1987), which may contribute to fluid accumulation in the placenta (hydrops amnii or allantois). These factors contribute to fetal death, and trigger abortions (Ellis *et al.*, 1985). Locoweed ingestion by pregnant sheep also results in skeletal contractual deformities (James *et al.*, 1976), but swainsonine has not been specifically implicated.

Lambs born to ewes fed locoweed were developmentally impaired (Astorga, 1992). They were slow to get up following birth, lacked the nursing instinct, and would not seek their dams. Parturition was abnormally long, even though the lambs were smaller than average. Without assistance, all lambs from locoed ewes would have died. However, intoxicated lambs recovered quickly (Pfister *et al.*, 1993).

Swainsonine is excreted in milk (Molyneux *et al.*, 1985). Nursing calves and lambs whose mothers were fed locoweed developed lesions, as did cats fed milk from cows that consumed locoweed (James and Hartley, 1977).

Locoweed reduces fertility in both males and females. Spermatogenesis and libido were impaired in rams (Panter *et al.*, 1989), and estrus and conception rates were suppressed in ewes (Balls and James, 1973).

Desert and semi-desert locoweeds (*A. pubentissimus*, *A. lentiginosus* and *A. mollissimus* var. *earlei*) germinate following autumn rains, and either remain green over winter, or they are the first plants to green up in the spring (Welsh, 1989). Livestock generally prefer the green-growing locoweeds to other forage that is dormant. Cattle readily grazed *O. sericea* and *A. mollissimus* var. *mollissimus* in the spring, ceasing only when warm-season grasses began active growth (Ralphs *et al.*, 1993). Cattle prefer the immature seed pods of *O. sericea* that occurs on mountain range lands (Ralphs, 1987). Dry, senescent stalks of *A. lentiginosus* var. *wahweapensis* on desert winter range were more nutritious than senescent grasses, and cattle grazed them in proportion to their availability (Ralphs *et al.*, 1988).

Management Strategies

Livestock should be denied access to locoweeds during critical periods when it is more palatable than the other forage. Most locoweed species are endemic, growing only in certain habitats or on specific soils. Reserving locoweed-free pastures for grazing in the critical periods of spring and fall can prevent locoweed poisoning.

Locoweed was thought to be addictive, but preference for locoweed was shown to be relative to availability and condition of other forage species (Ralphs *et al.*, 1991). Some animals learn to prefer locoweed. Social facilitation or peer pressure is a very strong influence inducing others to eat locoweed (Ralphs *et al.*, 1994). Livestock should be removed if they start eating locoweed to prevent progressive intoxication, and to prevent them from influencing others to graze it. Ranchers should not over-stock locoweed-infested ranges, but ensure that adequate forage is available. Conditioned food aversion can be used as a management tool to train animals to avoid eating locoweed (Ralphs *et al.*, 1997).

Herbicide control of locoweed (Ralphs and Ueckert, 1988) in strategic locations can also provide locoweed-free pastures for critical times, but long-lived seed in soil will germinate and reestablish when environmental conditions are favorable (Ralphs and Cronin, 1987).

References

Astorga, J.B. (1992) Maternal Ingestion of Locoweed: Effects on Ewe-Lamb Bonding and Behavior. PhD Dissertation, Utah State University, Logan, UT.

Balls, L.D. and James, L.F. (1973) Effect of locoweed (*Astragalus* spp.) on reproductive performance of ewes. *Journal of the American Veterinary Medical Association* 162, 291-292.

Ellis, L.C., James, L.F., McMillen, R.W. and Panter, K.E. (1985) Reduced progesterone and altered cotyledonary prostaglandin values induced by locoweed intoxication in range cattle. *American Journal of Veterinary Research* 46, 1903-1907.

James, L.F. (1976) Effect of locoweed (*Astragalus lentiginosus*) feeding on fetal lamb development. *Canada Journal of Comparative Medicine* 40, 394.

James, L.F. and Hartley, W.J. (1977) Effects of milk from animals fed locoweed on kittens, calves, and lambs. *Journal of American Veterinary Research* 38, 1263-1265.

James, L.F., Hartley, W.J. and Van Kampen, K.R. (1981) Syndromes of *Astragalus* poisoning in livestock. *Journal of American Veterinary Medical Association* 178, 146-150.

Molyneux, R.J., James, L.F. and Panter, K.E. (1985) Chemistry of toxic constituents of locoweed (*Astragalus* and *Oxytropis*) species. In: Seawright, A.A., Hegarty, M.P., James, L.F. and Keeler, R.F. (eds), *Plant Toxicology*. Queensland Poisonous Plant Committee, Yeerongpilly, QLD, Australia, pp. 266-278.

Molyneux, R.J., James, L.F., Panter, K.E. and Ralphs, M.H. (1989) The occurrence and detection of swainsonine in locoweeds. In: James, L.F., Elbein, A.D., Molyneux, R.J. and Warren, C.D. (eds), *Swainsonine and Related Glycosidase Inhibitors*. Iowa State University Press, Ames, IA, pp. 100-117.

Panter, K.E., Bunch, T.D., James, L.F., and Sisson, D.V. (1987) Ultrasonographic imaging to monitor fetal and placental developments in ewes fed locoweed (*Astragalus lentiginosus*). *American Journal of Veterinary Research* 48, 686-690.

Panter, K.E., James, L.F. and Hartley, W.J. (1989) Transient testicular degeneration in rams fed locoweed (*Astragalus lentiginosus*). *Veterinary and Human Toxicology* 31, 42-46.

Pfister, J.A., Astorga, J.B., Panter, K.E. and Molyneux, R.J. (1993) Maternal locoweed exposure *in utero* and as a neonate does not disrupt taste aversion learning in lambs. *Applied Animal Behavior Science* 36, 159-167.

Ralphs, M.H. (1987) Cattle grazing locoweed: influence of grazing pressure and palatability associated with phenological growth stage. *Journal of Range Management* 40, 330-332.

Ralphs, M.H. and Bagley, V.L. (1988) Population cycles of Wahweap milkvetch on the Henry Mountains and seed reserve in the soil. *Great Basin Naturalist* 48, 541-547.

Ralphs, M.H. and Cronin, E.H. (1987) Locoweed seed concentration in soil: longevity, germination potential and viability. *Weed Science* 35, 792-795.

Ralphs, M.H. and Ueckert, D.N. (1988) Herbicide control of locoweed: a review. *Weed Technology* 2, 460-465.

Ralphs, M.H., James, L.F., Nielsen, D.B., Baker, D.C. and Molyneux, R.J. (1988) Cattle grazing Wahweap milkvetch on the Henry Mountains. *Journal of Animal Science* 66, 3124-3130.

Ralphs, M.H., Panter, K.E. and James, L.F. (1991) Grazing behavior and forage preference of sheep with chronic locoweed toxicosis suggests no addiction. *Journal of Range Management* 44, 208-209.

Ralphs, M.H., Graham, D.L., Molyneux, R.J. and James, L.F. (1993) Seasonal grazing of locoweeds in northeastern New Mexico. *Journal of Range Management* 46, 416-420.

Ralphs, M.H., Graham, D.L. and James, L.F. (1994) Social facilitation influences cattle to graze locoweed. *Journal of Range Management* 47, 123-126.

Ralphs, M.H., Graham, D., Galyean, M.L. and James, L.F. (1997) Creating aversions to locoweed in naive and familiar cattle. *Journal of Range Management* 50, 361-366.

Stegelmeier, B.L., Ralphs, M.H., Gardner, D.R., Molyneux, R.J. and James, L.F. (1994) Serum α-mannosidase and the clinicopathologic alterations of locoweed (*Astragalus mollissimus*) intoxication in range cattle. *Journal of Veterinary Diagnostic Investigations* 6, 473-479.

Stegelmeier, B.L., Molyneux, R.J., Elbein, A.D. and James, L.F. (1995a) The comparative pathology of locoweed, swainsonine and castanospermine in rats. *Veterinary Pathology* 32, 289-298.

Stegelmeier, B.L., James, L.F., Panter, K.E. and Molyneux, R.J. (1995b) Serum swainsonine concentration and alpha-mannosidase activity in cattle and sheep ingesting *Oxytropis sericea* and *Astragalus lentiginosus*. *American Journal of Veterinary Research* 56, 149-154.

Welsh, S.L. (1989) *Astragalus* and *Oxytropis*: definitions, distributions and ecological parameters. In: James, L.F., Elbein, A.D., Molyneux, R.J. and Warren, C.D. (eds), *Swainsonine and Related Glycosidase Inhibitors*. Iowa State University Press, Ames, IA, pp. 3-13.

Chapter 7

Use of a Native Insect to Deter Grazing and Prevent Poisoning in Livestock

W.A. Jones[1], M.H. Ralphs[2] and L.F. James[2]

[1]*USDA-ARS, Beneficial Insects Research Unit, Subtropical Agricultural Research Center, 2413 East Highway 83, Weslaco, Texas 78596, USA;* [2] *USDA-ARS, Poisonous Plant Research Laboratory, Logan, Utah 84321, USA*

Introduction

Tall larkspurs *Delphinium barbeyi* Nutt. and *D. occidentale* S. Wats (Ranuculaceae) are responsible for more cattle deaths due to poisoning than any other plant on mountain rangeland in the western US (Kingsbury, 1964). These plants contain complex diterpenoid alkaloids that cause acute intoxication and death from respiratory paralysis (Olsen, 1978). Widespread losses have stimulated continuing effort to develop methods that can reduce poisonings. Recently imposed restrictions in the use of herbicides have brought further pressure to develop environmentally safe methods for reducing cattle deaths due to these plants.

Attempts to manipulate plant-feeding arthropods to manage plants poisonous to livestock have not previously been reported. The introduction and release of natural enemies of exotic competitive weeds have a successful history. The managing of native, coevolved insects for managing native weedy plants has rarely been investigated (Thompson and Richman, 1989; Holtkamp and Campbell, 1995).

Among an array of insects previously found associated with tall larkspurs, *Hoplomachus affiguratus* (Uhler) (Heteroptera: Miridae), the larkspur mirid, was selected for study in association with *D. barbeyi*. Nancy Peterson first observed stunting in infested *D. barbeyi* (Ralphs *et al.*, 1997); the affected plant populations were observed to decline over time. Fitz (1972) studied biological aspects of some of the more common insects related with tall larkspurs and recorded certain aspects of the biology and damage potential of *H. affiguratus*. Subsequently, it was noted that cattle prefer not to feed on plants infested by *H. affiguratus*. Unfortunately, the occurrence and population density of the mirid can fluctuate dramatically, and most larkspur sites do not harbor damaging *H. affiguratus* populations. Thus began a series of greenhouse and field experiments to measure the effects of insect feeding on plant growth, reproduction, alkaloid levels, and cattle feeding preference. The feasibility of establishing insect populations in uninfested sites was investigated. The results were encouraging and are summarized below.

The Larkspur Mirid

Hoplomachus affiguratus is a native insect associated with the tall larkspurs *D. occidentale* and *D. barbeyi* in the western US in wooded aspen and open sub-alpine rangeland communities. The phytophagous family Miridae composes the largest family of true bugs within the order Heteroptera (Henry and Wheeler, 1988). Several members of the genus cause enough damage to commercially relevant plant species to have themselves been targets for biological control (Hedlund and Graham, 1987). Others are insectivores and thus are important natural enemies of some pest insects (Wheeler *et al.*, 1975).

Several biological attributes of *H. affiguratus* have added to the original information observed by Fitz (1972). The insect is univoltine and overwinters as diapausing eggs embedded in old stems, which collapse and remain buried under the snowpack during winter. Eggs hatch in spring and first stage nymphs eclosing from the overwintered eggs crawl from the old stems onto new, developing stems emerging from the perennial rootstock. There are five nymphal instars. The immature insects cluster and feed together, mainly on the upper surface of leaves, then move to reproductive parts as plant and insect phenologies progress during the relatively short growing season. Feeding effects on plants appeared to be related to insect density, and visible effects can be readily discerned on both leaves and fruiting plant parts (unpublished data).

The larkspur mirid will likely not control larkspur since they are co-evolved. However, this native insect may be used as a biological tool to prevent cattle from eating larkspur and thus avoid poisoning. Early observations that cattle and sheep would not eat larkspur plants that were heavily damaged by the mirid were verified. The plants' chemical defenses against insect herbivores may differ from those produced against mammalian herbivores, and reallocation to insect defense could reduce toxic alkaloid content (Chew and Rodman, 1979; Baldwin, 1993).

With most insect species, maintaining a laboratory colony is restricted only by the ability to maintain their host. However, *H. affiguratus* goes through an obligatory winter diapause in the egg stage, and thus cannot be continuously reared. To obtain specific numbers of newly emerged first instars, a means had to be devised to store eggs successfully during winter, then program them to hatch at the desired time when test plants were ready to be infested.

To test winter storage methods, several hundred stems from heavily infested plants were collected before heavy snowfall and were divided into four storage treatments: three in cold storage and kept either wet, humid or dry; the fourth treatment consisted of placement under a snowbank in the shade of a building. Stems were removed in late spring and held at room temperature. To determine if females preferred different parts of the stem to deposit their eggs, a subsample of stems were separated into thirds, representing the upper, middle and lower sections. These stems were broken into small sections and placed into Petri dishes kept at room temperature and observed daily for hatching nymphs.

The results clearly illustrated that the greatest number of hatching nymphs were from stems stored outside in the snowbank. Insects emerged within the first 3d following removal from storage. Those stored under the snowpack began emerging first and completed emergence before the survivors from the other treatments. The stems averaged 22 mirid nymphs per inch; over twice that of the second-best storage method. Cold, dry conditions yielded the least number of nymphs. More nymphs hatched from the upper third of the stems, fewest from the lower third.

Another group of egg-infested stems was gathered and placed at a site that would be more accessible the following spring. The eggs were targeted for use in a greenhouse test relating density to damage. In June, 50-75 stems with eggs on them were placed at ambient laboratory temperature in a large box. Larkspur leaves were placed among the stems as food for emerging nymphs. Resulting nymphs were placed on potted *D. barbeyi* plants in the vegetative stage (about 0.3m tall) at zero, 50, 100 and 200 per plant, ten replications each.

The results clearly showed that the insects can have a severely damaging effect on larkspur reproduction, with all three infestation levels exceeding 90% aborted heads with few stems maturing to make seed. The extent of damage was essentially equivalent between plants with 100 and 200 nymphs.

Several field experiments were conducted over a 3yr period in natural *D. barbeyi* populations using different densities and insect stages (unpublished data). The results clearly demonstrated that an almost complete cessation of plant reproduction can readily be affected through transfer of insects to young plants.

Long- and short-term effects of insect feeding on alkaloid levels were tested. Toxic and total alkaloid concentrations were measured in two larkspur populations having established mirid populations and in two manually infested larkspur populations. In the naturally infested populations both undamaged and damaged leaves and flowering heads were harvested and analyzed. In the manually infested sites individual plants were divided with screen mesh, and mirids were placed on half of each plant. Leaves and heads from the damaged and undamaged halves were harvested 3wks later. There were no consistently significant differences in toxic or total alkaloid concentration between mirid-damaged and undamaged leaves in either study. These results confirmed that mirid damage does not affect either local or systemic alkaloid content or profile.There was no significant difference in toxic or total alkaloid concentration between larkspur populations with established mirid infestations and newly infested plants. A large degree of plant-to-plant variability masked any treatment differences. Other studies have implied reduced alkaloid in damaged plants, but the results have been inconsistent.

When present, the reproductive plant parts are preferred by cattle, while leaves are readily fed upon prior to initiation of flowering (Pfister *et al.*, 1988). Mirids concentrate their feeding on leaves but migrate toward blooms as they appear.

Two experiments were conducted to verify that cattle do not prefer insect-damaged plants. A feeding trial consisted of offering cattle a choice of equal amounts of mirid-damaged and undamaged flowering plants. The damaged plants contained no detectible insect residues. Preference for undamaged plants was highly significant, with cows consuming a mean of 0.8kg of undamaged plants per 10min feeding period, compared with 0.1kg of the insect-damaged plants (Ralphs *et al.*, 1997). A grazing trial took advantage of characteristic insect behavior to learn if cattle have a feeding preference under natural field conditions. The host plant has clusters of the gregarious *H. affiguratus* feeding on it. Such clustered feeding behavior and subsequent damage typically begins on the southeast quadrant of each plant, likely due to exposure to the warmth afforded by the rising sun. The net effect is that the insects infest and damage one side of the plant before eventually moving to the other side, thus providing damaged and undamaged portions of each plant at the initiation of flowering. When each plant was about half damaged, a heavily infested larkspur population was temporarily fenced and cows were allowed to graze freely for 15d. Cows did not consume mirid-damaged portions of the plants. As the insects moved to previously unfed portions of the plants, the proportion of larkspur feeding by the cattle declined correspondingly.

Tall larkspur typically occurs in geographically discrete patches or populations, and *H. affiguratus* populations are not always associated with the host plant. Damaging insect infestations may be the exception, and insect infestations ebb and flow over years. The bases for the insect population changes are unknown. The most obvious interim solution was to harvest insects from heavily infested sites and release them in uninfested sites. This was tried in various ways over a 4yr period.

In late fall when the insects had deposited most of their eggs, egg-infested stems were harvested from heavily infested plant populations and moved into known uninfested plant populations before snowfall. In other trials, large nymphs and adults were collected in August using insect nets, and released into plant populations containing no mirids. Each subsequent summer, these release sites were sampled to learn if the transfer resulted in new insect establishments.

The multi-year assessment of manual release sites gave mixed but promising results. Some previously uninfested sites became heavily infested, while others showed no infestation. Reasons for the successes and failures were not apparent.

Conclusions to Date and Recommendations for Future Research

Findings from this series of studies indicate that cattle prefer not to eat *D. barbeyi* damaged by the larkspur mirid, *H. affiguratus*. The larkspur mirid overwinters in the egg stage; egg-infested stems can be held during winter and can be programmed to emerge in spring. Attempts to establish damaging populations of the insect in previously uninfested plant populations was met with mixed results.

The array of toxic and total alkaloids in the foliage of *D. barbeyi* is extremely variable. There is some evidence that insect feeding may reduce alkaloid levels, but between-plant variability masks detection of a definite trend.

Life table studies are required to identify the key factors that drive the fluctuation in *H. affiguratus* populations from year to year. Previous studies of mirids show that natural enemies are major factors in adjusting populations (Clancy and Pierce, 1966; Jackson and Graham, 1983). Conventional wisdom has asserted that native plants cannot be regulated by native, coevolved herbivores. Simple manipulation of the system may not measurably reduce plant populations but still decrease cattle poisonings.

The unique possibility and appeal of the *H. affiguratus* system is that there is actually no "biological control" at all, but rather a resultant non-preference for cattle feeding due to the presence of a native insect. Future research should be focused on identifying the key factors that cause fluctuations in insect populations. Current studies include determining the role of natural enemies and weather events. Additional studies are needed to ascertain the factors involved in determining why some artificial insect populations can be established manually while others cannot.

References

Baldwin, I.T. (1994) Chemical changes rapidly induced by folivory. In: Bernays, E.A. (ed), *Insect-Plant Interactions*. CRC Press, Inc., Boca Raton, FL, pp.1-23.

Chew, F.S. and Rodman, J.E. (1979) Plant resources for chemical defense. In: Rosenthal, G.A. and Janzen, D.H. (eds), *Herbivores: Their Interactions with Secondary Plant Metabolites*. Academic Press, New York, NY, pp. 271-308.

Clancy, D.W. and Pierce, H.D. (1966) Natural enemies of some lygus bugs. *Journal of Economic Entomology* 59, 853-858.

Fitz, F. K. (1972) Some Ecological Factors Affecting the Distribution and Abundance of Two Species of Tall Larkspur in the Intermountain Region, with Thoughts on Biological Control. PhD Dissertation, Utah State University, Logan, UT.

Hedlund, R.C. and Graham, H.M. (eds), (1987) *Economic Importance and Biological Control of* Lygus *and* Adelphocoris *in North America*. USDA-ARS 64.

Henry, T.J. and Wheeler, A.G. Jr (1988) Family Miridae Hahn, 1833. In: Henry, T.J. and Froeschner, R.C. (eds), *Catalog of the Heteroptera, or True Bugs, of Canada and the Continental United States*. E.J. Brill Publishers, Leiden, The Netherlands, pp. 251-507.

Holtkamp, R.H. and Campbell, M.H. (1995) Biological control of *Cassinia* spp. (Asteraceae). In: Delfosse, E.S. and Scott, R.R. (eds), *Proceedings of the Eighth International Symposium on Biological Control of Weeds*. Lincoln University, Canterbury, NZ, pp. 447-449.

Jackson, C.G. and Graham, H.M. (1983) Parasitism of four species of *Lygus* (Hemiptera: Miridae) by *Anaphes ovijentatus* (Hymenoptera: Mymaridae) and an evaluation of other possible hosts. *Annals of the Entomological Society of America* 76, 772-775.

Kingsbury, J.M. (1964) *Poisonous Plants of the United States and Canada*. Prentice-Hall, Englewood Cliffs, NJ, pp. 131-140.

Olsen, J.D. (1978) Tall larkspur poisoning in cattle and sheep. *Journal of the American Veterinary Medicine Association* 173, 762.

Pfister, J.A., Ralphs, M.H. and Manners, G.D. (1988) Cattle grazing tall larkspur on Utah mountain rangeland. *Journal of Range Management* 41, 118-122.

Ralphs, M.H., Jones, W.J. and Pfister, J.A. (1997) Damage from the larkspur mirid deters cattle grazing of larkspur. *Journal of Range Management* 50, 371-373.

Thompson, D.C. and Richman, D.B. (1989) The role of native insects as snakeweed biological control agents. In: *Snakeweed: Problems and Perspectives*. New Mexico State University, Agricultural Experiment Station Bulletin No. 751, pp. 179-187.

Wheeler, A.G. Jr, Stinner, B.R. and Henry, T.J. (1975) Biology and nymphal stages of *Deraeocoris nebulosus* (Hemiptera: Miridae), a predator of arthropod pests on ornamentals. *Annals of the Entomological Society of America* 68, 1063-1068.

Chapter 8

Estimating Lethal Dose of Larkspur for Cattle

J.D. Olsen[1] and D.V. Sisson[2]
[1]*USDA-ARS, Poisonous Plant Research Laboratory, 1150 East 1400 North, Logan, Utah 84341, USA;* [2]*Agricultural Experiment Station, Utah State University, Logan, Utah 84322, USA*

Introduction

Larkspur poisoning of cattle is a serious economic problem for producers and managers of rangeland in the western US (Nielsen *et al.*, 1994). The amount of larkspur required to produce a given level of intoxication would be useful in management of the problem. A model was developed to estimate equally toxic doses of larkspur for a cattle population as a means of predicting the response by grazing cattle to intake of larkspur.

The model was developed under three working hypotheses. First, that the clinical response to larkspur can be measured within the range of two doses of plant, a maximal dose that causes no signs of poisoning (X_o) and a dose that causes death (X_d). Secondly, that intermediate doses cause proportionately intermediate signs of poisoning. And finally, that the relative toxic effect of various larkspur communities can be compared using equally effective doses as a basis.

Measurement of the Clinical Response of Cattle to Larkspur

Larkspur poisoning is a result of toxic alkaloids in the plant causing neuromuscular blockade (Olsen and Manners, 1989). As the collective effective toxic dose continuously increases, the strength of muscular activity continuously decreases.

Effective loss of muscular activity was defined, according to lifting distance and time of lifting of the cow's body, to form an array of stages (classes) of clinical signs (Table 8.1). Severity of poisoning increased in each class of signs, as did the

Table 8.1. Classes of clinical signs of larkspur poisoning in cattle and associated Response Index values.

Clinical Signs	Stage of Poisoning	Response Index
No signs of poisoning.		0
Repeated tremor and collapse.	I	1
Can lift body, cannot fully stand.	II	2
Maintains sternal position, but cannot lift body to stand.	III	3
Cannot lift body to sternal position.	IV	4
Death from asphyxiation.	V	5

associated Response Index. Classes of signs were unique and mutually exclusive.

Because muscular exertion by the cow exacerbated the effect of larkspur poisoning (Olsen and Sisson, 1991), a test protocol was followed to minimize bias due to extraneous exertion and to standardize aspects of biomechanical function during testing. The cattle were rested for at least 10min immediately before the response was measured. The standardized body position varied with stage of poisoning. A standing position was used for cattle up to and including Stage I. Animals were rested in sternal recumbency during Stages II and III. When necessary, cattle were helped to sternal position by the observer at the beginning of the rest period. A minimal electrical shock was used to stimulated response in order to ensure maximal effort by the cow to remain standing or to attempt to stand. When appropriate, subclasses of clinical signs were formed and related to smaller dose increments.

With this measure of response, poisoning was evaluated hourly following each single daily dose of tall larkspur in late-vegetative/early-bud stage of growth. The Daily Response was quantified by averaging the three hourly Response Indices recorded during maximal intoxication (5-8hrs after dosing). The effect of a particular daily dose of larkspur was not fully realized until it had been given for at least 3d (Olsen and Sisson, 1991). For each animal, daily doses were incrementally increased until lethal, and responses were thus measured over the dose interval between the last dose causing zero response (X_o) and the dose causing death (X_d).

Ruminal bloat occurred secondarily to larkspur poisoning of cattle. Up to 57L of gas/hr could be produced by cattle on a ration of hay or green alfalfa (Colvin *et al.*, 1957; Dougherty and Cook, 1962). When rumen gas could not escape, as has occurred with certain legumes, death ensued as soon as 1hr after eating (Waghorn, 1991). Sternal recumbency allowed eructation, but lateral recumbency did not.

At Response Index 4, cattle were laterally recumbent for 3hrs in response to larkspur poisoning. Lethal bloat was prevented in the laboratory because the cow was helped to sternal position during each hour according to the test protocol; otherwise, intake of larkspur sufficient to induce Response 4 would likely have been lethal in 100% of the cattle.

Estimating a Dose-Response Curve for Larkspur-poisoned Cattle

The quantitative relationship of increasing dose and increasing severity of response was described by a regression line resulting from statistical analysis (dose-response curve). Since the points on the regression line included sampling variation, the inversely estimated dose would also vary on replication, with the probability of occurrence distributed within the Confidence Interval limits for the regression.

Variation in the toxic response of individual animals was also incorporated into the modal. Equally effective toxic doses of a larkspur plant collection causing a Response 3 have been observed to differ by as much as a factor of five for individual cows (Olsen, unpublished data).

The cattle were originally tested in three groups, each dosed with a different collection of plant at similar stages of growth and from the same larkspur community. The linear regression dose-response curve for each cow originated from the highest dose of larkspur causing no measured response. The relative adjusted dose-response of each cow was collectively analyzed and collectively plotted for model development. Due to differences in individual susceptibility, the adjusted dose-response for each cow varied about a common dose (X_0), where X_0 was the average last dose causing no signs of toxicity for all cows (set by definition to equal zero in the Model plot). As a collective result, the dose-response curve, Confidence Interval Limits (95%) for regression, and Prediction Interval Limits (95%) were determined by linear regression analysis of the relative adjusted equally effective doses combined for 19 cattle. Response greater than Response 4 was not determined for all 19 cattle, but linear regression analysis gave Prediction Interval Limits that graphically estimated the sample Confidence Interval Limits $(X_L$ and $X_U)$ for the distribution of equally effective doses (Ostle, 1963).

The derived Model Gaussian distribution curves for random equally effective doses for Responses of magnitude 1, 2, 3 and 4, as shown in Fig. 8.2. The adjusted dose-response regression line was calculated for each cow and the equally effective dose associated with each level of response determined, then the sample group standard deviation was calculated for each set of equally effective doses for individual cows for each response level (Table 8.2).

Cumulative Relative Frequency Distribution curves for equally effective toxic doses for each Response level were plotted as a composite graph. The percentage of the population displaying particular signs of larkspur poisoning can be quantitatively predicted according to intake of larkspur, where dose X_0 of the particular mix of alkaloids in the plant has been characterized (Fig. 8.3).

Prediction of Response of Cattle to Intake of Larkspur Toxic Alkaloids Based upon Distributions of Equally Effective Doses

Collections of larkspur species have been obtained at phenological stages likely to be grazed by cattle. Benchmark toxic alkaloid profiles will be determined and corresponding dose-response curves (estimated for the interval between group mean alkaloid dose X_0 and dose X_3 at Response 3) obtained for cattle of known susceptibility. These equally effective alkaloid dose curves can then be applied, with the working model, to predict the percentage of the cattle population expected to have a particular response to intake of a set quantity of larkspur alkaloids.

Knowledge of the benchmark equally effective dose curves would be useful as a reference base to compare seasonally obtained larkspur collections from problem

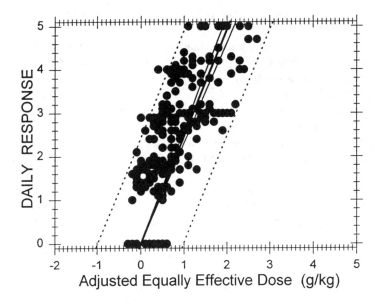

Fig. 8.1. Relative dose-response relationship for cattle (n=19) given incrementally increasing daily doses of larkspur. The adjusted equally effective dose was a coded value where Dose Zero equaled the group mean for the highest dose causing no measured response.

Table 8.2. Standard deviations for distributions of relative equally effective doses (g/kg) used to develop a working model for estimating population response to larkspur intake by cattle.

Daily	Individual Regression		Prediction Model	
Response	Mean	SD	Mean	SD
0	0	0.30	0	0.84
1.0	0.37	0.31	0.37	0.87
2.0	0.71	0.35	0.71	0.98
3.0	1.04	0.42	1.04	1.18
4.0	1.38	0.51	1.38	1.43

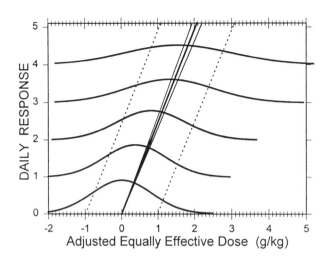

Fig. 8.2. Model Gaussian distribution curves for random equally effective doses at specific response levels for cattle. Mean and standard deviation (Table 8.2) for each curve. Variability increased as the response level increased due to forcing the regression line through Dose Zero.

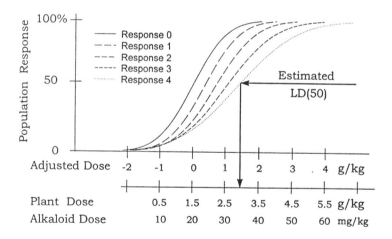

Fig. 8.3. Model equally effective dose-response curves estimating the percentage of the cattle population responding to a particular dose of larkspur. Adjusted dose is relative to the mean maximal dose (Dose X_0) causing no clinical signs of poisoning for the cattle population. Plant dose is an example where 1.5g/kg is known to be equivalent to Dose X_0. Alkaloid dose is an example where 20mg/kg is known to be equivalent to Dose X_0.

rangelands. The risk of poisoning (the percentage of the population responding to a corresponding bench mark equally effective dose) could be estimated according to intake by grazing cattle.

Some of the experimental cattle were predisposed to repeated, and at times intense, struggling when only moderately poisoned by larkspur (Stage II, or less), and unassisted, would have died from bloat. One reason cattle killed by larkspur poisoning have been reported to bloat rapidly may be that most cattle poisoned by larkspur were already bloated at the time of death. Therefore, the risk of death would be somewhat greater than otherwise predicted for the equally effective dose.

Acknowledgments

Thanks to J. Richard Schmid and David L. Turner for helpful discussions of statistics; to David L. Turner for help in preparing Fig. 8.3; and to Terrie Wierenga and Joyce Johnson for help in preparation of the manuscript.

References

Dougherty, R.W. and Cook, H.M. (1962) Routes of eructated gas expulsion in cattle: A quantitative study. *American Journal of Veterinary Research* 23, 997-1000.

Colvin, H.W. Jr, Wheat, J.D., Rhode, E.A. and Boda, J.M. (1957) Technique for measuring eructated gas in cattle. *Journal of Dairy Science* 40, 492-502.

Nielsen, D.B., Ralphs, M.H., Evans, J.O. and Call, C.A. (1994) Economic feasibility of controlling tall larkspur on rangelands. *Journal of Range Management* 47, 369-372.

Olsen, J.D. and Manners, G.D. (1989) Toxicology of diterpenoid alkaloids in rangeland larkspur (*Delphinium* spp.). In: Cheeke, P. (ed), *Toxicants of Plant Origin*, Vol. I, CRC Press, Boca Raton, FL, pp. 291-326.

Olsen, J.D. and Sisson, D.V. (1991) Description of a scale for rating the clinical response of cattle poisoned by larkspur. *American Journal of Veterinary Research* 52, 488-493.

Ostle, B. (1963) *Statistics in Research*, 2nd ed. Iowa State University Press, Ames, IA, pp. 30, 99-100, 176.

Waghorn, G.C. (1991) Bloat in cattle, No. 47: Relationships between intra-ruminal pressure, distension, and the volume of gas used to simulate bloat in cows. *New Zealand Journal of Agricultural Research* 34, 213-220.

Chapter 9

Control Technology Selection for Poisonous Plants with the EXSEL Expert System

W.T. Hamilton
Department of Rangeland Ecology and Management, Texas A&M University, College Station, Texas 77843, USA

Introduction

Professionals working with range livestock producers frequently encounter noxious plants that present problems to animal health. Often the problem is acute, requiring immediate removal of animals and destruction of the problem plants. Other times, noxious plants may influence range vegetation negatively by reducing desirable forage species and diminishing range nutritional value. Then the need for weed control may be less critical in the short term, but relate significantly to the overall health and productivity potential of livestock and wildlife species. Therefore, producers often ask for advice on weed-control methods. The objective for developing EXSEL was to assist in selection of appropriate control technology for a target species based on predicted efficacy of the treatment.

Methods

Brush- and weed-control factors necessary for development of EXSEL included target species, density, average mainstem diameter, average height, and soil moisture conditions, texture, depth, topography, amount of desirable vegetation present (to determine if seeding was necessary) and restrictions on use of aerial application of herbicides. Also, information on various brush- and weed-control methods available for use in Texas was obtained from the literature and expert opinions (Scifres, 1980; Scifres *et al.*, 1989; Welch, 1990). Then rules were formed to utilize this knowledge along with user input to select technically feasible alternatives for brush or weed management problems (Fig. 9.1).

Statements on expected response of forage target species to broadcast treatments were formulated. Treatment efficacy based on percent target plant mortality was assigned to each control method and plant in the species database. Caution or information statements on important environmental factors, or unique characteristics of treatments, including limitations critical to proper application were incorporated. Also, regulations that govern the use of brush and weed management practices, such as restrictions on herbicide applications, were retained as caution statements within the program when the practices were considered.

Mechanical technology in EXSEL is programmed *via* a forward chaining expert system using rules of inference to deduce conclusions. Chemical technology selection and burn feasibility sections use decision tree logic to determine specific recommendations. EXSEL is written in Microsoft C language 6.00a for the IBM PC. Some functions are written in Microsoft assembler to allow for responsive video output. The program is designed to operate with all popular IBM PC video adapters. The software is organized modularly into four areas: 1) user interface, 2) mechanical brush and weed management treatments, 3) chemical treatments and 4) prescribed burn feasibility. NASA's CLIPS system was used in the preliminary scheme, but was adapted into a smaller, more easily maintained and faster system.

Results and Discussion

Isolation of the user interface from the rest of the product has two advantages. If the user interface must change, it would minimally affect the rest of the application, and other interface modules could be developed so that the software might be more easily ported to other platforms.

Involvement of expert opinion is essential, and serves to assist the subject matter experts in their knowledge and identify areas where opinions differ so that the knowledge may be made homogeneous. Developing a rule base is complex. In testing, there were cases that the expert system would arrive at a conclusion with which the human expert would disagree. A rudimentary explanation feature was added to facilitate understanding the machine reasoning involved as an aid to debugging the deduced logic of the expert system.

Figure 9.2 is an example of the logic flow using EXSEL for selection of technically feasible mechanical treatments. The initial data entry screen prompts the user for identification of the target plant, which can be obtained from the plant database (Fig. 9.1). For mechanical practices, the county and soil type do not function in the process to determine possible treatments, but these may be necessary for chemical control selection. The entry for soil depth influences mechanical practice potential. If the soil depth had been "shallow," the practice of root-plowing would not have been selected. The user input of "moderately deep" does not preclude the use of root-plowing, but a caution statement is invoked so that users know that soil depth may be a limiting factor.

The number of plants per acre keys EXSEL to the practicality of individual plant treatment. Stem diameter determines if practices such as shredding are feasible. Plant height is significant to the use of ground chemical application equipment, but is not limiting in the selection of mechanical practices. There are four entries for soil moisture ranging from adequate throughout the soil profile to "low overall." The low soil moisture entry does not constrain the selection of mechanical practices, but does raise a caution statement that poor soil moisture can negatively affect the practice. The control method to be used can be specified by the user, or deduced by the expert system from data inputs. In Fig. 9.2, EXSEL was allowed to select between a broadcast or individual plant method and selected broadcast based on plant density. The user input of no restrictive topography prevents this consideration in the determination of mechanical treatment feasibility. If topography were restricting, this entry would inject a caution statement about the need for extreme care in proceeding with mechanical measures.

Treatment type refers to mechanical or chemical practices. In Fig. 9.2, the user selects mechanical, which forces the consideration of only these practices. When seeding is specified, the system will exhibit to the user potential problems with seed bed preparation in the form of caution statements.

If a prescribed burn is selected, there are additional input fields for fine fuel load, continuity and distribution. Those pertinent elements for possible mechanical treatments interact with the rule base and the treatments are deduced and reported.

EXSEL generates reports that can be sent to a file, the screen or the printer. Reports include a disclaimer statement and information entered by the user. It contains the specific treatment recommendations, both chemical and mechanical, for the target species. The report includes the recommended rates of application for each compound, as well as additives and specific application instructions. The expected results from the treatment are expressed as the level of efficacy: very high (76-100% mortality), high (56-75% mortality), moderate (36-55% mortality) and low (less than 35% mortality). When individual plant chemical treatments are recommended, the report will provide guidance to herbicide formulation for total volume of spray mix. Tables provide the amount of formulated herbicide to use for the concentration shown in the rate recommendations for individual plant and spot treatments and for a range of total spray volumes desired.

EXSEL provides expected projections of vegetation responses following application of selected technologies. A combination of predicted treatment efficacy and expected response provides information for use in economic analysis comparing alternative treatments (Conner *et al.*, 1990).

If the user is interested in prescribed fire, EXSEL determines the potential for use as a brush- and weed-control method. A detailed checklist of factors required for the safe implementation of prescribed burning is provided. A worksheet can be printed that prompts the user for necessary input information and provides a place to record entries that match the screen input requirements of the software.

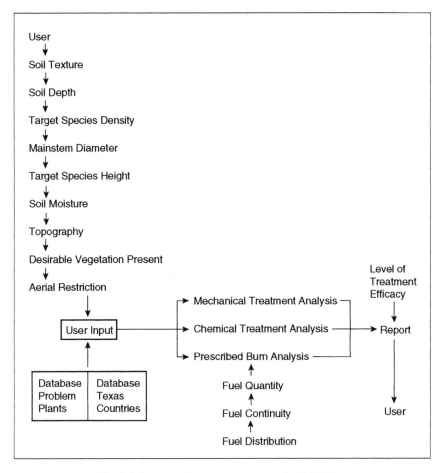

Fig. 9.1. Input and output components of EXSEL.

Conclusions

Field tests indicate that EXSEL successfully matches a specific weed or brush problem with the most technically feasible treatment alternatives. Agricultural Extension Agents evaluating the software have found the user's guide provides sufficient instruction to preclude formal training, making it user friendly. The addition of new herbicide use recommendations illustrated the ease of program updating. EXSEL is a powerful tool for rangeland producers, consultants and service agencies. It organizes the complexities of control measures and provides optimum control technology for a diversity of plant problems.

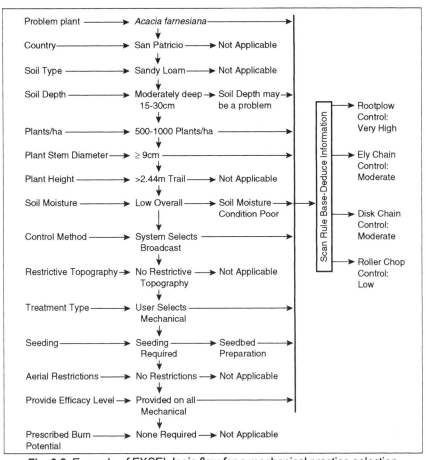

Fig. 9.2. Example of EXSEL logic flow for a mechanical practice selection.

References

Conner, J.R., Hamilton, W.T., Stuth, J.W. and Riegel, D.A. (1990) *ECON: An Investment Analysis Procedure for Range Improvement Practices.* MP-1717, Texas Agricultural Experiment Station, Texas A&M University, College Station, TX.

Scifres, C.J. (1980) *Brush Management: Principles and Practices for Texas and the Southwest.* Texas A&M University Press, College Station, TX.

Scifres, C.J., Koerth, B.H., Crane, R.A., Flinn, R.C., Hamilton, W.T., Welch, T.G., Ueckert, D.N., Hanselka, C.W. and White, L.D. (1989) *Management of South Texas Mixed Brush with Herbicides.* Bulletin 1623, Texas Agricultural Experiment Station, Texas A&M University, College Station, TX.

Welch, T.G. (1990) *Chemical Weed and Brush Control Suggestions for Rangeland.* Bulletin 1466, Texas Agricultural Extension Service, Texas A&M University, College Station, TX.

Chapter 10

Management and Edaphic Factors Related with the Incidence of Marsh Ragwort

I. McClements[1], A.D. Courtney[1] and F.E. Malone[1, 2]

[1]Applied Plant Sciences Division, Department of Agriculture for Northern Ireland, Newforge Lane, Belfast BT9 5PX, UK; [2]Veterinary Sciences Division, Department of Agriculture for Northern Ireland, 43 Beltany Road, Omagh, County Tyrone BT78 5NF, UK

Introduction

Ragwort is a noxious weed of grassland, reducing pasture productivity and poisoning livestock if eaten in sufficient quantities (Petrie and Logan, 1980-1). Two species of ragwort are found in County Fermanagh, Northern Ireland. The predominant species, marsh ragwort (*Senecio aquaticus* Hill), prefers wetter, heavier soils (Clapham *et al.*, 1989), whereas the common ragwort (*Senecio jacobaea* L.) is found on dry ground. No surveys have been conducted to determine the extent of the ragwort problem in Fermanagh.

Although Forbes (1977) achieved good control of *S. aquaticus* in Scotland, poor control has been achieved in Fermanagh. An understanding of herbicidal, plant and environmental factors could lead to the development of better control strategies for *S. aquaticus*.

Hopkins and Peel (1985) collated data from a number of Grassland Research Institute surveys in England and other data from Scotland and Northern Ireland to determine the incidence of weeds in permanent grassland. Docks, thistles and buttercup were identified as the major broad-leaved weeds of established grassland. Ragwort species (*S. jacobaea* and *S. aquaticus*) only infested 1% of swards in Scotland and Northern Ireland and 2% in England and Wales. Ragwort appears to be a localized problem, and the only two surveys of ragwort in the British Isles have concentrated on local regions (Davies, 1953; Forbes, 1974). However, Perring and Walters (1962) found *S. jacobaea* and *S. aquaticus* to be widespread throughout the British Isles. No attempt had been made to correlate ragwort incidence with soil environmental factors such as pH or soil nutrients and, more importantly, with the soil seed bank reserves responsible for the continued reinfestation of swards.

Materials and Methods

Two hundred farms in County Fermanagh (4% of the total) were surveyed during 1989-90 to assess the incidence of *S. jacobaea* and *S. aquaticus*. Two visits were made to each farm. On the first visit a detailed questionnaire was completed relating to the size of the farm, livestock numbers and incidence of ragwort poisoning. Two fields were selected at random from the farm map. The main factors believed to relate to the incidence of ragwort were recorded for each field.

The degree of infestation was assessed by placing twenty, $1m^2$ quadrats along a transect from one corner of the field to the other. Within each quadrat the number of ragwort seedlings or flowering plants was recorded. A soil sample was taken from each field and assessed for pH, phosphorus (P), potassium (K) and magnesium (Mg) (MAFF, 1981). These samples were analyzed for weed seed burden by placing approximately 100g of each dried soil sample in an unheated glasshouse between May and August. Germination was recorded every 3wks.

Data analysis involved calculating Pearson correlation coefficients and chi squared analyses. Plant numbers were stratified into the following categories when assessing the effects of each of the management factors: zero, 1-24, 25-49, 50-74, >75plants/$20m^2$. These were calculated using SPSS version 4.2 (McGraw Hill) on an ICL 2900.

Results

Ragwort was present in 51% of the fields surveyed (17.6% *S. jacobaea*, 74.1% *S. aquaticus*, and 8.3% mixed). The majority of infested farms (66.4%) had less than 0.5plants/m^2, while 18.3% had between 0.5-0.9plants/m^2 and 15.3% had more than 1plant/m^2. Only 2.5% of the farms related possible ragwort poisoning in livestock.

Sward management, grazing and conservation seemed to affect the incidence of ragwort. Sixty-seven percent of the fields were grazed by both cattle and sheep, 74% by cattle only and 2% by sheep only. There was a positive correlation ($P<0.05$, $r=0.4$) between total ragwort numbers on each farm and cattle density.

Thirty-seven per cent of the swards were used for silage compared to 7.5% for hay. Cutting for silage had a significant effect in reducing the incidence of ragwort, with ragwort present in 63.1% of the grazed swards, but in only 28.6% of the cut swards. Increased cutting frequency significantly reduced the incidence of ragwort. Both species showed a similar response to cutting frequency (Fig. 10.1). The apparent increase with three cuts may be because only five fields were involved. On sward age evaluation, 51% of the permanent pasture (swards over 10yrs old), 26% of 5-10yr-old swards and 37% of 1-5yr old swards were infested. Both ragwort species behaved in a similar way. In relation to drainage status the two species appeared to show divergent incidence, with the proportion of marsh ragwort increasing (19.7% to 58.6%) and common ragwort declining (42.2% to 35.6%) between free-draining and restricted-drainage situations. The incidence of ragwort increased as sward cover decreased: 34.3% of swards with less than 25% bare ground and 66.7% with more than 75% bare ground were infested (Fig. 10.2). Sandy soils were found to have low infestation rates (8.9% *S. jacobaea* and no *S. aquaticus*). Fields with clay soils had the largest *S. aquaticus* infestation (56.7%), while the largest *S. jacobaea* infestations (31.1%) were recorded on clay loams. There were no significant correlations between soil pH, K or Mg and total ragwort numbers. Although not statistically significant, low pH appeared to favor ragwort incidence. Increased soil P concentrations were found to have a significant effect in reducing ragwort infestations (Fig. 10.3).

The seedlings of 20 species germinating from the soil seed bank of 175 fields was

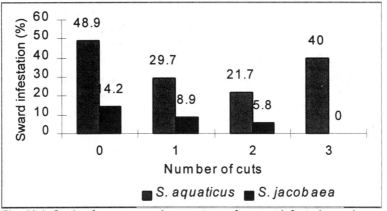

Fig. 10.1. Cutting frequency and percentage of ragwort infested swards.

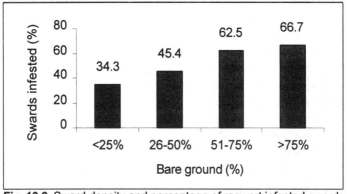

Fig. 10.2. Sward density and percentage of ragwort infested swards.

recorded. The largest proportion (77.7%) were grasses, followed by buttercup (*Ranunculus* spp.) and hairy bittercress (*Cardamine hirsuta*). *Senecio aquaticus* was the fifth most abundant species (1.7%) and *S. jacobaea* was one of the least abundant species (0.042%). Studies on population dynamics confirmed the high seed productivity of both species, with more seed being produced per plant by *S. jacobaea* (4794) than *S. aquaticus* (1369).

Discussion

Agricultural production in County Fermanagh is constrained by the nature of the soils and the heavy rainfall conditions (average 1,033mm/yr). Courtney (1973) reported that *S. jacobaea* was a problem of the drier coastal regions. Forbes (1974) found *S. aquaticus* to be more abundant than *S. jacobaea* where clay soils tend to become impermeable near the surface. In the present survey, drainage was frequently restricted. Forty-two percent of free-draining soils were infested with *S. jacobaea* and 19.7% with *S. aquaticus*. This contrasts with 35.6% of restricted-soils being infested with *S. jacobaea* and 58.6% with *S. aquaticus*. Clearly, drainage is a key factor in determining the incidence of either *S. aquaticus* or *S. jacobaea*.

Ragwort was not as prevalent in fields cut for silage, particularly in those cut more than once. Stock removal reduces the number of potential sites for ragwort invasion. High fertilizer applications associated with silage swards would also encourage grass competition. Watt (1987) showed that seedling emergence of *S. jacobaea* was more rapid in plots receiving high nitrogen fertilizer (200kg/ha). Forbes (1976) showed that it was possible to reduce *S. aquaticus* numbers significantly with an intensive management regime of heavy cattle stocking rates (2,250kg/ha) in a paddock grazing system with nitrogen fertilizer application of

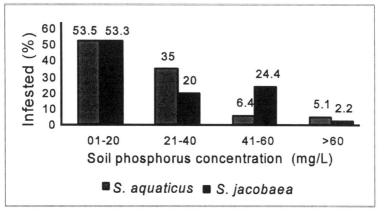

Fig. 10.3. Soil phosphorus concentration and percentage of ragwort infested swards.

300kg/ha. It may be possible to control ragwort infestation by more intensive grassland management in Fermanagh.

Seed production of the marsh ragwort is not as prolific as that of *S. jacobaea*. However, the higher prevalence of marsh ragwort in Fermanagh is reflected in the higher quantities of seed in the seed bank. The fifth most abundant species was *S. aquaticus*. This large seed burden ultimately ensures the potential for reinfestation.

References

Clapham, A.R., Tutin, T.G. and Moore, D.M. (1989) *Flora of the British Isles*, 3rd ed. Cambridge University Press, Cambridge, UK.

Courtney, A.D. (1973) Noxious weeds in grassland - docks, thistles and ragwort. *Agriculture in Northern Ireland* 48, 22-25.

Davies, A.J. (1953) The ragwort problem in Wales. In: *Proceedings of the 1st British Weed Control Conference* Cliftonville, Margate, UK, pp. 204-210.

Forbes, J.C. (1974) A survey of ragwort in Orkney. In: *Proceedings of a Symposium on Weed Control in the Northern Environment*. British Crop Protection Council Monograph 9, Edinburgh, UK, pp. 115-120.

Forbes, J.C. (1976) Influence of management and environmental factors on the distribution of marsh ragwort (*Senecio aquaticus* Huds) in agricultural grassland in Orkney. *Journal of Applied Ecology* 13, 985-990.

Forbes, J.C. (1977) Chemical control of marsh ragwort (*Senecio aquaticus* Huds) in established grassland. *Weed Research* 17, 247-250.

Hopkins, A. and Peel, S. (1985) Incidence of weeds in permanent grassland. In: Brockman, J.S. (ed), *Weeds, Pests and Diseases of Grasslands and Herbage Legumes*. British Crop Protection Council Monograph 29, pp. 93-103.

MAFF (1981) *The Analysis of Agricultural Materials*, 2nd ed. Ministry of Agriculture, Fisheries and Food, HMSO, London, UK.

Perring, F.H. and Walters, S.M. (1962) *Atlas of British Flora.* Botanical Society of the British Isles, Nelson and Sons, London, UK.

Petrie, L. and Logan, A. (1980-1) Ragwort poisoning. In: *Proceedings of the British Cattle Veterinary Association*, UK. pp. 167-170.

Watt, T.A. (1987) Establishment of *Senecio jacobaea* from seed in grassland and in boxed swards. *Weed Research* 27, 267-274.

Chapter 11

Management, Environmental and Livestock Interactions Impact on Perennial Ryegrass/*Neotyphodium*/Livestock Associations

W.M. Wheatley

Orange Agricultural College, The University of Sydney, Orange, New South Wales 2800, Australia

Introduction

Perennial ryegrass (*Lolium perenne* L.) is a highly regarded component of improved pastures in the Central Tablelands, and *Neotyphodium* endophyte is widespread (Wheatley, 1996). Climatic conditions in the central zone are characterized by cold winters with some snowfall, warm summers and non-seasonal rainfall. On average, rainfall is effective every month of the year across much of this zone. Dry, late summers/early autumns with little or no pasture growth are often experienced. The perennial ryegrass/*Neotyphodium*/livestock relationship affects grazing industries in a variety of ways (Fletcher *et al.,* 1990; Cunningham *et al.,* 1993; Familton *et al.,* 1995). A preliminary survey revealed that 91% of producers had never heard of endophyte (Wheatley, 1997) and only relate a problem to ryegrass staggers.

Set stocking/continuous grazing is commonly practiced, and animals can be exposed to toxin intake for considerable periods. This is particularly so following the "autumn break" when animals are forced to graze the basal stems of perennial ryegrass where the alkaloids, lolitrem B and ergovaline are concentrated (Familton *et al.,* 1995). Deferred grazing is sometimes practiced, where stock are excluded from perennial ryegrass dominant paddocks at the autumn break and not reintroduced until growth is considered to be tall enough to be "safe for grazing." This strategy is not always effective as perennial ryegrass/*Neotyphodium* associations are inconsistent in their impact on grazing animals, even when environmental and livestock management

45

systems appear to be the same. This chapter presents results from the first year of a research project examining these associations (Wheatley, 1997).

Methods

High endophyte (74% of seed at sowing) and nil endophyte lines of the perennial ryegrass cultivar Grasslands Lincoln were sown in August, 1995, across four randomly allocated replicates of 0.8ha each. Eight fine-wool Merino wether weaners, randomly allocated after ranking by weight, were continuously grazed on each treatment from March to August, 1996. Establishment/early growth in the pastures was measured by cutting five quadrats (0.5m²)/treatment. Five exclusion cages (1.0m²)/treatment were then established, and ungrazed herbage growth was harvested from a pre-trimmed area (0.5m²) at 2-3wk intervals, depending on rainfall. Dry matter (DM) and botanical composition were determined on the cut herbage.

Sheep were weighed every 14d when climatic conditions allowed, fecal samples were removed from the rectum of each animal immediately following weighing, and fecal moisture was determined. The incidence of ryegrass staggers was noted during the temporary stress of paddock mustering and weighing. Sheep that staggered and fell to the ground were transported to and from the weighing site. Wool samples from the midside of each sheep were taken at the beginning and again at the end of the grazing period, and were forwarded to the Australian Wool Testing Authority for micron and staple length testing. Internal parasites were controlled by a pregrazing anthelmintic drench and monitored by fecal egg counts.

Results

During the establishment and early growth stages, the high endophyte line of Grasslands Lincoln out-performed the nil endophyte line, having an average of 2,046kg/ha DM compared with 1,576kg/ha DM available as "feed on offer" when sheep were introduced. Despite rainfall being well below average for the entire period of grazing, subsequent variations in growth between the two lines were not significant. Botanical composition data indicated that in all the treatments feed on offer was close to 100% perennial ryegrass.

Body weights of sheep grazing the high endophyte treatments were significantly lower than the nil endophyte treatments during the 4wk period in which ryegrass staggers developed, but compensatory gain was rapid, and there was no significant difference in body weights between the high and the nil endophyte groups 60d later. There was a variation in body weights in the sheep when the high endophyte animals were grouped according to the degree in which they exhibited staggers. Body weight gains were greatest in animals that had no obvious clinical signs. Shaky sheep were intermediate, and those that staggered and dropped had the lowest gain. These differences were significant on one occasion well before the onset of staggers and again during the 4wk period of staggers. There was not a significant difference between the body weights of animals grouped this way 60d later. Fecal moisture content from the high endophyte treatment was significantly higher on one occasion immediately prior to the onset of ryegrass staggers. Ryegrass staggers were experienced over a 4wk period from late April to late May following the autumn break. There was variation in the development of ryegrass staggers between replicates (Table 11.1). Some sheep staggered and fell to the ground, and some exhibited shaking but were capable of moving, while others showed no obvious signs of ryegrass staggers. The ensuing 10wk period remained predominantly dry, with intermittent rainfall, and sheep continued to eat into the crown of the perennial ryegrass plants. Even when sheep were temporarily stressed during mustering, no signs of staggers were exhibited during this latter period. There was no significant difference in wool quality or quantity.

Discussion

The persistence of nil endophyte lines of perennial ryegrass is normally less than that of high endophyte lines (Cunningham *et al.*, 1993), so differences in the future production of perennial ryegrass may impact on the quantity of feed on offer. This would impact sheep weights and the quantity and quality of wool.

Providing sheep have sufficient feed on offer, the perennial ryegrass/ *Neotyphodium* association with sheep body weights appears to be temporary, and compensatory gain following staggers is rapid. Therefore, the higher nutrition from perennial ryegrass could be utilized to grow out Merino wether weaners.

There is some confusion over the conditions required to produce ryegrass staggers. Staggers occurred after a dry spell, when sheep were grazing short green pick, and for a period of 4wks. The 10wks following the period of 4wks of fresh green growth, initiated by rain after a dry spell, did not appear to be any different ecologically and climatically from the 4wks in which staggers were observed. Sheep were forced to eat into the crown of the perennial ryegrass, but there were no overt signs of staggers. It is unclear whether the lolitrem B levels fell sufficiently to enable

Table 11.1. Numbers of sheep affected by ryegrass staggers on high endophyte treatments.

Replicate	Died	Staggered (needed assistance)	Shaky	No effect
1	-	2	5	1
2	-	1	2	5
3	1	2	1	4
4	-	-	1	7

affected animals to recover or the animals developed an increased tolerance to the toxin, or a combination of both factors occurred.

The variation in the development/extent of ryegrass staggers between replicates needs further investigation. Many (43%) of the Central Tablelands producers surveyed had never experienced ryegrass staggers, even though perennial ryegrass across these properties ranged from moderate to dominant (Wheatley, 1997). Some producers' sheep experienced staggers even after a strategic deferred grazing management system had been implemented (Wheatley, 1997), and animals were not necessarily grazing the basal stems where lolitrem B was concentrated. The interactions between management, environmental factors and livestock obviously play a vital, but ill-defined, role in outbreaks of ryegrass staggers.

Acknowledgments

This work is funded by a University of Sydney Research Grant. Technical assistance from E. Hunt, P. Henry, N. Cother, R. Maxey, B. Upjohn, and B. McCarthy. Biometrical assistance from H. Nicol. Seed supply and testing from Wrightson Seeds (Aust) Pty Ltd, Sydney.

References

Cunningham, P.J., Foot, J.Z. and Reed, K.F.M. (1993) Perennial ryegrass (*Lolium perenne*) endophyte (*Acremonium lolii*) relationships - the Australian experience. *Agriculture, Ecosystems and Environment* 44, 157-168.

Familton, A.S., Fletcher, L.R. and Pownall, D.B. (1995) Endophytic fungi in grasses and their effect on livestock. *Proceedings of the 25th Sheep and Beef Cattle Seminar.* Massey University, Palmerston North, NZ, pp. 160-173.

Fletcher, L.R., Hoglund, J.H. and Sutherland, B.L. (1990) The impact of *Acremonium* endophytes in New Zealand: past, present and future. *Proceedings of the New Zealand Grassland Association* 52, 227-235.

Wheatley, W.M. (1996) Preliminary survey of the incidence of perennial ryegrass (*Lolium perenne*) staggers in Central Tablelands of NSW. *Proceedings of the Grassland Society of New South Wales.* Wagga Wagga, NSW, pp. 123-124.

Wheatley, W.M. (1997) Perennial ryegrass (*Lolium perenne*) staggers in the Central Tablelands, NSW, Australia: A survey of livestock producers. In: Bacon, C.W., and Hill, N.S (ed), *Neotyphodium/Grass Interactions*, Plenum Press, New York, pp. 447-449.

Chapter 12

Development of an Immunoassay for Corynetoxins

K.A. Than, Y. Cao, A. Michalewicz and J.A. Edgar

CSIRO Division of Animal Health, Australian Animal Health Laboratory, Private Mail Bag 24, Geelong, Victoria 3220, Australia

Introduction

Corynetoxins are extremely poisonous and cumulative toxins produced by a bacterium (*Clavibacter toxicum*), which colonizes nematode galls formed in the seedheads of annual ryegrass/wimmera ryegrass (*Lolium rigidum*), annual beard grass (*Polypogon monspeliensis*) and blown grass (*Agrostis avenacea*). They are responsible for annual ryegrass toxicity (ARGT) in Western Australia and South Australia (Vogel *et al.*, 1981; Edgar *et al.*, 1982), and floodplain staggers (FPS) in South Australia and New South Wales (Edgar *et al.*, 1994). Annually, ARGT kills 20,000-80,000 sheep annually in Western Australia (Roberts and Bucat, 1992). During a three-month period in 1990-91, 1,722 cattle and 2,466 sheep died from FPS in New South Wales (Bryden *et al.*, 1994). Corynetoxins have also caused poisoning of livestock feeding on contaminated hay (Roberts and Bucat, 1992).

The method most commonly used to estimate the toxicity of annual ryegrass paddocks involves counting the number of "bacterial" galls in pasture samples (Riley, 1992). An immunoassay for the bacterium is also available to detect the presence or absence of bacterium in grass samples (Riley and Mckay, 1991). Gall counts are sometimes augmented by estimates of the antimicrobial activity of the galls (Riley and Ophel, 1992), but this is at best a semi-quantitative indication of the toxin level. A simple high performance liquid chromatography (HPLC) method for quantifying the corynetoxins in "bacterial" galls is available (Cockrum and Edgar, 1985). Sample preparation for HPLC analysis of samples other than "bacterial" galls is, however, very labor intensive and time consuming. The cost of HPLC analysis is high and the method is insufficiently sensitive for many requirements. We have developed a more sensitive, lower cost, accurate and easily applicable immunoassay for measuring the corynetoxins in a variety of samples.

Antibody Production

Corynetoxins are closely related in structure to the tunicamycin antibiotics and other tunicaminyl-uracil toxins, differing only in regard to the series of fatty acids linked to the amino group of the central tunicamine unit (Frahn *et al.*, 1984). These toxins are of low molecular weight (800-900kD) and do not naturally induce an immune response in animals. Commercially available tunicamycins (Sigma T7765) were made immunogenic by conjugation to proteins, and injected into sheep in an oil adjuvant.

A radio-immunoassay (RIA) and an enzyme-linked immunosorbent assay (ELISA) were developed for assessing the production of antibodies in vaccinated sheep. Sera (0.1ml) were incubated with 3,000 counts per minute (cpm) of tritium-labeled tunicamycins, specific activity of 3,000cpm/8.9ng, with or without additional unlabeled tunicamycins, at 4°C overnight. Tunicamycins bound to antibodies and unbound toxins were separated by adding an equal volume of saturated $(NH_4)_2SO_4$ to precipitate the antibody fraction. Increased binding of tunicamycins to serum antibodies were detected in sheep after vaccination.

Tritium-labeled tunicamycins bound to the antibody fraction were displaced by adding unlabeled corynetoxins and tunicamycins, demonstrating competition between labeled and unlabeled tunicamycins for anti-toxin antibodies in vaccinated sheep sera (Fig. 12.1). After vaccination, increased binding of sheep serum IgG to microtiter plate wells coated with modified tunicamycin attached to bovine serum albumin (BSA) was detected using anti-sheep IgG conjugated to peroxidase.

Competitive ELISA for Toxin Detection

Serum IgG from vaccinated sheep was purified using a protein G sephrose 4 fast flow affinity column (Pharmacia). After dialysis against phosphate buffered saline, purified IgG was aliquoted and stored at -20°C. The tunicamycin standard solution was prepared by weighing crystalline tunicamycins and dissolving them in a known volume of methanol. The concentrations of Corynetoxin U17a and H17a standards, purified by preparative HPLC, were estimated by multiplication of optical density at 260nm by their respective molecular weights (858 and 876kD) and dividing by 9,650 (ϵ_{Max} at 260nm).

To analyze the cross-reactions of purified serum IgG with different tunicaminyl-uracil toxins, microtiter plates treated with glutaraldehyde were coated with chemically modified tunicamycin (4ng/well). The plates were washed and different amounts of free corynetoxin U17a, corynetoxin H17a, and tunicamycin standards were added to the wells in 100µl volumes along with the optimum dilution of purified IgG in 100µl. After a 2hr period of competition for antibody binding between the solid and liquid phase toxins, the unbound reagents were washed out. The antibodies bound to the wells were detected by the addition of anti-sheep IgG conjugated to

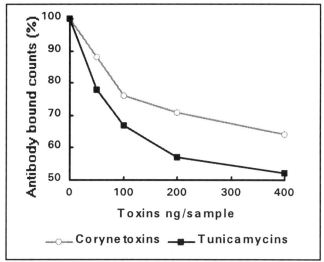

Fig. 12.1. Displacement of tritium-labeled tunicamycins in the serum of vaccinated sheep with different amounts of unlabeled corynetoxins and tunicamycins, showing competitive binding to the anti-toxin antibodies.

peroxidase. After 2hrs incubation, the plates were washed and 3,3',5,5'-tetramethylbenzidine (Sigma T2885) substrate was added and incubated for a further 20min before stopping solution was added. A comparison of the optical densities of the wells at 450nm for different amounts of corynetoxin U17a, corynetoxin H17a, and tunicamycins is shown in Fig. 12.2.

Cross-reactivity of different tunicaminyl-uracil toxins and compounds similar to or forming a part of the chemical structure of tunicaminyl-uracil toxins were compared at 80% and 50% inhibition of the binding of purified IgG to plates coated with 4ng of chemically modified tunicamycin (Table 12.1).

Discussion

Sheep injected with modified tunicamycin conjugated to proteins produced antibodies against tunicaminyl-uracil toxins (corynetoxin U17a, corynetoxin H17a and tunicamycins). Cross-reactions were very strong with all three of the tunicaminyl-uracil toxins tested; there was very little or no cross-reaction with compounds similar to or forming a part of the structure of tunicaminyl-uracil toxins (Table 12.1).

A number of approaches have been developed for detecting corynetoxins in field and laboratory samples. A bioassay using nursling rats (Stynes *et al.*, 1979) takes 5d

Than et al.

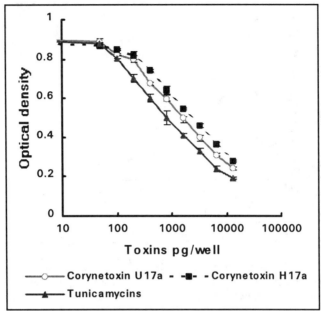

Fig. 12.2. Comparison of optical densities of the wells with different amounts of free corynetoxin U17a, corynetoxin H17a, and tunicamycin standards, showing competition for binding with purified vaccinated sheep serum IgG in microtiter plates coated with chemically modified tunicamycin. Bars indicate standard errors of the means of four replicates.

and is costly due to the requirement for laboratory animals. Methods currently used to estimate toxicity, such as counting the number of "bacterial" galls in pasture samples and hay (Riley, 1992), immunoassay for the presence of bacterium (Riley and Mckay, 1991) and estimates of the antimicrobial activity of the galls (Riley and Ophel, 1992), have provided a useful guide to predict the risk of toxicity. The number of "bacterial" galls or the level of bacterium does not always directly correlate with the production of corynetoxins. Production of corynetoxins increases dramatically as the grass matures (Stynes and Bird, 1983) and the toxin production appears to be correlated to infection of the bacterium with a bacteriophage (Ophel *et al.*, 1993). Weather conditions also appear to influence the likelihood of animals being poisoned.

Corynetoxins can be quantified by HPLC (Cockrum and Edgar, 1985). Chromatographic analysis of corynetoxins involves detection by UV absorption. UV-absorbing impurities, with similar retention times to those of corynetoxins in the samples, can interfere with the HPLC assay, reducing both specificity and sensitivity of the method.

The ELISA method developed in this lab for measuring corynetoxins and other tunicaminyl-uracil toxins overcomes previous limitations and offers several

Table 12.1. Cross-reactivity of different tunicaminyl-uracil toxins and compounds similar to or forming a part of the chemical structure of tunicaminyl-uracil toxins at 80% and 50% inhibition of the binding of purified IgG to 4ng/well of chemically modified tunicamycin.

	Compound	Cross-reaction	
		80%	50%
1	Corynetoxin U17a	100	100
2	Corynetoxin H17a	94.9	91.1
3	Tunicamycins	204	212
4	N-acetylglucosamine	NCR	NCR
5	D-glucose	NCR	NCR
6	Uracil	0.011	0.009
7	Uridine	0.045	0.036
8	Uridine 5'-diphosphate	0.003	0.004
9	Uridine 5'-diphosphate-N-acetylglucosamine	0.008	0.006

NCR = No cross-reactivity.

advantages over other screening methods. Sensitivity and specificity are high, no clean up of sample is required, sample throughput is rapid and the cost per assay and equipment costs are lower than other methods. As multiple samples can be tested simultaneously, the ELISA is suitable for routine monitoring of corynetoxins and other tunicaminyl-uracil toxins in large numbers of samples.

Acknowledgments

The authors would like to thank Dr Simon Stuart for generously providing the tritium-labeled tunicamycins and Peter Cockrum for HPLC-purified corynetoxins.

References

Bryden, W.L., Trengove, C.L., Davis, E.O., Giesecke, P.R. and Curran, G.C. (1994) Corynetoxicosis of livestock: a nematode-bacterium disease complex associated with different grasses. In: Colegate, S.M. and Dorling, P.R. (eds), *Plant-Associated Toxins: Agricultural, Phytochemical and Ecological Aspects.* CAB International, Wallingford, Oxon, pp. 410-415.

Cockrum, P.A. and Edgar, J.A. (1985) Rapid estimation of corynetoxins in bacterial galls from annual ryegrass (*Lolium rigidum* Gaudin) by high-performance liquid chromatography. *Australian Journal of Agricultural Research* 36, 35-41.

54   *Than* et al.

Edgar, J.A., Frahn, J.L., Cockrum, P.A., Anderton, N., Jago, M.V., Culvenor, C.C.J., Jones, A.J., Murray, K. and Shaw, K.J. (1982) Corynetoxins, causative agents of annual ryegrass toxicity; their identification as tunicamycin group antibiotics. *Journal of The Chemical Society Chemical Communications*, 222-224.

Edgar, J.A., Cockrum, P.A., Stewart, P.L., Anderton, N.A. and Payne, A.L. (1994) Identification of corynetoxins as the cause of poisoning associated with annual beard grass (*Polypogon monspeliensis* (L) Desf.) and blown grass (*Agrostis avenacea* C. Gemelin). In: Colegate, S.M. and Dorling, P.R. (eds), *Plant-Associated Toxins: Agricultural, Phytochemical and Ecological Aspects*. CAB International, Wallingford, Oxon, pp. 393-398.

Frahn, J.L., Edgar, J.A., Jones, A.J., Cockrum, P.A., Anderton, N. and Culvenor, C.C.J. (1984) Structure of corynetoxins, metabolites of *Corynebacterium rathayi* responsible for toxicity of annual ryegrass (*Lolium rigidum*) pastures. *Australian Journal of Chemistry* 37, 165-182.

Ophel, K.M., Bird, A.F. and Kerr, A. (1993) Association of bacteriophage particles with toxin production by *Clavibacter toxicus*, the causal agent of annual ryegrass toxicity. *Phytopathology* 83, 676-681.

Riley, I.T. (1992) Paddock sampling for management of annual ryegrass toxicity. *Western Australian Journal of Agriculture* 33, 51-56.

Riley, I.T. and Mckay, A.C. (1991) Inoculation of *Lolium rigidum* with *Clavibacter* sp., the toxigenic bacteria associated with annual ryegrass toxicity. *Journal of Applied Bacteriology* 71, 302-306.

Riley, I.T. and Ophel, K.M. (1992) Rapid detection of corynetoxins produced by *Clavibacter toxicum*. *Letters in Applied Microbiology* 14, 96-99.

Roberts, W.D. and Bucat, J. (1992) *The Surveillance of Annual Ryegrass Toxicity in Western Australia*. Miscellaneous publication No. 39/92, Department of Agriculture, Western Australia.

Stynes, B.A. and Bird, A.F. (1983) Development of annual ryegrass toxicity. *Australian Journal of Agricultural Research* 34, 653-660.

Stynes, B.A., Petterson, D.S., Lloyd, J., Payne, A.L. and Lanigan, G.W. (1979) The production of toxin in annual ryegrass, *Lolium rigidum*, infected with a nematode, *Anguina* sp. and *Corynebacterium rathayi*. *Australian Journal of Agricultural Research* 30, 201-209.

Vogel, P., Petterson, D.S., Berry, P.H., Frahn, J.L., Anderton, N., Cockrum, P.A., Edgar, J.A., Jago, M.V., Lanigan, G.W., Payne, A.L. and Culvenor, C.C.J. (1981) Isolation of a group of glycolipid toxins from seedheads of annual ryegrass (*Lolium rigidum* Gaud.) infected by *Corynebacterium rathayi*. *Australian Journal of Experimental Biology and Medical Science* 59, 455-467.

Chapter 13

[^{14}C]Senecionine Production in *Senecio vulgaris* Root Culture

P.R. Dorling[1], J. McComb[2], S.M. Colegate[1,3], K. Lambert[2,4] and S. Jawaheer[2,5]

[1]School of Veterinary Studies, and [2]School of Biological and Environmental Sciences, Murdoch University, Murdoch, Western Australia 6150, Australia; [3]Plant Toxins Unit, CSIRO Division of Animal Health, Australian Animal Health Laboratory, Private Bag 24, Geelong, Victoria 3220, Australia; [4]Faculty of Agriculture, The University of Western Australia, Western Australia 6709, Australia; [5]Faculty of Science, University of Mauritius, Reduit, Mauritius

Introduction

During the planning phase of a study into the mechanism of hepatogenous chronic copper poisoning, there was the need for a radiolabeled pyrrolizidine alkaloid (PA). Earlier work showed that hepatogenous chronic copper poisoning in sheep resulted from the synergistic activity of copper and PAs (Howell *et al.*, 1991) and studies were initiated to identify the point of interaction of these two toxins.

The most common source of PA toxicity in the field is from the plant *Heliotropium europeaum.* Hartmann and Toppel (1987) showed that putrescine and spermidine were precursors of PA synthesis in plant root cultures. Therefore, we attempted to grow root cultures from heliotrope plants and to radiolabel PA at specific sites in the necine base. Despite the fact that the roots of field-grown heliotrope plants contain PA, cultured roots produced negligible alkaloid. Similar results were obtained with roots from a number of *Crotalaria* species (Dorling and McComb, unpublished). Pyrrolizidine alkaloid production was successful in root cultures from *Senecio lautus* and *S. vulgaris*. This chapter describes the growth of root cultures from *S. vulgaris*, the production of PA and the specific labeling of the compound in these cultures.

Materials and Methods

Since [1,4-^{14}C]putrescine is commercially available and relatively inexpensive, it was the favored precursor for PA production. The incorporation was predicted to give a [3,5,8,9-^{14}C]pyrrolizidine.

Senecio vulgaris plants were collected near Perth, Western Australia. Leaves from the rosettes were sterilized in 2% sodium hypochlorite solution for 15min, washed three times in sterile water and placed on solid Murashige-Skoog medium (MS medium, 2% agar, pH 5.8) with 5μM indolebutyric acid (IBA) (Murashige and Skoog, 1962). Roots that regenerated from the leaf explants were removed and placed in 100ml flasks with 15ml liquid MS medium (pH 7.0) with 4% sucrose (Hartmann and Toppel, 1987). Upon further growth, the roots were transferred to 250ml flasks with 50ml of the same growth medium and subcultured every 15-20d. Flasks were sealed with a double layer of aluminum foil and shaken at 80 cycles/min in the dark at 25°C. Figure 13.1 shows the typical appearance of a culture grown for 15d.

To extract PA, roots were macerated in 0.05M H_2SO_4 and incubated for 2hrs at 25°C. The mixture was centrifuged or filtered and the orange supernatant collected. The sediment was re-extracted, and the combined supernatant was first partitioned with chloroform and then made up to 0.2M H_2SO_4 and stirred with excess zinc for 5hrs at 25°C. Filtration of the zinc reduction reaction yielded a pale yellow aqueous solution, which was extracted with chloroform and then basified by the addition of NH_4OH. The chloroform extract of the basic solution was dried with anhydrous Na_2SO_4, and evaporated to dryness under reduced pressure.

Fig. 13.1. A 250ml flask containing a root culture of *Senecio vulgaris* after 15d growth.

Crude PA extracts were subjected to radial chromatography to produce pure senecionine, examined by GC/MS, or redissolved in deuterated chloroform for NMR spectroscopic investigation. Senecionine content of the extracts was determined by GC (SE30 microbore column, 0.53mm x 15m, 250°C isothermal) using purified senecionine and heliotrine as internal standards.

The alkaloid production characteristics of *S. vulgaris* roots were measured by placing five root tips (1-2mg in total) from stock cultures into 15ml of standard growth medium in 100ml flasks, with three replicates harvested after 5, 10 ,15, 20 and 30d growth. Resulting root masses were blotted dry and the fresh weights recorded and then bulked for each time of harvest. Aliquots of the crude PA extracts were dried into the bottom of glass tubes. The PA content of these dried samples was determined by the Ehrlich reagent method of Mattocks (1967; 1968) using monocrotaline as a standard.

For [14C]putrescine incorporation studies, approximately 200mg of roots were placed in 50ml of growth medium in 250ml flasks with three replicates grown for 5, 10 and 20d before 10μCi of [1,4-14C]putrescine was added to each flask. Flasks were shaken and aliquots of the medium were taken 0, 2, 5, 8 and 24hr after addition and counted using a scintillation counter to assess 14C content. After this 24hr period, the roots were macerated, PA extracted as above and the level of 14C incorporation was determined.

The effect of temperature on 14C incorporation was determined by setting up cultures with 250mg roots in 50ml of medium in 250ml flasks and incubating for nine days at 25°C before exposure to 16, 20, 25 and 33°C for 24hrs. [14C]putrescine (50μCi) was then added and the cultures were incubated for a further 24hr at the respective treatment temperatures. Crude PA was extracted, samples taken for counting and determination of senecionine content by gas chromatography.

Results

Senecio vulgaris roots grow vigorously in culture, and the PA content of the tissue is about 0.05% (fresh wt.) in this early growth phase (Fig. 13.2). Root growth could be more than doubled by the inclusion of 1μM IBA in the growth medium, but roots grown in this way synthesized only trace amounts of PA (data not shown).

Almost total uptake of [14C]putrescine by root cultures occurred within the first 24hr of incubation with the most rapid rate occurring within the first 5hrs (Fig. 13.3). As expected, the rate of uptake was dependent on tissue mass. The crude PA fractions extracted from these cultures contained, on average, 17% of the added 14C with values ranging from 10.5-24.6%. This is not as high as the 33% reported by Toppel *et al.* (1987).

The effect of incubation temperature on 14C incorporation is shown in Table 13.1. The gas chromatograms of the alkaloidal extracts indicated that they were comprised mainly of one component, which co-chromatographed with authentic senecionine.

While growth rate was retarded at the extreme temperatures 16°C and 33°C, incorporation was most efficient at 16°C, resulting in a senecionine sample with a very high specific radioactivity. Following alkaline hydrolysis, all of the radioactivity could be accounted for within the retronecine base, which is consistent with the findings of Robins and Sweeney (1979).

The proton NMR spectrum of the major component was as expected for senecionine (Liddell and Stermitz, 1994) and was identical to the NMR spectrum of an authentic sample of senecionine. A comparison of the integration of the NMR doublet at δ5.5ppm and a contaminating doublet at δ5.42ppm indicates an approximate senecionine content of 90% in the crude extract (Fig. 13.4). A similar estimate was derived from other such comparisons of like resonances across the spectrum. The mass spectrum was as expected for senecionine. It indicated a molecular weight of 335 and showed a fragmentation pattern identical to that observed with an authentic sample of senecionine.

Table 13.1. The effect of temperature on growth and [1,4-^{14}C]putrescine incorporation.

Incubation Temperature (°C)	Weight of Roots (g)	Weight of Senecionine (µg)	Senecionine Content (%)	Specific Activity (cpm/mg)
16	3.75	595	0.016	1.9×10^7
20	6.60	534	0.008	1.5×10^7
25	5.72	642	0.011	1.5×10^7
33	3.43	602	0.018	0.3×10^7

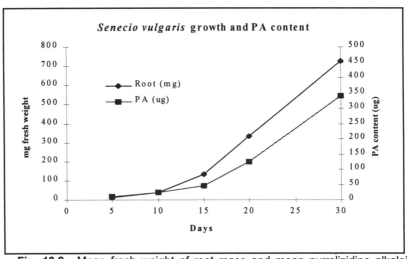

Fig. 13.2. Mean fresh weight of root mass and mean pyrrolizidine alkaloid content of roots following specified days of growth.

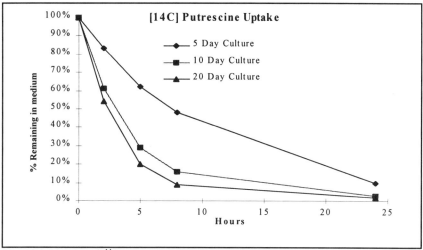

Fig. 13.3. Uptake of [^{14}C]putrescine from the medium over a period of 24hr by 5-, 10-, and 20-day-old root cultures.

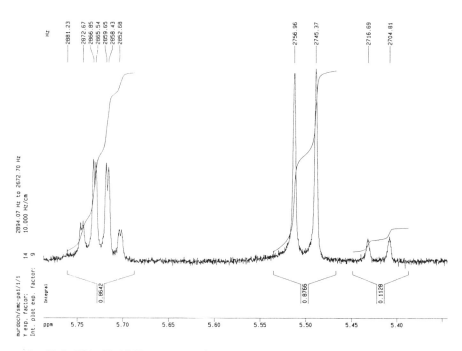

Fig. 13.4. 500mHz NMR spectrum of a crude, unlabelled pyrrolizidine alkaloid extract showing a contaminant doublet at δ5.42ppm and the expected senecionine doublet at δ5.5ppm and complex between δ5.70ppm and δ5.75ppm.

Discussion

Pyrrolizidine alkaloids like senecionine are metabolized by the hepatic P_{450} system and converted to toxic pyrrole esters. The metabolites are very unstable and spontaneously hydrolyze to form species that can undergo electrophilic reactions to alkylate a wide range of nucleophiles. Pyrrole di-esters can cross-link macromolecules of many types and form persistent adducts.

The advantages of producing labeled PA by this technique are the ease of production, safety and the ability to specifically label that part of the molecule which forms the tissue adduct as part of the toxic process. All of the label will be present as part of the adduct, not a small proportion as would be expected when using uniformly labeled PA.

Root culture is a relatively simple technology and any laboratory able to grow animal cells in culture would have little trouble growing plant roots. Since the whole process is held within the limits of simple glassware, radioactivity within the product and wastes is easily confined and disposed of (Segall, 1985). The product from this process was at least 90% senecionine, which is probably pure enough for use in adduct studies without the imposition of various forms of chromatographic purification and the accompanying risks of radioactive contamination. The relative percentage of senecionine in the total PA content of a given culture may vary with root culture line and growth conditions. It could be important to establish a number of lines and to optimize the growth conditions of the most promising of these.

From the figures in Table 13.1, incorporation at 16°C produced about 600μg of senecionine containing about 11% of the isotope of the original putrescine. This sample had a specific activity of 2.9mCi/mM. If 1mCi of labeled putrescine had been used instead of 50μCi, and if the growth conditions had allowed 20% incorporation, then a specific activity in excess of 100mCi/mM could be expected.

Ideally, a final tissue mass of about 4g is required for extraction and further processing. To obtain this, growth from a 200-250mg stock for about 10d is required. During this period, the culture is producing unlabeled senecionine, which will "dilute" labeled alkaloid produced during a 24hr labeling period. If extremely "hot" alkaloid is required for a specific purpose, it is suggested that roots be "grown up" in an IBA-containing medium, promoting rapid root growth but little senecionine production, and then placed in IBA-free medium for labeling. In this way, an even higher specific activity should be obtained. The manufacture of specifically labeled natural products in plant tissue culture in general, and roots in particular, will prove to be a very useful technique in toxic plant research.

References

Hartmann, T. and Toppel, G. (1987) Senecionine N-oxide, the primary product of pyrrolizidine alkaloid biosynthesis in root cultures of *Senecio vulgaris*. *Phytochemistry* 26, 1639-1643.

Howell, J.McC., Deol, H.S. and Dorling, P.R. (1991) Experimental copper and *Heliotropium europeaum* intoxication in sheep: clinical syndromes and trace element concentrations. *Australian Journal of Agricultural Research* 42, 979-992.

Liddell, J.R. and Stermitz, F.R. (1994) Pyrrolizidine alkaloids from *Trollius laxus*. In: Colegate, S.M. and Dorling, P.R. (eds), *Plant-Associated Toxins: Agricultural, Phytochemical and Ecological Aspects*. CAB International, Wallingford, Oxon, pp. 217-220.

Mattocks, A.R. (1967) Spectrophotometric determination of unsaturated pyrrolizidine alkaloids. *Analytical Chemistry* 39, 443-447.

Mattocks, A.R. (1968) Spectrophotometric determination of pyrrolizidine alkaloids - some improvements. *Analytical Chemistry* 40, 1749-1750.

Murashige, T. and Skoog, F. (1962) A revised medium for rapid growth and bioassays with tobacco tissue cultures. *Physiologia Plantarum* 15, 473-497.

Robins, D.J. and Sweeney, J.R. (1979) Pyrrolizidine alkaloids: evidence for the involvement of spermidine and spermine in the biosynthesis of retronecine. *Journal of the Chemical Society Chemical Communications*, 120-121.

Segall, H.J. (1985) The biosynthesis of ^{14}C macrocyclic pyrrolizidine alkaloids and identification of pyrrolizidine metabolites. In: Seawright, A.A., Hegarty, M.P., James, L.F. and Keeler, R.F. (eds), *Plant Toxicology*. Dominion Press, Hedgers and Bell, Melbourne, Australia, pp. 219-226.

Toppel, G., Witte, L., Riebesehl, B., von Borstel, K. and Hartmann, T. (1987) Alkaloid patterns and biosynthetic capacity of root cultures from some pyrrolizidine alkaloid producing *Senecio* species. *Plant Cell Reports* 6, 466-469.

Chapter 14

Hematological Changes in Sheep with Experimentally Produced Acute and Chronic Lupinosis

J.G. Allen and D.V. Cousins

Animal Health Laboratories, Agriculture Western Australia, 3 Baron-Hay Court, South Perth, Western Australia 6151, Australia

Introduction

Based on field observations, Gardiner (1961; 1967) reported that sheep with subacute and chronic lupinosis developed anemia and leukopenia. Hematological changes observed in sheep in which lupinosis was experimentally produced by the administration of an extract of *Diaporthe toxica* (anamorph *Phomopsis* spp.), the fungus that causes lupinosis (Williamson *et al.*, 1994), are first reported here.

Materials and Methods

Three experiments were conducted using Merino hogget wethers individually penned in an animal house. The sheep were given selenium and cobalt pellets and drenched with an anthelmintic when first put into the pens 2wks prior to the experiment starting. Throughout the experiment, the sheep were provided with commercial sheep cubes and water *ad libitum*, unless otherwise stated. Lupinosis was produced using a crude toxic extract of *D. toxica* prepared as described by Allen (1985). The toxic bioactivity of the extract (phomopsin A concentration) was determined using a nursling rat bioassay (Peterson, 1978; Petterson *et al.*, 1985). The gross and microscopic pathology observed in intoxicated sheep when they were euthanized confirmed the production of lupinosis.

For Experiment 1, four sheep were selected for intoxication and four of a similar weight served as pair-fed controls. On day zero, sheep in the intoxication group were given an intraperitoneal (i.p.) injection of 30g phomopsin A/kg. Daily feed intakes of the intoxicated sheep were measured, and the quantity consumed was offered to the respective pair-fed controls the next day. All intoxicated sheep became moribund on day five and were terminated. Their pair-fed controls were euthanized the next day. Blood samples were collected on days zero to five.

Experiment 2 was a second acute lupinosis trial with a single i.p. dose of phomopsin A, using six sheep for intoxication and three of a similar weight as controls. Apart from the controls not being pair-fed, this trial was conducted in an identical manner to the first.

Finally, in Experiment 3 using multiple i.p. doses of phomosin A a chronic lupinosis was created. Nine sheep were selected for intoxication and nine of similar weight served as pair-fed controls. On day zero, sheep in the intoxication group were given an i.p. injection of 6g phomopsin A/kg. Weekly feed intakes of the intoxicated sheep were measured, and the pair-fed controls were offered each day one seventh of the quantity of feed consumed by the respective intoxicated pair in the previous week. After day zero toxin was given weekly, and the amount dosed to each sheep varied depending upon their feed consumption and weight change in the previous week. The dose rates of phomosin A varied between 0-12g/kg, based on weight at the time. The aim was to produce a number of episodes of moderate liver damage over an extended period, as might occur under field conditions in sheep that develop chronic lupinosis. All sheep were weighed and had blood samples collected from them on day zero, then again weekly and immediately prior to being euthanized for necropsy. Six of the intoxicated sheep became moribund and were terminated during the experiment, while the remaining three were euthanized 18wks after intoxication commenced. The pair-fed controls were each euthanized 1wk after the death of their respective intoxicated pair.

All blood samples were collected in the mornings prior to carrying out other procedures. Collection was by jugular venipuncture directly into tubes containing EDTA anticoagulant. Hematological methods used were those described by Benjamin (1961). Total red and white cells were counted with a Royco (Fisher) 920A cell counter. Hemoglobin (Hb) was assayed using the cyanomethemoglobin method and reading the color reaction on a Linson photometer, and packed cell volume (PCV) was measured using the microhematocrit method. The various values recorded in the intoxicated sheep were compared with those in the control sheep and the normal ranges for sheep reported by Schalm *et al.* (1975). Student's *t* test was used for the detection of statistical significance.

Results

Experiment 1

The red cell concentrations in the intoxicated sheep were never significantly greater than in the pair-fed controls (Fig. 14.1), but the increase in red cell concentrations from day zero to day five in intoxicated sheep was significantly (*P*<0.05) greater than in the controls. Also, intoxicated sheep had significantly (*P*<0.025) greater red cell concentrations on day five than on day zero. Although remaining within normal ranges, the PCVs and Hb concentrations in the intoxicated sheep became significantly greater than those in the controls (Fig. 14.1). The mean total white cell and neutrophil concentrations, and the percentages of neutrophils, in the intoxicated sheep all became significantly greater than those in the controls (Fig. 14.2). All intoxicated sheep developed a leukocytosis and relative and absolute neutrophilias.

The intoxicated sheep had a significantly (*P*<0.05) lower concentration of lymphocytes on day five than day zero, and on day five they had a significantly lower concentration than the pair-fed controls (Fig. 14.3). However, although the intoxicated sheep developed a relative lymphocytopenia, the concentration of lymphocytes remained within the normal range.

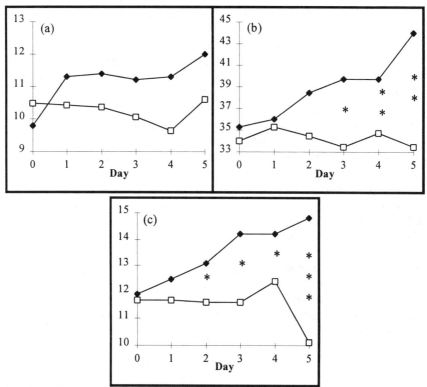

Fig. 14.1. Mean total red cell concentrations (a; 10⁶cells/ml), packed cell volumes (b; %) and hemoglobin concentrations (c; g/dl) in intoxicated (♦) and pair-fed control (□) sheep in Experiment 1. Asterisks indicate values that were significantly different (*P*<0.05, **P*<0.01, ***P*<0.001).

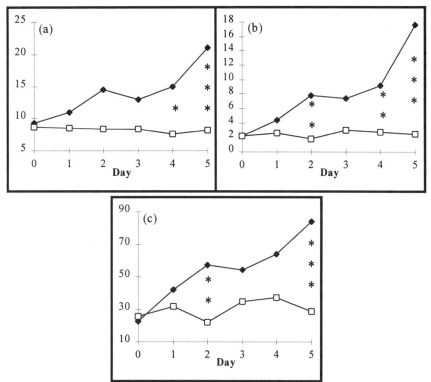

Fig. 14.2. Mean total white cell concentrations (a; 10³cells/ml), neutrophil concentrations (b; 10³cells/ml) and neutrophil percentages (c; %) in intoxicated (◆) and pair-fed control (□) sheep in Experiment 1. Asterisks indicate values that were significantly different (*P<0.05, **P<0.01, ***P<0.001).

Experiment 2

The results were similar to those of Experiment 1. One intoxicated sheep had a high terminal red blood cell concentration (17.6 x 10⁶cells/ml). Although the mean total red cell and Hb concentrations increased in the intoxicated sheep from day zero to day five (11.7-14.1 x 10⁶cells/ml and 11.7- 13.1g/dl, respectively), the increases were not significant and the values were never significantly different to those in the controls. The increase in mean PCVs in the intoxicated sheep from day zero to day five was significant (33-40%, P<0.05), but the values remained within the normal range and never significantly greater than those for the controls.

As in Experiment 1, the mean total white cell and neutrophil concentrations and the percentages of neutrophils in the intoxicated sheep all became significantly elevated above those in the controls. Three of the intoxicated sheep developed a leukocytosis and relative and absolute neutrophilias.

The mean lymphocyte concentrations in the intoxicated sheep remained within the normal range and never significantly differed from the controls, but two sheep on day four and one on day five had a lymphocytosis (9.4, 10.1 and 11.5 x 10^3cells/ml, respectively). In spite of this, the intoxicated sheep developed a mean relative lymphocytopenia and a significantly reduced lymphocyte percentage compared to the controls, but the decreases in lymphocyte percentages from day zero to day five for both groups of sheep were not significantly different.

Experiment 3

Two intoxicated sheep had become anemic (<8 x 10^6cells/ml) by the eighth week, and before the end, seven of the nine sheep were anemic. Packed cell volumes and Hb concentrations in the intoxicated sheep were also reduced, but in only three and two sheep, respectively, were the levels reduced to abnormally low levels (<24% and 8g/dl). The mean red cell concentrations, PCVs and Hb concentrations in the intoxicated sheep were significantly less than those in their pair-fed controls on most occasions from the eighth week onwards (Fig. 14.4), and the terminal mean values for these parameters were also significantly ($P<0.05$) less in the intoxicated sheep (7.7 *vs.* 12.3 x 10^6red cells/ml, 27 *vs.* 36% PCV and 8.9 *vs.* 12.0g Hb/dl).

Three intoxicated sheep developed a leukocytosis just prior to their death. On each occasion it was due to a neutrophilia. One other sheep developed a neutrophilia, but not a leukocytosis. However, three others developed a leukopenia leading up to their death, and one other had a gradual decrease in its white cell concentration from 8.1- 4.7 x 10^3/ml over the 8wks prior to its death. Two of the sheep with leukopenia had neutrophil and lymphocyte concentrations in the lower normal range, while the other had a neutropenia. At no time did the mean white cell and neutrophil concentrations in the intoxicated sheep (terminal mean values of 8.3 x 10^3cells/ml and 4.5 x 10^3cells/ml, respectively) differ significantly from those in the pair-fed

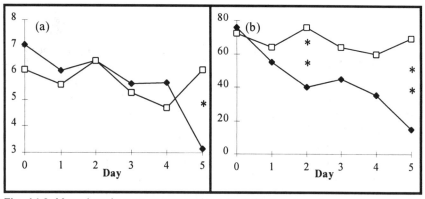

Fig. 14.3. Mean lymphocyte concentrations (a; 10^3cells/ml) and percentages (b; %) in intoxicated (♦) and pair-fed control (□) sheep in Experiment 1. Asterisks indicate values that were significantly different (*$P<0.05$, **$P<0.01$).

controls (terminal mean values of 7.5 x 10^3cells/ml and 2.9 x 10^3cells/ml, respectively).

Discussion

Experiment 3 confirms Gardiner's (1967) observation that sheep with chronic lupinosis develop anemia, with reductions in PCV and Hb concentration. Pathogenesis of the anemia is uncertain. Gardiner (1961) reported that bone marrow became only mildly hypoplastic in lupinosis, and that there were no toxic effects on hematopoiesis. A long-term reduced feed intake and accompanying gradual weight loss were ruled out because the pair-fed controls in this experiment did not develop these changes. Gardiner (1967) concluded that intravascular hemolysis was at least partially responsible for the anemia.

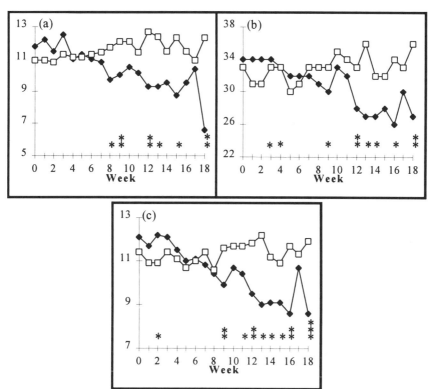

Fig. 14.4. Mean total red cell concentrations (a; 10^6cells/ml), packed cell volumes (b; %) and hemoglobin concentrations (c; g/dl) in intoxicated (♦) and pair-fed control (□) sheep in Experiment 3. Asterisks indicate values that were significantly different (*P<0.05, **P<0.01, ***P<0.001).

Results of Experiment 3 do not support Gardiner's (1961; 1967) observation that sheep with chronic lupinosis consistently develop a leukopenia, and associated depressed lymphocyte concentration. There was no consistent change in the white cells of sheep in Experiment 3: some developed a leukocytosis and others a leukopenia.

The changes in red blood cell parameters, and white cell and neutrophil concentrations, observed in Experiments 1 and 2, may have resulted from dehydration. The acutely intoxicated sheep stopped eating almost completely, and it is not known whether there was a similar reduction in water intake. The changes were quite remarkable, considering that substantial hemorrhage into the gastro-intestinal tract is one of the features of acute lupinosis (Allen, 1989).

The leukocytosis and neutrophilia seen in the acute lupinosis produced in Experiments 1 and 2 do not support the suggestion by Gardiner (1967) that acute lupinosis is not associated with any leukocytic changes. Apart from dehydration, uremia and/or acidosis, common features of acute lupinosis (Allen, 1989) may have contributed to these changes in the white cells (Benjamin, 1961).

References

Allen, J.G. (1985) The toxicity of phomopsin in a crude extract of *Phomopsis lepto-stromiformis* to sheep and pigs. In: Seawright, A.A., Hegarty, M.P., James, L.F. and Keeler, R.F. (eds), *Plant Toxicology*. Queensland Poisonous Plants Committee, Yeerongpilly, Qld, pp. 596-603.

Allen, J.G. (1989) *Studies of the Pathogenesis, Toxicology and Pathology of Lupinosis and Associated Conditions*. PhD Thesis, Murdoch University, Perth.

Benjamin, M.M. (1961) *Outline of Veterinary Clinical Pathology*, 2nd ed. Iowa State University Press, Ames, IA.

Gardiner, M.R. (1961) Lupinosis, an iron storage disease of sheep. *Australian Veterinary Journal* 37, 135-140.

Gardiner, M.R. (1967) Lupinosis. *Advances in Veterinary Science* 11, 85-138.

Peterson, J.E. (1978) *Phomopsis leptostromiformis* toxicity (lupinosis) in nursling rats. *Journal of Comparative Pathology* 88, 191-203.

Petterson, D.S., Peterson, J.E., Smith, L.W., Wood, P.McR. and Culvenor, C.C.L. (1985) Bioassay of the contamination of lupin seed by the mycotoxin phomopsin. *Australian Journal of Experimental Agriculture* 25, 434-439.

Schalm, O.W., Jain, N.C. and Carroll, E.J. (1975) *Veterinary Haematology*, 3rd ed. Lea and Febiger, Philadelphia, PA.

Williamson, P.M., Highet, A.S., Gams, W., Sivasithamparam, K. and Cowling, W.A. (1994) *Diaporthe toxica* sp. nov., the cause of lupinosis in sheep. *Mycological Research* 98, 1364-1368.

Chapter 15

Atypical Pneumonia Associated with Ryegrass Staggers in Calves

L.L. Blythe, C.B. Andreason, E.G. Pearson and A.M. Craig
College of Veterinary Medicine, Oregon State University, Corvallis, Oregon 97331, USA

Introduction

Turf quality perennial ryegrass *(Lolium perenne)* is frequently infected with an endophyte fungus (*Acremonium lolii*), since it increases resistance to insects and drought (Porter, 1995). In 1997, 160,000 acres in Oregon were planted with perennial ryegrass, approximately 80% of which was endophyte infected. The fungus produces varying quantities of tremorgenic toxins including lolitrem B, which has been associated with ryegrass staggers. This is a neurologic syndrome in sheep, cattle and horses that graze pastures or eat the straw fields containing endophyte infected perennial ryegrass (Galey *et al.*, 1991; Plumlee and Galey, 1994; Tor-Agbidye *et al.*, 1994). Clinical signs include fine tremors of the head and neck, spasticity, stiffness and incoordination of gait (Galey *et al.*, 1991; van Essen, 1995). Clinical signs increase when the animals are excited or forced to move and diminish when feeding with infected straw is discontinued. Ryegrass staggers, not usually related to abnormal respiratory signs, is typically regarded as a transient condition alleviated as soon as the animals are removed from the source of the toxin. This chapter relates a study of 17 recently weaned feeder calves that were introduced to perennial ryegrass feed and developed both ryegrass staggers and a severe acute onset of atypical interstitial pneumonia (Pearson *et al.*, 1996).

Farm Cases

In Oregon, 17 Hereford/Red Angus male and female calves weighing 136-181kg were weaned off a grass pasture and placed in drylot confinement. They were fed perennial ryegrass straw and started on a small grain ration, which was increased over a 10d period to 1.82kg/calf/d. Some of the calves begin to exhibit head tremors and a staggering, stiff gait 7d after initiation of perennial ryegrass feeding. The farmer treated them with mineral oil drenches, procaine penicillin G and 20mg dexamethasone intramuscularly (i.m.) since he suspected a grain overload. The treatment was repeated 10d and 13d later. A veterinarian was called when one calf died on day 23 of ryegrass feeding, and a tentative diagnosis of ryegrass staggers was made. The calves were removed from the perennial ryegrass feed and given alfalfa hay. In addition to the classic neurological signs, the affected calves had abnormal respiratory signs, which consisted of labored, rapid, and shallow respirations. At necropsy, the dead calf had emphysematous lungs.

Over the next few days, three more calves died, and four of the remaining 13 were transported to the College of Veterinary Medicine at Oregon State University for complete evaluation. One of the transported calves died en route and was later necropsied. Lesions were limited to the lungs, which were firm, diffusely gray, and emphysematous, with numerous bullae ranging from 2-10cm in diameter. Histologically, numerous interstitial neutrophils and mononuclear cells were seen, with small foci of fibrin and neutrophils in the alveoli. Mild, diffuse type-II pneumocytic hyperplasia was present, and the diagnosis was diffuse interstitial pneumonia with purulent inflammation and pulmonary bullae. Ruminal ingesta and a sample of the perennial ryegrass straw were submitted for HPLC analysis for lolitrem B (Tor-Agbidye *et al.*, 1994). The straw had 3,711±243µg/kg lolitrem B while the ruminal ingesta contained 185µg/kg of the toxin even though the calf had not eaten the endophyte-infected straw for 5d previously.

The three live calves had similar clinical signs of ryegrass staggers with obvious incoordination and head and body tremors. Thoracic and pelvic limbs had increased muscle tone and the calves sometimes assumed a wide-base stance. All other cranial nerve and neurological functions appeared normal. All three animals had labored, shallow respirations (60 breaths/min), groaned on expiration, and stood with extended heads and open mouths to breathe. Temperatures ranged from 39.4-41.3°C. Blood work revealed that the calves were hypoxic and hypercapnic and had elevated muscle-related enzymes. Thoracic radiographs of 2/6 calves revealed a hyper-inflated thorax and well-marginated, lucent foci in the lungs compatible with emphysematous bullae. A mixed alveolar and interstitial infiltrate, suggestive of interstitial lung disease or edema, was present. The calves were treated for atypical interstitial pneumonia with atropine and dexamethasone, and 2/3 improved and were discharged 8d after admission. The third calf died 7d after admission and necropsy findings were similar to those of the other calves.

Ten days later, the farmer elected to euthanize and necropsy six of the most severely affected calves including the aforementioned two calves, due to persistent respiratory signs and poor weight gains. These calves were trucked to the University, with one dying en route. The neurological signs were minimal in these calves, but the respiratory difficulties (>66 breaths/min) were still evident, with increased effort on expiration. Gross and histologic lesions were similar in all the calves. Lolitrem B concentrations in the rumen contents from the five calves ranged from 106-183µg/kg even though the calves had not ingested perennial ryegrass straw for 38d. Pearson *et al.* (1996) reported the pathological findings.

The differential diagnosis for the acute atypical interstitial pneumonia seen in these calves included the syndrome seen in adult cattle when transferred from poor pasture to lush pasture (Carlson *et al.*, 1972; Hammond *et al.*, 1979), inhalation of *Micropolysporum faeni* or ingestion of *Fusarium solani* from sweet potatoes, purple mint or other fungally infected plants (Hibbs *et al.*, 1981; Pearson, 1984), and insect fogger pneumonitis (Pearson, 1984). Cases of pulmonary edema and emphysema of unknown cause have been reported in weaned calves (Hibbs *et al.*, 1981; Johnson *et al.*, 1982). Other causes of pneumonia including bovine herpesvirus-1, parainfluenza-3, bovine respiratory syncytial virus, and bovine viral diarrhea virus were ruled out, although one calf was positive for parainfluenza-3 virus and one calf was positive for bovine viral diarrhea virus. No likely pathogens were isolated from specimens of lung or tracheobronchial lymph nodes.

Experimental Cases

An effort to reproduce the dual clinical syndromes with the remaining hay from the farm was attempted. Four clinically healthy 4-6mos-old mixed breed heifer calves (118-165kg) were used. Blood, cerebrospinal fluid and videotapes of the gait were obtained prior to initiation of *ad libitum* feeding of the perennial ryegrass with the high lolitrem B concentration and the same grain supplementation that the farmer had used. Within 4d, one calf developed the classic neurological syndrome described above, and by day five all four calves showed abnormal neurologic signs. By days 29 and 32, two of the calves were recumbent, unable to rise, and were euthanized and necropsied. The other two calves remained severely neurologically affected until 6d after the feed was switched to alfalfa hay. At that time they appeared neurologically normal and were returned to the perennial ryegrass straw diet for 11 and 16d. Neurological signs reappeared after 9d on the perennial ryegrass, but the calves never reached the initially observed severity. At no time did any of the four calves have clinical respiratory signs that were present in all of the calves of the field case. Pathological evaluation of the calves supported this clinical impression, with no lesions in the lungs. One calf did have a number of degenerate Purkinje cells in the cerebellum. Elevations of creatine kinase activity in the CSF was seen in 3/4 calves with values greater than 400IU/L, which was a 14 to 90-fold increase over values obtained before feeding the perennial ryegrass straw.

Summary

The factor that triggered the acute atypical interstitial pneumonia is still unknown. During that year, reports from four other farms indicated that the pneumonic problems were being seen in association with ryegrass staggers, but in-depth investigations were not conducted. This study does support the earlier reports (Tor-Agbidye *et al.*, 1994) that perennial ryegrass straw with lolitrem B levels greater than 2,000µg/kg produces the neurological syndrome of ryegrass staggers. In the experimental feeding trial, the second intoxication took longer for the calves to develop the neurologic signs and they were of lesser severity. This raises the possibility of increased resistance to the toxin with repeated exposure, possibly due to induction of detoxifying ruminal microbes. This theory remains to be proven.

References

Carlson, J.R., Yokoyama, M.T. and Dickinson, E.O. (1972) Induction of pulmonary edema and emphysema in cattle and goats with 3-methylindole. *Science* 176, 298-299.

Galey, F.D., Tracy, M.L., Craigmill, A.L., Barr, B.C., Markegard, G., Peterson, R. and O'Conner, M. (1991) Staggers induced by consumption of perennial ryegrass in cattle and sheep from northern California. *Journal of the American Veterinary Medical Association* 199, 466-470.

Hammond, A.C., Bradley, B.J. and Yokoyama, M.T. (1979) 3-Methylindole and naturally occurring acute bovine pulmonary edema and emphysema. *American Journal of Veterinary Research* 40, 1398-1401.

Hibbs, C.M., Frey, M.L. and Bohlender, R. (1981) Pulmonary edema and emphysema in feeder calves. *Modern Veterinary Practice* 62, 381-383.

Johnson, J.L., Pommer, J.L. and Hudson, D.B. (1982) Pulmonary emphysema in weaned calves: laboratory diagnosis. *Proceedings of the American Association of Bovine Practitioners* 14, 122-126.

Pearson, E.G. (1984) Diagnosis of causes of respiratory diseases in cattle. *Modern Veterinary Practice* 65, 204-209.

Pearson, E.G., Andreasen, C.B., Blythe, L.L. and Craig, A.M. (1996) Atypical pneumonia associated with ryegrass staggers in calves. *Journal of the American Veterinary Medical Association* 209, 1137-1142.

Porter, J.K. (1995) Analysis of endophyte toxins: fescue and other grasses toxic to livestock. *Journal of Animal Science* 73, 871-880.

Plumlee, K.H. and Galey, F.D. (1994) Neurotoxic mycotoxins: a review of fungal toxins that cause neurological disease in large animals. *Journal of Veterinary Internal Medicine* 8, 49-54.

Tor-Agbidye, J., Blythe, L.L. and Craig, A.M. (1994) Correlation of quantities of ergovaline and lolitrem B toxins in clinical cases of tall fescue toxicosis and perennial ryegrass staggers. In: Colegate, S.M. and Dorling, P.R. (eds), *Plant-Associated Toxins: Agricultural, Phytochemical and Ecological Aspects*. CAB International, Wallingford, Oxon, pp. 369-374.

van Essen, G.J. (1995) Ryegrass cramps in horses. *Tijdschrift voor Diergeneeskunde* 120, 710-711.

Chapter 16

Dissimilatory Metabolism by Ruminal Microbes: Impact on Ruminant Toxicoses

M.A. Rasmussen[1] and R.C. Anderson[1,2]

[1]*USDA-ARS, Metabolic Diseases and Immunology Research Unit, National Animal Disease Center, Ames, Iowa 50010, USA;* [2]*Milk Specialties BioScience Division, Dundee, Illinois 60118, USA*

Introduction

Microbial activities in the rumen are predominately anaerobic, fermentative processes, which metabolize carbohydrates and proteins into volatile fatty acids and provide energy for microbial growth through substrate-level phosphorylation, providing nutrients to the host animal and a means of protection against toxicants of dietary origin. The metabolic processes of anaerobic fermentation produce a surplus of reducing equivalents that must be disposed of in an environment that has very little oxygen available as a terminal electron acceptor. Therefore, in the rumen, fermentation is intricately linked to anaerobic respiration or dissimilatory metabolism, in which microbes use compounds other than oxygen as a terminal electron acceptor. Dissimilatory metabolism contributes to the reduction of many oxidized minerals (Lovely, 1993). Recent studies have documented the important role that these fermentative microbes play in terrestrial mineral cycles and their contribution to the degradation of organic matter in anoxic environments.

Dissimilatory Metabolism

Dissimilatory metabolism is a process whereby electrons generated from organic compounds or hydrogen are shifted to an oxidized "electron acceptor" like sulfate or nitrate. Oxygen-linked respiration provides the largest quantity of free energy for growth, but other less energetically favored substrates replace O_2 in anaerobic environments (Table 16.1). All forms of dissimilatory metabolism conserve energy by variations of a membrane bound electron transport chain. Higher-energy electron acceptors yield more energy for microbial growth (NO_3>SeO_4), and thus are preferred by microbes capable of using more than one substrate. Regardless of the electron acceptor, these "respiratory" pathways are used to power membrane transport, generate transmembrane gradients, and produce ATP.

The term "dissimilatory" is used primarily to distinguish this energy yielding mechanism from processes used for the assimilation of minerals needed in the synthesis of cellular components (assimilatory metabolism). It is also important to distinguish dissimilatory metabolism from other reductive mechanisms that microbes employ strictly for purposes of detoxification. Whereas dissimilatory metabolism yields energy for microbial growth, microbes that detoxify minerals like selenium for their own survival must expend energy to do so.

Nitrogen

Nitrate is a high-potential electron acceptor that is toxic to ruminants only when microbial activity rapidly dissimilates it to nitrite in the rumen. In contrast, nitropropanol and nitropropionic acid are toxic nitroalkanes, which upon reduction are converted to their nontoxic amines, aminopropanol and β-alanine, respectively (Anderson *et al.*, 1993).

In carbon-rich environments like the rumen, the predominant end-product of nitrate reduction is ammonia, where environments low in carbon reduce oxidized forms of nitrogen to molecular nitrogen (Cole, 1988). Rumen microbes capable of dissimilatory reduction of oxidized nitrogen compounds include *Desulfovibrio* spp. and *Wolinella succinogenes*. Other more recent isolates are discussed in Chapter 32 (Anderson *et al.*, 1993). Enterobacteriaceae also possess the capacity for dissimilatory reduction of nitrate and nitrite, but this group is not prominent in ruminal populations (Cole, 1996). Nitrate-reducing bacteria contain b-type cytochromes and molybdenum-coupled nitrate reductases. High tungsten concentrations interfere with reductase activity, and this has been exploited as a treatment for nitrate intoxication. The presence of b-type cytochromes in other ruminal bacteria (*Anaerovibrio lipolytica*, *Selenomonas ruminantium* and *Veillonella alcalescens*) suggests that they

Table 16.1. A hierarchy of some common electron acceptors.

Electron Acceptor[1]	Reduced Product
O_2	CO_2
NO_3	NH_4
Fe(III)	Fe(II)
Mn(IV)	Mn(II)
Cr(VI)	Cr(III)
SeO_4	SeO_3
AsO_4	AsO_3
SO_4	SO_3
CO_2	CH_4

[1]In decreasing order of free energy, $\Delta G^{o'}$

too contain a respiratory nitrate reductase, although this has not yet been proven.

Rapid rates of nitrate or nitrite reduction require an abundant supply of reducing equivalents that arise from rapid carbohydrate fermentation. Many substrates including formate, hydrogen, lactate, succinate, citrate and malate can support nitrate reduction. Ruminal nitropropanol and nitropropionic acid reduction are supported by hydrogen and formate, but not by carbohydrate additions (Anderson *et al.*, 1996).

Iron

Ferric iron is a relatively insoluble, nontoxic mineral that undergoes microbial dissimilation to the more soluble and bioavailable ferrous form. Ferric iron is a relatively high-energy electron acceptor, but in the rumen its reduction is restricted due to its solubility. The formation of insoluble ferrous sulfide restricts iron's bioavailability. The chemistry of non-microbial iron transformations in the rumen is better understood than the microbial pathways. A recent study of sediments indicated that iron reduction can be carried out by the sulfate-reducing bacteria including *Desulfovibrio desulfuricans* (Nealson and Saffarini, 1994). Microbial reduction of iron is stimulated by chelation agents that make iron more available for dissimilation.

The importance of iron in the degradation of organic matter in underground aquifers has only recently been appreciated. Iron is a major factor influencing the quality of ground water for human and agricultural use, both providing a dissimilatory mechanism for the degradation of organic contamination and lowering water quality by the presence of dissolved ferrous iron.

Sulfur

Sulfur in the form of sulfate, sulfite and sulfur-containing amino acids is reduced by microbial activity to sulfide, which occurs as insoluble sulfide precipitates (cupric thiomolybdate) or as toxic hydrogen sulfide. Although sulfur compounds have a relatively low free-energy potential (Table 16.1), they are rapidly reduced to sulfide because of their solubility. Sulfur compounds can divert electron flow from carbon dioxide (CO_2) reduction and thus inhibit methane formation.

The microbes responsible for the reduction of sulfur compounds are some of the best-known species of dissimilatory bacteria. In the rumen they include *D. desulfuricans*, *D. sapovorans* and *Desulfotomaculum ruminis*. The sulfate-reducing bacteria have a wide range of metabolic capabilities and can metabolize fatty acids and simple aromatic compounds in addition to substrates like formate and molecular hydrogen (H_2). The production of fatty acids and aromatic species tends to be slower than the rumen's rate of turnover, which limits the degradation of fatty acids that can be used for energy production by the host animal.

Arsenic

A mixed rumen inoculum can reduce arsenate to arsenite, but does not carry the reduction to the fully reduced arsine. The toxicity of arsenic increases as reduction increases its solubility and availability, and it strongly inhibits rumen fermentation, gas production and bacterial growth. The ruminal bacterium *Megasphaera elsdenii* displays a relatively high resistance to arsenite inhibition, but the basis for this resistance is unknown. However, inferences can be made based upon other anoxic habitats like lake sediment (Dowdle *et al.*, 1996). Arsenate reduction in sediments can be increased with lactate and H_2 additions whereas the reaction can be inhibited with respiratory uncouplers, indicating that arsenic reduction follows a true dissimilatory mechanism. Arsenate reduction is inhibited by nitrate additions but is unaffected when sulfate is added as a competing substrate.

Interspecies Hydrogen Transfer

Interspecies H_2 transfer is a critical interaction between rumen microbes linking fermentative and dissimilatory metabolism. Reducing equivalents in the form of H_2 are transferred from H_2-producing species (those that degrade complex organic compounds) to H_2-utilizing species (sulfate reducers, nitrate reducers and methanogens). Hydrogen is a significant regulatory feedback molecule, and its removal promotes efficient fermentation, increased organic matter degradation, greater ATP synthesis and increased microbial cell yields, which benefit the host animal.

Impact of Dissimilatory Metabolism on the Livestock Industry

Dissimilatory metabolism impacts livestock health and production in terms of nutrition, toxicology and the environment. Nutritionally, dissimilatory metabolism is responsible for many of the mineral transformations that occur in the rumen, either increasing or decreasing bioavailability of a mineral (Jonnalagadda and Prasada Rao, 1993). There is a growing interest in the use of organic trace minerals in the livestock industry (Spears, 1996), but the stability of these compounds during rumen fermentation is uncertain and requires further research.

The toxicological properties of some minerals can be influenced by dissimilatory microbes. Those minerals that become less soluble after reduction (selenium, chromium) tend to be less toxic than their oxidized counterparts (Ohtake and Silver, 1994; Daniels, 1996). Those elements that are reduced to more available forms (sulfur and arsenic) tend to be more toxic. Toxic intermediates like nitrite and selenite can also accumulate as a result of dissimilatory metabolism in the rumen.

Dissimilatory metabolism also has an impact upon environmental issues pertinent to the ruminant livestock industry. Methane production by ruminants has been identified as one source of this greenhouse gas, and its production has instigated many methanogenesis research programs. Methanogenesis is one of the least energetically favorable dissimilatory pathways for the rumen flora, but it predominates because of a scarcity of acceptors with higher energy potentials. One strategy for reducing ruminant methane production is to divert the rumen fermentation's reducing equivalents to environmentally benign or energetically productive electron acceptors (Tamminga, 1996). Further progress awaits a better understanding of rumen dissimilatory processes.

References

Anderson, R.C., Rasmussen, M.A. and Allison, M.J. (1993) Metabolism of the plant toxins nitropropionic acid and nitropropanol by ruminal microorganisms. *Applied and Environmental Microbiology* 59, 3056-3061.

Anderson, R.C., Rasmussen, M.A. and Allison, M.J. (1996) Enrichment and isolation of a nitropropanol metabolizing bacterium from the rumen. *Applied and Environmental Microbiology* 62, 3885-3886.

Cole, J. (1988) Assimilatory and dissimilatory reduction of nitrate to ammonia. In: Cole, J.A. and Ferguson, S.J. (eds), *The Nitrogen and Sulphur Cycles*. Cambridge University Press, Cambridge, UK, pp. 281-329.

Cole, J. (1996) Nitrate reduction to ammonia by enteric bacteria: Redundancy, or a strategy for survival during oxygen starvation. *FEMS Microbiology Letters* 136, 1-11.

Daniels, L.A. (1996) Selenium metabolism and bioavailability. *Biological Trace Element Research* 54, 185-199.

Dowdle, P.R., Laverman, A.M. and Oremland, R.S. (1996) Bacterial dissimilatory reduction of arsenic (V) to arsenic (III) in anoxic sediments. *Applied and Environmental Microbiology* 62, 1664-1669.

Jonnalagadda, S.B. and Prasada Rao, P.V.V. (1993) Toxicity, bioavailability and metal speciation. *Comparative Biochemistry and Physiology* 106C, 585-595.

Lovely, D.R. (1993) Dissimilatory metal reduction. *Annual Review of Microbiology* 47, 263-269.

Nealson, K.H. and Saffarini, D. (1994) Iron and manganese in anaerobic respiration. *Annual Review of Microbiology* 48, 311-343.

Ohtake, H. and Silver, S. (1994) Bacterial detoxification of toxic chromate. In: Chaudhry, G.R. (ed), *Biological Degradation and Bioremediation of Toxic Chemicals*. Dioscorides Press, Portland, OR, pp. 403-415.

Spears, J.W. (1996) Organic trace minerals in ruminant nutrition. *Animal Feed Science Technology* 58, 151-163.

Tamminga, S. (1996) A review on environmental impacts of nutritional strategies in ruminants. *Journal of Animal Science* 74, 3112-3124.

Chapter 17

Yew (*Taxus* spp.) Poisoning in Domestic Animals

W.R. Hare

National Animal Poison Control Center, Department of Veterinary Biosciences, College of Veterinary Medicine, University of Illinois, Urbana, Illinois 61801, USA

Introduction

Yew (*Taxus* spp.) poisoning generally occurs due to ingestion of decorative shrubs or their trimmings, and continues to be an occasional problem in domestic animals. Sudden death is often the only clinical sign reported. There is no antidote or specific treatment for yew poisoning. Diagnosis is based on exposure history and finding the characteristic foliage in the animal's rumen or gastric contents. Prompt and aggressive decontamination measures may be beneficial.

Yews are rapidly growing evergreen shrubs with worldwide distribution. They are commonly used in landscape architecture, which has resulted in a widespread naturalized population of yews throughout the US.

The yew belongs to the plant family Taxaceae. Members of this family include the European yew *(Taxus baccata)*, Western yew *(T. brevifolia)*, American yew *(T. canadensis)*, Chinese yew *(T. chinensis)* and Japanese yew *(T. cuspidata)*. The European yew and the Japanese yew have both become well established in the US as ornamental shrubs. These two species grow as bushes and seldom reach greater than 6m in height. The Chinese yew grows as a tree or shrub, reaching 15m in height. The Western yew is an evergreen tree up to 23m tall, ranging from California to Montana and North to Alaska. The American yew, commonly referred to as ground hemlock, grows as a spreading shrub, up to 1-2m in height. The yew is characterized by alternate, stiff, flat to needle-like leaves, 1-2.5cm long and 1-3mm wide, with upper dark-green and lower pale-green or yellow-green surfaces, a prominent mid-rib, and inconspicuous, unisexual flowers. Their fruit is a bright scarlet, ovoid, fleshy, cupped-berry (aril), 1-2cm long which surrounds a very small, hard, single, dark-brown nutlet. The trunk and branches are covered by a thin, flaky, reddish-brown bark.

Toxicology

Yews are known to contain numerous toxic substances, including at least ten alkaloids, nitriles (cyanogenic glycoside esters), ephedrine and an essential oil (oil of yew) (Kingsbury, 1964; Clarke and Clarke, 1967; Beal, 1975; Khan and Parveen, 1987; Ogden, 1988). The principle toxin is taxine, a non-irritating alkaloid found in all parts of the plant except for the red, fleshy aril. Taxine has been studied for over 125yrs, and is multiple similar alkaloids (Miller, 1980), named taxine A, B, C, etc. Taxine B is the most significant and remains intact in dried foliage. It disrupts sodium and calcium currents in cardiac myocytes, blocking myocardial conduction (Tekol, 1991; Leikin and Paloucek, 1996), which results in cardiac arrest. Taxine is rapidly absorbed, metabolized, conjugated in the liver, and eliminated as conjugated hippuric acid by the kidney (Clarke and Clarke, 1967). Herbivores have the ability to convert large quantities of benzoic acid to hippuric acid, and so may have a greater ability to eliminate taxine.

Oil of yew is a known intestinal irritant, and is most likely responsible for the colic and diarrhea reported in animals exhibiting a subacute clinical syndrome (Kingsbury, 1964; Clarke and Clarke, 1967). There are variations in toxicity of yew. The Japanese yew and European yew are considered to be the most toxic species (Kingsbury, 1964) and male plants are more toxic than female plants (Ogden, 1988). More mature portions of the plant contain higher concentrations of toxins (Thomson and Barker, 1978; Maxie, 1991).

Reports on the harmful affects of yew ingestion are well documented (Alden *et al.*, 1977; Thomson and Barker, 1978; Kerr and Edwards, 1981; Ogden, 1988; Maxie, 1991; Tekol, 1991; Helman *et al.*, 1996) and poisoning has been reported in humans, deer, cattle, sheep, goats, pigs, horses, burros, alpacas, llamas, dogs, chickens, rabbits, rats and mice. Cattle are the most common animal poisoned.

Ingestion of green foliage is toxic to equines at 0.5g/kg, to other monogastrics at 1g/kg, and to ruminants at approximately 5g/kg (Kingsbury, 1964). A few animals have been reported to recover spontaneously, even after the development of severe adverse clinical signs. Poisoned dogs have survived with medical treatment (Evans and Cook, 1991). There are reports of some wildlife, particularly deer, having ingested the plants without untoward effects.

Yew poisoning is usually peracute and the animal is found dead. When the dose of toxin is smaller, clinical signs are uneasiness, trembling, dyspnea, staggering, weakness and diarrhea. Cardiac arrhythmias increase and are eventually fatal (Leikin and Paloucek, 1996). Necropsy findings are unremarkable and nonspecific. Pulmonary, hepatic and splenic congestion are generally present, and a mild inflammation of the upper gastrointestinal tract may also be found in animals dying subacutely (Kingsbury, 1964; Clarke and Clarke, 1967).

There is no specific treatment for yew poisoning. Induction of emesis or gastric lavage, followed by activated charcoal and saline cathartic may prevent absorption of a fatal dose if taxine. Rumenotomy and removal of rumen contents has been effective in some cases (Clarke and Clarke, 1967). Treatment with atropine to counteract bradycardia (Kingsbury, 1964; Ogden, 1988) is not always effective (Leikin and Paloucek, 1996), and slows peristalsis, prolonging elimination of the toxic material. Other supportive measures could include intravenous fluids, electrolyte maintenance, norepinephrine, calcium gluconate, and/or tripelennamine (Evans and Cook, 1991; Barragry, 1994; Leikin and Paloucek, 1996). Seizure activity may be controlled with diazepam or pentobarbital.

References

Alden, C.L., Fosnaugh, C.J., Smith, J.B. and Mohan, R. (1977) Japanese yew poisoning of large domestic animals in the Midwest. *Journal of the American Veterinary Medical Association* 170, 314-316.

Barragry, T.B. (1994) *Veterinary Drug Therapy.* Lea and Febiger, Malvern, PA, p. 213.

Beal, J. (1975) Poisonous properties of *Taxus.* International Taxus Symposium, Ohio Agricultural Research and Development Center, Wooster, OH.

Clarke, E.G.C. and Clarke, M.L. (1967) *Garner's Veterinary Toxicology,* 3rd ed. Williams and Wilkins Co., Baltimore, MD, pp. 11-12, 399-401.

Evans, K.L. and Cook, J.R. (1991) Japanese yew poisoning in a dog. *Journal of the American Animal Hospital Association* 27, 300-302.

Helman, R.G., Fenton, K., Edwards, W.C., Panciera, R.J. and Burrows, G.E. (1996) Sudden death in calves due to *Taxus* ingestion. *Agricultural Practice* 17, 16-18.

Kerr, L.A. and Edwards, W.C. (1981) Japanese yew: A toxic ornamental shrub. *Veterinary Medicine; Small Animal Clinics* 76, 1339-1340.

Khan, N. and Parveen, N. (1987) The constituents of the genus *Taxus. Journal of Scientific Industrial Research* 46, 512-516.

Kingsbury, J.M. (1964) *Poisonous Plants of the United States and Canada.* Prentice Hall, Englewood Cliffs, NJ, pp. 121-123.

Leikin, J.B. and Paloucek, F.P. (1996) *Poisoning and Toxicology Handbook,* 2nd ed. Lexi-Comp Inc., Hudson, OH, pp. 1159-1160.

Maxie, G. (1991) Another Japanese yew poisoning. *Canadian Veterinary Journal* 32, 370.

Miller, R.W. (1980) A brief survey of *Taxus* alkaloids and other taxane derivatives. *Journal of Natural Products* 43, 425-437.

Ogden, L. (1988) *Taxus* (Yew): A highly toxic plant. *Veterinary and Human Toxicology* 30, 563-564.

Tekol, T. (1991) Acute toxicity of taxine in mice and rats. *Veterinary and Human Toxicology* 33, 337-338.

Thomson, G.W. and Barker, I.K. (1978) Japanese yew *(Taxus cuspidata)* poisoning in cattle. *Canadian Veterinary Journal* 19, 320-321.

Chapter Eighteen

Pasture Mediated Acute Bovine Pulmonary Emphysema and Edema in British Columbia

W. Majak[1], L. Stroesser[1], R.E. McDiarmid[1] and G.S. Yost[2]
[1]Agriculture and Agri-Food Canada, Range Station, 3015 Ord Road, Kamloops, British Columbia V2B 8A9, Canada; [2]Department of Pharmacology and Toxicology, University of Utah, Salt Lake City, Utah 84112, USA

Introduction

Seasonal losses of range cattle associated with non-infectious, acute respiratory distress have been known in the interior of British Columbia (BC) since the early years of the twentieth century (Brink *et al.*, 1976). The clinical signs are similar to those manifested in acute bovine pulmonary emphysema and edema (ABPE), a naturally occurring disease of adult cattle characterized by sudden respiratory distress and increased respiration (Dickinson and Carlson, 1978; Carlson and Yost, 1989). Death can occur within 48hrs of initial signs, especially after exertion. The disease occurs in many parts of the world, including the US, Canada and Europe. The clinical abnormalities appear within 2-4d following a drastic change from a dry, low-energy diet to a relatively lush, rapidly growing feed. The former is typical of the late summer forage in the dry interior forest range and the latter is typical of the fall regrowth on grassland range, especially when there is adequate moisture during September and October.

A 4yr (1993-1996) study was conducted near Kamloops, BC, to determine the etiology of the ABPE syndrome in the interior of BC and to test the efficacy of the monensin controlled-release capsule (Rumensin CRC®, Elanco, Division of Eli Lilly Canada, Inc.) for the prevention of ABPE. Earlier studies in the US showed that monensin was effective in the prevention of ABPE in cattle when given either intraruminally (Carlson *et al.*, 1983) or as a feed supplement (Potchoiba *et al.*, 1992) at 200mg/head/d. The Rumensin CRC® (21mm orifice) releases monensin at a rate of 272mg/head/d (Merrill and Stobbs, 1993). A number of studies have shown that monensin reduces the ruminal putrefaction of tryptophan to 3-methylindole (MI), the putative pneumotoxic agent in ABPE (Potchoiba *et al.*, 1992; Honeyfield *et al.*, 1985). 3-Methylindole causes selective damage to pulmonary tissue, and is bioactivated to toxic metabolites by cytochrome P_{450} enzymes (Yost, 1989). A highly specific enzyme-linked immunosorbent assay (ELISA) was recently developed for the detection of adducts of MI to proteins in tissues (Kaster and Yost, 1997). An increase in plasma concentrations of MI-protein adducts would be consistent with an increase in circulating levels of bioactivated MI. Our objective was also to determine MI levels in the rumen of cattle treated with Rumensin CRC® and in untreated controls.

Materials and Methods

Two groups of ten Hereford cows (2-11yrs old) were used each year. The cattle grazed a typical forest range near Pass Lake, BC, during summer. At the end of September they were trailed to Field 3 on the upper grasslands. Animals grazing Field 3 had a long history of ABPE (Table 17.1). One group of cows were given Rumensin CRC® while the other group served as controls. The capsules were administered while cattle were still on the summer range, 10d prior to exposure to Field 3. Cattle were stomach pumped for rumen contents after grazing Field 3 for 5d in 1993, 3d in 1994 and 1995, and 2d in 1996. In 1994 and 1995, the microbial capacity to generate MI from indoleacetic acid (IAA) was examined using rumen fluid samples from 40 cattle on the grassland pasture. The incubations were conducted under anaerobic conditions with freshly collected rumen contents (40ml) using 5mM IAA as the substrate in a 6hr incubation at 39°C.

Each sample of rumen contents (50ml, unfiltered) collected for MI evaluation was preserved with 2ml 5% $HgCl_2$. A 2ml aliquot was extracted with hexane (2x20ml) using a vortex mixer, and the extract was concentrated under vacuum (34°C, 200mbars) just to dryness and reconstituted in a final volume of 0.5ml hexane. 3-Methylindole was determined by GC carried out on DB-17 (0.25mm x 30m fused silica capillary column, J&W Scientific) using flame ionization detection (FID), split injection, helium as the carrier gas and 1-methylindole as the internal standard (injector 300°C, oven 150°C, dectector 330°C, Fig. 18.1). The quantitative detection level for MI in rumen fluid was 1ppm, which is well within the range of values reported for cattle after an abrupt change to lush forage (Carlson *et al.*, 1983).

Blood samples were collected from 20 animals on test in 1996, before and after the abrupt change to the lush forage on the grassland pasture, and were subjected to the ELISA.

Table 18.1. Incidence of acute bovine pulmonary edema (ABPE) in cattle on field three in the upper grasslands near Kamloops, British Columbia, from 1979 to 1996.

Year	Number affected[a]
1979	2
1980	2
1981	3
1982	1
1983	1
1986	5
1988	5
1989	6
1992	4
1993	2
1994	4
1995	2
1996	9

[a]Sixteen animals grazed the field from 1979 to 1993, and 20 from 1994 to 1996.

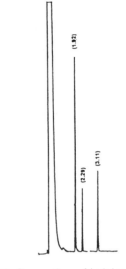

Fig. 18.1. Separation of indoles by GC on DB-17 (retention time in brackets): 1-methylindole (1.92 min), indole (2.29 min) and MI (3.11 min).

Results

The number of ABPE cases in the experimental groups over time is listed in Table 18.1. No MI was detected in the rumen contents of affected animals (<1ppm). Rumen content samples from 40 animals on test produced no MI when incubated with IAA under appropriate conditions. The blood samples subjected to the ELISA test gave negative results. No increase in plasma MI-protein adducts was detected (Table 18.2).

Discussion

Eight cases of ABPE occurred from 1993-95, only one of which was in the CRC group, but in general the clinical signs were not severe. In four of the cases, the disease affected 2yr-old first-calf heifers. In the fall of 1996, the moisture conditions were ideal for a potential outbreak of ABPE. Nine severe cases of ABPE occurred in 1996 (Table 18.1) and in seven of the cases, the disease affected 2yr-old first-calf heifers. Classical ABPE is considered to be a disease of mature cattle, those that are more than 2-3yrs old (Dickinson and Carlson, 1978). In 1996, five of the nine cases occurred in the CRC group, indicating that Rumensin CRC® may not be effective in the prevention of the local syndrome of ABPE. These results suggested that the disease was induced by a pneumotoxin other than MI.

A 4yr search for MI in the rumen fluid of 80 cattle on test proved to be fruitless. Our GC detection limits for MI in rumen fluid were adequate (<1ppm), as were the recoveries of synthetic MI from spiked rumen fluid (>80%). These results further suggested the presence of an unknown pneumotoxin.

Formation of MI in the rumen requires the conversion of tryptophan to IAA, which is then decarboxylated by strains of *Lactobacillus* to produce MI (Honeyfield and Carlson, 1990). In rumen content samples collected in 1994-95, there was no sign (<1ppm) of MI production, suggesting the absence in the rumen of the specific strains of *Lactobacillus* required for MI production.

A small but statistically significant decrease was observed in the CRC group (Table 18.2), but this may not have therapeutic meaning since one-half of the treated animals were affected by the disease.

In summary, (1) the relatively high incidence of ABPE among young cattle; (2) the apparent inefficacy of Rumensin CRC® for the prevention of the syndrome; (3) the failure to detect MI in rumen contents or to generate it *in vitro* from IAA in rumen fluid from cattle grazing fall regrowth; and (4) the negative results with the serum ELISA test for MI-protein adducts strongly indicate the action of a pneumotoxic agent yet to be identified.

Table 18.2. Levels of plasma 3-methylindole adducts in cattle before and after exposure to the grassland forage in 1996. Comparisons were made within treatments groups or within affected groups.

Group	MI protein adducts (absorbance/µg)		SE	*P*
	Before	After		
Control	58.9	58.8	0.9	n.s.
Rumensin-CRC®	60.7	56.3	1.3	<0.01
Affected by ABPE	59.0	57.3	1.6	n.s.
Unaffected by ABPE	60.8	57.9	1.1	n.s.

Acknowledgments

The authors would like to thank Keith Ogilvie for care and handling of livestock and Diane Lanza for technical assistance with the ELISA studies. We also thank Elanco and Dr John Merrill for supplying Rumensin CRC®.

References

Brink, V.C., Clapp, J.B., Salisbury, P.J. and Stovell, P.J. (1976) *Pasture Mediated Bovine Pulmonary Emphysema on British Columbia Ranges.* University of British Columbia, Department of Plant Science, Vancouver, BC.

Carlson, J.R. and Yost, G.S. (1989) 3-Methylindole-induced acute lung injury resulting from ruminal fermentation of tryptophan. In: Cheeke, P.R. (ed), *Toxicants of Plant Origin, Vol III: Proteins and Amino Acids.* CRC Press, Boca Raton, FL, pp. 107-123.

Carlson, J.R., Hammond, A.C., Breeze, R.G., Potchoiba, M.J. and Heinemann, W.W. (1983) Effect of monensin on bovine ruminal 3-methylindole production after abrupt change to lush pasture. *American Journal of Veterinary Research* 44, 118-122.

Dickinson, E.O. and Carlson, J.R. (1978) Acute respiratory distress of rangeland cattle. In: Keeler, R.F., Van Kampen, K.R. and James L.F. (eds), *Effects of Poisonous Plants on Livestock.* Academic Press, New York, NY, pp. 251-259.

Honeyfield, D.C. and Carlson, J.R. (1990) Effect of indoleacetic acid and related indoles on *Lactobacillus* sp. strain 11201 growth, indoleacetic acid catabolism, and 3-methyl-indole formation. *Applied and Environmental Microbiology* 56, 1373-1377.

Honeyfield, D.C., Carlson, J.R., Nocerini, M.R. and Breeze, R.G. (1985) Duration of inhibition of 3-methylindole production by monensin. *Journal of Animal Science* 60, 226-231.

Kaster, J.K. and Yost, G.S. (1997) Production and characterization of specific antibodies: utilization to predict organ- and species-selective pneumotoxicity of 3-methylindole. *Toxicology and Applied Pharmacology* 143, 324-327.

Merrill, J.K. and Stobbs, L.A. (1993) The effect of the monensin controlled release capsule on the incidence and severity of bloat and average daily gain in animals grazing legumes. *Journal of Animal Science* 71(Suppl. 1) Abstr., 181.

Potchoiba, M.J., Carlson, J.R., Nocerini, M.R. and Breeze, R.G. (1992) Effect of monensin and supplemental hay on ruminal 3-methylindole formation in adult cows after abrupt change to lush pasture. *American Journal of Veterinary Research* 53, 129-133.

Yost, G.S. (1989) Mechanisms of 3-methylindole pneumotoxicity. *Chemical Research in Toxicology* 2, 273-279.

Chapter 19

A Study of Persin, the Mammary Cell Necrosis Agent from Avocado (*Persea americana*), and its Furan Derivative in Lactating Mice

P.B. Oelrichs[1], S. Kratzmann[1], J.K. MacLeod[2], L. Schäffeler[2], J.C. Ng[1] and A.A. Seawright[1]

[1]*The National Research Centre for Environmental Toxicology, 39 Kessels Road, Coopers Plains, Queensland 4108, Australia;* [2]*The Research School of Chemistry, Australian National University, Canberra, Australia*

Introduction

Lactating livestock that eat a Guatemalan variety of avocado (*Persea americana*) leaves may develop a non-infectious mastitis and agalactia associated with extensive coagulative necrosis of the secretory epithelium (Kingsbury, 1964). Frequently, complete loss of milk production occurs in cattle for that lactation. The disorder affects all domestic animals and rabbits, and recently laboratory tests have shown mice are suitable test animals (Sani *et al.*, 1994). Mice were used as test animals to monitor the isolation of the active principle termed "persin" from avocado leaves. This was identified as (Z,Z)-1-(acetyloxy)-2-hydroxy 12,15-heneicosadien-4-one (Fig. 19.1, 1). Persin when dosed to mice was found to produce identical lesions to leaves from the avocado (Oelrichs *et al.*, 1995a). Apart from the action of persin on the mammary gland, it is cardioactive when administered in high doses. Persin was screened for activity against human cell lines including breast cancer cell lines. In these tests a differential sensitivity was shown by human breast cancer cell lines suggestive that persin has specific activity against breast cancer cells *in vitro* (Oelrichs *et al.*, 1995b).

Because mammary tissue necrosis has only been observed in animals that are dosed orally with persin, it seemed reasonable to assume that mild acid treatment of persin in the gut to form its furan derivative may be essential for its activity. It was therefore decided to determine in what form the active principle occurred in the target organs at various times after dosing with persin. To establish the stereochemistry of persin, both the R and S isomers were synthesized and tested for activity, as were a number of closely related compounds.

Methods and Materials

Milled, freeze-dried Guatemalan avocado leaf (200g) was extracted in a Soxhlet extractor with chloroform ($CHCl_3$) for 18hrs (Oelrichs *et al.*, 1995a). After removal of the solvents under reduced pressure, the dried residue was purified by silica gel and Florisil chromatography. Fractions collected were monitored and the active fractions combined. The active fractions were further purified by reverse phase and gel permeation chromatography, and finally preparative silica HPLC.

Initial testing of crude extracts in the isolation procedure was carried out by mixing powdered rodent diet with the test samples and drying under reduced pressure. The dried mixture was fed to lactating mice with pups 4d old. Alternatively, samples were administered as a suspension in Tween 80 or Intralipid. In either case loss of milk production was indicated by failure of the pups to gain weight. The animals were euthanized after 3d and the mammary gland removed for pathological examination.

The structure of persin (Fig. 19.1,1) was determined following extensive spectroscopic, spectrometric and chemical methods (Oelrichs *et al.*, 1995a). The stereochemistry at C_2 has been confirmed by enantioselective synthesis of both the R and S enantiomers. Closely similar compounds to persin were synthesized to determine if the activity is affected by a small structural change (Fig. 19.1, 2-8).

In earlier work, Kashman *et al.* (1969) showed that compounds with a terminal 1-acetyloxy-2-hydroxy-4-one group in acidic medium form stable furan derivatives. This reaction could be explained by the initial elimination of water to form an intermediate keto-allylic derivative, followed by the departure of the acetate group to form an allylic cation. Cis-trans isomerism could then occur, and the furan derivative was formed by ring closure and proton abstraction. This reaction was used to determine if persin or its furan derivative were to be found in the mammary glands, heart or blood of lactating mice dosed with persin.

Lactating Quackenbush mice were dosed with persin by oral gavage at the rate of 100mg/kg, and euthanized after set times with carbon dioxide (CO_2). Tissues were immediately removed, weighed then homogenized (2x) with ethanol (EtOH). The homogenates were filtered and the filtrates dried under reduced pressure. The residues were extracted with 20ml ethyl ether (Et_2O) and dried by adding anhydrous sodium sulfate (Na_2SO_4). The solution of Et_2O was divided into Samples A and B. To half (10ml) the Et_2O solution was added 2ml p-toluene sulfonic acid (5mg/ml) in Et_2O, and the solution was incubated at room temperature in the dark for 48hrs. After this, the Et_2O solution was washed (2x) with water and applied to a 10g neutral aluminum oxide (Al_2O_3) column and eluted with 20ml Et_2O, the eluate concentrated to dryness

Oelrichs et al.

Fig. 19.1. Structures of persin and synthetic derivatives.

and the residue resuspended in 2ml n-hexane - Sample A. The remaining 10ml of Et₂O solution from above was concentrated to dryness, and resuspended with 2ml n-hexane - Sample B. The EI mass spectra of Sample A and Sample B were compared using a Varian Saturn 4D GC/MS. The samples were applied using a direct insertion technique and the resultant spectra compared with that from a standard furan solution.

Results and Discussion

The yield of persin from the isolation procedure was about 1%. The R and S enantiomers of persin were tested for activity required to induce widespread lactating mammary gland necrosis in mice at the dose rates of 50-100mg/kg. These studies indicated conclusively that only the R isomer was active. The S isomer was inactive even when administered in excess of 200mg/kg.

Only the mammary gland displayed any effects of persin intoxication at a gross or microscopic level. Single doses of persin in excess of 100mg/kg have produced cardiac lesions (Sani *et al.*, 1994), but no ill effects were noted when a series of small doses totaling more than 100mg/kg were administered at 4hr intervals.

None of the synthetic persin analogs (Fig. 19.1, 2-8) showed any activity on the mammary gland. These were compounds that replaced the acetyloxy group at C_1 with a butoxy or phenoxy group, changed the ester linkage to an ether linkage, or replaced the ketone group at C_4 with a hydroxyl.

Further evidence was given from the two compounds isolated by Kashman *et al.* (1970) from avocado fruit and seeds. These have the same 1-(acetyloxy)-2-hydroxy-4-one component as persin, but differ in chain length. Neither of these compounds had the effect on the mammary gland reported for persin. Both these compounds yielded furan derivatives by the addition of p-toluene sulfonic acid to their ether solutions, as persin did.

A comparison of spectra from Samples A and B showed no persin furan derivative in Sample B for any tissue. The lack of furan in solution B (not treated with p-toluene sulfonic acid) meant that no conversion of persin to furan had taken place either in the gut or tissues tested. This evidence strongly indicated that persin *per se* was the active agent in the tissues tested, and not its furan derivative.

The maximum concentration of persin was higher in the mammary gland than in the cardiac muscle or blood. The concentration of persin reached its highest point in the blood after about 2hrs, and in the mammary gland at 6-7hrs after dosing. Levels in both tissues declined rapidly after their respective peaks. The concentration of persin in the heart was too small to measure accurately. No detectable persin or its furan derivative could be found in any tissue after 24hrs.

Interestingly, some persin or its furan derivative could be recovered from previously extracted (2x with EtOH) mammary tissue by 6N hydrochloric acid hydrolysis. In future work, it is hoped to determine if this constitutes persin that has reacted with the tissue to form a stable bond. Further work to identify the site of action of persin in the mammary gland and its biochemical mode of action will be essential before any future therapeutic trials can be carried out.

References

Kashman, Y., Neéman, I. and Lifshitz, A. (1969) New compounds from avocado pear. *Tetrahedron* 25, 4617-4631.

Kashman, Y., Neéman, I. and Lifshitz, A. (1970) New compounds from avocado pear. *Tetrahedron* 26, 1943-1951.

Kingsbury, J.M. (1964) *Poisonous Plants of the United States and Canada.* Prentice-Hall, Englewood Cliffs, NJ, p. 124.

Oelrichs, P.B., Ng, J.C., Seawright, A.A., Ward, A., Schäffeler, L. and MacLeod, J.K. (1995a) Isolation and identification of a compound from avocado (*Persea americana*) leaves which causes necrosis of the acinar epithelium of the lactating mammary gland and the myocardium. *Natural Toxins* 3, 344-349.

Oelrichs, P.B., Ng, J.C., Seawright, A.A., O'Brien, G.P., Ward, A., Schäffeler, L., MacLeod, J.K., Hurst, T. and Ward, B. (1995b) Persin, a novel agent from avocado (*Persea americana*) which has the potential to suppress lactation and destroy mammary tumor cells *in vivo*. In: Kotsaki-Kovatsi, V.-P. and Vafiadiou, A.J. (eds), *Aspects on Environmental Toxicology*. Bookshop "Erasmos" St. Tsiamita, Ippodromiou 18, 546 Thessaloniki, Greece, pp. 121-125.

Sani, Y., Seawright, A.A., Ng, J.C., O'Brien, G. and Oelrichs, P.B. (1994) The toxicity of avocado leaves *(Persea americana)* for the heart and lactating mammary gland of the mouse. In: Colegate, S.M. and Dorling, P.R. (eds), *Plant-Associated Toxins: Agricultural, Photochemical and Ecological Aspects*. CAB International, Wallingford, Oxon, pp. 552-556.

Chapter 20

Bioassay-directed Isolation of Active Principle from *Lavendula stoechas*

K. Aftab[1], Atta-ur-Rahman[2], S.I. Ahmed[1] and K. Usmanghani[3]

[1]*Dr HMI Institute of Pharmacology and Herbal Sciences, Hamdard University;*
[2]*HEJ Research Institute of Chemistry, Department of Pharmacology and*
[3]*Department of Pharmacognosy, University of Karachi, Karachi-75270, Pakistan*

Introduction

Lavendula stoechas belongs to the family Labiatae, and is known in Karachi, Pakistan as "Ustukhudus" (Nasir and Ali, 1972). *Lavendula stoechas* has been used as a drug in the indigenous system of medicine for the treatment of chest complaints and to relieve biliousness. It has been considered to be cephalic, deobstruent and carminative (Sastri, 1962; Watt and Breyer-Bandwijk, 1962). Muslim physicians consider it to be the broom for cleaning the brain and it is given as a cerebral tonic. It is used in traditional medicine as a diuretic and for healing coronary disorders. It is a stimulant, aromatic, hypotensive, diaphoretic, expectorant, antiphlogistic and emmenagogue. An essential oil is distilled from the flowers and is used in colic and chest complaint, and is applied locally to relieve nervous headaches. Fomentation with the flowers relieves rheumatic and neuralgic pain.

Isolation of a smooth muscle relaxant principle identified as 7-methoxycoumarin is reported. Lingipinen derivatives are ursolic acid, β-sitosterol, flavonoids luteolin, acacetin, vitexin and coumarin. The oil is reported to contain 51 components including fenchone, pinocarvyl acetate, camphor, eucalyptol and myrthenol (Hoppe, 1975). The aqueous extract of flowers and stem of *L. stoechas* have hypotensive effects blocked by atropine in dogs, and negative chronotropic effects on isolated atria and spasmodic effect on guinea pig ileum.

Materials and Methods

The following reference materials were obtained from the sources specified:

acetylcholine chloride, atropine sulfate, chlorphenramine maleate, histamine phosphate, indomethacin, isoproterenol, norepinephrine hydrochloride, phentolamine hydrochloride, phenylephrine, propranolol, verapamil hydrochloride, glibenclamide (RBI), ranitidine (Ciba), potassium chloride, calcium chloride (E. Merck, Darmstadt, Germany) and pentothal sodium (Abbott Laboratories, Pakistan). All drugs were dissolved in distilled water and dilutions were made fresh in normal saline (0.9% NaCl) on the day of the experiment.

Animals were housed at the Animal House of HEJ Research Institute, University of Karachi at 23-25°C and fed standard diets with tap water *ad libitum*.

In vivo experiments

Wistar rats of either sex weighing 200-250g were anesthetized with 50mg/kg pentothal sodium injected intraperitoneally. The experiment followed the method described by Magos and Vidrio (1991), with some modifications.

An anesthetized rat was fixed in a supine position on a dissecting table. A longitudinal midtracheal incision approximately 2cm long exposed the trachea, the right jugular vein and both common carotid arteries. The trachea was cannulated with a polyethylene tube (2.75mm diameter) to maintain a free airway. The right jugular vein was cannulated with a saline-filled polyethylene tube (1mm diameter), and the exposed surface covered with cotton wool moistened in warm saline.

Systemic blood pressure was recorded at the left carotid arterial cannula connected to a physiological pressure transducer and displayed on a Grass 7D Polygraph. Cannulation of the carotid artery was performed like that of the jugular vein, and the cannula (1mm diameter) was filled with heparin sodium in saline solution. The heart rate was measured with a Grass tachograph. The temperature of the animal was maintained at 37°C by use of a heated table and overhead lamp.

Blood pressure (BP) and heart rate were monitored until steady baseline levels were obtained. Drugs were then administered by intravenous injection and flushed in with 0.2ml saline. Response of mean blood pressure (diastolic BP plus ⅓ of pulse width) and heart rate were expressed as per cent change from the control measurement taken immediately before injection. The preparation was allowed to equilibrate for at least 30min before the experiment (Gilani and Aftab, 1992).

In vitro studies

Guinea pigs of either sex weighing 400-600g were terminated by cervical dislocation and exsanguination. The heart was quickly removed and the atria were dissected. The tissue was immersed in Kreb's solution at 35°C and bubbled with a mixture of 95% oxygen (O_2) and of 5% carbon dioxide (CO_2). In all experiments, the spontaneously beating paired atria were used (Edinburgh, 1970).

A cotton thread was tied around the tip of each atrium and the atrial pair was mounted in a 10ml tissue bath maintained at 35°C. The force and rate of contraction were recorded on a Grass Polygraph *via* a force displacement isometric transducer under a resting tension of 1g.

The rabbit aorta was prepared based on the rabbit thoracic aorta strip preparation described by Furchgott and Bhadrakom (1953). New Zealand white rabbits of either sex weighing 2-3kg were euthanized and the descending thoracic aorta was quickly removed and placed in Kreb's solution, maintained at 37°C temperature and aerated with O_2 containing 5% CO_2. The aorta was trimmed of connective tissue and cut into rings 2-3mm wide, some of which were cut perpendicular to the axis of the cylindrical vessel to make strips.

The aortic ring preparations with undamaged endothelium were used for studies involving muscarinic receptors, and the strip preparations were used for other studies, where possible damage to the endothelial surface was of little importance. Special care was taken to protect the endothelial layer, as unintentional rubbing of the initial surface can result in failure to respond to the relaxant effect of acetylcholine (Furchgott and Zawadzki, 1980; Furchgott *et al.*, 1981). Aortic preparations used immediately or stored overnight at 4°C were found to be equally sensitive. A maximum of four preparations were obtained from a single rabbit. A resting tension of 2g was applied to each tissue and changes in isometric tension of the strips were measured with a force-displacement transducer connected to a Grass polygraph. During the initial equilibration period, contractile responses to a submaximal concentration of norepinephrine (NE) were measured until constant responses were elicited.

When mounting the uncut ring preparations, a stiff steel wire attached to a rigid supporting frame was placed intraluminally. Another stainless steel wire was inserted into the aortic lumen and attached by a cotton thread to an isometric transducer. The experimental procedure was similar to that described for strip preparations, except that phenylephrine was utilized to induce tone rather than NE.

Results

In vivo experiments

In anesthetized rats, the crude extract of *L. stoechas* and ursolic acid caused a dose-dependent fall in systolic, diastolic and mean arterial blood pressure (Table 20.1). A slight decrease in heart rate was also observed. The hypotensive effect was brief, returning to normal within 1-2min. Acetylcholine at a dose of 1μg/kg produced a considerable drop in blood pressure, and pretreatment of animals with atropine (1mg/kg) abolished this effect. However, atropine pretreatment did not alter the hypotensive effect of ursolic acid. Pretreatment of animal with ranitidine (10mg/ kg) or chlorpheniramine (10mg/kg) did not block the hypotensive response of ursolic acid, while vasodilator effect of histamine (10μg/kg) was completely blocked by ranitidine with chloropheniramine. Pretreatment of animal with propranolol (1mg/kg) did not abolish the hypotensive effect of ursolic acid, where the vasodilator effect of isoprenaline (1μg/kg) was completely abolished by propranolol. Norepinephrine

(1µg/kg) produced a pronounced increase in arterial blood pressure that was completely blocked by phentolamine (1mg/kg) but was not affected by pretreatment with ursolic acid.

In vitro studies

Ursolic acid at a concentration range of 10-300µg/ml caused a progressive decrease in force and rate of atrial contractions. The inhibitory responses were concentration related with plateau levels achieved within 2min, and were reversible on wash out. Pretreatment of tissue with atropine (1µM) did not abolish the inhibitory responses to the compound, whereas inhibitory responses to acetylcholine (0.1µM) were completely abolished.

In isolated aorta preparations, NE (1µM) and K^+ (30 and 80mM) produced sustained contractions; ursolic acid was then added, inducing vasorelaxation. Ursolic acid produced relaxation of K^+-induced contractions at similar concentrations (10-300µg/ml) to that of NE.

Discussion

The crude extract of *L. stoechas* and pure ursolic acid produced hypotensive effects in anesthetized rats. Acetylcholine produced a similar vasodilator response, which was blocked by atropine. Pretreatment with atropine failed to abolish the hypotensive responses of ursolic acid, suggesting that this effect was not mediated through acetylcholine receptors. Pretreatment of animals with antihistaminic agents (ranitidine and chlorpheniramine) and a beta-adrenergic blocker (propranolol) failed to abolish the cardiovascular response of ursolic acid, indicating that these effects were not mediated through mechanisms similar to those of histamine or isoprenaline. Norepinephrine is a potent vasoconstrictor, mediating this effect through activation of alpha-1 adrenoceptors (Gilman *et al.*, 1990). Norepinephrine plays a role in maintaining the normal tone of blood vessels (Ganong, 1981), and alpha-1 blocking drugs produce hypotension. Vasoconstrictor response to NE was not abolished by

Table 20.1. Effect of the crude extract of *Lavendula stoechas* and ursolic acid on mean arterial blood pressure (MABP) and heart rate (HR) in anesthetized rats.

Dose (mg/kg)	# Obs	Crude extract		Ursolic acid	
		MABP* (mmHg)	HR (beats/min)	MABP (mmHg)	HR (beats/min)
1	5	13.40+5.55	6.10+2.55	16.72+2.15	8.66+2.45
3	5	23.40+4.10	10.11+2.07	28.62+3.56	16.06+3.97
10	5	45.40+3.23	14.09+3.99	59.70+3.67	20.65+3.03
30	5	60.00+6.95	22.25+3.75	-	-

* % decrease; values shown represent mean + standard error of the mean.

pretreatment with ursolic acid, ruling out the possibility of adrenoceptor involvement.

The crude extract of *L. stoechas* and ursolic acid both produced a decrease in the force and rate of atrial contractions similar to that induced by acetylcholine. However, the inhibitory action of ursolic acid was not blocked by atropine, a competitive blocker of acetylcholine at a muscarinic receptor site (Arunlakhshana and Schild, 1959; Gilani and Cobbin, 1987). This suggests that the mechanism of ursolic acid is independent of muscarinic receptor activation.

Contraction of cardiac and smooth muscles are dependent upon an increase in the concentration of cytoplasmic free calcium ion (Ca^{++}) which activates the contractile elements. The source of Ca^{++} may be extracellular or intracellular (Meisheri *et al.*, 1981). Ursolic acid inhibits the contraction of isolated atria, a tissue dependent on extracellular Ca^{++} influx and intracellular Ca^{++} release.

Ursolic acid appears to interfere with the contraction process beyond the cell membrane. It inhibits the contractions of aortic muscle, which depend on extra-cellular Ca^{++} influx (Van Breeme, 1971), and it abolishes the contractions of aorta in Ca^{++}-free solution.

Ursolic acid inhibited the NE and potassium ion-induced contractions in a concentration dependent fashion in the aortic rings and strips, regardless of presence or absence of endothelium. In the aortic ring, vascular endothelial cells release several vasoactive factors including endothelium-dependent relaxing factors that regulate the tone of the underlying smooth muscle. Ursolic acid does not relax the aorta through endothelium-dependent mechanism (Olah and Rahwan, 1988).

Vasodilation by ursolic acid is not due to stimulation of beta-adrenoceptors because propranolol did not abolish the inhibition of the norepinephrine-induced contraction by ursolic acid, and isoprenaline-induced relaxation was completely inhibited (Yano *et al.*, 1991). These results suggest that the hypotensive and bradycardiac effects of *L. stoechas* are due to its ursolic acid constituent.

References

Arunlakshana, O. and Schild, H.O. (1959) Some quantitative uses of drug antagonists. *British Journal of Pharmacology* 14, 48-58.

Edinburgh, Staff, Department of Phamacology at the University of Edinburgh (ed), (1970) Experiments with other smooth muscle preparations. In: *Pharmacological Experiments on Intact Preparations*, 2nd ed. E & S Livingston Ltd., Edinburgh, pp. 89-112.

Furchgott, R.F. and Bhadrakom, S. (1953) Reactions of rabbit aorta to norepinephrine, isopropylarterenol, sodium nitrite and other drugs. *Journal of Pharmacology and Experimental Therapeutics* 108, 129-143.

Furchgott, R.F. and Zawadzki, J.V. (1980) The obligatory role of endothelial cell in the relaxation of arterial smooth muscle by acetylcholine. *Nature* 288, 273-376.

Furchgott, R.F., Zawadzki, J.V. and Cherry, P.D. (1981) Role of endothelium in vasodilator response to acetylcholine. In: Vanhout, P.M. and Leusen, I. (eds), *Vasodilatation*. Raven Press, New York, NY, pp. 49-66.

Ganong, W.F. (1981) *Review of Medical Physiology*. Asian edition. Lange Medical Publications, Singapore, p. 494.

Gilani, A.H. and Aftab, K. (1992) Pharmacological actions of *Cuscuta reflexa*. *International Journal of Pharmacognosy* 30, 296-302.

Gilani, A.H. and Cobbin, L.B. (1987) The interaction of himbacine with carbachol at muscarinic receptors in heart and smooth muscle. *Archives Internationales de Pharmacodynamie et de Therapie* 290, 46-53.

Gilman, A.G., Rall, T.W., Nies, A. and Taylor, P. (1990) *The Pharmacological Basis of Therapeutics*, 8th ed. Pergamon Press, New York, NY.

Hoppe, H.A. (1975) *Drogenkunde, Vol 1: Angiosperms*, 8th ed. Walter de Gruyter, Berlin, p. 1028.

Magos, G.A. and Vidrio, H. (1991) Pharmacology of *Casimiroa edulis*: Part I. Blood pressure and heart rate effects in the anaesthetized rats. *Planta Medica* 57, 20-24.

Meisheri, K., Hwang, O.L. and Van Breemen, C. (1981) Evidence for two separate Ca^{++} pathways in smooth muscle plasmalemma. *Journal of Membrane Biology* 59, 19-25.

Nasir, E. and Ali, S.I. (1972) *Flora of Pakistan*. Fakhri Printing Press, Karachi, Pakistan.

Olah, M.E. and Rahwan, R.G. (1988) Differential effects in calcium channel blockade and intracellular calcium antagonists on endothelium-dependent responses of the rat aorta to drugs. *Pharmacology* 37, 305-320.

Sastri, B.N. (1962) *The Wealth of India*. Council of Scientific and Industrial Research, New Delhi, India.

Van Breeme, V. (1971) Calcium requirement for activation of intact aortic smooth muscle. *Journal of Physiology* 272, 317-329.

Watt, J.M. and Breyer-Brandwijk, M.G. (1962) *The Medicinal and Poisonous Plants of Southern and Eastern Africa*. E&S Livingstone, Edinburgh and London.

Yano, S., Horiuchi, H., Horie, S., Aimi, N., Sakai, S. and Watanabe, K. (1991) Ca++ channel blocking effects of hirsutine, an indole alkaloid from *Uncaria genus*, in isolated rat aorta. *Planta Medica* 57, 403-405.

Chapter 21

Analysis of Gossypol in Whole Cottonseed and Rumen Fluid by HPLC

Ismartoyo[1] and T. Acamovic[2]

[1]Department of Animal Science, Hasanuddin University, Ujungpandang 90245, Indonesia; [2]Department of Biochemistry and Nutrition, SAC, Ayr KA6 5HW, UK

Introduction

Gossypol is a polyphenolic compound that can be toxic to animals, microbes and cells (Ismartoyo and Acamovic, 1994; Ismartoyo, 1997). The accurate and specific quantitation of gossypol is therefore important.

Whole cottonseed (WCS) contains gossypol at about 4-20g/kg glanded cotton-seed (Botsoglou, 1992; Acamovic, 1994). During processing, the aldehyde groups of gossypol react with free amino groups of proteins, forming "bound gossypol" (Jones, 1979; Calhoun et al., 1991; Risco et al., 1997). Gossypol reacts with other constituents to form condensation products that are soluble in aqueous acetone and represent "soluble bound gossypol." Some gossypol may be oxidized and/or degraded to various unidentified products. The remaining gossypol and gossypol-related compounds soluble in aqueous acetone that have an aldehyde moiety represent "unbound gossypol" (Botsoglou and Kufidis, 1990; Botslogou, 1992).

Determination methods for gossypol in WCS and cottonseed meal have often been non-specific and lacking in sensitivity. In this study, an HPLC method was developed for specific determination of unbound and acetone-soluble bound gossypol in WCS and unbound gossypol in the liquor. The method was based on that by Botsoglou (1992).

Materials and Methods

The solvents were HPLC-grade methanol, acetone, chloroform, acetonitrile, hydrochloric acid and phosphoric acid. Aqueous acetone-ascorbic acid solution was prepared by diluting 2.5g ascorbic acid in 150ml distilled water (H_2O), which was then mixed with 350ml acetone. Ascorbic acid solution was prepared by dissolving 1.5g ascorbic acid in 500ml distilled H_2O.

Mobile phase A was prepared as follows: 1ml phosphoric acid (Sigma Ltd., Poole, Dorset) was added to 25ml double distilled H_2O in a volumetric flask and made to 1L with double distilled H_2O. Mobile phase B was prepared by adding 1ml phosphoric acid to 1L HPLC-grade methanol in a volumetric flask. Both solutions were filtered through a 0.45μm membrane filter to remove particulate material.

A stock solution of gossypol was prepared by dissolving 6mg gossypol in 25ml acetonitrile. Working gossypol solutions of 0.5-8μg/ml were prepared.

Sample preparation and extraction

A laboratory mill (1mm screen) was used to grind WCS. The ground WCS sample (2g) was weighed accurately, transferred to a 250ml glass stoppered Erlenmeyer flask, and acetone-ascorbic acid solution (100ml) was added. The flask was stoppered and shaken for 1hr. The extract was filtered through a Whatman No. 40 filter paper and three separate 25ml aliquots transferred into flasks for the determination of unbound and acetone-soluble bound gossypol (USB), unbound gossypol (U), and unbound and acetone-soluble lipophilic form of bound gossypol (USL).

For determination of USB, the WCS extract (25ml) was transferred into a 50ml conical flask and 0.05ml concentrated HCl added. The flask was stoppered and heated in a water bath for 1hr at 65°C. The flask was removed and cooled to room temperature, then the extract was transferred into a separating funnel. Chloroform (50ml), 100ml ascorbic acid solution and 1ml concentrated HCl were added to the funnel, which was shaken for 3min and allowed to stand for 5min. The lower organic layer was filtered through anhydrous Na_2SO_4 in a Whatman No. 40 filter paper into a 100ml flask. The Na_2SO_4 was rinsed with chloroform and the filtrate evaporated to dryness by rotary evaporation at 30°C. The residue was dissolved in 25ml acetonitrile for HPLC analysis. The WCS extract (25ml) for U was transferred directly into a separating funnel and then processed as in USB, but heating with HCl was omitted.

For USL determination, the WCS extract (25ml) was transferred to a separating funnel (250ml) and processed as in U, but instead of reconstitution in acetonitrile, the residue was dissolved in 25ml acetone-ascorbic acid solution and then processed as in USB.

Chromatography

Analysis was performed on a Varian HPLC System consisting of a 9010 pump with a 9065 polychromator diode array detector (DAD). Data was handled using a Desk Pro 486 workstation with Varian Star software Version 4.0. The chromatographic column used was a stainless steel Spherisorb ODS2 (S5 ODS2, 250x4.6mm; Phase

Separation, Ltd., Deeside Industrial Park, Clwyd, Wales). The DAD was set to collect data from 190-365nm. The mobile phase flow rate was 1.5ml/min. The mobile phase composition was 80:20 (mobile phase B:mobile phase A). The temperature was maintained at 30°C using a column heater, and injection of both sample and working standard solutions (25µl) was achieved using a 9100R autosampler. Values of gossypol were determined by relation to the standards.

The calculated values from USB, U and USL allowed further calculation of a "soluble bound gossypol" (lipophilic and hydrophilic form of bound gossypol). The acetone-soluble lipophilic form of bound gossypol (SLB) was determined by subtracting U from USL, or SLB=USL-U. The acetone-soluble hydrophilic bound gossypol (SHB) was calculated by subtracting USL from USB, or SHB=USB-USL.

Gossypol determination in rumen fluid

Selected frozen rumen samples from sheep fed basal diet grass hay (1kg/d) supplemented with 500g WCS/d were thawed at room temperature to 20°C overnight and centrifuged at 2,000xg for 10min. The supernatant was transferred to HPLC vials for the determination of free gossypol. No gossypol peaks were found in any rumen liquor samples from sheep that consumed cottonseed except the gossypol spiked controls. This suggests that the concentration of any free gossypol in the rumen was very low (less than 0.5µg/ml).

Analysis of standard gossypol solutions by HPLC

A single pure peak of gossypol was obtained for the standard as determined by spectral comparison and statistical analyses of the spectra across the peaks. Gossypol was eluted at about 6min (5.9±0.015min).

The regression line of the standard gossypol solutions against peak area at 234nm was linear (r^2=0.98). Thus quantitation of gossypol to a concentration as low as 0.5µg/ml was attainable. Detection at 287nm gave a linear response and a more stable baseline, but was less sensitive by a factor of about three. The average peak purity parameter was 245.73±0.54nm.

Conclusions

Gossypol analysis by HPLC was rapid, reproducible, specific and sensitive for the gossypol in WCS with the limit of detection of 0.5µg/ml, which is similar to that of Botslogou (1992). The absence of, or very low, free gossypol concentration in the rumen samples indicated that a large proportion of free gossypol may have been bound to protein and possibly to other components in the rumen such as lipids (Reyes *et al.*, 1984) and iron salts (Waldrup, 1981; Risco *et al.*, 1997). Jones, (1985) reported that during processing, free gossypol is bound to cottonseed protein, resulting in bound gossypol and unavailable amino acids; lysine is considered to be the primary amino acid to which free gossypol is bound (Baliga and Lyman, 1957;

Martin, 1990). It has been suggested that not all free gossypol is bound in the rumen and that some may be released during digestion (Calhoun *et al.*, 1991). This incident needs further detailed investigation, particularly the mechanism of how the free gossypol binds and its possible release during digestion.

References

Acamovic, T. (1994) The advantages and disadvantages of xenobiotics in plant foods and feeds. In: Weitzner, M.I. (ed), *Development and Ethical Considerations in Toxicology.* Royal Society of Chemistry, London, pp.129-138.

Baliga, B.P. and Lyman, C.M. (1957) Preliminary reports on the nutritional significance of bound gossypol in cottonseed meal. *Journal of the American Oil Chemist Society* 34, 21.

Botsoglou, N.A. (1992) Liquid chromatographic determination of unbound and acetone-soluble bound gossypol in cottonseed meals and mixed feeds. *Journal of AOAC International* 75, 815-822

Botsoglou, N.A. and Kufidis, D.C. (1990) Determination of total gossypol in cottonseed and cottonseed meals by derivative UV spectrophotometer. *Journal of AOAC International* 73, 447-451.

Calhoun, M.C., Huston, J.E., Calk, C.B., Baldwin, B.C. Jr and Kuhlman, S.W. (1991) Effects of gossypol on digestive and metabolic function of domestic livestock. *Journal of Animal Science* 69 (Supp.1), 534(A).

Ismartoyo (1997) Studies *in vitro* and *in vivo* on the Nutritive Value of Whole Cottonseed (*Gossypium* sp.) for Sheep. PhD Thesis, Aberdeen University, Scotland.

Ismartoyo and Acamovic, T. (1994) The effect of gossypol on the growth and total protein of animal cells in culture. In: Colegate, S.M. and Dorling, P.R. (eds), *Plant-Associated Toxins: Agricultural, Phytochemical and Ecological Aspects.* CAB International, Wallingford, Oxon, pp. 201-206.

Jones, L.A. (1979) Gossypol and some other terpenoids, flavanoids, and phenols that affect quality of cottonseed protein. *Journal of the American Oil Chemist Society* 56, 727-730.

Jones, L.A. (1985) Gossypol chemistry and plant distribution. In: Lobi, T.J. and Hafez, E.S.E. (eds), *Male Fertility and its Regulation.* MTP Press Ltd., Lancaster, pp. 93-110.

Martin, S.D. (1990) Gossypol effects in animal feeding can be controlled. *Feedstuffs* 62 ,14.

Reyes, J., Allen, J., Tanphaichitr, N., Bellve, A.R. and Benos, D.J. (1984) Molecular mechanism of gossypol action on lipid membranes. *Journal of Biological Chemistry* 259, 9607.

Risco, C.A., Chase, C.C. and Robertson, J. (1997) Gossypol. In: D'Mello, J.F.P. (ed), *Plant and Fungal Toxicants.* CRC Press, Boca Raton, FL, pp. 243-252.

Waldroup, P.W. (1981) Cottonseed meal in poultry diets. *Feedstuffs* 53, 21.

Chapter 22

The Excretion of Minerals from Broilers Fed Tannic Acid, Polyethylene Glycol and Protein

B. Mansoori[1,2] and T. Acamovic[1]

[1]AFT Department, SAC and [2]Department of Agriculture, University of Aberdeen, MacRobert Building, 581 King Street, Aberdeen AB24 5UD, UK

Introduction

Tannins have molecular masses between 500 and 3,000 and are able to form strong soluble and/or insoluble complexes with proteins, carbohydrates, nucleic acids and alkaloids (Artz *et al.*, 1987; Mueller-Harvey and McAllan, 1992). Tannins can adversely influence digestibility and absorption of nutrients such as proteins and amino acids, carbohydrates and lipids, and also the activity of digestive enzymes in monogastrics (Elkin *et al.,* 1978; Horigome *et al.,* 1988; Longstaff and McNab, 1991; Ortiz *et al.*, 1993). Some studies reveal that tannins also affect mineral metabolism in man and animals. Most tannin molecules contain adjacent phenolic hydroxyl moieties, which form stable chelates with many metal ions including iron (III), copper (II) and zinc (II), or reduce their solubility (McDonald *et al.*, 1996; Chapter 23). Tannins can also disturb the absorption of minerals through the gastrointestinal (GI) tract and/or increase the endogenous losses of the minerals such as calcium (Ca), magnesium (Mg) and phosphorus (P) (Chang *et al.*, 1994; Mansoori and Acamovic, 1996; 1997). It has been reported that feeding gelatin to poultry that have ingested tannic acid reduced the endogenous losses of these macrominerals (Mansoori and Acamovic, 1997). The following experiment was designed to examine the effect of feeding gelatin and/or polyethylene glycol on the excretion of endogenous Ca, Mg and P from broiler cockerels in the presence of tannic acid.

Materials and Methods

Tannic acid, gelatin and polyethylene glycol 4,000 (PEG) were commercially purchased. Ninety-six 7wk-old broiler cockerels (2.25±0.15kg) were randomly grouped into 12 treatments with eight replicates each. Birds were individually housed in raised floor battery cages with controlled environmental conditions and had *ad libitum* access to water throughout the experiment. The tube feeding experiment was carried out based on the method of Sibbald (1976) and modified method of McNab and Blair (1988). After 24hr starvation, each bird was tube fed 60ml warm (40°C) saturated glucose solution (SGS). This was followed 24hrs later with 50ml warm (40°C) SGS or liquid gelatin alone (containing 6 or 12g gelatin), or with 4.5g tannic acid (TA) dissolved in 15ml water, and/or 2g PEG dissolved in 10ml water. In all cases tannic acid was administered subsequent to SGS, gelatin or PEG. Total collection of excreta was carried out for 48hrs immediately after tube feeding of the birds. The collected excreta were kept frozen (-18°C) until analysis. All birds were weighed before tube feeding of the test materials and at the end of the experiment. All remained healthy and survived the experimental treatments.

Frozen samples of excreta were freeze-dried and weighed as total dry matter output (DM output). Each sample (about 1.5g) was ashed at 500°C for 8hrs to determine the ash. The ash was dissolved in hydrochloric acid (6M), dried and diluted in distilled deionized water (AOAC, 1980). Calcium and Mg were measured by atomic absorption spectrometry, and P by a Technicon Auto Analyzer.

Analysis of data was carried out by ANOVA (Minitab Statistical Package, Minitab, Inc). Tukey's Highest Significant Difference method was used to obtain confidence intervals for all pairwise differences between level means.

Results

Control birds (SGS) lost about 6% of their original body weight during the experiment. There was no effect ($P>0.05$) of feeding 6g or 12g gelatin and/or 2g of PEG on the weight loss compared to the control. Weight loss significantly ($P<0.01$) increased in the SGS fed birds when TA was ingested. In the presence of TA, there was no significant difference in weight loss between SGS-fed birds and birds fed 6g gelatin, and birds fed 12g gelatin had significantly ($P<0.01$) smaller weight loss compared to the birds fed SGS or 6g gelatin. The adverse effect of TA on weight loss was alleviated by PEG in all treatments.

There was a significant ($P<0.05$) increase in DM output due to feeding 6g or 12g of gelatin compared to the control. Feeding PEG did not increase the amount of DM output significantly in control or 6g-gelatin-fed birds, but it increased ($P<0.05$) DM output from the birds fed 12g gelatin. Tannic-acid-treated birds had highly significant ($P<0.05$) increases in DM output. It was assumed that all TA and/or PEG was excreted during the 48hrs of excreta collection, and this was subtracted from the amount of DM excreted. In this case, PEG had no effect on DM output. Tannic acid significantly ($P<0.01$) increased the amount of DM output in the SGS-fed birds and birds fed 6g gelatin but not ($P>0.05$) in the birds fed 12g gelatin. In all cases, PEG could completely alleviate the adverse effect of TA on DM output ($P<0.01$).

Gelatin and PEG had no effect ($P>0.05$) on the excretion of the ash compared to the control birds, while TA dramatically increased ($P<0.01$) the amount of ash excreted in SGS-fed birds. Feeding 6g and 12g of gelatin significantly ($P<0.01$) reduced the influence of TA on the ash excreted. In all cases, PEG completely alleviated ($P<0.01$) the adverse effect of TA on the excretion of ash.

Gelatin had no effect on the excretion of Ca, P and Mg compared to the control birds. Although PEG significantly reduced ($P<0.01$) the amount of Ca excreted in 6g-gelatin-fed birds, there was no effect on the excretion of Ca in SGS-fed birds and birds fed 12g gelatin. Tannic acid increased dramatically ($P<0.01$) the amount of Ca excreted compared to SGS-fed birds. Feeding 6g or 12g gelatin reduced significantly ($P<0.01$) but not completely the adverse effect of TA on the excretion of Ca. The effect of TA in gelatin-fed birds was completely alleviated by PEG ($P<0.01$), but only partially ($P<0.01$) in SGS-fed birds.

Tannic acid significantly ($P<0.05$) increased the amount of Mg excreted in SGS-fed birds. Feeding 6g of gelatin had no effect in birds fed TA, and 12g of gelatin could only partially ($P>0.01$) reduce the adverse effect of TA on the excretion of Mg. Polyethylene glycol could completely ($P<0.01$) alleviate the adverse effect of TA on Mg in birds fed 6g gelatin, and partially but not significantly in birds fed SGS or 12g gelatin.

Tannic acid dramatically increased ($P<0.01$) the amount of P excreted in SGS-fed birds. Feeding 6g of gelatin could partially but not significantly ($P<0.05$) reduce the adverse effect of TA, but 12g gelatin reduced ($P<0.01$) the effect of TA on excretion of phosphorus. Polyethylene glycol alleviated ($P<0.01$) the adverse effect of TA on the excretion of phosphorus in birds fed SGS and 6g gelatin, but not ($P>0.05$) in birds fed 12g gelatin.

Discussion

Endogenous ash, Ca, Mg and P excretion were not affected by feeding gelatin compared to the control birds, which is in agreement with other works where birds were fed 18g gelatin (Mansoori and Acamovic, 1997). Tannic acid heavily influenced the amount of total endogenous ash, Ca, Mg and P excreted in SGS birds compared to control. Mansoori and Acamovic (1997) reported that TA dramatically increased the amount of total endogenous ash, Ca, Mg and P excreted from water-fed birds and was positively correlated to the amount of TA fed. Kubena *et al.* (1983) reported that diets containing TA alone or with ochratoxin decreased the serum Ca levels significantly, suggesting that this might have been due to impaired GI absorption, increased urinary excretion and tissue deposition secondary to the injury. Mitjavila *et al.* (1977) reported a 40% increase in fecal excretion of Ca, but a decrease in Mg, in rats fed diet containing 10g TA/kg diet. Chang *et al.* (1994) mentioned that the apparent Ca absorption was reduced when tea tannins and cowpea tannins were fed to rats, but there was no apparent effect on Mg absorption. The contrast between this paper and two other reports might be due to different susceptibility of each species to the tannin, experimental methods, or the levels of tannins fed. The results obtained here are in agreement with other work (Mansoori and Acamovic, 1997). There are a number of reports that the addition of protein or amino acids reduced the antinutritional effects of tannins (Fuller *et al.,* 1967; Rayudu *et al.,* 1970). Rayudu *et al.* (1970) found that adding Tween 80 and polyvinylpyrrolidone (PVP) reduced the growth-depressing effect of TA in chicks. Hewitt and Ford (1982) and Garrido *et al.* (1991) reported that the depression of the nutritional quality of protein caused by field bean tannins could be reversed by addition of PEG 4,000 and PVP to the diets of chicks and rats.

Conclusion

Diets containing high amounts of tannic acid, particularly for chickens, might cause a significant increase in loss of nutrients, specially minerals, of which a large part is likely to be due to endogenous losses. Polyethylene glycol and PVP may reduce the endogenous losses associated with tanniniferous diets.

References

AOAC (1980) *Official Methods of Analysis,* 13th ed. Association of Official Analytical Chemists, Arlington, VA.

Artz, W.E., Bishop, P.D., Dunker, A.K., Schanus, E.G. and Swanson, B.G. (1987) Interaction of synthetic proanthocyanidin dimer and trimer with bovine serum albumin and purified bean globulin fraction G-1. *Journal of Agricultural and Food Chemistry* 35, 417-421.

Chang, M.J., Bailey, J.W. and Colins, J.L. (1994) Dietary tannins from cowpeas and tea transiently alter apparent calcium absorption but not absorption and utilisation of protein in rats. *Journal of Nutrition* 124, 283-288.

Elkin, R.G., Featherston, W.R. and Rogler, J.C. (1978) Investigations of leg abnormalities in chicks consuming high tannin sorghum grain diets. *Poultry Science* 57, 757-762.

Fuller, H.L., Chang, S.I. and Potter, D.K. (1967) Detoxification of dietary tannic acid by chicks. *Journal of Nutrition* 91, 477-481.

Garrido, A., Cabrera, A.G., Guerrero, J.E. and van der Meer, J.M. (1991) Effects of treatment with polyvinylpyrrolidone and polyethylene glycol on faba bean tannins. *Animal Feed Science and Technology* 35, 199-203.

Hewitt, D. and Ford, J.E. (1982) Influence of tannins on the protein nutritional quality of food grains. *Proceedings of Nutrition Society* 4, 7-17.

Horigome, T., Kumar, R. and Okamoto, K. (1988) Effects of condensed tannins prepared from leaves of fodder plants on digestive enzymes *in vitro* and in the intestine of the rats. *British Journal of Nutrition* 60, 275-285.

Kubena, L.F., Philips, T.D., Creger, C.R., Witzel, D.A. and Heidelbaugh, N.D. (1983) Toxicity of ochratoxin A and tannic acid to growing chicks. *Poultry Science* 62, 1786-1792.

Longstaff, M.A. and McNab, J.M. (1991) The inhibitory effects of hull polysaccharides and tannins of field beans (*Vicia faba* L.) on the digestion of amino acids, starch and lipid and on digestive enzyme activities in young chicks. *British Journal of Nutrition* 65, 199-216.

McNab, J.M. and Blair, J.C. (1988) Modified assay for true and apparent energy based on tube feeding. *British Poultry Science* 29, 697-707.

Mansoori, B. and Acamovic, T. (1996) The effect of tannic acid on endogenous calcium, magnesium, and phosphorus losses in broilers. *British Poultry Science* 67-68.

Mansoori, B. and Acamovic, T. (1997) The excretion of minerals from broilers fed tannic acid with and without gelatine. In: *Proceedings of Spring Meeting of the World Poultry Science Association.* World Poultry Association, pp. 25-26.

McDonald, M., Mila, I. and Scalbert, A. (1996) Precipitation of metal ions by plant polyphenols: optimal conditions and origin of precipitation. *Journal of Agricultural and Food Chemistry* 44, 599-606.

Mitjavila, S., Lacombe, C., Carrera, G. and Derache, R. (1977) Tannic acid and oxidised tannic acid on the functional state of rat intestinal epithelium. *Journal of Nutrition* 107, 2113-2121.

Mueller-Harvey and McAllan, A.B. (1992) Tannins: Their biochemistry and nutritional properties. *Advances in Plant Cell Biochemistry and Biotechnology* 1, 151-217.

Ortiz, L.T., Centeno, C. and Tervino, J. (1993) Tannins in faba bean seeds: effects on the digestion of protein and amino acids in growing chicks. *Animal Feed Science and Technology* 41, 271-278.

Rayudu, G.V.N., Kadirvel, R., Vohra, P. and Kratzer, F.K. (1970) Effect of various agents in alleviating the toxicity of tannic acid for chickens. *Poultry Science* 49, 1323-1326.

Sibbald, I.R. (1976) A bioassay for true metabolizable energy in feedstuffs. *Poultry Science* 55, 303-308.

Chapter 23

The Influence of Tannic Acid on Amino Acid Digestibility in Broilers

B. Mansoori[1,2] and T. Acamovic[1]

[1]AFT Department, SAC and [2]Department of Agriculture, University of Aberdeen, MacRobert Building, 581 King Street, Aberdeen AB24 5UD, UK

Introduction

Tannins are water-soluble polyphenolic compounds that form hydrogen and hydrophobic bonds with macromolecules such as proteins (Artz et al., 1987; Mueller-Harvey and McAllan, 1992). Hydrolyzable tannins are oxidized and polymerized to form insoluble products, and are hydrolyzed by certain types of microbes and enzymes (Bickley, 1992).

Tannins have some beneficial effects in human and animal foodstuff (Vinson et al., 1995), but are known as antinutrients, complexing proteins by interaction between hydroxyl and carbonyl groups (Hagerman and Butler, 1978; Jansman, 1993). Tanniniferous components in the diet reduced the apparent digestibility of crude protein and increased excretion of nitrogen in rats (Moseley and Griffiths, 1979; Shahkhalili et al., 1990), pigs (Cousins et al., 1981; Jansman et al., 1993), and chickens (Nelson et al., 1975). Tannins have adverse effects on enzymatic activity in the gastrointestinal tract (GI) (Griffiths and Moseley, 1980; Lizardo et al., 1995). Digestibility of amino acids (AA) is adversely affected by inclusion of high amounts of dietary tannins in feed (Ford and Hewitt, 1979; Ortiz et al., 1993), but little is known about how tannins influence endogenous losses of nitrogen and AA. This study compares protein digestibility and endogenous losses of AA with and without tannic acid (TA) and gelatin.

Materials and Methods

Tannic acid and gelatin were purchased. Eighty 9wk-old broiler cockerels (3-3.6kg) were randomly grouped in ten treatments of eight replicates each. Birds were individually kept in raised floor battery cages with controlled environmental conditions and *ad libitum* access to water during the trial. The tube feeding trial was done according to Sibbald (1976). After 24hrs starvation, each bird was tube fed 60ml of warm (40°C) saturated glucose solution. This was followed 24hrs later with 60ml warm (40°C) liquid gelatin alone (containing 18g gelatin), or with 1.5, 3, 4.5, or 6g TA dissolved in 15ml water, 50ml water (control birds) or 50ml water containing 1.5, 3, 4.5, and 6g tannic acid. Tannic acid was given subsequent to gelatin or water. Total collection of excreta was for 48hrs after tube feeding the birds, and the material was frozen (-18°C) until analysis. Birds were weighed before tube feeding and at the end of the experiment. All birds remained healthy and survived the experimental treatments.

Samples of excreta were lyophilized, weighed, and assayed for uric acid (Marquardt, 1983). Gelatin and excreta were analyzed for nitrogen (AOAC, 1990), and AA contents were determined by weighing (100mg) into a 20ml screw-capped glass hydrolysis tube and hydrolyzing with HCl (10ml, 6M) in a heating block (110°C) for 24hrs (AOAC, 1984). The hydrolysate was filtered, dried in a shaking water bath (43°C) for 18hrs and dissolved in HOAc (25mM). Solutions were stored at 4°C until analysis. The methionine and cysteine content of samples were determined by weighing (100mg) into a 20ml screw-capped hydrolysis tube, to which was added cold performic acid (2ml), and stored at 4°C overnight. While the tube was still cold, sodium metabisulfite (350mg) was added, and then the solution was hydrolyzed with HCl (8ml, 7.5M) for 24hrs (AOAC, 1984). The hydrolysates was treated as above.

All AA were derivatized with *ortho*-phthaldialdehyde and 2-mercapto-ethanol (AOAC, 1984) and separated by HPLC. Statistical analysis was by ANOVA (Minitab Statistical Package, Minitab Inc). Tukey's Highest Significant Difference method was used to obtain confidence intervals on pairwise differences between level means.

Results

Total dry matter output (DMO) was greatly influenced ($P<0.05$) by tube feeding TA in water to birds, and positively correlated ($R^2=0.903$) to the dose of TA. Assuming that all of the TA was excreted within the 48hrs of collection, the amount of DMO excreted from endogenous origin was still highly influenced ($P<0.05$, $R^2=0.698$) by TA. Feeding gelatin (18g) elevated the excretion of DMO from 8.6g to 13.5g, but there was no significant difference in DMO feeding gelatin and TA up to 4.5g. Apparent nitrogen digestibility of gelatin was not affected by 1.5 or 3g of TA, but was significantly different ($P<0.01$) when TA was increased to 4.5 and 6g. Total endogenous loss of nitrogen was affected ($P<0.05$) in water-fed birds when TA was fed, and it correlated positively ($R^2=0.906$) with the amount of TA. Feeding 18g of gelatin increased ($P<0.05$) total nitrogen excreted compared to the control birds. Only the higher levels of TA (4.5 and 6g) increased ($P<0.05$) the total nitrogen excreted compared to the gelatin-fed birds. The amount of endogenous nitrogen excreted in the form of uric acid ($P<0.05$) was positively correlated ($R^2=0.74$) to the amount of TA administered. Feeding gelatin increased ($P<0.05$) the excretion of uric acid compared with control birds. TA had a small ($P>0.05$) influence on the excretion of uric acid in gelatin-fed birds.

Excretion of endogenous AA was affected ($P<0.05$) by the administration of tannic acid to the water-fed birds. The amount of AA excreted increased with increasing TA, but the response was different for each of the AA. Methionine, histidine and lysine were most affected in water-fed birds with TA (6g) compared to the control birds. The least affected were isoleucine and glycine. Feeding gelatin (18g) increased total excretion of glycine, alanine, and aspartic acid, compared to the control birds, but there was little or no effect on excretion of other AA. The excretion of AA in gelatin-fed birds with TA (6g) and gelatin-fed birds without TA was not as big as that in water-fed birds with and without TA.

Mueller-Harvey and McAllan (1992) reported that tannins including TA depressed DMO, protein and amino acid digestibility in poultry. Increase in DMO and nitrogenous compounds might be due to increases in secretion of GI fluid, which is high in protein, and/or increase in catabolism of endogenous proteins. There is evidence that ingested tannin causes the hypersecretion of mucin, and hypertrophy and hyperplasia of the goblet cells in the intestinal tract (Mitjavila *et al.* 1977; Sell *et al.*, 1985; Ortiz *et al.*, 1994). Mucin forms a protective layer, reducing the effect of tannins on intestinal brush border membrane and/or making mucin-tannin complex in order to deactivate tannin. There are reports indicating that TA is degraded and absorbed from the GI tract, conjugated with glutathione and excreted (Booth *et al.*, 1959; Potter and Fuller, 1968; Hagerman *et al.*, 1992), causing an increase in endogenous losses of AA. Methionine, histidine and arginine were the most excreted in TA-fed birds, possibly due to their high affinity for tannins. Adding methionine and arginine to the diet alleviated the adverse effects of added TA or sorghum tannins on the chicks (Fuller *et al.*, 1967; Armstrong *et al.*, 1973; Elkin *et al.*, 1991). An increase in excretion of histidine may indicate a breakdown of muscle proteins, causing a release of this amino acid and/or its methylated form, or its dimer forms such as balenine or anserine (Fuller, 1994). Feeding gelatin increased the amount of nitrogen and uric acid excreted. Gelatin is digested by GI tract enzymes and absorbed. Since the birds were fasted, a large proportion of the absorbed AA were likely to be catabolized to produce energy, so that the amount of excreted uric acid and nitrogen should have been elevated compared to the control birds. Apparent dry matter digestibility of the diet was adversely affected by the amount of TA ingested, but real dry matter digestibility did not change up to 4.5g of TA, which could be due to the amount of uric acid excreted in each treatment. Real nitrogen digestibility was significantly reduced when TA was ingested, but there was no difference ($P>0.05$) between different levels of TA. This reduction was mainly due to the excretion of uric acid nitrogen. Results obtained from apparent and true digestibility of gelatin amino acids showed that, in general, TA had an adverse effect ($P<0.05$) on digestibility for each amino acid. It has been reported that methionine is specifically required by the chick to detoxify dietary TA (Fuller *et al.*, 1967). Gelatin contains negligible amounts of tyrosine and cysteine, therefore the only source of these is endogenous, and calculating their digestibility is not possible. On the other hand, gelatin-fed birds with TA had lower endogenous losses of these AA compared to water-fed birds with TA. If this fact is expanded to other AA, it can be suggested that protein could reduce the adverse effect of TA on the excretion of AA from endogenous origin.

Acknowledgments

B. Mansoori is in receipt of a scholarship from the IR of Iran and this is gratefully acknowledged.

References

AOAC (1984) *Official Methods of Analysis*, 4th ed. Association of Official Analytical Chemists, Arlington, VA.

AOAC (1990) *Official Methods of Analysis*, 15th ed. Association of Official Analytical Chemists, Arlington, VA.

Armstrong, W.D., Featherston, W.R. and Rogler, J.C. (1973) Influence of methionine and other dietary additions on the performance of chicks fed bird resistant sorghum grain diets. *Poultry Science* 52, 1592-1599.

Artz, W.E., Bishop, P.D., Dunker, A.K., Schanus, E.G. and Swanson, B.G. (1987) Interaction of synthetic proanthocyanidin dimer and trimer with bovine serum albumin and purified bean globulin fraction G-1. *Journal of Agricultural and Food Chemistry* 35, 417-421.

Bickley, J.C. (1992) Vegetable tannins and tanning. *Journal of the Society of Leather Technologists and Chemists* 76, 1-5.

Booth, A.N., Morse, M.S., Robbins, D.J., Emerson, O.H., Jones, F.T. and DeEds, F. (1959) The metabolic fate of gallic acid and related compounds. *Journal of Biological Chemistry* 234, 3014.

Cousins, B.W., Tanksley, T.D., Knabe, D.A. and Zebrowska, T. (1981) Nutrient digestibility and performance of pigs fed sorghums varying in tannin concentration. *Journal of Animal Science* 53, 1524-1537.

Elkin, R.G., Rogler, J.C. and Sullivan, T.W. (1991) Differential response of ducks and chicks to dietary sorghum tannins. *Journal of Science of Food and Agriculture* 57, 543-553.

Ford, J.E. and Hewitt, D. (1979) Protein quality in cereals and pulses. 3. Bioassays with rats and chickens on sorghum (*Sorghum vulgare* Pers.), Barley and field beans (*Vicia faba* L.). Influence of polyethylene glycol on digestibility of proteins in high tannin grain. *British Journal of Nutrition* 42, 325-340.

Fuller, H.L., Chang, S.I. and Potter, D.K. (1967) Detoxification of dietary tannic acid by chicks. *Journal of Nutrition* 91, 477-481.

Fuller, M.F. (1994) Amino acid requirements for maintenance, body protein accretion and reproduction in pigs. In: D'Mello, J.P.F. (ed), *Amino Acids in Farm Animal Nutrition*. CAB International, Wallingford, Oxon, pp. 155-184.

Griffiths, D.W. and Moseley, G. (1980) The effect of diets containing field beans of high and low polyphenolic content on the activity of digestive enzymes in the intestines of rats. *Journal of Science of Food and Agriculture* 31, 255-259.

Hagerman, A.E. and Butler, L.G. (1978) Protein precipitation method for the quantitative determination of tannins. *Journal of Agricultural and Food Chemistry* 26, 809-812.

Hagerman, A.E., Robbins, C.T., Weerasuriya, Y., Wilson, T.C. and McArthur, C. (1992) Tannin chemistry in relation to digestion. *Journal of Range Management* 45, 57-62.

Jansman, A.J.M. (1993) Tannins in feedstuffs for simple stomached animals. *Nutrition Research Reviews* 6, 209-236.

Jansman, A.J.M., Verstegen, M.W.A. and Huisman, J. (1993) Effects of dietary inclusion of hulls of faba beans (*Vicia faba* L.) with a low and high content of condensed tannins on digestion and some physiological parameters in piglets. *Animal Feed Science and Technology* 43, 239-257.

Lizardo, R., Peiniau, J. and Aumaitre, A. (1995) Effect of sorghum on performance, digestibility of dietary components and activities of pancreatic and intestinal enzymes in the weaned piglet. *Animal Feed Science and Technology* 56, 67-82.

Marquardt R.R. (1983) A simple spectrophotometric method for the direct determination of uric acid in avian excreta. *Poultry Science* 62, 2016-2108.

Mitjavila, S., Lacombe, C., Carrera, G. and Derache, R. (1977) Tannic acid and oxidized tannic acid on the functional state of rat intestinal epithelium. *Journal of Nutrition* 107, 2113-2121.

Moseley, G. and Griffiths, D.W. (1979) Varietal variation in the anti-nutritive effects of field beans (*Vicia faba* L.) when fed to rats. *Journal of Science of Food and Agriculture* 30, 772-778.

Mueller-Harvey, I. and McAllan, A.B. (1992) Tannins, their biochemistry and nutritional properties. *Advances in Plant Cell Biochemistry and Biotechnology* 1, 151-217.

Nelson, T.S., Stephenson, E.L., Burgos, A., Floyd, J. and York, J.O. (1975) Effect of tannin content and dry matter digestion on energy utilization and average amino acid availability of hybrid sorghum grains. *Poultry Science*, 54, 1620-1623.

Ortiz, L.T., Centeno, C. and Tervino, J. (1993) Tannins in faba bean seeds: effects on the digestion of protein and amino acids in growing chicks. *Animal Feed Science and Technology* 41, 271-278.

Ortiz, L.T., Alsueta, C., Tervino, J. and Castano, M. (1994) Effects of faba bean tannins on the growth and histological structure of the intestinal tract and liver of chicks and rats. *British Poultry Science* 35, 743-754.

Potter, D.K. and Fuller, H.L. (1968) Metabolic fate of dietary tannins in chickens. *Journal of Nutrition* 96, 187-191.

Sell, D.R., Reed, W.M., Chrisman, C.L. and Rogler, J.C. (1985) Mucin excretion and morphology of the intestinal tract as influenced by sorghum tannins. *Nutrition Reports International* 31, 1369-1374.

Shahkhalili, Y., Finot, P.A., Hurrell, R. and Fern, E. (1990) Effects of foods rich in polyphenols on nitrogen excretion in rats. *Journal of Nutrition* 120, 346-352.

Sibbald, I.R. (1976) A bioassay for true metabolizable energy in feedstuffs. *Poultry Science* 55, 303-308.

Vinson, J.A., Dabbagh, Y.A., Serry, M.M. and Jang, J. (1995) Plant flavonoids, especially tea flavonols, are powerful antioxidants using an *in vitro* oxidation model for heart disease. *Journal of Science of Food and Agriculture* 43, 2800-2802.

Chapter 24

Effect of Polyethylene Glycol on the Chelation of Trace Elements by Tannins

C. Kainja, L. Bates and T. Acamovic
Animal and Feed Technology Department, SAC, 581 King Street, Aberdeen AB24 5UD, UK

Introduction

Tannins are among the most abundant secondary polyphenolic metabolite compounds found in plants and thus are frequently found in animal feeds and human foods. They are a complex mixture of water-soluble polymeric polyphenolic compounds, which have been demonstrated to have adverse effects on mammals, birds, insects, fish and microbes (Brune *et al.*, 1989; Scalbert, 1991; McAllister *et al.*, 1994; Muhammed *et al.*, 1994; Muhammed, 1997; Makkar *et al.*, 1995; Salawu *et al.*, 1997a, b, c). Hydrolyzable tannins and their degradation products have been reported toxic to animals (Reed, 1995; Chapter 107) often by absorption of these compounds through the gastrointestinal epithelium. Condensed tannins (proanthocyanidins) are not absorbed through the healthy gut epithelium although they exert adverse effects on animals (Reed, 1995; Salawu *et al.*, 1997a, b, c). The adverse effects of tannins have generally been attributed to their interaction with proteins (Hagerman, 1992; Dietz *et al.*, 1994; Reed, 1995). Higher concentrations of tannins tend to be associated with harsh environments and/or increased herbivory of growing plants. Anecdotal evidence from Zimbabwe suggests that when cattle were forced to consume a high proportion of tanniniferous plants during periods of drought, addition of small quantities of a commercially prepared polymeric tannin-binding agent (essentially polyethylene glycol, PEG) to the diet caused large improvements in performance. The farmers' recommendation was about 1g/steer/d, calculating to a concentration of PEG in a 50L rumen of about 5µM. It would be surprising if such low concentrations exerted their effects by releasing dietary protein. Both *in vitro* and *in vivo* studies have demonstrated that PEG has beneficial effects in tanniniferous feeds (Khazaal and Orskov, 1994; Makkar *et al.*, 1995; Salawu *et al.*, 1997a, b, c). Work with poultry has shown that dietary tannins increase the excretion of major minerals, and cause increased excretion of protein (Chapters 22, 23). Polyethylene glycol seems to have some ameliorative effect on the excretion of minerals. Cattle grazing high-tannin feeds when on a diet marginal or deficient in trace elements may suffer exacerbated trace element deficiency caused by chelation of the trace elements. The addition of PEG to the diet could alleviate this. Relatively large responses in performance may occur with the administration of small quantities of PEG to release tannin-chelated trace elements. The work reported here investigated the chelating effect of tannins on copper (Cu) and cobalt (Co) ions in the absence and presence of PEG 4,000 in buffers at different pH values.

Materials and Methods

Quebracho and mimosa tannins and were obtained from Hodgson Chemicals Ltd., Beverley, North Humberside, England, UK. Reverse-phase HPLC of these compounds showed that they were complex mixtures, although the major peak areas indicated that the main components were proanthocyanidins. Solutions of tannin were prepared in distilled water at 10g/L. Solutions of 0.25-5g/L Cu(II) and 0.20-4.2g/L Co(II) were prepared from the respective sulfates. Phosphate-citrate and phosphate-acetate buffers at pH2-6 were used to prevent basic precipitation of the tannins above pH6 and PEG (4,000g/mole) prepared in distilled water (20g/L).

Solutions of tannins, Cu, Co and PEG were mixed to yield different tannin:metal ion concentrations at pH2-6. The mass ratios of tannin:metal ion ranged from about 0.7-20 metal ion:1,000 tannin. These ratios are much higher than would be expected in the rumen fluid of an animal consuming feedstuffs with a tannin content of about 200g/kg dry matter (DM), but the concentrations were selected because good spectra could not be obtained at lower concentrations. The spectra of the mixtures were obtained in microcuvettes reading 290-900nm using a Kontron Instruments double-beam scanning spectrophotometer. Data was collected by computer and λ_{max} determined by computer software.

Results

In all cases, addition of tannin caused a hypochromic shift of the λ_{max} of the metal ions, which increased with increasing relative concentration of tannin. The response tended to be linear, with correlation coefficients varying from 0.6-0.97. Poorer correlations were associated with chelation of Cu ion, especially at pH3.

At all pH values, PEG addition produced a bathochromic shift and, in contrast to untreated solutions of metal ions, gave yields equivalent to the total concentrations of the ions present in the solution. This is due to release of the metal ions from their tannin chelates in the presence of PEG. Results obtained support the hypothesis that trace elements are chelated by tannins within the gastrointestinal tract of ruminants, and the presence of PEG releases the metal ion from the chelate, allowing a more efficient absorption of those elements that form tannin chelates.

Discussion

The above results agree well with other work on the chelation effects of various flavonoids and flavonoids with mineral elements, and suggest that as the chelating agent increases in relative concentration to the metal ions, oligodentate chelates are produced as supported by the continued hypochromic shift (Nemcova *et al.*, 1996). This is in agreement with other recent work by McDonald *et al.* (1996), who found that various polyphenolic monomers and tannins chelated and precipitated Cu (II) and Zn (II) ions from solution at pH5. These workers proposed structures for metal ion:polyphenolic compound chelates, and confirmed the time periods for precipitation of metal ions by tannins and other polyphenolics varied from 2-48hrs, which is within the time frame for chelation (and precipitation) of trace elements within the ruminant gastrointestinal tract. Others implicated the chelating effects of polyphenolics as inhibitory to attack of plants by microorganisms (Scalbert, 1991). The presence of tannins in microbial cultures may reduce the effectiveness of these microbes by depleting them of essential mineral elements (McAllister *et al.*, 1994; Muhammed *et al.*, 1994; Makkar *et al.*, 1995; Salawu *et al.*, 1997b, c). Chelation of trace elements may contribute to their direct or indirect adverse effects in the nutrition of animals, affecting the function of gastrointestinal microflora.

Polyethylene glycol appears to reduce the chelating effects of tannins and thus may contribute to the amelioration of these effects. This supports the idea that PEG may alleviate trace element deficiency in animals. Furthermore, this work suggests that the addition of tannins (especially proanthocyanidins) to animal diets may be a suitable method of reducing trace element toxicity in animals.

Acknowledgments

This work was conducted as part of a European Union funded project (Project No TS*3-CT93-0211). The Scottish Agricultural College is Funded by the Scottish Office, Agriculture Environment and Food Department.

References

Brune, M., Rossander, L. and Hallberg, L. (1989) Iron absorption and phenolic compounds: importance of different phenolic structures. *European Journal of Clinical Nutrition* 43, 547-558.

Dietz, B.A., Hagermann, A.E. and Barrett, G.W. (1994) Role of condensed tannin on salivary tannin-binding proteins, bioenergetics, and nitrogen digestibility in *Microtus pennsylvanicus*. *Journal of Mammalogy* 75, 880-889.

Hagerman, A.E. (1992) Tannin-protein interactions. In: Ho, C.T., Lee, C.Y. and Huang, M.T. (eds), *Phenolic Compounds in Food and Their Effects on Health I: Analysis, Occurrence and Chemistry*. ACS Symposium Series, 506.

Khazaal, K.J. and Orskov, E.R. (1994) Assessment of phenolic-related antinutritive effects in Mediterranean browse: a comparison between the *in vitro* gas technique with or without insoluble polyvinylpyrrolidone or nylon bag. *Animal Feed Science and Technology* 49, 133-149.

Makkar, H.P.S., Blummel, M. and Becker, K. (1995) Formation of complexes between polyvinylpyrrolidones or polyethylene glycol and tannins, and their implication in gas production and true digestibility in *in vivo* techniques. *British Journal of Nutrition* 73, 897-913.

McAllister, T.A., Bae, H.D., Yanke, L.J. and Cheng, K.J. (1994) Effect of condensed tannin from birdsfoot trefoil on endoglucanase activity and the digestion of cellulose filter paper by ruminal fungi. *Canadian Journal of Microbiology* 40, 298-305.

McDonald, M., Mila, I. and Scalbert, A. (1996) Precipitation of metal ions by plant polyphenols: optimal conditions and origin of precipitation. *Journal of Agriculture and Food Chemistry* 44, 599-606.

Muhammed, S. (1997) Antinutrient Effects of Plant Phenolic Compounds. PhD Thesis, Aberdeen University, Aberdeen.

Muhammed, S., Stewart, C.S. and Acamovic, T. (1994) Effect of tannic acid on cellulose degradation, adhesion and enzyme activity of rumen micro-organisms. *Proceedings of the Society of Nutritional Physiology* 3, 174.

Nemcova, I., Cermakova, L. and Gasparic, J. (1996) *Practical Spectroscopy Series 22*. Marcel Dekker, pp. 57-134.

Reed, J.D. (1995) Nutritional toxicology of tannins and related polyphenols in forage legumes. *Journal of Animal Science* 73, 1516-1528.

Salawu, M., Acamovic, T., Stewart, C.S. and Hovell, F.D.DeB. (1997a) Quebracho tannins with or without Browse Plus (a commercial preparation of polyethylene glycol) in sheep diets: effect on digestibility of nutrients *in vivo* and degradation of grass hay *in sacco* and *in vitro*. *Animal Feed Science and Technology* 69, 67-78.

Salawu, M., Acamovic, T., Stewart, C.S., Hovell, F.D.DeB. and McKay, I. (1997b) Assessment of the nutritive value of *Calliandra calothyrsus*: *in sacco* degradation and *in vitro* gas production in the presence of Quebracho tannins with or without Browse Plus. *Animal Feed Science and Technology* 69, 219-232.

Salawu, M., Acamovic, T., Stewart, C.S. and Maasdorp, B. (1997c) Assessment of the nutritive value of *Calliandra calothyrsus*: its chemical composition and the influence of tannins, pipecolic acid and polyethylene glycol on *in vitro* organic matter digestibility. *Animal Feed Science and Technology* 69, 207-217.

Scalbert, A. 1991. Antimicrobial properties of tannins. *Phytochemistry* 30, 3875-3883.

Chapter 25

The Use of Copper Boli for Cattle Grazing High-molybdenum Forage

W.G. Gardner[1], D.A. Quinton[1], J.D. Popp[2], Z. Mir[2], P.S. Mir[2] and W.T. Buckley[3]

Agriculture and Agri-Food Canada Research Station, [1]Kamloops, British Columbia V2B 8A9, Canada; [2]Lethbridge, Alberta T1J 4B1, Canada; [3]Brandon, Manitoba R7A 5Y3, Canada

Introduction

Molybdenum (Mo) is a required element for both plants and animals and is essential in several enzyme systems in the animal body. The term molybdenosis is often used to describe a secondary copper (Cu) deficiency that occurs in animals consuming feeds high in Mo. The clinical signs of a Mo-induced Cu deficiency are severe diarrhea, weight loss, color loss in hair (achromotrichia), lameness with a characteristic stiff gait, and sometimes death (NRC, 1980; Ward, 1994). Tolerance to high dietary levels of Mo varies with the species and age of the animal (Underwood, 1971), with ruminants being the most sensitive and cattle being the most susceptible of any ruminant species (Ward, 1994).

In terms of metabolism, studies have shown that Mo is readily and rapidly absorbed from most diets (Underwood, 1971). As Mo concentrations in the diet increase, the molybdate (MoO_4^{-2}) and sulfide (S^{-2}) interact in the rumen of the animal to form thiomolybdates (TMs), which may then bind with Cu and form unabsorbable Cu-thiomolybdate complexes (Cu-TM) (Allen and Gawthorne, 1987). Some TMs may be absorbed (Suttle, 1991) and can react with plasma albumin to change Cu binding in circulation (Woods and Mason, 1987), impairing absorption at the tissue level. Precipitation of the plasma with trichloroacetic acid (TCA) gives a measure of TCA-insoluble Cu, which indicates of the presence of TMs in plasma (Mason, 1986).

The 1984 National Research Council guidelines recommend a maximum level of 5-6ppm Mo in feeds for beef cattle. The Cu:Mo ratio is critical in maintaining Cu availability, and if it falls below 2:1, Cu deficiency symptoms may arise (Miltimore and Mason, 1971). Dietary levels of sulfur (S), manganese, zinc, iron, lead, tungsten, ascorbic acid, methionine, cystine and protein all influence Mo metabolism in the ruminant.

High concentrations of Mo have been identified in forages from several areas of British Columbia (BC), particularly those grown on mine wastes. A 3yr grazing study was conducted using beef cow/calf pairs grazed on Mo-rich forage (21-44ppm) in order to address concerns on animal health and productivity.

Materials and Methods

A grazing trial was conducted at Highland Valley Copper Mine near Logan Lake, BC, on 55ha of reclaimed mine tailings. The trial ran from July to September, 1994-96. The pasture vegetation consisted of crested wheatgrass (*Agropyron cristatum*), pubescent wheatgrass (*Agropyron trichophorum*), smooth bromegrass (*Bromus inermis* Leyss.), orchardgrass (*Dactylis glomerata*), red fescue (*Festuca rubra*), alfalfa (*Medicago sativa*), sainfoin (*Onobrychis viciaefolia*), timothy (*Phleum pratense*), shore pine (*Pinus contorta*), Canada bluegrass (*Poa compressa*), Russian wild rye (*Elymus junceus*), yarrow (*Achillea millefolium*), bentgrass (*Agrostis* spp.), willowherb (*Epilobium* spp.) and Kentucky bluegrass (*Poa pratensis*). The site was stocked continuously at 0.85animal units/ha.

In each year, 32 Hereford x Angus cow/calf pairs were randomly assigned to treatment groups. Half of the animals received All-Trace mineral boli (Agrimin Ltd.), while the remaining animals served as controls. All-Trace boli are designed to release Cu, other vitamins and minerals continuously over an 8mo period. The grazing herd was also offered free-choice cobalt (Co)-iodized salt. Blood and milk samples and animal weights were collected from all animals at the start of each grazing season and every 3wks thereafter. Initially, and then every 6wks, liver samples were collected from half of the animals in each treatment group using the "tru-cut biopsy technique" (Davies and Jebbeth, 1981) with a liver biopsy instrument for large animals (Buckley *et al.*, 1986). Serum, milk and liver tissue samples were digested in nitric acid and hydrogen peroxide and analyzed for Cu and Mo by inductively coupled plasma atomic emission spectrophotometry (Thermo Jarrell Ash ICAP61). The instrumentation was equipped with an ultrasonic nebulizer for the liver determinations. Hand-collected forage samples to represent the animals' diet (Edlefsen *et al.*, 1960) were collected throughout each grazing season. Samples were dried at 60°C for 24hrs, ground to pass through a 1mm steel sieve using a Wiley Mill and analyzed for Cu (Allan, 1961; Baker and Smith, 1974), Mo (Marczenko, 1976), sulfur (S) and crude protein (CP) (AOAC, 1990).

Results

The concentrations of Mo in the forage decreased from an average of 44ppm for the 1994 grazing season to an average of 21ppm for the 1996 grazing season. The average overall Cu value was 15ppm, resulting in a Cu:Mo ratio of 0.44:1. However, even though the Cu:Mo ratio fell below the recommended guideline of 2:1 (Miltimore and Mason, 1971), no signs of adverse health effects were noted. Both the cows with the All-Trace boli and the animals acting as controls maintained body weight and condition. The weight gain for the calves for all three seasons averaged 91kg for both the treatment and control group, and was similar to gains of cattle grazing elsewhere in the Kamloops region (Quinton, 1987).

The results of the liver biopsy data show that the liver Mo increased linearly for both treatment groups (Fig. 25.1). Even with the high levels of Mo fed, the liver Cu concentrations remained within the normal range of 88-525ppm on a dry matter (DM) basis (Puls, 1980). The results showed significantly higher liver Cu values at 6 and 12 wks for the treatment animals *vs.* the controls (Fig. 25.2). This indicates that the bolus was effective in supplying additional Cu to the animals, as the liver Cu concentrations reflect the Cu status of the animal (Hidiroglou *et al.*, 1990).

The serum Cu and Mo levels for both cows and calves did not differ between treatment groups. The serum Cu levels stayed in the normal range of 0.7-1.5µg/ml (Gengelbach *et al.*, 1994; Smart *et al.*, 1981) while the serum Mo levels increased linearly over the season. The combined milk Mo values also increased linearly over time, as Mo content of milk in cattle responds rapidly and directly to increased Mo intake (Huber *et al.*, 1971; Vanderveen and Keener, 1964), with normal milk containing 0.02-0.12µg/ml (Mills and Davis, 1987). There were no differences between the treatment and control groups in terms of milk Mo content, and the overall average milk Mo concentration being 0.56µg/ml.

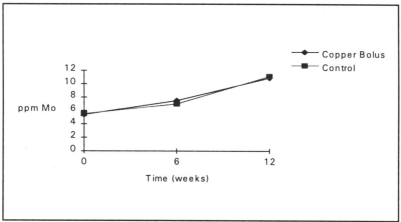

Fig. 25.1. Liver molybdenum concentrations of cows grazing high-molybdenum forage.

Conclusions

Grazing of high Mo herbage from the reclaimed mine site did not lead to any adverse effects on animal health or productivity. Molybdenum accumulated in the liver, serum and milk, but there was no negative impact on liver or serum Cu. Although a Cu:Mo ratio of 2:1 may be a useful guideline on other high-Mo rangeland in BC, it is not valid on the reclaimed mining site at Highland Valley. The All-Trace bolus was effective in providing Cu to liver tissues of the animals in the treatment group, although all animals had normal serum and liver Cu ranges. It is possible that other factors, such as the form of Mo in the forage, the high level of Cu in the diet or the interaction of those minerals with sulfur, may have prevented a Mo-induced Cu deficiency from occurring.

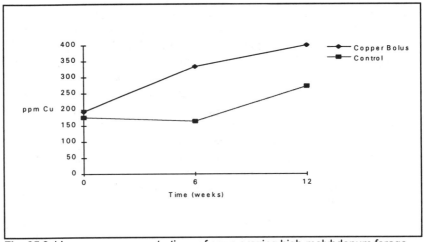

Fig. 25.2. Liver copper concentrations of cows grazing high-molybdenum forage.

Acknowledgments

The authors would like to thank Highland Valley Copper Mine and the British Columbia Cattlemen's Association for their participation in this study.

References

AOAC (1990) *Official Methods of Analysis*, 15th ed. Association of Official Analytical Chemists, Arlington, VA.
Allan, J.E. (1961) The determination of copper by atomic absorption spectroscopy. *Spectrochimica Acta* 17, 459–466.

Allen, J.D. and Gawthorne, J.M. (1987) Involvement of the solid phase rumen digesta in the interaction between copper, molybdenum and sulfur in sheep. *British Journal of Nutrition* 58, 265-276.

Baker, A.A. and Smith, R.L. (1974) The preparation for atomic absorption spectroscopy. *Analyst* 83, 655-661.

Buckley, W.T., Eigendorf, G.K. and Dorward, W.J. (1986) A liver biopsy instrument for large animals. *Canadian Journal of Animal Science* 66, 1137-1140.

Davies, D.C. and Jebbeth, I.H. (1981) Liver biopsy of cattle. *In Practice* 3, 14-16.

Edlefsen, J.L., Cook, C.W. and Blake, J.T. (1960) Nutrient content of the diet as determined by hand plucked and esophageal fistula samples. *Journal of Animal Science* 19, 560-567.

Gengelbach, G.P., Ward, J.D. and Spears, J.W. (1994) Effect of dietary copper, iron, and molybdenum on growth and copper status of beef cows and calves. *Journal of Animal Science* 72, 2722-2727.

Hidiroglou, M., Ivan, M. and McDowell, L.R. (1990) Copper metabolism and status in cattle. In: *Proceedings of the 16th World Buiatrics Congress*. Interlink Consultoria and Eventos, Salvador, Bahia, Brazil, pp. 1247-1252.

Huber, J.T., Price, N.O. and Engel, R.W. (1971) Response of lactating dairy cows to high levels of dietary molybdenum. *Journal of Animal Science* 32, 364-367.

Marczenko, Z. (1976) *Spectrophotometric Determination of Elements*. John Wiley and Sons, Toronto, Ontario, pp. 360-362.

Mason, J. (1986) Thiomolybdates: mediators of molybdenum toxicity and enzyme inhibitors. *Toxicology* 42, 99-109.

Mills, C.F. and Davis, G.K. (1987) Molybdenum. In: *Trace Elements in Human and Animal Nutrition, Vol 1*. Academic Press, New York, NY, p. 429.

Miltimore, J.E. and Mason, J.L. (1971) Copper to molybdenum ratio and molybdenum and copper concentration in ruminant feeds. *Canadian Journal of Animal Science* 51,193-200.

National Research Council (1980) Copper, molybdenum, mercury, lead and cadmium. In: *Mineral Tolerance of Domestic Animals*. National Academy Press, Washington, DC, pp. 107, 169-170, 264-265, 334-335, 312-313.

Puls, R. (1980) *Veterinary Trace Mineral Deficiency and Toxicity Information*. Province of British Columbia Ministry of Agriculture, British Columbia.

Quinton, D.A. (1987) Beef cattle nutrition and performance on seeded clearcuts in southern interior of British Columbia. *Canadian Journal of Animal Science* 67, 919-928.

Smart, M.E., Gudmundson, J. and Christensen, D.A. (1981) Trace mineral deficiencies in cattle: a review. *Canadian Veterinary Journal* 22, 372.

Suttle, N.F. (1991) The interactions between copper, molybdenum and sulfur in ruminant nutrition. *Annual Review of Nutrition* 11, 121-140.

Underwood, E.J. (1971) Molybdenum. In: *Trace Elements in Human and Animal Nutrition*, 3rd ed. Academic Press, New York, NY, pp. 116-140.

Vanderveen, J.E. and Keener, H.A. (1964) Effects of molybdenum and sulfate sulfur on metabolism copper in dairy cattle. *Journal of Dairy Science* 47, 1224.

Ward, G.M. (1994) Molybdenum requirements, toxicity and nutritional limits for man and animals. In: Braithwaite, E.R. and Haber, J. (eds), *Molybdenum: An Outline of its Chemistry and Uses*. Elsevier Science, The Netherlands, pp. 452-476.

Woods, M. and Mason, J. (1987) Spectral and kinetic studies on the binding of trithio-molybdate to bovine and canine serum albumin *in vitro*: the interaction with copper. *Journal of Inorganic Biochemistry* 30, 261-272.

Chapter 26

A Newly Discovered Toxic Plant, *Stemodia kingii*, in Western Australia

J.G. Allen[1] and A.A. Mitchell[2]

[1]*Animal Health Laboratories and* [2]*Natural Resource Management Services, Agriculture Western Australia, 3 Baron-Hay Court, South Perth, Western Austalia 6151, Australia*

Introduction

During a rangeland survey of the Pilbara region of Western Australia, pastoralists drew attention to a plant responsible for the death of an unknown number of sheep each year. This plant was *Stemodia kingii* F. Muell., and it was not recorded as being toxic. Sheep did not normally eat it, but would do so if feed was short and they were hungry (Housen, 1995). A feeding trial that established its toxicity is reported here.

Stemodia kingii is a dicotyledonous angiosperm in the family Scrophulariaceae. It is found only in the Pilbara region of Western Australia, within an area bounded by Pannawonica in the south, the De Grey River in the north, Wittenoom in the east and the coast in the west. It is a short-lived (2-5yrs) shrub that grows to 0.25m tall, has blue flowers, and bright, glossy, bushy, green foliage. The plant grows only on crabhole plains of red, cracking clay soils, with a variety of tussock grasses (Roebourne Plains grass, *Eragrostis xerophila*). In dense grasslands, *S. kingii* is a minor member of the vegetation, but in heavily grazed areas it becomes dominant, increasing in above average rainfall years and almost disappearing in dry years.

Materials and Methods

Several entire plants (voucher specimen A.A. Mitchell PRP260, 2/4/95, 1km north of Mulga Downs Homestead) were collected from Mulga Downs Station near Wittenoom in September, 1995. Plants were air-dried, the base of the main stem and roots removed, and the remaining material hammermilled to provide a powder with a maximum particle length of 2mm. This was stored dry until required.

Six Merino wether hoggets (24.6-42.5kg) were kept in individual pens in an animal house. They were given *ad libitum* access to water and provided with 300g of a commercial sheep pellet and 200g of good quality oaten chaff each day, for 2wks prior to, and throughout, the experiment. Daily feed consumption was measured. To test the toxicity of the plant, the milled material was mixed with sufficient water to form a free-flowing slurry and administered to sheep by an intraruminal tube. Dose rates are shown in Table 26.1.

Blood samples (into lithium heparin) and rumen samples were collected immediately prior to dosing, then where possible at 6hrs, and 1, 2, 3 and 4d after initial dosing, and immediately prior to death. Plasma was assayed for gamma glutamyl transferase, glutamate dehydrogenase, aspartate amino transferase and creatine kinase activities, and total bilirubin, creatinine, urea, protein, albumin, calcium, magnesium and phosphate concentrations, using commercial kits and an automatic chemistry analyzer. The pH of the rumen samples was measured within 30min of collection using a pH meter. Sheep were euthanized when they were profoundly depressed, or 4d after they were initially dosed, and post mortems conducted. A complete range of tissues was collected for histopathology.

Results

Sheep 1 died overnight, and all others except Sheep 4 were terminated in a depressed state. Sheep 4 appeared to be recovering 4d after dosing. The survival periods of the sheep are shown in Table 26.1. Sheep 1 stopped eating immediately after being dosed, Sheep 2 within 4hrs, and the rest, except for Sheep 4, within 24hrs. Sheep 4 ate 15, 80 and 141g on days two, three and four after dosing. Rumen pH dropped from 7.00-7.05 to 5.85-6.15 within 8hrs of the initial dosing. In surviving sheep it remained at a similar level for about a day then started to rise. It did not completely recover in any sheep before study's end.

The leading clinical sign was a severe watery, and sometimes mucoid, diarrhea, that developed within 24hrs of dosing. There was frequent urination. The sheep were

Table 26.1. Dose rates of *Stemodia kingii* administered and the survival periods for the sheep.

Sheep no.	Dose rate (g/kg)[1]	Survival period
1	12.0	18hrs
2	3.0	30hrs
3	2.5	26hrs
4	2.0	4d
5	2.5 + 2.0	4d
6	1.5 + 2.0	2.5d

[1] Two dose rates, the second was given 48hrs after the first.

quiet but remained alert until about 3-4hrs before they were terminated. In the last few hours they rapidly became depressed and lay down. The watery feces and urine continued to be passed while they were recumbent. In the terminal stages it was difficult to collect a blood sample due to low blood pressure and dehydration.

Plasma creatinine and urea concentrations were elevated in Sheep 2 (355µM/L and 14.02mM/L, respectively) and Sheep 3 (255µM/L and 11.74mM/L) just prior to death. Other plasma constituents had no significant changes in any of the sheep.

Gross pathological changes in all except Sheep 4 were similar, and were mainly restricted to the gastrointestinal tract. Rumen contents were dry, there was usually hyperemia of the ventral sac, and papillae over large areas of the internal surfaces of both the ventral and dorsal sacs were swollen and whitish in color. In Sheep 5 and 6 (dosed twice) the mucosa of many of the reticular folds, and the floors of the "honeycomb" cells between them, was thickened, darkly discolored and furry. Apart from this the reticulum and the omasum appeared normal. The mucosa of the abomasum was mildly to moderately hyperemic, particularly along the tips of the folds in the fundus. Hyperemia was evident throughout the small and large intestines and the cecum, especially in the ileum. Contents of the entire intestinal tract were moderately voluminous and extremely watery, with no formed fecal pellets anywhere in the large intestine. Yet, there was no unusual smell. Two of the sheep had creamy colored discoloration of the adrenal cortex and one had subendocardial hemorrhages over the papillary muscles in the left ventricle.

In Sheep 4 the only gross change was that the contents of the small and large intestine and the cecum were mucoid in the duodenum and jejunum, and very watery elsewhere. There were no formed fecal pellets in the large intestine.

Microscopically, rumen changes were present in all sections, even those taken from areas that had no gross pathology. Grossly swollen papillae had extensive degenerative, necrotic and inflammatory changes along the lengths and over the tips of the papillae. The earliest change was a pronounced intercellular edema within the stratum spinosum that progressed to liquefactive necrosis with serous exudation and intense polymorphonuclear (PMN) inflammatory cell infiltration into the necrotic areas. This necrotic process often spread, causing loss of the stratum cylindricum and disruption of the basement membrane, with cells and the inflammatory exudate spilling into the lamina propria. In sections taken from areas where gross changes were not evident, there were many large foci of layers of small punched-out holes containing pyknotic nuclei and degenerating inflammatory cells restricted to the strata granulosum and corneum. Occasionally there was liquefactive necrosis and PMN infiltration into the adjacent outer stratum spinosum. Intercellular edema in the stratum spinosum was not seen. There was regenerative response of the stratum cylindricum as early as 26hrs after dosing.

Similar microscopic changes were in the reticulum and omasum. In the former the widespread liquefactive necrosis with intense PMN infiltration of the stratum spinosum was restricted mainly to the mucosa on the floors of the "honeycomb" cells. The darkly discolored, furry appearance in parts of the reticulum of the sheep given two doses was due to the toxic rumenitis becoming a bacterial rumenitis. There were multiple foci of complete necrosis of the mucosa, loss of the basement membrane, heavy bacterial invasion and PMN infiltration. Every omasal fold examined had multiple degenerative foci that had variable mixes of pronounced intercellular edema, liquefactive necrosis, serous exudation and PMN infiltration of the stratum spinosum. Intercellular edema was so evident in some foci that intraepithelial vesicles were formed. In many of the foci there was also loss of the stratum cylindricum and disruption of the basement membrane.

Over the fundic folds in the abomasum, and particularly over the tips of the folds, were many necrotic foci gastric pit epithelium, with a light infiltration of PMNs below the necrotic areas. There was also congestion of blood vessels throughout the lamina propria and submucosa. Similar focal changes were present throughout the duodenum and jejunum. More substantial changes were present in the ileum, where there was extensive necrosis of villi epithelium and loss of villi, with large numbers of degenerate epithelial and PMN cells located in the lumen. There was also marked congestion of the blood vessels in the lamina propria and submucosa, some hemorrhage in the lamina propria, and dilation of submucosal lymphatics. Over almost the entire surface of the cecum there was necrosis and sloughing of surface epithelial cells, with an accompanying passing of PMN cells into the lumen. Blood vessels in the lamina propria and submucosa were congested. Changes in the colon were similar to those in the cecum, but focal in distribution.

Throughout the myocardium in the left and right ventricles and the inter-ventricular septum were multiple degenerative foci of variable size. These consisted of small aggregates of acutely necrotic myocytes that were swollen, stained strongly eosinophilic and had pyknotic nuclei, and these were often surrounded by vacuolated myocytes. Occasionally several foci merged together.

Other changes were marked apoptosis of lymphoid cells, manifesting as scattered pyknotic chromatin, in the germinal centers of the splenic corpuscles in the spleen; vacuolar change in periportal hepatocytes and dense, eosinophilic, "glassy" masses within the cytoplasm of periacinar and midzonal hepatocytes (probably reduplication of the smooth endoplasmic reticulum (Cheville, 1983)) in the liver; dilation of the sinusoids in the lower zona fasciculata, the zona reticularis and the outer medulla of the adrenal glands, with a pink proteinaceous material in many of these sinusoids; scattered focal necrosis of renal epithelial cells, primarily in the distal convoluted and collecting tubules, and occasionally some proteinaceous casts in the collecting tubules in the medulla; and in the lungs congestion of alveolar capillaries caused alveolar walls to appear thickened.

The mucosa of the forestomach of Sheep 4 had foci where the cytoplasm of cells in the strata cylindricum and spinosum was contracted from the cell walls, and foci of mild intracellular edema where spaces between the cells in the same two strata were just visible. There was no necrosis, and most of the mucosa seemed normal. The other change in the gastrointestinal tract was scattered focal necrosis of the surface epithelium in the cecum, colon and rectum. The myocardium had some foci of a few necrotic myocytes encircled by macrophages and fibroblasts.

Discussion

Stemodia kingii is toxic and consumption of a very small amount results in fatal intoxication. Pathological features are inflammation of the entire gastrointestinal tract, resulting in a watery diarrhea, dehydration and a cardiomyopathy. In the Pilbara region of Western Australia, it is now known to avoid this plant if possible.

The gross changes in the rumen were subtle and would not have been identified if the rumen contents had not been removed and the internal surface washed. No gross changes were noticed in the reticulum or omasum of Sheep 3 and Sheep 5 respectively, despite there being very severe degeneration of the mucosa of these organs. Only in the sheep that received two doses of the toxic material were there obvious gross changes in the reticulum. This poor correlation between gross and microscopic changes in the rumen is not uncommon. Recently it has been observed in black soil blindness, cereal grain induced ruminal acidosis and lupin seed induced ruminal acidosis (J.G. Allen and D.C. Main, personal observations). It is recommended that each of the forestomachs be regularly sampled for microscopic examination when post mortems for diagnostic purposes are conducted.

Stemodia florulenta W.R. Barker (formerly *Morgania floribunda* Benth. (Barker, 1990)), of the same genus is reported to be toxic (McKenzie *et al.*, 1996). In a feeding trial sheep developed similar clinical signs to those reported here, and had a gastroenteritis (Hurst, 1942). McKenzie *et al.* (1996) suspected that *S. florulenta* contains cardiac glycosides. The toxic agent(s) in *S. kingii* is not known, but the clinical signs, and some of the clinical chemistry and pathological changes observed in this experiment are consistent with intoxication by cardiac glycosides (Seawright, 1989). However, pathological changes were observed that have not previously been associated with cardiac glycoside poisoning.

References

Barker, W.R. (1990) New taxa, names and combinations in *Lindernia, Peplidium, Stemodia* and *Striga* (Scrophulariaceae) mainly of the Kimberley region, WA. *Journal of the Adelaide Botanic Gardens* 13, 79-93.

Cheville, N.F. (1983) *Cell Pathology*, 2nd ed. Iowa State University Press, Ames, IA, pp. 415-417.

Housen, S. (1995) Personal communication, Manager, Mulga Downs Station.

Hurst, E. (1942) *The Poisonous Plants of New South Wales*. Poisonous Plants Committee of New South Wales, Sydney, p. 381.

McKenzie, R.A., Blood, D.C., Larcombe, M.T. and Brightling, P. (1996) *PHYTOX. An International Computerised Guide to the Clinical Signs, Necropsy Lesions and Toxins of Plants, Fungi and Cyanobacteria Poisonous to Animals*. Animal Management Pty., Werribee.

Seawright, A.A. (1989) *Animal Health in Australia, Vol. 2: Chemical and Plant Poisons*, 2nd ed. Australian Government Publishing Service, Canberra, pp. 22-25.

Chapter 27

Site and Sequence Selectivity of DNA Binding by Divalent Cadmium: Evidence for Inhibition of BPDE-induced DNA Lesions in the Major Groove but Not in the Biologically Relevant Minor Groove

A.S. Prakash, S.R. Koyyalamudi and C.T. Dameron
National Research Centre for Environmental Toxicology, 39 Kessels Road, Coopers Plains, Queensland 4108, Australia

Introduction

Cadmium (Cd^{2+}) is a widely distributed metallic pollutant and a known carcinogen (Kazantzis, 1987). Human exposure to Cd^{2+} is increasing (The Food Standard, 1995) through foods grown in contaminated soils and manufacturing processes. The concentration of Cd^{2+} is elevated in smokers' tissues compared to those of non-smokers, and may play a role in the pathology and carcinogenesis experienced by many smokers (Hahn *et al.*, 1987). Cadmium has been detected in both neoplastic and adjacent non-neoplastic lung tissue of smokers (Waalkes and Obserdorster, 1990; Hart *et al.*, 1993). In addition to Cd^{2+}, smoke contains polycyclic aromatic hydrocarbons (PAH), nitroso compounds and aldehydes that contribute to lung cancer. Alkylation of DNA is believed to be the first step in the initiation of some chemically induced carcinogenesis. Modified bases in specific codons of a proto-oncogene lead to point mutations, which in turn lead to protein products with deleterious effects. Smoke-derived DNA adducts have been detected in human lung tissues from cigarette smokers (Paakko *et al.*, 1989; Van Schooten *et al.*, 1990). Benzo(a)pyrene (BP), a PAH found in cigarette smoke, when activated to its ultimate metabolite (±) *trans*-7,8-dihydroxy-anti-9,10-epoxy-7,8,9,10-tetrahydrobenzo(a)-pyrene (BPDE), reacts with the exocyclic amino group of guanine (G) in DNA (Rill and Marsch, 1990). The n-7G are also alkyated, but the former is considered to be the mutagenic event (Sage and Hasteltine, 1984). Increasing exposure to Cd^{2+} is of concern because of its association with renal toxicity and its carcinogenicity. Mandel and Ryser (1987) demonstrated that cadmium chloride ($CdCl_2$) is a weak mutagen in *Salmonella typhimurium*, but had a strong co-mutagenic (synergistic) effect with the methylating agents. To test the hypothesis that selective binding of Cd^{2+} to DNA modulates the reactivity of DNA towards electrophilic carcinogens, the interaction of Cd^{2+} with DNA and its effect on the reactivity of BPDE were studied.

Materials and Methods

Fotemustine (FM) and BPDE were obtained from Servier Labs Australia and Midwest Research Institute (USA), respectively. Bromoethylphenol (BEP) was a kind gift from Dr David Young, Griffith University, Brisbane.

A 375 base pair *BamH1/EcoR1* fragment of pBR322 was 3'-end-labeled at the *EcoR1* site and isolated (Prakash *et al.*, 1990). The labeled DNA was incubated in presence and absence of $CdCl_2$ (0.5-10µM) with different concentrations of chemical probes (BPDE-0.4µM; FM-125µM and BEP-125µM) in triethanolamine (TEA) buffer (10µM, pH 6.7) for 1hr at 37°C. Incubation was stopped by ethanol (EtOH). The BEP-treated DNA pellet was resuspended in 100ml of TEA buffer (pH 7.4) and incubated at 60°C for 1hr to induce quantitative depurination at alkylated sites. The DNA was precipitated and all reaction samples were dissolved in 100µl of 1M piperidine and subjected to heat treatment at 90°C for 10min. The samples were run on a sequencing gel (Prakash *et al.*, 1990).

Calf thymus DNA (35µg) was incubated with and without $CdCl_2$ or magnesium chloride ($MgCl_2$) (10µM) with BPDE (2µM) for 1hr at 37°C. At the end of the incubation, the reaction products were precipitated with ammonium acetate and EtOH. The supernatant (Sup I) contained mainly tetrol from unreacted BPDE and was saved for fluorescence measurements. The DNA pellet was dissolved in TEA buffer and heated at 90°C for 10min and precipitated as before. The resulting supernatant (Sup II) contained mainly depurinated adducts and intercalated tetrol from the precipitated DNA. The DNA pellet was dissolved in 0.1N HCl and heated at 90°C for 1hr followed by neutralization with NaOH to pH 7-7.5 before precipitation. The supernatant (Sup III) contained mainly tetrol released from N^2-adducts. The emission spectrum of each supernatant fraction was obtained using a Perkin Elmer Luminescence Spectrometer Model LS50B at λ_{ex}=350nm.

Results

Lanes one through four (Fig. 27.1) show the piperidine-induced strand cleavage pattern obtained from BPDE-modified DNA with and without various concentrations of $CdCl_2$. BPDE in the absence of the metal alkylates all Gs, with a preference for runs of G residues (lane four). In the presence of Cd^{2+} the N7-G alkylation is strongly inhibited even at 0.5μM (lanes one through three). Fotemustine (FM), a chloroethylnitrosourea, generates piperidine-sensitive N7 alkyl guanines with a preference for Gs occurring in runs of Gs (bases 80-81, 121-122 and 129-131) (Shelton *et al.*, 1996). Lanes eight and nine show the alkylation pattern produced by FM with and without Cd^{2+}. Divalent Cd inhibits bands corresponding to GG sequences more intensely than isolated Gs (bases 70, 74 and 78). The divalent cation Mg^{2+} has much less effect (lane ten).

The second chemical probe used (BEP) alkylates N3 of adenine (N3-A) in the minor groove and N7-G in the major groove (White *et al.*, 1995). Lanes three and four show the alkylation pattern of BEP (Fig. 27.2) with and without Cd^{2+}. Guanines are strongly inhibited (bases 80-81, 121-123, 129-131) whereas the As are only marginally affected (bases 46-48, 57, 62, 66, 87 and 95). Isolated Gs are not quenched as strongly as the runs of Gs (bases 50-51, 68 and 70). Again Mg^{2+} has only a marginal effect (lane five).

When treated with acid, the minor groove adduct releases the hydrolysis product of BPDE, a tetrol with a high fluorescent quantum yield (Prakash *et al.*, 1988; Alexandrov *et al.*, 1992). This reaction was monitored to determine the effect of Cd^{2+} on the minor groove alkylation by BPDE (Fig. 27.3). Supernatant I contains mainly hydrolyzed BPDE catalyzed by acidic DNA. The intensity is greatest in this supernatant because only a small fraction of BPDE covalently binds to DNA. The intensities of the peaks are about the same with and without Cd^{2+}. The weak fluorescence observed from Sup II with and without Cd^{2+} is mainly from the intercalated tetrol that precipitated with DNA in the first step. The depurinated adduct does not contribute to the intensity of the peaks (Prakash *et al.*, 1988). The fluorescence from Sup III containing the tetrol released from the acid-treated minor groove adduct shows that Cd^{2+} has marginally increased the fluorescence intensity.

Discussion

The chemical probe FM and BPDE alkylate N7-Gs in the major groove, which is sensitive to piperidine cleavage. In a normal sequencing gel assay, these chemicals produce characteristic band patterns that correspond to Gs, with hot spots occurring in sequences containing runs of Gs (Fig. 27.1 lanes four and eight). The presence of Cd^{2+} alters the alkylation patterns of both chemicals. The alkylation of contiguous Gs is dramatically inhibited, while those in other sequences are either weakly or not at all affected. In contrast, the Mg^{2+} at the same concentration has little effect on the alkylation pattern of these chemicals. These results clearly demonstrate that Cd^{2+} preferentially binds to N7-G occurring in runs of Gs. Similar results have been obtained for major groove binding.

To determine if Cd^{2+} could bind in the minor groove, BEP was used; BEP is a minor groove probe that alkylates N3-As in the minor groove, and to a small extent N7-Gs in the major groove (Shelton *et al.*, 1996; Fig. 27.2, lane three). It preferentially alkylates (White *et al.*, 1995) As which occur in runs of As and those

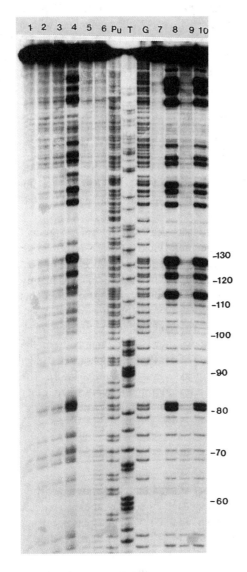

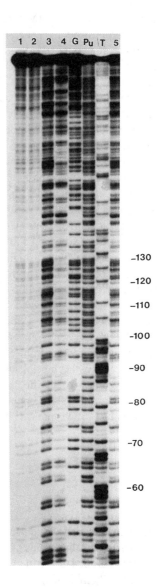

Fig. 27.1. Autoradiogram of alkylation patterns of BPDE and FM with and without Cd²⁺ or Mg²⁺. Lanes one through three, BPDE (0.4µM) + Cd²⁺ (5.0, 1.0, 0.5µM); lane four, BPDE (0.4µM); lane five, Cd²⁺ (5.0µM); lane six, control; lane seven, Mg²⁺ (5.0µM); lane ten, FM + Mg²⁺ (5.0µM). Pu, T, and G are sequencing lanes.

Fig. 27.2. Autoradiogram of the alkylation pattern of BEP with and without 5.0µM Cd²⁺ or Mg²⁺. Lane one control; lane two, Cd²⁺; lane three, BEP + Cd²⁺; lane five, BEP + Mg²⁺. G, Pu, and T are sequencing lanes.

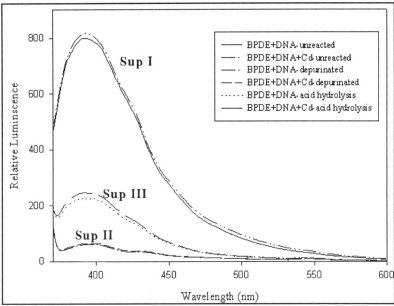

Fig. 27.3. Fluorescence spectra of BP tetrol from DNA-BPDE reaction products.

having contiguous thiamines at the 5' end. The presence of Cd^{2+} strongly inhibited G alkylation, especially in contiguous Gs, but not A alkylation (lane four). It is clear from these two gel experiments that Cd^{2+} is a major groove binder and it inhibits other molecules from interacting with DNA at the site of its binding.

The majority (>90%, Sage and Hasteltine, 1984) of BPDE alkylation occurs at exocyclic amino group of G in the minor groove, and this is considered to be the key lesion leading to mutation and subsequent initiation of cancer. Thus any process that may enhance this lesion will act as a co-carcinogen. The minor groove lesion is not susceptible to chemical cleavage. The fluorescence assay shows that under the same conditions used in the gel studies, Cd^{2+} has a marginal enhancing effect on the alkylation at the N2-G by BPDE. The small increase observed may be due to the fact that only a small fraction of BPDE reacts in the major groove. Divalent Cd has the potential to enhance mutagenic responses of carcinogens through modulation of the reactivity of cellular DNA.

References

Alexandrov, K., Rojas, M., Geneste, O., Castegnaro, M., Camus, A., Petruzzelli, S., Giuntini, C. and Bartsch, H. (1992) An improved fluorometric assay for dosimetry of BPDE-DNA adducts in smoker's lung: Comparisons with total bulky adducts and aryl hydrocarbon hydroxylase activity. *Cancer Research* 52, 6248-6253.

Hahn, R., Ewers, U., Jermann, E., Freier, J., Brockhaus, A. and Schlipkoter, H.W. (1987) Cadmium in kidney cortex of inhabitants of North-West Germany: Its relationship to age, sex, smoking and environmental pollution by cadmium. *International Archives of Occupational and Environmental Health* 59, 165-176.

Hart, B.A., Voss, G.W. and Vacek, P.M. (1993) Metallothionein in human lung carcinoma. *Cancer Letters* 75, 121.

Kazantzis, G. (1987) Cadmium. In: Fishbein, L., Furst, A. and Meehlman, M.A. (eds), *Advances in Modern Environmental Toxicology Vol. XI.* Princeton, NJ, pp. 127-143.

Mandel, R. and Ryser, H.J. (1987) Mechanism of synergism in the mutagenicity of cadmium and N-methyl-N-nitrosourea in *Salmonella typhimurium*: the effect of pH. *Mutation Research* 176, 1-10.

Paakko, P., Kokkone, P., Anttila, S. and Kalliomaki, P.L. (1989) Cadmium and chromium as markers of a smoking in human lung tissue. *Environmental Research* 49, 197-207.

Prakash, A.S., Harvey, R.G. and Lebreton, P.R. (1988) Differences in the influences of π physical binding interactions with DNA on the reactivity of bay versus k-region hydrocarbon epoxides. In: Cooke, M. and Dennis, J. (eds), *Polynuclear Aromatic Hydrocarbons: A Decade of Progress.* Battelle Press, Columbus, OH, pp. 699-710.

Prakash, A.S., Denny, W.A., Gourdie, T.A., Valu, K.K., Woodgate, P.D. and Wakelin, L.P.G. (1990) DNA-directed alkylating ligands as potential antitumor agents: Sequence specificity of alkylation by intercalating aniline mustards. *Biochemistry* 29, 9799-9807.

Rill, R.L. and Marsch, G.A. (1990) Sequence preferences of covalent DNA binding by anti-(+)- and anti(-)-benzo(a)pyrene diol epoxides. *Biochemistry* 29, 6050-6058.

Sage, E. and Hasteltine, W.A. (1984) High ratio of alkali-sensitive lesions to total DNA modification induced by BPDE. *Journal of Biological Chemistry* 259, 11098-11102.

Shelton, C.J., Harding, M.M. and Prakash, A.S. (1996) Enzymatic and chemical footprinting of anthracycline antitumor antibiotics and related saccharide side chains. *Biochemistry* 35, 7974-7982.

The Food Standard: The Regular Bulletin of the National Food Authority (1995) Cadmium Levels: Should the Australian Standard be Changed? Issue 14, ISSN 1038-300X.

Van Schooten, F.J., Hillebrand, M.J., Van Leeuwen, F.E., Lutgerink, J.T., Van Zandwijik, N., Jansen, H.M. and Kriek, E. (1990) Polycyclic aromatic hydrocarbon-DNA adducts in lung tissue from cancer patients. *Carcinogenesis* 11, 1677-1681.

Waalkes, M.P. and Obserdorster, G. (1990) Cadmium Carcinogenesis. In: Foulkes, E.C. (ed), *Biological Effects of Heavy Metals Vol II: Metal Carcinogenesis.* CRC Press, Boston, MA.

White, R.H., Parsons, P.G., Prakash, A.S. and Young, D.J. (1995) The *in vitro* cytotoxicity and DNA alkylating ability of the simplest functional analogues of the sec CC-1065 alkylating subunit. *Bioorganic and Medicinal Chemistry Letters* 5, 1869-1874.

Chapter 28

Monitoring of Physiological and Pathological Changes in Turkey Poults Fed Leaves of Potentially Cardiomyotoxic *Nerium oleander* and *Persea americana*

A. Shlosberg[1], D.G. Ohad[2], M. Bellaiche[1] and S. Perl[1]

[1]*Kimron Veterinary Institute, PO Box 12, 50250 Bet Dagan, Israel;* [2]*Koret School of Veterinary Medicine, Hebrew University of Jerusalem, PO Box 12, 76100 Rehovot, Israel*

Introduction

One of the most perplexing clinical syndromes facing a veterinary diagnostician is that of "sudden death," particularly marked in plant toxicoses, where diagnosis is often difficult. Peracute plant toxicoses are best diagnosed by finding undigested identifiable parts of the toxic plant in the upper gastrointestinal tract, as is done frequently in yew (*Taxus baccata*) or oleander (*Nerium oleander*) toxicoses (Kingsbury, 1964; Galey *et al.*, 1996). As more time passes between the ingestion of the toxic plant and death, the plant undergoes digestion and becomes unidentifiable without specialized microscopic procedures. In acute or subacute toxicoses, therefore, the diagnostic emphasis shifts to examining urine or rumen contents for suspect plant constituents or metabolites (Galey *et al.*, 1996), or to investigating physiological, biochemical or pathological changes which do occur in this longer time frame and which might characterize ingestion of a specific plant. Two cosmopolitan cultivated plants with cardiotoxic potential, and which are not infrequent causes of toxicosis in domestic animals, are oleander (*N. oleander*) and avocado (*Persea americana*). Oleander is a well-known toxicant, with all parts of the plant containing very toxic cardiac glycosides (Rezakhani and Maham 1994; Oryan *et al.*, 1996). Avocado fruit has caused toxicosis in some avian species (Hargis *et al.*, 1989), but the leaves are generally more toxic in other species (Grant *et al.*, 1991; McKenzie and Brown, 1991; Sani *et al.*, 1991; Stadler *et al.*, 1991; Burger *et al.*, 1994). The target organ in avocado leaf toxicosis is usually the heart (Grant *et al.*, 1991; Sani *et al.*, 1991; Burger *et al.*, 1994). Persin, a cardiomyotoxic constituent of the leaves, has recently been investigated (Oelrichs *et al.*, 1995).

This present work was planned to compare some potential physiological, biochemical and pathological changes in subacute toxicoses induced by feeding leaves of oleander and avocado, with an aim of determining whether such changes together might characterize one or both of the toxicoses.

Materials and Methods

Leaves from mature varieties of a pink *N. oleander*, and from a Hass strain of *P. americana* were collected, washed, freeze dried and blended. The resulting powdered materials were stored at -20°C until use.

Fifteen 21d-old turkey poults were allocated at random to three groups. All birds received turkey poult mash feed and water *ad libitum* throughout the 5d trial. The control group received no treatment. Birds in the oleander group received on days one and three a gelatin capsule containing 0.25g/kg of the *N. oleander* powder, and the avocado group received feed containing 10% by weight of the *P. americana* powder for 4d and 20% by weight for 1d. Birds that died were replaced by new birds. Feed intake of each group was measured daily.

Blood was taken from the poults on the fourth day and examined for serum sodium (Na), potassium (K), chloride (Cl), triglycerides, albumin, total protein, uric acid, lactic dehydrogenase (LDH), creatine phosphokinase (CPK) and aspartate aminotransferase (AST) by standard autoanalyzer techniques. All birds euthanized, and those that died during the trial, were examined for signs of gross pathology, and samples of heart, liver, kidney, brain, sciatic nerve and skeletal muscle were placed in buffered formalin for standard histopathological processing.

Birds were observed daily for clinical signs. Electrocardiograms (ECG) were recorded (Boulianne *et al.*, 1992) and oximetry (oxygen saturation of arterial blood, and pulse rate) was performed daily.

Results

Twenty-four hours after the first administration of oleander, one poult was moribund and died shortly thereafter, and two others were very depressed. One of these birds was dead the next morning. The three remaining birds in this group appeared mildly depressed. Two new poults were added to this group and all five birds then received a second dose. On the next day all birds appeared mildly depressed and, on the day after, one of the new birds and one of the original birds were dead. Two of the three remaining poults seemed very depressed, with occasional clonic convulsions, and on the last day, all oleander birds were moderately depressed. Poults in the avocado and control groups seemed healthy throughout. Feed intake in the control and the 10% avocado groups were similar, but the 20% avocado feed caused a 30% reduction in feed intake. The oleander-dosed group had a severe reduction in daily feed intake, ranging daily from 6-60% of that of the control group.

There were no treatment-related significant ($P<0.05$) changes between the groups for values of Na, K, Cl, LDH, CPK, AST, although the enzymes in the oleander group were elevated. The oleander group showed (Table 28.1) a reduction in albumin and total protein, and an elevation in uric acid; the avocado group had an elevation in triglycerides. Necropsy and histopathological examinations did not reveal any abnormal change that could be ascribed to treatment.

In the ECG, significant findings were only manifested in poults in the oleander group. An incremental, dose-dependent response was seen (Fig. 28.1) in heart rates of oleander-fed poults, decreasing from 376±36 beats/min (bpm) prior to oleander dosing, to 199±45bpm on the last experimental day ($P=0.0023$). As the cumulative oleander dose increased, there were progressive prolongations in the PR-interval (from 0.06±0.010 to 0.08±0.004s, $P>0.05$), the QRS-duration (from 0.02±0.006 to 0.05±0.011s, P=0.003), and the QT-interval (from 0.09±0.01 to 0.16±0.01s, $P=0.03$) in oleander-fed birds. Electrocardiograms recorded in this group on the fifth experimental day (Fig. 28.1) clearly demonstrated a first-degree atrio-ventricular conduction block, expressed as a prolonged PR-interval. Although this difference did not reach statistical significance when compared to other experimental days in the same group, it was significant by comparison between the three treatment groups on a day-to-day basis (Table 28.2). The kinetic pattern of the heart rate appeared to reflect a "mirror-image" of the PR, QRS, and QT intervals kinetics in this group (Fig. 28.1). Based on cyclic variation recorded on the baseline, the respiratory rate was 64±10, 103±57, and 58±5 cycles per minute in the control, oleander and avocado groups, respectively.

Discussion

The aim of this work was to monitor and compare some physiological, biochemical and pathological changes in subacute toxicoses induced by feeding leaves of oleander and avocado to determine whether such changes together might characterize one or both of the toxicoses. Unfortunately, very few changes were actually recorded. This is particularly surprising in the oleander group as, out of seven poults, one died after one dose, three died after two doses, and the remaining birds showed clinical signs. The fact that no pathological changes were seen may be ascribed to the relatively short period between dosage and death (at maximum, 5d), which may not have sufficed to permit visualization of changes in cardiac integrity. However, no significant changes were seen in the levels of the blood enzymes LDH, CPK, or AST, which are often more sensitive than histopathology as indicators of cellular damage in cardiac or skeletal muscle, and in liver. The reductions in total protein and albumin may have been due to liver damage (Oryan *et al.*, 1996), but are more likely

Table 28.1. Some of the biochemical and physiological parameters with significant changes between groups, in an oleander- and avocado-dosed poult trial.

Group	Biochemical parameters (mg %)				Heart rate based on oximetry (beats/min)	
	Protein	Albumin	Uric acid	Triglycerides	Day 4	Day 5
Control	3.56[a]	1.70[a]	5.40[b]	69.6[b]	320[a]	308[a]
Oleander	2.42[b]	1.20[b]	10.83[a]	62.0[b]	202[b]	205[b]
Avocado	3.20[a]	1.48[a]	2.94[b]	138.6[a]	342[a]	352[a]

[a,b] Means within columns with no common superscript differ significantly ($P < 0.05$).

Table 28.2. Electrocardiographic differences recorded on day five in values calculated from the control and oleander groups (mean ± SD, based on a one-way analysis of variance, n=5); s=seconds.

Parameter	Control	Oleander	*P*-value
Heart rate (bpm)	380.5 ± 32	199 ± 45	0.040
PR-interval (s)	0.056 ± 0.006	0.083 ± 0.004	0.033
QRS-interval (s)	0.022 ± 0.004	0.048 ± 0.011	0.008
QT-interval (s)	0.09 ± 0.017	0.162 ± 0.009	NS (0.09)

a consequence of the severe reduction in daily feed intake; the elevation in uric acid may have been due to nephrotoxicosis (Oryan *et al.*, 1996), but was probably secondary to cachexia and renal hypoperfusion, secondary to bradycardia. The cause of the rise in triglycerides in the avocado group is unknown; in contrast to these findings, decreases in triglycerides were seen in hypercholesterolemic humans receiving avocado fruit (Lopez *et al.*, 1996).

The birds' respiratory rates, as evident from their ECG tracings, differed between the oleander group and the two other groups. This difference is possibly due to passive pulmonary vascular congestion accompanying the slow heart rate that was progressively demonstrated in this group.

Based on clinical and pathologic signs reported in the literature, avocado toxicity may provoke congestive heart failure, mainly of the right-sided type (Hargis *et al.*, 1989; Burger *et al.*, 1994). Such findings might have been expected in the present study, but were not supported by electrocardiographic findings typical of cardiac chamber dilatation, probably due to the short-term nature of this study. The arterial oxygen saturation results did not reveal any differences between the groups, and this would exclude any severe cardiopulmonary malfunction.

Conclusions

In comparison with other species, turkey poults (250mg/kg) were not very sensitive to oleander. Lethal doses in horses, donkeys and calves were 30-50mg/kg (Rezakhani and Maham, 1994; Oryan *et al.*, 1996).When compared with untreated turkey poults, oleander-dosed poults showed surprisingly few deleterious physiological, biochemical or pathological changes, despite 4/7 birds dying and clear signs of toxicosis. Heart rate and interval duration abnormalities were observed in ECG tracings rather than rhythm or amplitude changes. The poults were also relatively insensitive to avocado leaf toxicosis (30g/kg), compared with lethal doses of 3-9g/kg in goats (Sani *et al.*, 1991), and 30g/kg in ostriches (Burger *et al.*, 1994). The turkey poult is an unsatisfactory model for avocado leaf toxicosis.

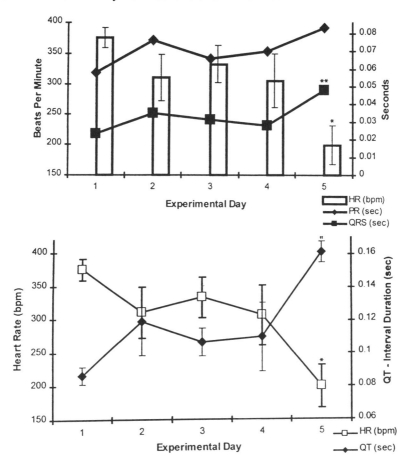

Fig. 28.1. Heart rate (HR) and interval durations measured from oleander-fed poults during five consecutive beats (mean ± SEM). Oleander was fed on days one and three. Based on a one-way analysis of variance in repeated measures, the HR at day five (*) differed significantly from HR during previous days (*P*=0.0023). A) Heart rate, PR- and QRS-interval durations. The QRS-duration at day 5 (**) differed significantly (*P*=0.003) from QRS-duration in previous days. B) Heart rate and QT-interval duration. The QT-interval at day five (″) significantly differed from the QT-interval during previous days (*P*=0.03, one-way analysis of variance in repeated measures).

References

Boulianne, M., Hunter, D.B., Julian, R.J., O'Grady, M.R. and Physick-Sherad, P.W. (1992) Cardiac muscle mass distribution in the domestic turkey and relationship to electrocardiogram. *Avian Diseases* 36, 582-589.

Burger, W.P., Naudé, T.W., Van Rensburg, I.B., Botha, C.J. and Pienaar, A.C. (1994) Avocado (*Persea americana*) poisoning in ostriches. In: Colegate, S.M. and Dorling, P.R. (eds), *Plant-Associated Toxins: Agricultural, Phytochemical and Ecological Aspects.* CAB International, Wallingford, Oxon, pp. 546-551.

Galey, F.D., Holstege, D.M., Plumlee, K.H., Tor, E., Johnson, B., Anderson, M.L., Blanchard, P.C. and Brown, F. (1996) Diagnosis of oleander poisoning in livestock. *Journal of Veterinary Diagnostic Investigation* 8, 358-364.

Grant, R., Basson, P.A., Booker, H.H., Hofherr, J.B. and Antonissen, M. (1991) Cardiomyopathy caused by avocado (*Persea americana* Mill) leaves. *Journal of the South African Veterinary Association* 62, 21-22.

Hargis, A.M., Stauber, E., Casteel, S. and Eitner, D. (1989) Avocado (*Persea americana*) intoxication in caged birds. *Journal of the American Veterinary Medical Association* 194, 64-66.

Kingsbury, J.M. (1964) *Poisonous Plants of the United States and Canada.* Prentice-Hall, New York, NY, pp. 121-123, 264-266.

Lopez, L.R., Frati, M.A.C., Hernandez, D.B.C., Cervantes, M.S., Hernandez, L.M.H., Juarez, C. and Moran, L.S. (1996) Monounsaturated fatty acid (avocado) rich diet for mild hypercholesterolemia. *Archives of Medical Research* 27, 519-523.

McKenzie, R.A. and Brown, O.P. (1991) Avocado (*Persea americana*) poisoning of horses. *Australian Veterinary Journal* 68, 77-78.

Oelrichs, P.B., Ng, J.C., Seawright, A.A., Ward, A., Schaffeler, L. and McLeod, J.K. (1995) Isolation and identification of a compound from avocado (*Persea americana*) leaves which cause necrosis of the acinar epithelium of the lactating mammary gland and the myocardium. *Natural Toxins* 3, 344-349.

Oryan, A., Maham, M., Rezakhani, A. and Maleki, M. (1996) Morphological studies on experimental oleander poisoning in cattle. *Journal of Veterinary Medicine A* 43, 625-634.

Rezakhani, A. and Maham, M. (1994) Cardiac manifestations of oleander poisoning in cattle and donkeys. In: Colegate, S.M. and Dorling, P.R. (eds), *Plant-Associated Toxins: Agricultural, Phytochemical and Ecological Aspects.* CAB International, Wallingford, Oxon, pp. 534-537.

Sani, Y., Atwell, R.B. and Seawright, A.A. (1991) The cardiotoxicity of avocado leaves. *Australian Veterinary Journal* 68, 150-151.

Stadler, P., Van Rensburg, I.B.J. and Naudé, T.W. (1991) Suspected avocado (*Persea americana*) poisoning in goats. *Journal of the South African Veterinary Association* 62, 186-188.

Chapter 29

Enzootic Geophagia and Hepatitis of Calves and the Possible Role of Manganese Poisoning

J.A. Neser[1], M.A. De Vries[2], M. De Vries[2], A.J. Van der Merwe[3], A.H. Loock[3], H.J.C. Smith[3], J.H. Elsenbroek[4], F.H. Van der Vyver[5] and R. Delport[6]

[1]*Pathology Section, Onderstepoort Veterinary Institute, Private Bag X5, Onderstepoort 0110, South Africa;* [2]*State Veterinary Laboratory, Private Bag X36, Vryburg 8600, South Africa;* [3]*Institute for Soil, Climate and Water, Private Bag X6\79, Pretoria 0001, South Africa;* [4]*Council for Geoscience and Geological Survey, Private Bag X112, Pretoria 0001, South Africa;* [5]*Directorate of Animal Health, Private Bag X369, Pretoria 0001, South Africa;* [6]*Department of Chemical Pathology, Faculty of Medicine, University of Pretoria, PO Box 2034 Pretoria 0001, South Africa*

Introduction

Geophagia, or the deliberate ingestion of soil, has been classified as a form of pica (Ammerman *et al.*, 1965; Halsted, 1968; Kreulen and Jager, 1984; Johns and Duquette, 1991). While pica is usually defined as the compulsive ingestion of inappropriate or foreign material, geophagia may be beneficial when it is selective for certain sites and soil types, often termed natural licks, in order to exploit particular nutrient elements, and can therefore be symptomatic for deficiencies of several elements, notably sodium (Na), magnesium (Mg), phosphorus (P), sulfur (S), copper (Cu), cobalt (Co) and manganese (Mn) (Kreulen and Jager, 1984). However, more recent reports have also suggested that geophagia may be an instinctive behavioural response to gastrointestinal disturbances caused by the ingestion of certain toxic plants or irritant substances (Kreulen, 1985; Johns and Duquette, 1991; Reid, 1992).

A specific enzootic form of geophagia has been identified in young cattle on Mn-rich soil derived from the weathering of superficial dolomitic (carbonate) rock formations in a restricted area in northern South Africa (SA). Affected animals are characterized by marked icterus, large amounts of ingested black to dark brown soil, and severe chronic cholangiohepatitis, with a high mortality rate in untreated cases. The pathological lesions observed in organs sporadically referred to the Onderstepoort Veterinary Institute (OVI) since 1972 were at first suspected to be due to a plant toxicosis or a chronic infection of the liver. The condition was referred to as "Vryburg hepatosis," after the district in which it primarily occurred.

Epidemiology

Field observations and a questionnaire survey indicate that the disease was limited to farms with outcrops of dolomitic rock formations reported to be rich in Mn, and an association between the disease and Mn poisoning was suspected. The farms are all situated on an extensive area known as the Ghaap Plateau, which is underlain by Fe-containing Mn-rich dolomitic or carbonate rock with a unique geological composition of limestone, dolomite and chert. Although superficial manganiferous dolomitic rock formations occur in other areas of South Africa, the geological composition is different according to the Official Geological Map of SA, and it is not known if cases of enzootic geophagia have been reported from other areas.

The surface topography of the affected farms is characterized by numerous superficial outcrops of black to brown and grey-white dolomitic rocks and poorly drained extensive low-lying areas. The soil contains many densely distributed small round to ovoid black-grey Fe-containing Mn-rich concretions of about 1-10mm diameter derived from the superficial weathering of parent rock under moist warm conditions in shallow pans.

Clinical Signs

On the affected farms, young calves of both sexes and all breeds developed an insatiable appetite for the Mn-rich soil. Some calves habitually licked Fe poles. Marked icterus, cholangiohepatitis, constipation developed and death occurred in a high percentage of cases. The problem apparently only occurred among calves less than 2mos old, with the highest frequency in the 7-14d age group. The uniform symptoms usually started with intermittent, progressive geophagia, followed by constipation, dehydration and death within about 7-10d in untreated cases. The morbidity and mortality rates were difficult to estimate accurately, since farmers used the preventative measure of rearing calves in pens covered with a thick layer of dung to prevent them from licking the underlying soil. On severely affected farms, about 50-75% of calves developed geophagia. The mortality rate in untreated advanced cases was estimated by farmers at close to 100%.

According to farmers, the treatment of calves by the parenteral injection of commercial Fe-dextran compounds and vitamin B_{12} at 1-2d after birth and at 14d at the registered therapeutic doses appeared to have a significant preventative effect on the occurrence of geophagia.

Pathological Examinations

Blood and serum parameters, including Fe levels, were within normal limits with the exception of very high serum bilirubin levels that corresponded to the severe icterus and the histopathological lesions of cholangiohepatitis. The levels of homocysteine in plasma were also raised in cows from the affected area, thereby indicating a possible vitamin B_{12} or Co deficiency (Jubb *et al.*, 1994).

Macroscopical pathological changes consisted of moderate to severe icterus, and a markedly enlarged and yellow liver. The stomach and intestinal contents contained variable amounts of dark brown to black soil, largely composed of small black to dark brown Mn-rich dolomitic concretions about 1-5mm in diameter. Histopathology revealed a marked subacute to chronic cholangiohepatitis with hyperplasia of bile duct epithelium and periductular fibrosis.

Mineral Analysis

Atomic Absorption Spectrometer (AA) analysis of liver specimens from 23 calves with soil in their digestive tracts or a history of geophagia revealed concentrations of Mn ranging from 10-1,800ppm wet mass, considerably higher than the normal range of 2-3ppm (Hurley and Keen, 1986; Graham *et al.*, 1994). Liver specimens from four full-term dead bovine fetuses, four new-born calves that had not yet ingested milk, 11 yearling steers and 15 culled cows from affected farms showed no significant deviation from the normal values for Mn.

Milk specimens from representative numbers of cows with calves of about 1-12wks of age from three severely affected farms were examined by the AA for their Mn and Fe content, and compared with values determined from cows under similar grazing conditions at the Onderstepoort experimental farm (control site), where the Mn content of the soil is marginal. The levels of Mn in the milk of cows on affected farms were higher and the Fe lower when compared with control values. Although there was considerable variation in the data, the differences were statistically significant at $P< 0.01$ for Mn and $P< 0.001$ for Fe. The levels of Co were not determined.

Rectal fecal specimens from the same cows were collected at the same time as the milk samples, and analyzed for the same elements by the AA and Inductively Coupled Plasma Mass Spectrometer (ICP-MS) methods. Levels of Mn, Fe and Co were significantly higher in the cows from the affected farms ($P< 0.001$).

Specimens from a palatable grass (red grass, *Themeda triandra*) and an edible shrub (rosyntjiebos, *Grewia flava*) were collected from grazing areas on affected farms near Vryburg as well as in the control area at Onderstepoort during autumn 1995 and 1997 and analyzed for their Mn, Fe and Co content. Specimens from affected farms contained grossly higher levels of Mn, and the ratio of Mn to Fe was higher in the plants from the affected areas (Table 29.1).

Surface soil specimens (n=2,882) were collected over an extensive area of the Ghaap Plato, including most of the affected farms at a sampling density of 1/km². These were dry sieved and analyzed by X-ray fluorescence spectrometry. Samples were also collected from selected sites on the Onderstepoort experimental farm, where the Mn content in the soil is considered to be marginal, and the Fe content high. The mean levels of Mn were grossly higher in specimens from the affected area (0.54 *vs.* 0.25%), compared to control specimens (n=4). The levels of both Fe and Co were lower (3.88 *vs.* 9.30% and 9 *vs.* 65ppm, respectively), than the control specimens from the Onderstepoort area where soil Fe and Co levels are considered to be high.

The average Mn content of unsieved surface soil samples, including concretions (i.e. the soil usually ingested by affected calves) collected from selected sites on affected farms, was grossly higher (7.8 *vs.* 0.39%) than control specimens. Soil specimens containing similar concretions collected from the stomachs of affected calves contained comparable concentrations of Mn (5.9 *vs.* 7.8%).

Feeding and Dosing Experiments

Geophagia was reproduced experimentally at both the State Veterinary Laboratory at Vryburg and at the OVI by offering Mn-rich soil from affected farms, as well as ordinary red soil from unaffected farms, to new-born calves. The calves were kept in enclosures on concrete floors and fed on colostrum and unsupplemented whole milk according to conventional rearing methods. Geophagia of ordinary soil was of brief duration, whereas ingestion of the black Mn-rich soil became progressively worse until calves become seriously ill with typical symptoms. This serious form of geophagia was aggravated by dosing animals with $Mn(SO_4)_2$ at 0.5g/kg/d for 7-10d. The experimental animals developed the characteristic histopathological changes of subacute to chronic cholangitis as in field cases of enzootic geophagia.

Table 29.1. Mean Mn, Fe and Co concentrations in specimens of selected sites in the affected and contol areas during the autumn seasons of 1995 and 1997.

Plant Species	T. triandra			G. flava		
Element	Mn%	Fe%	Co ppm	Mn%	Fe%	Co ppm
Affected farms	0.059	0.068	0.454	0.017	0.034	0.228
Control farm	0.007	0.042	0.235	0.007	0.037	-

Discussion

While deficiency of Mn has been cited as a cause of geophagia in domestic ruminants (Kreulen and Jager, 1984), a form of geophagia is described that apparently results from the excessive ingestion of Mn. The initial stimulus for the intake of soil by calves and lambs is unknown. Routine necropsies on young calves and lambs frequently reveal small amounts of soil in the digestive tract. It has been shown that calves reared in close confinement voluntarily ingest limited amounts of soil from both affected and unaffected farms. Soil licking may be an instinctive mechanism to secure necessary mineral nutrients in the soil, especially Fe, as calves on unsupplemented whole milk diets have developed Fe deficiency in the absence of soil (Graham *et al.*, 1994). Alternatively, it has been suggested that geophagia may be an instinctive adaptive behavioral response to gastrointestinal disturbances (Kreulen, 1985; Johns and Duquette, 1991; Reid, 1992), or represent displaced appetitive behavior due to hunger while the cows graze far away for long periods of the day. Calves on the affected farms, temporarily deserted while their dams are pursuing lush green growth, were observed to start licking small amounts of soil that may often include Mn-rich concretions. It is clear from these experiments that Mn-rich soil, or even $Mn(SO_4)_2$, aggravates ingestion of soil, to produce fatal geophagia in a high percentage of young calves. The tendency apparently disappears as soon as the calf develops a functioning rumen.

Analysis of edible plants and rectal feces indicated that cows on the affected farms ingested relatively large amounts of Mn, and secreted relatively high levels of Mn in the milk. Although enzootic geophagia could be reproduced in calves that had not been conceived and born in the enzootic areas, the possible influence of the maternal animal in initiating or aggravating this condition through her milk cannot be discounted. Although transplacental transmission of Mn with accumulation in the liver has been reported in calves, rabbits and mice (Hurley and Keen, 1986), these findings suggested that Mn accumulated in the liver of affected calves after the ingestion of Mn-rich soil or milk.

Disease production by a high intake of Mn appears as a complex entity. Manganese interferes with Fe, Co and Zn uptake in the digestive tract (Pfander *et al.*, 1966; Thomson and Valberg, 1972; Hurley and Keen, 1986). The observations imply that treatment of calves with Fe-dextran and vitamin B_{12} preparations reduce the prevalence and intensity of geophagia.

The main feature of the disease is liver damage, although the ingestion of soil also leads to variable degrees of constipation and colic that probably cause the calf to stop suckling. The characteristic clinical signs and pathological lesions could be induced by dosing with relatively large amounts of $Mn(SO_4)_2$. These lesions correspond with the changes induced by experimental administration of $MnCl_2$ and $Mn(SO_4)_2$ to laboratory rodents (Marshall Findlay, 1924; Witzleben *et al.*, 1968). Although the mechanism by which Mn causes geophagia may not yet be fully understood, there can be no doubt that the characteristic lesions in the liver can be attributed to a subacute to chronic form of Mn poisoning.

References

Ammerman, C.B., Loggins, P.E. and Wing, J.M. (1965) Pica - a craving for unnatural food or feeds. *Foodstuffs* 37, 26-27.

Graham, T.W., Thurmond, M.C., Mohr, F.C., Holmberg, C.A., Anderson, M.L. and Keen, C.L. (1994) Relationships between maternal and fetal liver copper, iron, manganese and zinc concentrations and fetal development in California Holstein dairy cows. *Journal of Veterinary Diagnostic Investigation* 6, 77-87.

Halsted, J.A. (1968) Geophagia in man: Its nature and nutritional effects. *American Journal of Clinical Nutrition* 21, 1384-1393.

Hurley, L.A.S. and Keen, C.L. (1986) Manganese. In: Mertz, W. (ed), *Trace Elements in Human and Animal Nutrition, Vol 1,* 5th ed. Academic Press, San Diego, CA, pp. 185-223.

Johns, T. and Duquette, M. (1991) Detoxification and mineral supplementation as functions of geophagy. *American Journal of Clinical Nutrition* 53, 448-456.

Jubb, K.V.F., Kennedy, P.C. and Palmer, N. (1994) Anaemia. In: *Pathology of Domestic Animals, Vol 3,* 4th ed. Academic Press, San Diego, CA, p. 172.

Kreulen, D.A. and Jager, T. (1984) The significance of soil ingestion in the utilization of arid rangelands by large herbivores, with special reference to natural licks on the Kalahari pans. In: Gilchrist, F.M.C. and Mackie R.I. (eds), *Herbivore Nutrition in the Subtropics and Tropics.* The Science Press, Craighall, South Africa, pp. 204-221.

Kreulen, D.A. (1985) Lick use by large herbivores: a review of benefits and banes of soil consumption. *Mammal Review* 15, 107-123.

Marshall Findlay, G. (1924) The production of biliary cirrhosis by salts of manganese. *British Journal of Experimental Pathology* 5, 92-106.

Pfander, W.H., Beck, H. and Preston, R.L. (1966) The interaction of manganese, zinc and cobalt in ruminants. *Federation Proceedings* 25, 1362.

Reid, R.M. (1992) Cultural and medical perspectives on geophagy. *Medical Anthropology* 13, 337-351.

Thomson, A.B.R. and Valberg, L.S. (1972) Intestinal uptake of iron, cobalt and manganese in the iron deficient rat. *American Journal of Physiology* 223, 1327-1329.

Witzleben, C.L., Pitlick, P., Bergmeyer, J. and Benoit, R. (1968) Acute manganese overload - A new experimental model of intrahepatic cholestasis. *American Journal of Pathology* 53, 409-403.

Chapter 30

The Feeding of Lupin Grain Can Cause Rumen Acidosis and Rumenitis

J.G. Allen[1], G.D. Tudor[2] and D.S. Petterson[1]

[1]*Agriculture Western Australia, 3 Baron-Hay Court, South Perth, Western Australia 6151, Australia;* [2] *Agriculture Western Australia, Bunbury, Western Australia 6230, Australia*

Introduction

The use of lupin grain (referred to as lupins from here) as a feed and feed supplement is widespread in Australia. Lupins contain less than 3% starch (Evans, 1994), the main fermentable carbohydrate involved in rumen acidosis when cereal grains are fed to ruminants. For this reason lupins have generally been regarded as a completely safe feed for sheep and cattle, and required no gradual introduction (Rowe, 1995). This belief was supported by the lack of any problems associated with the feeding of lupins to livestock over more than 20yrs up to 1995. Up to this time most lupins were fed by trailing or scattering them on the ground.

During the summer/autumn of 1995, 13 cases of lupin-grain-associated deaths were reported to the Animal Health Laboratories of Agriculture Western Australia. Four of these involved sheep, with mortality rates of 1/6, 4/14, 17/130 and 50/900, and nine involved cattle. Mortality rates for the cattle varied from 1/6 to 14/150 and 25/430. Other cases have subsequently been reported during the summer/autumn periods of 1996 and 1997, with one involving the deaths of over 200 sheep. Most cases involved hungry stock on a falling plane of nutrition gaining access to large quantities of lupins, often provided in self-feeders. With cattle, the lupins had usually been coarsely milled.

Insufficient samples were obtained from field cases in 1995 to establish a consistent pathogenesis. A severe rumenitis existed in cases in which the rumen was examined, but since lupins contain so little starch, acidosis was not considered a probable cause. Lupins contain high concentrations of protein (Petterson and Mackintosh, 1994), and ammonia toxicity was the favored diagnosis.

Animal House Experiment

An animal house trial involving ten Merino wethers (26.9-37.1kg at dosing) was conducted to determine the pathogenesis of this newly emerged disease. Table 30.1 details the treatments. "Chaff" pretreatment involved feeding only chaff for the 4wks prior to dosing. The sheep lost 3.8-6.5kg over this time. The "normal" sheep were fed a maintenance ration of sheep cubes and chaff for the same period and gained 0.5-0.6kg. Whole lupin grains (*Lupinus angustifolius* cv. Gungurru, no detectable starch) were hammer-milled and given as an aqueous slurry *via* an intraruminal tube. The amount given was based on a preliminary feeding trial with a 50kg Merino wether. This animal had been fed a maintenance ration of sheep cubes and chaff for several weeks before consuming 2.794kg of lupins (55.8g/kg) within 24hrs of being given *ad libitum* access to them. It then stopped eating, but resumed its normal ration 5d later. The "processed" lupins were lupin kernels (seed coat removed from whole grain) which had the protein, some water, some ash, the oligosaccharides and the oil removed by wet-milling. The resultant material represented 41% of whole lupins, and was accounted for in the amount given.

During the experiment, "chaff" sheep were offered 500g chaff each day and "normal" sheep 300g of sheep pellets and 200g chaff. Daily feed consumption was measured. Blood and rumen samples were collected at regular intervals and subjected to various tests. The sheep were euthanized at different times, as indicated in Table 30.1, to determine the pathological changes in the forestomachs. Daily feed intakes and clinical chemistry results are shown in Figs. 30.1 and 30.2.

Gross and Microscopic Pathology

Table 30.1 is a summary of the pathological changes in the forestomachs. The most noticeable gross change was the sloughing of the mucosa over variable sized areas. Where sloughing was not apparent, the mucosa could be easily dislodged. Occasionally, rumen papillae were swollen and had pale tips. There was no hyperemia of the forestomachs. In the sheep with fungal rumenitis there were several areas of depigmentation and apparent scarring, occasionally with small foci of ulceration, throughout the rumen.

Microscopically there was microvesiculation within, and necrosis and inflammatory cell infiltration of, the outer half of the mucosa. In many areas this necrotic layer was lifting off the intact epithelial cells in the mucosa. This lifting material was seen sloughing, or easily dislodged, at necropsy. There was a lot of mitotic activity in the germinal layer cells, and mild inflammatory cell infiltration of the submucosa. This resembled a "burning" of the outer layer of the mucosa.

Such a superficial lesion would normally be expected to regenerate quickly, but many areas had foci of deep mucosal necrosis, sometime extending to the germinal layer and the basement membrane. Intense inflammatory cell infiltration into the submucosa and bacteria in the necrotic tissues was evident.

Table 30.1. Experimental design, and pathological changes seen in sheep dosed with milled or processed lupins.

Pre-dosing treatment	Treatment	Number of sheep	Dose rate (g/kg)[1]	Forestomach pathology[2] at			
				2d	3d	4d	7d
Chaff	Lupins	6	48.1-64.7 (1740-1980g)	1 Sheep Ru, Re, O	1 sheep Ru, O		1 sheep Re
				1 Sheep Ru, O			1 sheep Ru, Re
							1 sheep Ru(f)
Normal	Lupins	2	46.9-49.6 (1740g)		1 Sheep NPC		1 Sheep NPC
Chaff	Processed lupins	2	24.6 (710-750g)			1 sheep Ru, Re, O	
						1 Sheep NPC	

[1] bracketed figures=total amount dosed; [2] Ru=Rumenitis, Ru(f)=Fungal rumenitis, Re=Reticulitis, O=Omasitis, NPC=No pathological changes. One sheep dosed with the processed lupins had no changes in clinical chemistry and NPC in its forestomachs after 4d. This animal's results are not included.

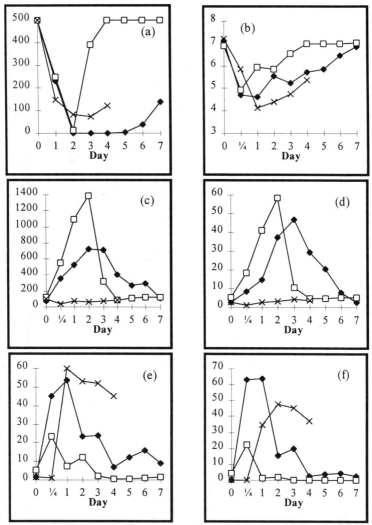

Fig. 30.1. Mean daily feed intake (a; g/hd/d), and rumen pH (b), ammonia (c; mg/L), urea (d; mM/L), L-lactate (e; mM/L) and D-lactate (f; mM/L) concentrations in sheep fed chaff prior to being dosed milled (♦) or processed lupins (✕), or fed a normal maintenance ration prior to being dosed with milled lupins (□).

In the sheep with a fungal rumenitis, scattered fungal hyphae were apparent, and there were multiple foci of fibrous and pyogranulomatous reaction, mainly in the submucosa, but also within the muscularis. Most of the overlying mucosa was normal, except for a thin layer of necrosis, vacuolar change and inflammatory cell infiltration on the surface. In sections through the grossly evident ulcers, there was full depth mucosal necrosis and destruction of the basement membrane. The ulcers merged into submucosal pyogranulomatous lesions.

Discussion

Transient rumen acidosis and lactic acidemia, with accompanying damage to the mucosa of the forestomachs, was produced by dosing milled lupins to six sheep that had been on a submaintenance ration. The resultant pH in the rumen was consistent with levels considered to result in a chemical rumenitis (rumen pH 4-5, Barker *et al.*, 1993). Pathological examinations revealed damage to the mucosa of the forestomachs was generally superficial, and in most cases affected sheep should recover. Yet, there were foci where full depth necrosis of the mucosa did occur, allowing for entry of bacteria or fungi to the deeper layers of the forestomachs. This presumably happened in field cases, resulting in death.

Dosing milled lupins to sheep that had been on a maintenance ration resulted in a reduced rumen pH. It was not as low as, and it returned to normal levels more

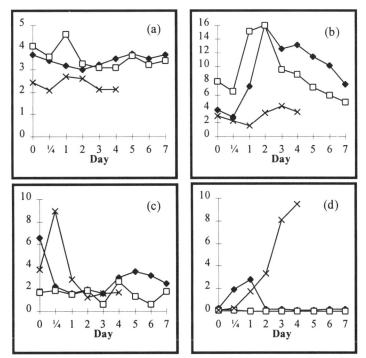

Fig. 30.2. Mean daily plasma ammonia (a; mg/L), urea (b; mM/L), L-lactate (c; mM/L) and D-lactate (d; mM/L) concentrations in sheep fed chaff prior to being dosed milled (◆) or processed lupins (✕), or fed a normal maintenance ration prior to being dosed with milled lupins (□).

quickly than, the sheep fed a submaintenance ration. These changes reflected lower rumen lactate levels. The lack of pathological changes in the forestomachs of the sheep fed the maintenance ration established that the rumen environment was not sufficiently altered to result in chemical rumenitis. Clearly, rumen microbiological activity at the time of first access to lupins influences whether or not a substantial acidosis will occur.

All sheep dosed with the milled lupins developed extremely high concentrations of ammonia and urea in the rumen. Although there were substantial, but transient, increases in their plasma urea concentrations, plasma concentrations of ammonia remained normal. This is not surprising, because at these low pH levels most of the ammonia would be in the form of the poorly absorbed ammonium ions (Seawright, 1989). These results indicate that ammonia toxicity is not part of this pathogenesis.

The finding that dosing processed lupins to a sheep fed a submaintenance ration resulted in rumen acidosis, high concentrations of rumen lactic acid, a substantial lactic acidemia, and damage to the mucosa of all forestomachs, indicated that the non-starch polysaccharides in the dietary fiber of the lupin kernel were the substrate for fermentation that leads to this condition. Evans (1994) found that this dietary fiber was readily fermentable to volatile fatty acids in the hind gut of the rat, causing a reduction in pH. The lack of protein in the processed lupins was further evidence that ammonia toxicity was not involved in this disease.

This experiment confirms that sheep on a falling plane of nutrition may develop rumen acidosis and damage to the forestomachs if given sudden access to large quantities of lupins. The recent adoption of the use of self-feeders increases the likelihood that such a situation may occur.

References

Barker, I.K., van Dreumel, A.A. and Palmer, N. (1993) The alimentary system. In: Jubb, K.V.F., Kennedy, P.C. and Palmer, N. (eds), *Pathology of Domestic Animals, Vol. 2*, 4th ed. Academic Press, New York, NY, p. 47.

Evans, A.J. (1994) The carbohydrates of lupins, compositions and uses. In: Dracup, M. and Palta, J. (eds), *Proceedings of the First Australian Lupin Technical Symposium*. Western Australian Department of Agriculture, South Perth, pp. 110-114.

Petterson, D.S. and Mackintosh, J.B. (1994) *The Chemical Composition and Nutritive Value of Australian Grain Legumes*. Grain Research and Development Corporation, Canberra, p. 10.

Rowe, J.B. (1995) New horizons for using cereal grain in the nutritional management of sheep. In: *Proceedings of the Australian Sheep Veterinary Society*. Australian Sheep Veterinary Society, Artarmon, pp. 78-83.

Seawright, A.A. (1989) *Animal Health in Australia, Vol 2: Chemical and Plant Poisons*, 2nd ed. Australian Government Publishing Services, Canberra, pp. 305-307.

Chapter 31

Effects of Dietary Cottonseed Meal and Cottonseed Oil on Nutrient Digestibility and Rumen Protozoa

M.N. Islam[1], J.R. Scaife[2] and T. Acamovic[3]

[1]*Department of Dairy Science, Bangladesh Agricultural University, Mymensingh 2202, Bangladesh;* [2]*Department of Agriculture, University of Aberdeen and* [3]*Scottish Agricultural College, 581 King Street, Aberdeen AB24 5UD, UK*

Introduction

Whole cottonseed (WCS) and cottonseed meal (CSM) have been widely used as dietary supplements in ruminant feeds. However, their use may be limited by gossypol, which can adversely affect rumen microflora. Ismartoyo (1997) and Ismartoyo and Acamovic (1994) reported that gossypol adversely affected rumen fungal attachment to cellulose and grass fermentation *in vitro*. Others (Calhoun *et al.*, 1990; Huston *et al.*, 1990) have shown gossypol to have negative effects on reproductive capacity and even to kill sheep. Ismartoyo *et al.* (1993) and Ismartoyo (1997) demonstrated that dry matter loss and gas production during *in vitro* incubation of WCS with sheep rumen liquor was lower than observed from other oil seeds. Subsequent studies have shown that 300g/d of WCS can be in sheep diets with no negative effects on feed intake, nutrient digestibility and rumen fermentation characteristics (Ismartoyo *et al.*, 1994). However, inclusion of 500g/d of WCS severely depressed rumen bacteria and protozoa (Ismartoyo *et al.*, 1995). These effects may have been caused by gossypol at 2g/kg in WCS, or the high dietary oil intake (200g/d) when WCS nears 500g/d. Previous studies (Moore *et al.*, 1986; Zinn, 1989) have shown that when fat concentrations in ruminant diets exceed 50-70g/kg dry matter (DM), fiber and organic matter digestibilities are reduced. Sheep were fed CSM or cottonseed oil (CSO) to determine the effects of CSO and gossypol on rumen microflora and nutrient digestibility.

Materials and Methods

Mature sheep (n=5) fitted with rumen cannulas were fed either CSM (n=3) or CSO (n=2) with good quality hay. Two levels of CSM (240 and 400g/d) and two levels of CSO (60 and 100g/d) were fed as constituents of maintenance diets. Diets were fed in three phases. In Phase 1, animals were fed a diet consisting of hay only. In Phase 2, sheep received hay plus either 240g/d CSM or 60g/d CSO. In Phase 3, the levels of CSM and CSO were increased to 400g/d and 100g/d respectively. At the start of Phases 2 and 3, CSM and CSO were introduced gradually and hay was reduced to maintain similar dry matter and energy contents of the rations. Digestibility was measured conventionally in each phase of the study, and rumen samples were collected to observe changes in the protozoal population. During Phase 3, animals receiving CSO lost appetite and were withdrawn from the study.

Effects on Nutrient Digestibility

Dry matter (DM), organic matter (OM), crude protein (CP), acid detergent fiber (ADF) and neutral detergent fiber (NDF) digestibilities during the different phases of the study are presented in Tables 31.1 and 31.2. There was no significant difference in nutrient digestibility between the treatment groups during Phase 1.

During Phase 2, digestibility values for all nutrients measured were significantly lower ($P<0.05$) in sheep fed the CSO diet (60g/d) compared to those fed CSM (240g/d) (Table 31.1). There was no significant change in the digestibility of DM, OM, ADF and NDF in the CSM group in this study, but digestibility of CP significantly increased ($P<0.01$) from 52.4% in Phase 1 to 74.4% in Phase 3 (Table 31.2).

This study's findings for CSO group agree with the work of Moore *et al.* (1986), who found that fat additions of 63-80g/kg, either as free fats or WCS, reduced ADF digestibility. Addition of more than 80g/kg fat reduced digestibility of DM, OM and gross energy. They also reported that oil fed as WCS caused a depression of digestibility similar to that observed when an equivalent amount of free fat was fed. In the present study, the fat content of the diet offered to the CSO group during Phase 2 was approximately 60g/kg of the total DM content of the diet, equivalent to the inclusion of WCS at 300g/kg diet. Data from this study suggests that CSO added as a free oil had more adverse effects on nutrient digestibility than when added at a comparable level in the form of WCS.

The increased protein digestibility in the CSM group was apparently due to the high protein intake of sheep fed the CSM diet, and suggested increased microbial protein synthesis in the rumen. Ismartoyo *et al.* (1995) also observed high protein digestibility when WCS was fed to sheep at 500g/d. At that level, the digestibility of DM, OM, ADF and NDF were reduced, and it was suggested that gossypol might be responsible for the decreases. In the present study, the level of CSM (400g/d) in Phase 3 was more than equivalent to that fed as WCS by Ismartoyo *et al.* (1995), and nutrient digestibilities did not decrease. The concentration of free gossypol was very low in CSM (32.3µg/g) compared to that in WCS (301.2µg/g), and no gossypol was detected in rumen liquor samples taken from any sheep. Huston *et al.* (1991) reported that essentially gossypol in WCS is free, while much is bound in CSM. The epsilon amino groups in lysine are considered an important site for binding gossypol to protein during processing of WCS to CSM and digestion in the rumen. The current results indicate that the amount of free gossypol in CSM is very low, and has no adverse effect on nutrient digestibility.

Table 31.1. Comparison of nutritional parameters (DM, OM, CP, ADF, NDF)[†] digestibility (%) in CSM and CSO groups (Mean ± SED).

Study phase	Parameters[†]	Groups of animals		Significance
		CSM group	CSO group	
Phase 1	DM	67.2 ± 1.30	66.9 ± 0.20	NS
	OM	68.7 ± 1.50	68.2 ± 0.46	NS
	CP	52.5 ± 0.88	51.32 ± 2.96	NS
	ADF	61.5 ± 5.25	61.6 ± 2.09	NS
	NDF	63.9 ± 5.32	64.5 ± 1.74	NS
Phase 2	DM	67.3 ± 2.20	62.2 ± 1.70	*
	OM	68.3 ± 1.85	63.5 ± 1.06	*
	CP	68.7 ± 2.16	44.4 ± 1.17	**
	ADF	62.1 ± 3.11	52.8 ± 0.38	*
	NDF	61.1 ± 1.26	48.5 ± 4.51	*
Phase 3	DM	68.0 ± 3.40	-	-
	OM	69.6 ± 2.99	-	-
	CP	74.4 ± 3.50	-	-
	ADF	56.4 ± 3.40	-	-
	NDF	64.7 ± 3.15	-	-

[†] DM=Dry matter; OM=Organic matter; CP=Crude protein; ADF=Acid detergent fiber; NDF=Neutral detergent fiber; NS=Not significant; *=*P*<0.05; **=*P*<0.01.

Table 31.2. Comparison of parameters (DM, OM, CP, ADF, NDF)[†] digestibility (%) between treatment phases for sheep fed CSM or CSO (mean ± SED).

Group	Parameter[†]	Phase of study			Significance
		1	2	3	
CSM	DM	67.2 ± 1.30	67.3 ± 2.20	68.0 ± 3.40	NS
	OM	68.7 ± 1.50	68.3 ± 1.85	69.2 ± 2.71	NS
	CP	52.4 ± 0.88	68.7 ± 2.16	74.4 ± 3.50	**
	ADF	61.5 ± 5.25	62.1 ± 3.11	56.4 ± 3.40	NS
	NDF	63.9 ± 5.32	61.1 ± 1.26	64.7 ± 3.15	NS
CSO	DM	66.9 ± 0.20	62.2 ± 0.28	-	**
	OM	68.2 ± 0.46	63.5 ± 1.06	-	*
	CP	51.3 ± 0.96	44.4 ± 1.17	-	*
	ADF	61.6 ± 2.09	52.8 ± 0.38	-	*
	NDF	64.4 ± 1.74	48.5 ± 4.51	-	*

[†] DM=Dry matter; OM=Organic matter, CP=Crude protein; ADF=Acid detergent fiber; NDF=Neutral detergent fiber;NS=Not significant; *=*P*<0.05; **=*P*<0.01.

Effect on Rumen Protozoa

Types and total count of protozoa in rumen liquor in the different phases of this study are shown in Table 31.3. The protozoal population was similar in both groups during Phase 1. However, no protozoa were detected in rumen liquor samples taken from the CSO group during Phase 2. Protozoal types and numbers were unaffected in the CSM group during Phases 2 and 3. Ismartoyo *et al.* (1995) found drastically reduced numbers of protozoa in the rumen liquor of sheep fed 500g/d WCS, and concluded that gossypol may inhibit growth of rumen protozoa. Here, CSM with low levels of free gossypol did not affect rumen protozoa, but CSO fed at a level equivalent to approximately 300g WCS/d rapidly eliminated protozoa from the rumen environment. Similar effects of oil on rumen protozoa have been reported in cattle fed diets supplemented with 500g rapeseed oil/d (Tesfa, 1993). The form in which the oil is fed, extracted or contained within the natural matrix of the whole oil seed, determines the level at which it can be fed without incurring a reduction in the digestibility of other nutrients. For cottonseed, the studies of Ismartoyo *et al.* (1994; 1995) suggest that, when contained in the matrix of WCS, CSO has no adverse effects on rumen microflora and fauna until WCS intakes reach 500g/d (equivalent to 100g CSO). When fed as the extracted oil, intakes of 60g/d are sufficient to defaunate the rumen. Since WCS contains a much higher concentration of free gossypol than CSM, effects of gossypol on rumen protozoa cannot be ruled out at high intakes of WCS.

The results of this study suggest that inclusion of extracted CSO in ruminant diet should be kept below 60g/kg diet to avoid rumen defaunation and reduction in nutrient digestibility. CSM at levels of up to 400g/d appears to have no adverse effect on rumen function or nutrient digestibility.

Table 31.3. Numbers of rumen protozoa in the rumen liquor samples taken from sheep fed CSM or CSO.

Phase of study	Group	TC* ($\times 10^4$)	Ent ($\times 10^4$)	Polypl ($\times 10^3$)	Diplopl ($\times 10^3$)	Dasy ($\times 10^3$)
				Protozoa		
Phase 1	CSM	7.77	6.50	0.95	1.60	3.10
	CSO	7.50	6.75	1.10	2.38	-
Phase 2	CSM	8.93	8.03	0.77	2.97	1.56
	CSO	NP	NP	NP	NP	NP
Phase 3	CSM	8.19	7.18	1.57	1.67	2.10
	CSO	Animals withdrawn from the trial				

*TC=Total counts, Ent=Entodinia, Polypl=Polyplastron, Diplopl=Diploplastron, Dasy=Dasytrichia; NP=none present.

Acknowledgments

Dr M.N. Islam was a Research Fellow supported by the Commonwealth Scholarship

Commission in the UK. The authors gratefully acknowledge their financial support for this study. SAC is financially supported by SOAEFD.

References

Calhoun, M.C., Huston, J.E., Baldwin, B.C., Kuhlmann, S.K., Engdahl, B.S. and Bales, K.W. (1990) Performance of lambs fed diets containing cottonseed meal processed by different methods. *Journal of Animal Science* 70, Suppl 1, 530A.

Huston, J.E., Calhoun, M.C., Kuhlmann, S.K., Baldwin, B.C., Engdahl, B.S. and Bales, K.W. (1990) Comparative toxicity of gossypol acetic acid and free gossypol in cottonseed meal and cottonseed to lambs. *Journal of Animal Science* 68, Suppl 1, 401A.

Huston, J.E., Calk, C.B., Baldwin, B.C. and Kuhlmann, S.W. (1991) Effects of gossypol on digestive and metabolic function of domestic livestock. *Journal of Animal Science* Supplement 1, 534A.

Ismartoyo (1997) Studies *in vitro* and *in vivo* on the nutritive value of whole cottonseed for sheep. PhD Thesis, Aberdeen University, Aberdeen.

Ismartoyo and Acamovic, T. (1994) The effect of gossypol on the growth and total protein of animal cells in culture. In: Colegate, S.M. and Dorling, P.R. (eds), *Plant-Associated Toxins: Agricultural, Phytochemical and Ecological Aspects.* CAB International, Wallingford, Oxon, pp. 201-206.

Ismartoyo, Acamovic, T. and Stewart, C.S. (1993) The effect of gossypol on the rumen microbial degradation of grass hay under consecutive bath culture. *Animal Production* 56, 462.

Ismartoyo, Stewart, C.S. and Acamovic, T. (1994) *In vitro* rumen microbial degradation of a selection of oil seeds and legume seeds under consecutive bath culture. *Animal Science* 58, 453.

Ismartoyo, Acamovic, T., Stewart, C.S., Richardson, A.J., Duncan, S.H. and Shand, W.J. (1995) The effect of feeding WCS (whole cotton seed) as a supplement for sheep fed GH (grass hay) on the rumen fermentation and defaunation of rumen microorganisms. Conference on Tropical Forage Utilization, held in the University of Zimbabwe, Aug. 27 to Sep. 1, 1995.

Moore, J.A., Swingle, R.S. and Hale, W.H. (1986) Effect of whole cottonseed, cottonseed oil or animal fat on digestibility of wheat straw by steers. *Journal of Animal Science* 63, 1267-1273.

Tesfa, A.T. (1993) Effect of rapeseed oil supplementation on digestion, microbial protein synthesis and duodenal microbial amino acid composition in ruminants. *Animal Feed Science and Technology* 41, 312-328.

Zinn, R.A. (1989) Influence of level and source of dietary fat on its comparative feeding value in finishing diets for feed lot steers: Metabolism. *Journal of Animal Science* 67, 1038-1049.

Chapter 32

Detoxification Potential of a New Species of Ruminal Bacteria that Metabolize Nitrate and Naturally Occurring Nitrotoxins

R.C. Anderson[1,2], W. Majak[3], M.A. Rasmussen[1] and M.J. Allison[1]

[1]*USDA-ARS, National Animal Disease Center, Metabolic Diseases Research Unit, Ames, Iowa 50010, USA; [2]BioScience Division of Milk Specialties Company, Dundee, Illinois 60118, USA; [3]Range Research Station, Agriculture and Agri-Food Canada, Kamloops, British Columbia V2B 8A9, Canada*

Introduction

Nitropropanol (NPOH) and nitropropionate are respiratory toxins found in various milkvetches consumed by livestock (Pass, 1994). More than 450 species and varieties of *Astragalus* are known to synthesize either glycosides of NPOH or glucose esters of nitropropionate (Williams and Barneby, 1977a, b). Certain species of *Coronilla*, *Indigofera* and *Lotus* also synthesize esters of nitropropionate (Majak and Pass, 1989). In North America, poisoning occurs mainly in the west, and it is estimated that up to 5% of range cattle exposed to *Astragalus miser* var. *serotinus* in British Columbia are poisoned annually (Majak and Cheng, 1983). Nitropropionate is also synthesized by certain species of fungi belonging to the genera *Penicillium*, *Aspergillus* and *Arthrinium* and human poisoning caused by mildewed sugar cane has been reported (Alston *et al.*, 1985; Ludolph *et al.*, 1991). There is no known cure for nitrotoxin poisoning.

Livestock are poisoned by high levels of dietary nitrate (NO_3), which, while relatively nontoxic *per se*, is reduced within the alimentary tract to nitrite. When absorbed into the bloodstream, nitrite complexes to hemoglobin, inhibiting oxygen transport (Pfister, 1988). Plants that accumulate high levels of NO_3 occur world-wide (Pfister, 1988), and modern agricultural practices often promote accumulation of high concentrations of NO_3 in feed and water (Schelcher *et al.*, 1992).

Ruminal microbes can metabolize NPOH and nitropropionate to amino-propanol and β-alanine, respectively (Anderson *et al.*, 1993). Nitrite, the product of NO_3

reduction, is further reduced to ammonia (Lewis, 1951; Kaspar and Tiedje, 1981). The rumen thus possesses a detoxification potential that can be enhanced by selection for competent detoxifying microbes, and gradual adaptation of populations of ruminal microbes to increasing toxin concentrations. The ruminal detoxification potential can be enhanced *via* inoculation with competent microbes.

Microbial Detoxification and Adaptation

Conjugated nitrotoxins consumed by ruminants are rapidly hydrolyzed by microbial enzymes, freeing the nitroalkane for further ruminal metabolism. Certain strains of bacteria of the genera *Bacteroides, Clostridium, Coprococcus, Desulfovibrio, Lactobacillus, Megasphaera, Peptostreptococcus, Ramibacterium, Selenomonas* and *Veillonella* are capable of metabolizing the nitrotoxins (Majak and Cheng, 1981; 1983). The specific contributions of these bacteria to ruminal detoxification is unclear, as amounts of toxin metabolized and rates of toxin metabolism by the pure cultures are much less than those by mixed ruminal populations. Several species of the above genera also reduce NO_3 and/or nitrite (Majak and Cheng, 1983; Cheng *et al.*, 1988). Another ruminal NO_3 and nitrite-reducing bacterium, *Wolinella succinogenes*, differs from the above genera in that it obtains energy for growth solely *via* anaerobic respiration (Wolin *et al.*, 1961).

Rates of NO_3 and nitrite detoxification increased following adaptation of ruminal populations to high NO_3 concentrations (Alaboudi, 1982; Allison and Reddy, 1984). Three groups of NO_3-reducing bacteria were isolated from adapted populations (Allison and Reddy, 1984). Two of the predominant groups were presumptively identified as *Selenomonas* spp., and the third group was unclassified. Species of *Butyvibrio, Clostridium, Peptococcus, Propionibacterium* and *Selenomonas* were the most often isolated NO_3-reducing bacteria (Alaboudi, 1982). Isolates belonging to *Propionibacterium* and *Clostridium* also reduced nitrite. *Propionibacterium* were atypical in that nitrous oxide was the product of nitrite reduction (Kaspar, 1982). Ruminal adaptation to NO_3 resulted in increased rates of NO_3 and nitrite reduction and an increased rate of NPOH detoxification (Cheng *et al.*, 1985). Metabolism of NPOH was increased by gradual adaptation to sublethal amounts of the β-D-glycoside miserotoxin and nitroethane (Majak *et al.*, 1986; Majak, 1992), a nontoxic analog of NPOH. Following an *in vitro* adaptation to NPOH, enrichment of competent NPOH-metabolizing bacteria was achieved (Anderson *et al.*, 1996). Rates of detoxification by adapted populations increased 8x and numbers of competent NPOH-metabolizing bacteria increased from 10^4/ml to more than 10^8/ml. Hydrogen was a more important source of reducing equivalents than carbohydrates for the enriched populations, implying that organisms other than those discussed were involved. A novel nitrotoxin-metabolizing bacterium, strain NPOH1, was isolated and a new genus and species is needed to include the properties of this bacterium (Anderson *et al.*, 1997).

Strain NPOH1 is a gram-positive bacterium that obtains energy exclusively *via* anaerobic respiration, oxidizing hydrogen, formate or lactate for the reduction of appropriate electron acceptors, including nitrotoxins and nitroethane. NPOH1 reduces NO_3 to ammonia. The requirement for an appropriate electron acceptor may be a major component of the selective pressures allowing populations to adapt to increasing levels of NO_3 or the nitrotoxins. This dissimilatory metabolism by strain NPOH1, which can reduce more than 20mM NPOH or nitropropionate within 24-48hrs, implicates these and other such organisms as major nitrotoxin detoxifiers. Less than one-tenth of NPOH or nitropropionate is metabolized by other known nitrotoxin metabolizing bacteria (Majak and Cheng, 1981).

Strain NPOH1 may be responsible for enhanced rates of nitrite detoxification observed following adaptation to NO_3. However, four strains similar to NPOH1 have since been isolated and the ability to reduce NO_3 and nitrite appears to be unique to strain NPOH1 (unpublished). All four strains are genotypically quite similar (>99% rRNA sequence homology). When ruminal populations from goats adapted to NO_3 were examined, 9/90 bacterial isolates reduced NO_3 and nitrite (unpublished), but none reduced NPOH, which further disputes our earlier hypothesis. Rates of NO_3 and nitrite reduction by the mixed populations increased 1.5-2x after adaptation of the goats to high levels of NO_3, but the rate of NPOH metabolism was not increased. Populations adapted to NO_3 in this study were different from the NO_3-adapted populations of Cheng *et al.* (1985), which had enhanced rates of NPOH metabolism as well.

Ruminal Inoculation

Ruminal inoculation may be a practical way to enhance nitrotoxin and nitrite detoxification, particularly when consumption of the poison increases abruptly. An animal's age and absence of previous toxin exposure may contribute to increased susceptibility to poisoning, in which cases inoculation may be especially effective. An inoculant for ruminal NO_3 poisoning has been reported (Muirhead, 1992).

Strain NPOH1 is a good candidate for development into a ruminal inoculant because of its unique ability to detoxify the nitrotoxins and nitrite and because it is a normal ruminal inhabitant. When tested *in vitro*, ruminal populations inoculated with strain NPOH1 reduced NO_3 6x more rapidly and reduced NPOH 2x as rapidly as populations not inoculated (unpublished). Inoculating ruminal populations with strain NPOH1 provides a shortcut to gradual adaptation. An additional advantage to using strain NPOH1 as an inoculant is that its population may be maintained at a high level by supplementing animal diets with nitroethane.

The feasibility of using strain NPOH1 as a ruminal inoculant to enhance ruminal nitrotoxin detoxification was assessed *in vivo*. Twenty cross-bred first-calf heifers nursing calves grazed on a pasture with a high timber milkvetch density. This was the first exposure of the year to abundant timber milkvetch. After being placed on pasture 4d, the mean±SE rate of NPOH metabolism determined from 15 of the heifers was 180±27μM/L/hr. Rates of NPOH metabolism were determined by *in vitro* incubation of the ruminal fluid that was collected by stomach pump. After 12d on pasture, ten heifers were inoculated with 70ml of a NPOH1 cell suspension, equal to approximately 3.6g of wet packed cells harvested from batch cultures. The suspension was prepared by reconstituting cells (from batch cultures grown with NO_3 or nitroproprionate and lyophilized or frozen in 25% glycerol) in ruminal fluid collected from a steer maintained on alfalfa hay containing no known nitrotoxin. The experiment was concluded 20d after NPOH1 inoculation, and rates of NPOH metabolism were measured as before. At no time did any animal exhibit symptoms of toxicity. All 15 animals measured on both occasions had increased rates of NPOH detoxification, indicating a natural selection of nitrotoxin-metabolizing microbes. Rates for those inoculated with strain NPOH1 (n=9) were higher on both occasions, increasing from 192μM/L/hr to 567μM/L/hr. Rates for those not inoculated (n=5) increased from 159μM/L/hr to 446μM/ L/hr on the final determination. When rates calculated on the later occasion were subjected to an unpaired t test, those inoculated with NPOH1 (567±35μM/L/hr, n=9) were higher ($P<0.05$) than those not inoculated (416±35μM/L/hr, n=10). Four of the animals in the inoculated group had the highest rates (exceeding 608μM/L/hr). Ruminal inoculation of strain NPOH1 provides some protection that was noticeable even after the occurrence of a natural *in vivo* enrichment.

Acknowledgments

We thank Rudy Karlen and the XH Ranch for providing cattle and pastures. The technical assistance of Monica Durigon, Ruth McDiarmid and Herbert Cook is greatly appreciated. Leo Stroesser and Keith Ogilvie handled livestock.

References

Alaboudi, A.R. (1982) PhD Dissertation. University of Saskatchewan, Saskatoon, Saskatchewan.

Allison, M.J. and Reddy, C.A. (1984) Adaptation of gastrointestinal bacteria in response to changes in dietary oxalate and nitrate. In: Klug, M.J. and Reddy, C.A. (eds), *Proceedings of the 3rd International Symposium on Microbial Ecology.* American Society for Microbiology, Washington, DC, pp. 248-256.

Alston, T.A., Porter, D.J.T. and Bright, H.J. (1985) The bioorganic chemistry of the nitroalkyl group. *Bioorganic Chemistry* 13, 375-403.

Anderson, R.C., Rasmussen, M.A. and Allison, M.J. (1993) Metabolism of the plant toxins nitropropionic acid and nitropropanol by ruminal microorganisms. *Applied and Environmental Microbiology* 59, 3056-3061.

Anderson, R.C., Rasmussen, M.A. and Allison, M.J. (1996) Enrichment and isolation of a nitropropanol metabolizing bacterium from the rumen. *Applied and Environmental Microbiology* 62, 3885-3886.

Anderson, R.C., Rasmussen, M.A., DiSpirito, A.A. and Allison, M.J. (1997) Characteristics of a nitropropanol metabolizing bacterium from the rumen. *Canadian Journal of Microbiology* 43, 617-624.

Cheng, K.J., Phillippe, R.C., Kozub, G.C., Majak, W. and Costerton, J.W. (1985) Induction of nitrate and nitrite metabolism in bovine rumen fluid and the transfer of this capacity to untreated animals. *Canadian Journal of Animal Science* 65, 647-652.

Cheng, K.J., Phillippe, R.C. and Majak, W. (1988) Identification of rumen bacteria that anaerobically degrade nitrite. *Canadian Journal of Microbiology* 34, 1099-1102.

Kaspar, H.F. (1982) Nitrite reduction to nitrous oxide by propionibacteria: detoxification mechanism. *Archives of Microbiology* 133, 126-130.

Kaspar, H.F. and Tiedje, J.M. (1981) Dissimilatory reduction of nitrate and nitrite in the bovine rumen: nitrous oxide production and effect of acetylene. *Applied and Environmental Microbiology* 41, 705-709.

Lewis, H. (1951) The metabolism of nitrate and nitrite in the sheep. 1. The reduction of nitrate in the rumen of sheep. *Biochemical Journal* 48, 175-180.

Ludolph, A.C., He, F., Spencer, P.S., Hammerstad, J. and Sabri, M. (1991) 3-Nitropropionic acid-exogenous animal neurotoxin and possible human striatal toxin. *Canadian Journal of Neurological Sciences* 18, 492-498.

Majak, W. (1992) Further enhancement of 3-nitropropanol detoxification by ruminal bacteria in cattle. *Canadian Journal of Animal Science* 72, 863-870.

Majak, W. and Cheng, K.J. (1981) Identification of rumen bacteria that anaerobically degrade aliphatic nitrotoxins. *Canadian Journal of Microbiology* 27, 646-650.

Majak, W. and Cheng, K.J. (1983) Recent studies on ruminal metabolism of 3-nitropropanol in cattle. *Toxicon* Supplement 3, 265-268.

Majak, W. and Pass, M.A. (1989) Aliphatic nitro compounds. In: Cheeke, P.R. (ed), *Toxicants of Plant Origin, Vol II: Glycosides.* CRC Press, Boca Raton, FL, pp. 143-159.

Majak, W., Cheng, K.J. and Hall, J.W. (1986) Enhanced degradation of 3-nitropropanol by ruminal microorganisms. *Journal of Animal Science* 62, 1072-1080.

Muirhead, S. (1992) *Propionibacterium* appears capable of reducing nitrate, nitrite toxicities. *Feedstuffs* 64, 12-13.

Pass, M.A. (1994) Toxicity of plant-derived aliphatic nitrotoxins. In: Colegate, S.M. and Dorling, P.R. (eds), *Plant-Associated Toxins: Agricultural, Phytochemical and Ecological Aspects.* CAB International, Wallingford, Oxon, pp. 541-545.

Pfister, J.A. (1988) Nitrate intoxication of ruminant livestock. In: James, L.F., Ralphs, M.H. and Nielson, D.B. (eds), *The Ecology and Economic Impact of Poisonous Plants on Livestock Production.* Westview Press, Boulder, CO, pp. 233-259.

Schelcher, F., Valarcher, J.F. and Espinasse, J. (1992) Abnormal ruminal digestion in cattle with dominantly non-digestive disorders. *Deutsche Tierarztliche Wochenschrift* 99, 175-182.

Williams, M.C. and Barneby, R.C. (1977a) The occurrence of nitro-toxins in North American *Astragalus* (Fabaceae). *Brittonia* 29, 310-326.

Williams, M.C. and Barneby, R.C. (1977b) The occurrence of nitro-toxins in Old World and South American *Astragalus* (Fabaceae). *Brittonia* 29, 327-336.

Wolin, M.J., Wolin, E.A. and Jacobs, N.J. (1961) Cytochrome-producing anaerobic vibrio, *Vibrio succinogenes* sp. nov. *Journal of Bacteriology* 81, 911-917.

Chapter 33

Vitamin K1 Therapy of *Ferula communis* variety *brevifolia* Toxicosis in Sheep

N. Tligui, M. El Haouzi and H. El Himer
Department of Pathology, Institut Agronomique et Vétérinaire Hassan II, BP 6202, Rabat-Instituts, Rabat, Morocco

Introduction

Ferula communis (giant fennel) is a perennial weed of the Umbelliferae family, growing in several Mediterranean countries. Several varieties of this plant are known to cause a hemorrhagic syndrome (ferulosis) in various species. The Moroccan *F. communis* var. *brevifolia*, from which ferulenol, a 4-hydroxy-coumarin, has been extracted (Lamnaouer, 1987), was confirmed to cause coagulopathy by substantially reducing the activity of vitamin-K-dependent clotting factors without affecting platelets or the liver (Tligui and Ruth, 1994). This study investigated the effect that route and dosage of vitamin K1 administration have on the reversal of *F. communis*-induced coagulopathy in sheep.

Materials and Methods

Leaves of *F. communis* were harvested at the stage of early growth from the Maamoura forest, dried in the sun, powdered and stored in plastic bags until used. Four clinically normal mixed-breed wethers, 11-15 mos old (38-45kg) were used. They were dewormed with albendazole (5mg/kg) and acclimated to the environment and to a diet of mixed hay and grain for 2wks before the study began. The animals were examined twice daily for appetite, attitude, and clinical evidence of bleeding (e.g. epistaxis, melena, hematuria, pale mucous membranes). Rectal temperature, heart rate and respiratory rate were recorded daily. Each sheep served as its own control during the study.

Blood samples were collected into sterile evacuated glass tubes containing EDTA anticoagulant. Routine manual laboratory procedures were used to determine total

RBC, WBC, platelet count, hemoglobin concentration and packed cell volume. The erythrocytic indices (MCV and MCHC) were calculated.

Blood samples were collected into 5ml sterile silicone-coated tubes containing 3.8% buffered sodium citrate solution (one volume of anticoagulant to nine volumes of blood). Platelet poor plasma (PPP) was used to determine prothrombin time (PT) using a coagulometer. Blood samples were collected at 18, 6 and 0hrs before intoxication was induced. The means of these values were used as baseline values, to which results of the PT during the study were compared with and expressed as a ratio for each individual sheep (treatment-to-pretreatment ratio). Data were analyzed pairwise using Student's *t* test.

In all experiments, each animal was given powdered *F. communis* mixed in water (600ml) *via* stomach tube at a rate of 2.5g/kg at 8hr intervals until the PT was prolonged to approximately 1.5x the pre-dosing baseline value. Treatment was then discontinued and the PT was monitored at hourly intervals for the first 12hrs, then every 12hrs for 4d (108hrs). There was a 2wk rest period between experiments. First, no treatment was applied. In the three successive experiments, when plant dosing was discontinued each animal was treated with 0.5mg/kg vitamin K (phytomenadione; Roche Neuilly-sur-Seine) subcutaneously (s.c.), intramuscularly (i.m.) or intravenously (i.v.). For the final three trials, after plant dosing a treatment of 1.5mg/kg was applied s.c., i.m. or i.v.

Results

In each experiment, *F. communis* (2.5g/kg) was administered 5x at 8hr intervals to each animal before the PT reached about 1.5x baseline values, 33-36hrs after the initial dosing and 1-4hrs after the last dose was given. In all trials, no significant ($P<0.05$) changes from predosing ranges were seen for clinical variables and for RBC, WBC, platelet count, packed cell volume, MCV and MCHC. Response to the same dose and route of vitamin K1 varied by individual.

The PT ratio was significantly increased in all wethers at 16hrs after the initial dose of *F. communis*. The PT was most prolonged 12hrs afer the last dose in the experiment with no vitamin K treatment, and began to decrease 24hrs after the last plant dose. The PT returned to baseline 72hrs after *F. communis* administration was discontinued (Fig. 33.1).

Vitamin K application at a dose of 0.5mg/kg s.c., i.m. and i.v. allowed the PT to recover to return to baseline levels 72, 9 and 7hrs after intervention, respectively (Fig. 33.2). Vitamin K treatment at a dose of 1.5mg/kg s.c., i.m. and i.v. aided in the recovery of PT to baseline levels in 9, 9 and 7hrs after application, respectively (Fig. 33.3). In both i.v. trials, the induction of PT recovery occurred significantly sooner than in any of the other treatments.

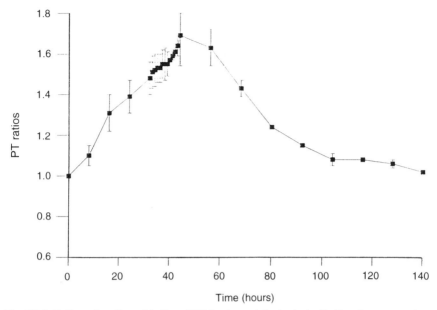

Fig. 33.1. Ratios of prothrombin time (PT) in sheep intoxicated with *Ferula communis*. No treatment has been used.

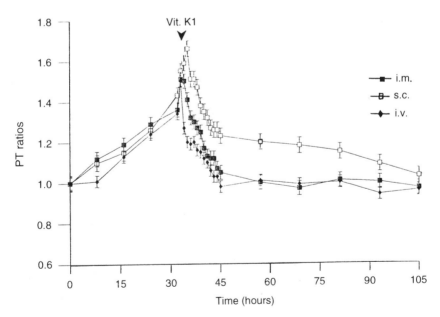

Fig. 33.2. Ratios of prothrombin time (PT) in sheep intoxicated with *Ferula communis* and treated with vitamin K1 (0.5mg/kg) by subcutaneous (s.c.), intramuscular (i.m.) and intravenous (i.v.) routes.

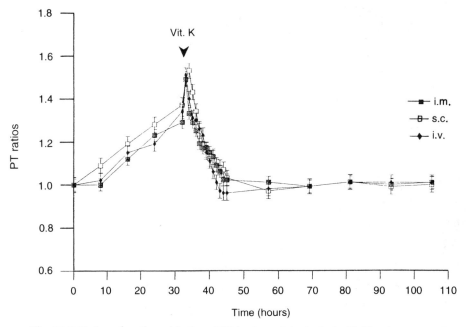

Fig. 33.3. Ratios of prothrombin time (PT) in sheep intoxicated with *Ferula communis* and treated with vitamin K1 (1.5mg/kg) by s.c., i.m. and i.v. routes.

Discussion

Administration of *F. communis* at the dose of 2.5g/kg 5x at 8hrs intervals induced a marked prolongation of PT. This dosage was chosen to induce only transient clinical pathologic changes without killing the animals. With the exception of treatment with vitamin K_1 *via* s.c. route at the dose of 0.5mg/kg, the use of vitamin K_1 *via* different routes (0.5 and 1.5mg/kg) was effective in rapidly returning the PT to normal (7-9hrs) compared to non-treated animals (72hrs). This finding is consistent with earlier results that the hypocoagul-ability induced with *F. communis* is related to decreased activity of the vitamin-K-dependent coagulation factors (II, VII, IX and X) (Tligui and Ruth, 1994; Tligui *et al.*, 1994). These results are consistent with the conclusion that ferulenol, a 4-hydroxycoumarinic component of *F. communis*, acts similarly to the anticoagulant rodenticides by diminishing synthesis of vitamin-K-dependent factors through inhibition of hepatic vitamin K_1 epoxide reductase activity (Bell and Matschiner, 1972).

In non-treated animals, approximately 3d were required for the PT to return to normal after *F. communis* was discontinued and the PT was 1.5x the baseline values. Similar results (3-4d) were obtained by Byars *et al.* (1986) using warfarin administration in horses until the PT reached 1.5x the baseline values. Thus, the duration of anticoagulation produced by the active compound of *F. communis*, ferulenol, seems to be closer to that caused by warfarin and its related first-generation anticoagulant rodenticides, which are considered to have a short-acting anticoagulant effect (Mount, 1988) compared with second-generation anticoagulant rodenticides, such as brodifacoum and difenacoum, which induce a longer coagulopathic effect (Park and Leck, 1982; Mount *et al.*, 1986).

Consistent with the short action of the toxic hydroxycoumarinic compound of *F. communis*, a single dose of vitamin K1 at the lowest dose recommended for large animals (0.5mg/kg) (Mount *et al.*, 1982) *via* i.m. or i.v. routes was sufficient to return the PT to normal a few hours after treatment. In dogs poisoned with brodifacoum, vitamin K_1 at the dose of 0.83mg/kg for a minimum of 5d has been advised (Woody *et al.*, 1992). When the low dose of vitamin K_1 (0.5mg/kg) was given s.c., the PT required a long time to return to normal, but the recovery began only 5hrs after treatment, compared to 24hrs in non-treated animals. The dose was effective, but this route is not recommended in urgent cases (Mount *et al.*, 1982).

With the exception of the s.c. route, the increase of vitamin K_1 doses (1.5mg/kg) did not reduce the time required for the PT to return to normal. The recovery of the PT depends not only on the vitamin K, but on the availability of inactive precursors of vitamin-K-dependent factors, and the latent time required for synthesis of new precursor factors in liver. Normalization of these factors in human beings previously treated with vitamin K antagonists takes 10-12hrs after oral administration of vitamin K (Williams, 1983).

The therapeutic response to vitamin K_1 was rapid after i.v. administration (7hrs) in comparison to the s.c. and i.m. routes (9hrs). The rapid response of the i.v. route was also observed by Byars *et al.* (1986) in treated horses. However, according to the same author, the s.c. route was quicker than i.m. route. In our study, no adverse effects (anaphylaxis) of i.v. administration were observed in the four treated sheep as were described in treated horses (Byars *et al.*, 1986).

This study demonstrates that the anticoagulant effect produced by ferulenol is similar to the action of the first-generation anticoagulant rodenticides. The response of sheep to i.v. administration of vitamin K_1 was rapid, and no adverse effects were observed (which should still be confirmed in other trials). However, the use of vitamin K_1 at the dose of 1.5mg/kg s.c. seems to be an effective and safe one for treatment of intoxicated animals, even though similar effect was obtained *via* i.m. route with a dose of 0.5mg/kg.

Acknowledgments

The authors are very grateful to the International Foundation of Science (IFS, Sweden) for financial support (Grant, B/1635-2F), and to Dr G. Ruth at the University of Minnesota for his valuable cooperation and interest.

References

Bell, R.G. and Matschiner, J.T. (1972) Warfarin and the inhibition of vitamin K activity by an oxide metabolite. *Nature* 237, 32-33.

Byars, T.D., Greene, C.E. and Kemp, D.T. (1986) Antidotal effect of vitamin K1 against warfarin-induced anticoagulation in horses. *American Journal of Veterinary Research* 47, 2309-2312.

Lamnaouer, D. (1987) *Ferula communis* L. Recherches sur la Toxicologie et la Composition Chimique des Taxons Marocains. Doct Es-Sciences Agronomiques I.A.V. Hassan II, Rabat, Morocco.

Mount, M.E. (1988) Diagnosis and therapy of anticoagulant rodenticide intoxications. *Veterinary Clinical North American [Small Animal Practice]* 18, 115-130.

Mount, M.E., Feldman, B.F. and Buffington, T. (1982) Vitamin K and its therapeutic importance. *Journal of American Veterinary Medical Association* 180, 1354-1356.

Mount, M.E., Woody, B.M. and Murphy, M.J. (1986) The anticoagulant rodenticides. In: Kirk R.D. (ed), *Current Veterinary Therapy IX. Small Animal Practice*. WB Saunders, Philadelphia, PA, pp. 156-165.

Park, B.K., and Leck, J.B. (1982) A comparison of vitamin K antagonism by warfarin, difenacoum and brodifacoum in the rabbit. *Biochemistry Pharmacology* 31, 3635-3639.

Tligui, N. and Ruth, G.R. (1994) *Ferula communis* brevifolia intoxication in sheep. *American Journal of Veterinary Research* 55, 1558-1563.

Tligui, N., Ruth, G.R. and Felice, L.J. (1994) Plasma ferulenol concentration and activity of clotting factors in sheep with *Ferula communis* variety brevifolia intoxication. *American Journal of Veterinary Research* 55, 1564-1569.

Williams, J.W. (1983) Production of plasma coagulation factors. In: Williams, J.W., Beutler, E., Erslev, A.J. and Lichtman, M.A. (eds), *Hematology*, 3rd ed. McGraw-Hill, New York, NY, pp. 1222-1230.

Woody, B.J, Murphy, M.J, Ray, A.C. and Robert, A.G. (1992) Coagulation effects and therapy of brodifacoum toxicosis in dogs. *Journal of Veterinary Internal Medicine* 6, 23-28.

Chapter 34

Development of a Vaccine Against Annual Ryegrass Toxicity

K.A. Than, Y. Cao, A. Michalewicz and J.A. Edgar

CSIRO Division of Animal Health, Australian Animal Health Laboratory, Private Mail Bag 24, Geelong, Victoria 3220, Australia

Introduction

Annual ryegrass toxicity (ARGT) is caused by corynetoxins produced by a bacterium (*Clavibacter toxicum*) which colonizes galls formed by the nematode *Anguina funesta* in the seedheads of annual (Wimmera) ryegrass (*Lolium rigidum*) (Vogel *et al.*, 1981; Edgar *et al.*, 1982). The bacterium also infects nematode galls formed by *Anguina* spp. in annual beard grass (*Polypogon monspeliensis*) and blown grass (*Agrostis avenacea*) and causes, in these cases, a form of corynetoxin poisoning called floodplain staggers (FPS) (Bryden *et al.*, 1994; Edgar *et al.*, 1994). In addition to stock deaths, field evidence suggests lambing losses (Davies *et al.*, 1996) and adverse effects on wool growth (Davies *et al.*, 1997). The number of new farms affected by ARGT is increasing each year (Roberts and Bucat, 1992). Farm management can reduce stock losses; however, management procedures are time-consuming and costly (Australian Bureau of Statistics, 1989). This chapter investigates the possibility of vaccinating sheep against ARGT and FPS.

Production of Vaccine and Measurement of Antibody Levels

Corynetoxins are structurally related to the tunicamycin antibiotics and other tunicaminyl-uracil toxins, which vary only in the fatty acids linked to the amino group of the central C_{11} amino-sugar, tunicamine. With low molecular weights (800-900), tunicaminyl-uracil toxins do not naturally induce an immune response in animals. By coupling the toxins as haptens to carrier proteins to form immunogens, an immune response to tunicaminyl-uracil toxins could be obtained.

Commercially available tunicamycins were chemically modified and linked as haptens to proteins, then injected into sheep with an oil adjuvant. Radioimmunoassay (RIA) and enzyme-linked immunosorbent assay (ELISA) methods were developed for detecting and estimating toxin antibodies levels in the sera of vaccinated sheep. Binding of radiolabeled and unlabeled tunicamycins and unlabeled corynetoxins did increase in the sera of vaccinated sheep. It was also found that there was very little or no binding with compounds similar to or forming a part of the structure of corynetoxins and tunicamycins (Chapter 12).

Induction of an immune response against the ARGT and FPS toxins did not imply that vaccinated animals were protected from poisoning. Resistance to poisoning must be correlated with the immune responsiveness of the animal to the vaccine. To determine if vaccinated sheep were protected, two pen trials were conducted, one simulating a chronic challenge and the other an acute challenge.

Challenge with Toxins

Chronic challenge

Ten vaccinated and ten unvaccinated sheep were individually housed. Mean body weights of vaccinated and unvaccinated sheep were, respectively, 30.6 ± 1.3 and 30.7 ± 1.3kg before vaccination and 32.6 ± 1.5 and 33.6 ± 0.9kg at the beginning of the challenge (mean$\pm$SE). No significant differences in body weights nor any adverse affects due to vaccination were observed in the sheep.

All sheep were injected subcutaneously with 1μg tunicamycins/kg/d, 5d/wk for 7wks. Feed intake was monitored daily. Blood samples were collected and body weight changes were recorded weekly.

After 7wks, surviving sheep had received an accumulated dose of 35μg toxins/kg, which approximates an LD_{50} dose (Jago and Culvenor, 1987). Five of the unvaccinated group (50%) showed convulsions, starting as early as 19μg toxins/kg, compared to only one vaccinated sheep (10%), which first convulsed after receiving 30μg toxins/kg. Four of the unvaccinated group (40%) died (at 20, 21, 23 and 25μg toxins/kg), but there were no deaths in the vaccinated group.

Acute challenge

Ten vaccinated and ten unvaccinated sheep were used in the trial. Mean body weights of vaccinated and unvaccinated sheep were 28.6 ± 0.5 and 28.7 ± 0.4kg, respectively, before vaccination and 36.0 ± 0.6 and 34.9 ± 1.0kg before the beginning of oral dosing. No significant differences in body weights or any adverse affects due to vaccination were observed in sheep.

Ground and homogenized annual ryegrass seedheads, infected with bacterium *C. toxicum*, were orally administered at 170g/d/sheep, for 6d during the first week and for 5d during the second week. Corynetoxin concentration in the ground seed was estimated by HPLC to be 83μg/gm. After 12d, all the surviving sheep had received a total of 155mg of corynetoxins. Feed intakes were monitored daily, blood sampling and body weight changes were recorded weekly. The sheep were monitored for 3wks after dosing ceased.

Nine of the unvaccinated group (90%) showed convulsions starting as early as 6d after initiation of dosing, compared to only one vaccinated sheep at 11d. Nine unvaccinated sheep died, starting at 7d, compared to one death in the vaccinated group on day 13, 2d after dosing ceased. All remaining vaccinated sheep and the one surviving control animal showed no clinical signs of poisoning during the experiment or during 3wks of subsequent monitoring.

Discussion

Following subcutaneous administration of a total of 35µg pure tunicamycins/kg over 7wks, at a rate of 1µg/kg/d, 5d/wk, 50% of the unvaccinated sheep displayed convulsions and 40% died. This result agrees with the quantitative toxicity measurements of Jago and Culvenor (1987), and reinforces their demonstration of a cumulative effect. Following the same level of exposure, only one sheep convulsed in the vaccinated group and none died.

When a total dose of 3.44mg corynetoxins/kg (155mg/sheep) was administered orally to unvaccinated sheep over 12d, 90% displayed convulsions and 90% died. This result compares well with Jago and Culvenor's estimate of an lethal oral dose between 3.2-5.6mg/kg in sheep. Vaccinated sheep displayed only 10% convulsions and 10% deaths when exposed to the same oral dose of corynetoxins.

The two vaccinated sheep that showed clinical signs of poisoning were found to have not responded as well as the other sheep to the vaccine (data not shown), indicating that protection is correlated with antibody titer.

Conclusion

A significant degree of protection was achieved in sheep vaccinated with a synthetic immunogen against ARGT/FPS. Protection appeared to be correlated with antibody titer.

Acknowledgments

The authors would like to thank Drs Steve Colegate, Neil Anderton, Peter Cockrum and Phil Stewart for preparing toxic ryegrass seedheads, estimating the level of corynetoxins by HPLC and for their support during the oral dosing.

References

Australian Bureau of Statistics (1989) *Ryegrass toxicity in Western Australia.* Catalogue No. 7421.5.

Bryden, W.L., Trengove, C.L., Davis, E.O., Giesecke, P.R. and Curran., G.C. (1994) Corynetoxicosis of livestock: a nematode-bacterium disease complex associated with different grasses. In: Colegate, S.M. and Dorling, P.R. (eds), *Plant-Associated Toxins: Agricultural, Phytochemical and Ecological Aspects.* CAB International, Wallingford, Oxon, pp. 410-415.

Davies, S.C., White, C.L., Williams, I.H., Allen, J.G. and Croker, K.P. (1996) Sublethal exposure to corynetoxins affects production of grazing sheep. *Australian Journal of Experimental Agriculture* 36, 649-655.

Davies, S.C., Williams, I.H., White, C.L. and Hocking Edwards, J.E. (1997) Tunicamycin reduces wool growth by slowing the mitotic activity of wool follicles. *Australian Journal of Agricultural Research* 48, 331-336.

Edgar, J.A., Frahn, J.L., Cockrum, P.A., Anderton, N., Jago, M.V., Culvenor, C.C.J., Jones, A.J., Murray, K. and Shaw, K.J. (1982) Corynetoxins, causative agents of annual ryegrass toxicity; their identification as tunicamycin group antibiotics. *Journal of The Chemical Society Chemical Communications.* 4, 222-224.

Edgar, J.A., Cockrum, P.A., Stewart, P.L., Anderton, N.A. and Payne, A.L. (1994) Identification of corynetoxins as the cause of poisoning associated with annual beard grass (*Polypogon monspeliensis* (L) Desf.) and blown grass (*Agrostis avenacea* C. Gemelin). In: Colegate, S.M. and Dorling, P.R. (eds), *Plant-Associated Toxins: Agricultural, Phytochemical and Ecological Aspects.* CAB International, Wallingford, Oxon, pp. 393-398.

Jago, M.V. and Culvenor, C.C.J. (1987) Tunicamycin and corynetoxin poisoning in sheep. *Australian Veterinary Journal* 64, 232-235.

Roberts, W.D. and Bucat, J. (1992) The surveillance of annual ryegrass toxicity in Western Australia. *Miscellaneous publication No. 39/92* Western Australia Department of Agriculture, South Perth, WA.

Vogel, P., Petterson, D.S., Berry, P.H., Frahn, J.L., Anderton, N., Cockrum, P.A., Edgar, J.A., Jago, M.V., Lanigan, G.W., Payne, A.L. and Culvenor, C.C.J. (1981) Isolation of a group of glycolipid toxins from seedheads of annual ryegrass (*Lolium rigidum* Gaud.) infected by *Corynebacterium rathayi. Australian Journal of Experimental Biology and Medical Science* 59, 455-467.

Chapter 35

Conditioned Food Aversions to Oxalic Acid in the Food Plants of Sheep and Goats

A.J. Duncan[1], P. Frutos[1] and I. Kyriazakis[2]
[1]Macaulay Land Use Research Institute, Craigiebuckler, Aberdeen, AB25 15X UK; [2]Scottish Agricultural College, West Mains Road, Edinburgh, UK

Introduction

Many food plants available to free-ranging herbivores are toxic when consumed in quantity. However, ruminants usually select an appropriate diet with minimal toxicity problems. The means by which toxic plants are avoided are unclear, but several recent trials have shown that conditioned food aversions (CFAs) may be involved. These CFAs involve the learned avoidance of harmful foods by association of the sensory properties of particular foods with their post-ingestive effects (Provenza *et al.*, 1992). Such aversions have been shown in ruminants by dispensing the emetic drug, lithium chloride (LiCl) intraruminally while allowing animals access to novel feeds and measuring subsequent preference (du Toit *et al.*, 1991). Some trials have been conducted using natural secondary compounds (Pfister *et al.*, 1980; Kronberg *et al.*, 1993), but the majority of work has involved LiCl as the aversive stimulus, and the relevance of CFAs to the avoidance of natural aversive stimuli such as the secondary plant compounds is questionable.

Oxalic acid depicts a well-studied toxic plant component for investigating the wider applicability of CFAs to plant/animal interactions. Oxalic acid, sometimes in high concentrations, is found in many food plants often consumed by ruminants. For example, *Halogeton glomeratus* and *Oxalis cernua* contain 10-30% oxalate as soluble salts (Libert and Franceschi, 1987). *Rumex* spp. may contain oxalic acid at notable concentrations. Consumption of plants rich in oxalic acid elicits toxic symptoms in ruminants as a result of the formation of insoluble precipitates of calcium oxalate in capillary beds and in the renal tubules. Toxicity can result from lesions caused by calcium oxalate crystal formation in other soft tissues, reducing concentrations of plasma calcium which may lead to tetany (Sanz and Reig, 1992). Oxalic acid is broken down in the rumen by *Oxalobacter formigenes* following a period of adaptation (Allison *et al.*, 1977). The rate of oxalic acid breakdown changes with the differences in rates of administration and has not been quantified. Ruminal breakdown of oxalic acid will reduce the amount of oxalic acid absorbed. Thus, adapted animals are likely to experience less-severe post-ingestive effects when offered oxalic acid containing foods than non-adapted animals. This may alter their perception of the toxicity of such foods and may influence the strength of avoidance behavior. Trials were conducted to (1) quantify the ruminal response to oxalic acid administration, (2) investigate learned avoidance of oxalate-related foods and (3) study the physiological adaptation to oxalic acid on diet selection.

Rumen Degradation of Oxalic Acid

The effect of oral administration of oxalic acid on the rate of oxalic acid degradation in the rumen of sheep and goats was studied by dosing ten female sheep (Scottish Blackface) and ten female goats (Scottish Cashmere) 2x daily for 3wks with one of five levels of oxalic acid (0.0, 0.3, 0.6, 0.9 and 1.2mM/kg/d). The dose level was gradually increased over the first 5d to avoid toxicity problems related to abrupt exposure. Samples of ruminal fluid were collected by tube once before the start of dosing and at weekly intervals in the dosing period. Rates of oxalic acid degradation were measured by the method of Allison *et al.* (1977). Rates of oxalic acid degradation increased rapidly (within 1wk) in proportion to the level of administration (Fig. 35.1) and were higher in goats than in sheep (9.38 *vs.* 5.95µM/ml/d; standard error of difference (SED), 0.966; $P<0.05$).

Conditioned Aversions to Flavored Hays with Oxalic Acid

An experiment was conducted to determine whether sheep were able to form conditioned food aversions to oxalic acid and whether the strength of aversion was dependent on the level of oxalic acid exposure. In this experiment, 48 sheep were allocated to one of four treatment groups each receiving the conditioning stimulus (oxalic acid) at a different level (0.67, 1.33, 2.0 and 2.67mM/kg, n=12). Sheep were allowed to form CFAs during four consecutive conditioning periods of 8d each. During each conditioning period animals were offered flavored hays (orange or aniseed) for 4hrs on days one and two, 2hrs after which oxalic acid or a placebo was administered. The process was repeated with the opposing flavor and conditioning stimulus on days five and six in completely balanced fashion. At the end of each conditioning period, preference for the two flavors was measured by offering both hays simultaneously for 20min and recording food intake.

Preference ratios for the oxalic-acid-related hay tended to diminish with rising dose levels of oxalic acid ($P<0.1$). The strength of avoidance increased as the trial progressed ($P<0.05$; Table 35.1), with preference ratios for oxalic-acid-related hays

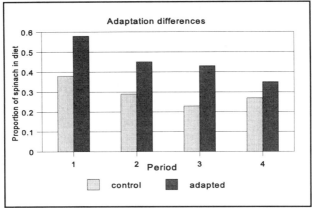

Fig. 35.1. Effect of rate of oxalic acid administration on its rate of breakdown under microbial action in the rumen of sheep and goats (means across species; n=4).

being approximately 0.50 (meaning no preference for either feed) in Period 1 and gradually diminishing in subsequent periods. Mean preference ratios for flavored hays were significantly lower when paired with oxalic acid than when paired with placebos (overall mean of 0.44 significantly different from 0.50; $P<0.05$). The results of the trial showed that sheep were able to relate oxalic acid administration with the flavored hay being consumed at the time and subsequently learned to modify their diet choice appropriately.

Rumen Adaptation to Oxalic Acid Relating to Diet Choice

The influence of rumen adaptation to oxalic acid on food choice of oxalic-acid-containing plants (spinach) and oxalic-acid-free plants (cabbage) was investigated. Six goats were dosed daily with 0.6mM/kg/d oxalic acid while six goats received empty gelatin capsules. This promoted an adapted rumen microbial population in the adapted animals, which was verified by measuring rates of intraruminal oxalic acid degradation (Allison *et al.*, 1977). After an 8d adaptation period goats grazed a paddock containing a mixture of spinach [neutral detergent fiber (NDF), 371g/kg DM, crude protein (CP), 103g/kg DM] and cabbage (NDF, 213g/kg DM; CP, 135g/kg DM) in a series of preference trials, which lasted 2d each, conducted at weekly intervals. Goats were moved to grass pasture during the intervening periods. Dietary proportions of spinach and cabbage were measured using the n-alkane technique (Dove and Mayes, 1996). Dietary proportions were estimated from fecal and herbage n-alkane concentrations using a least squares optimization procedure (Dove and Mayes, 1996).

Goats receiving daily oxalic acid capsules showed higher rates of intraruminal

Duncan et al.

Table 35.1. Mean preference ratios for flavored hays related with oxalic acid dosing measured in 20min, two-choice preference tests. Preference ratio was defined as weight of oxalic-acid-associated hay as a proportion of total intake over 20min. Standard error (SE) for dose level means was 0.0772; SE for conditioning period means was 0.0516.

| | Oxalic acid dose level (mM/kg) | | | | |
	0.67	1.33	2.00	2.67	Mean
Period 1	0.628	0.623	0.475	0.314	0.510
Period 2	0.481	0.582	0.414	0.449	0.482
Period 3	0.400	0.550	0.280	0.405	0.409
Period 4	0.465	0.337	0.330	0.274	0.352
Mean	0.494	0.523	0.375	0.360	0.438

oxalic acid degradation as expected. Average rates of oxalic acid degradation over the whole trial were 12.84μM/ml/d for the adapted group and 5.81μM/ml/d for the control group. Animals with an adapted rumen microbial population selected a higher proportion of spinach in their diet than non-adapted animals. This event was apparent during the first choice test and persisted throughout the trial (Fig 35.2).

The preference for the higher proportion of spinach in the diet of adapted animals was not explained by the relative nutritive value of the two foods available. The oxalic-acid-adapted animals may have selected more spinach to avoid the secondary compounds (S-methyl cysteine sulfoxide and glucosinolates) present in cabbage when the constraint of oxalic acid toxicity in spinach was reduced. The results indicate that goats that were physiologically adapted to oxalic acid were able to increase the proportion of this component in their diet, and that diet selections were made rapidly.

Discussion

Rumen biotransformation is an important mechanism for the detoxification of certain plant secondary compounds (Smith, 1992). A period of physiological adaptation to new foods is usually required, after which the adapted rumen microbial population may protect the host animal from the otherwise detrimental effects of secondary compounds present in the new food source. The current series of experiments was conducted to determine whether changes in the state of adaptation of the rumen microbial population might influence food selection.

First it was verified that discrete daily doses of free oxalic acid would elicit changes in rates of intraruminal oxalic acid degradation. Then the ability of ruminants to use CFAs to avoid oxalic-acid-associated feeds was confirmed.

The final phase directly addressed whether rumen adaptation to oxalic acid might alter perceptions of toxicity in grazing ruminants and so influence food selection. Results indicated that adapted animals selected a higher proportion of the potentially toxic plant in their diet. Differences in diet selection patterns were evident at the first choice test, while in the pen-fed test CFAs took time to develop.

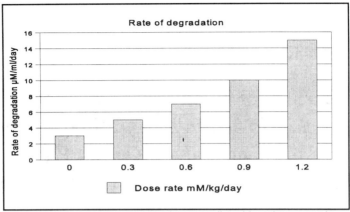

Fig. 36.2. Effect of rumen adaptation to oxalic acid on the proportion of spinach in the diet of goats offered a mixture of spinach and cabbage.

References

Allison, M.J., Littledike, E.T. and James, L.F. (1977) Changes in ruminal oxalate degradation rates associated with adaptation to oxalate ingestion. *Journal of Animal Science* 45, 1173-1179.

Dove, H. and Mayes, R.W. (1996) Plant wax components: a new approach to estimating intake and diet composition in herbivores. *Journal of Nutrition* 126, 13-26.

Kronberg, S.L., Muntifering, R.B., Ayers, E.L. and Marlow, C.B. (1993) Cattle avoidance of leafy spurge: A case of conditioned aversion. *Journal of Range Management* 46, 364-366.

Libert, B. and Franceschi, R. (1987) Oxalate in crop plants. *Journal of Agricultural and Food Chemistry* 35, 926-938.

Pfister, J.A., Provenza, F.D. and Manners, G.D. (1980) Ingestion of tall larkspur by cattle. Separating effects of flavour from post-ingestive consequences. *Journal of Chemical Ecology* 16, 1697-1700.

Provenza, F.D., Pfister, J.A. and Cheney, C.D. (1992) Mechanisms of learning in diet selection with reference to phytotoxicosis in herbivores. *Journal of Range Management* 45, 36-45.

Sanz, P. and Reig, R. (1992) Clinical and pathological findings in fatal plant oxalosis: a review. *American Journal of Forensic Medicine and Pathology* 13, 342-345.

Smith, G.S. (1992) Toxification and detoxification of plant compounds by ruminants: an overview. *Journal of Range Management* 45, 25-30.

du Toit, J.T., Provenza, F.D. and Nastis, A. (1991) Conditioned taste aversions: how sick must a ruminant get before it learns about toxicity in foods? *Applied Animal Behavior Science* 30, 35-46.

Chapter 36

Fungi Microbiota and Oxalate Formation on *Brachiaria brizantha* and their Possible Toxic Effects on Calves Kept at Pasture without Supplementation of Calcium and Magnesium

E.O. da Costa, S.L. Gorniak, N.R. Benites, B. Correa and E.L. Ortolani
Faculdade de Medicina Veterinaria e Zootecnia, Universidade de São Paulo, Brazil, Av Prof Dr Orlando Marques de Paiva, 87, CEP 05508-900, São Paulo-SP, Brazil

Introduction

Oxalate production by various species of commonly encountered fungi was known as early as 1887 (Foster, 1946). Extensive experiments on factors affecting oxalate formation, and the possibility of it being elaborated in dangerous quantities in moldy feeds, has not been seriously considered. Early studies showed that various filamentous fungi, namely *Aspergillus* spp., *Rhizopus* spp., *Mucor* spp., and 23 *Penicillium* species (e.g. *P. oxalicum*), were capable of degrading food substrates to oxalate (Foster, 1946; Wilson and Wilson, 1961). A possible relationship between oxalate-positive mold and several epizooties of mycotoxicosis was supported by comparison of clinical signs with those of halogeton poisoning and other hypocalcemic states. This link, if established, would indicate that oxalate is another toxic substance to be included in certain cases of moldy feed poisoning.

Materials and Methods

Twenty Zebu calves (*Bos indicus*) were ranked according to age, weight and general health, then randomly assigned to one of four groups. One experimental group and one control group were used in each of two 1yr pasture trials. The control groups, G1 in 1994 and G1A in 1995, were five calves each that received calcium (Ca) and magnesium (Mg) supplementation over the period of observation. The experimental groups, G2 in 1994 and G2A in 1995, were five calves without mineral supplementation. The pasture was largely composed of *Brachiaria brizantha*, a grass known to be low in Ca. The grass was evaluated monthly for oxalate content and fungal microbiota, and the calf weights were recorded monthly. The oxalate levels were determined by the method of Moir, and the fungal microbiota was determined by the method of Busta *et al.* (1984).

Results

The total oxalate ranged from 1.4% in winter to 3.1% in summer over the first experimental period and from 0.87% (winter) to 5.5% (summer) during the second experimental period (Fig 36.1).

Fusarium spp., *Aspergillus* spp., *Rhizopus* spp., *Mucor* spp., *Cladosporium* spp., *Helminthosporium* spp. and *Geothricum* spp., were the fungi detected. *Aspergillus* spp. and *Rhizopus* spp. were the oxalate-producing fungi isolated in high levels, ranging from 1×10^2cfu/ml (colony-forming units) to 1.1×10^5cfu/ml during the 1994 seasons and from 1×10^3cfu/ml to 4.2×10^5cfu/ml during the 1995 seasons (Fig. 36.2).

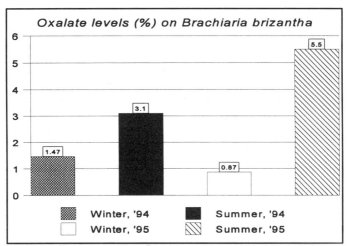

Fig. 36.1. Oxalate levels on *Brachiaria brizantha* during the winter and summer months of 1994 and 1995.

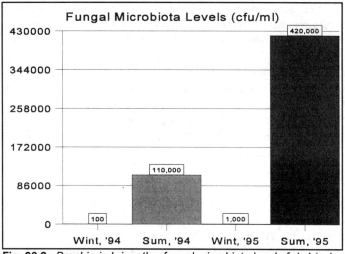

Fig. 36.2. *Brachiaria brizantha,* fungal microbiota level cfu/ml (colony-forming units/ml) during winter and summer, 1994 and 1995.

The weight gain (g/d) was statistically higher ($P<0.001$) in the control groups (supplemented) *vs.* the experimental groups: 171 ± 19.4(G1) and 125.6 ± 25.9 (G2) in 1994 (Fig. 36.3) *vs.* 135.6 ± 13.1(G1A) and 78.2 ± 22.0(G2A) in 1995 (Fig. 36.4).

Discussion

Wilson and Wilson (1961) demonstrated that an appreciable fungal oxalate production was observed at both 26°C and 35°C. These observations were very important, since they cover a range that includes ambient temperatures both in temperate and in sub-tropical zones. Adverse effects of high oxalate foods have been demonstrated by several investigators. Low milk output by Puerto Rican cattle may be traced to regular grazing on grasses containing more than 1.6% oxalic acid. Wilson and Wilson (1961) also reported that one species, *Pennistum merckerii,* contained about 5% total oxalate and was responsible for a negative Ca balance in cattle.

As to the mechanism of oxalate toxicity in ruminants, Talapatra *et al.* (1948) suggested that ingested soluble oxalates were degraded to alkali carbonates before reaching the true stomach. These compounds were believed to be responsible for a severe alkalosis interfering with Ca absorption. Pre-formed Ca oxalate was excreted unchanged in the feces. Whatever the mechanism of action, the available evidence indicates that continued ingestion of excess oxalate inevitably leads to a negative Ca availability. In addition to lowering the Ca availability, fungal growth necessarily decreases the caloric value of feed. In order to compensate for a caloric deficiency,

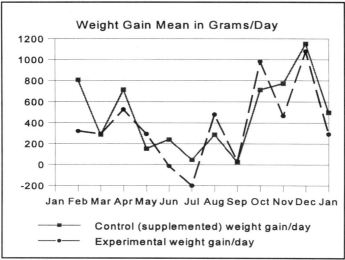

Fig. 36.3. Weight gain mean values in g/d of ten Zebu calves (*Bos indicus*) kept on *B. brizantha* pasture during 1994 with and without mineral supplementation.

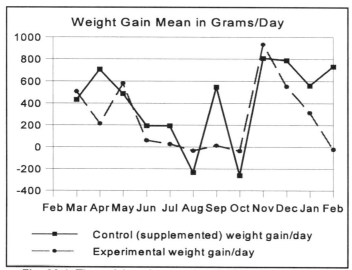

Fig. 36.4. The weight gain mean in g/d of ten Zebu calves (*Bos indicus*) kept on *B. brizantha* pasture during 1995 with and without mineral supplementation.

the amount of feed is increased, augmenting the intake of oxalate. Absorption of sublethal quantities of soluble oxalate from the gastrointestinal tract can form Ca oxalate in body tissues, leading to renal damage due to formation of calculi. Alteration of the calcium/phosphorus intake ratio and the destruction of food vitamin content must be considered as possible mechanisms of oxalate poisoning.

Symptoms of oxalate poisoning include albuminuria, urinary calculi, unsteady gait, falling with inability to rise, depression, hyperexcitability, tetany, paralysis and convulsions leading to death. In cases of winter tetany of cattle where clinical tests were performed and hypocalcemia and hypomagnesemia were detected, injections of therapeutic preparations of Ca and Mg evoked a prompt cessation of signs.

References

Busta, F.F., Peterson, E.H., Adams, D.M. and Johnson, M.G. (1984) Colony count methods. In: Speck, M.L. (ed), *Compendium of Methods for the Microbiological Examination of Foods.* American Public Health Association, Washington, DC, pp. 62-83.

Foster, J.W. (1946) *Chemical Activities of Fungi.* Academic Press, New York, NY, pp. 326-350.

Talapatra, S.K., Ray, S.C. and Sen, K.C. (1948), Calcium assimilation in ruminants on oxalate-rich diet. *Journal of Agricultural Science* 38, 163.

Wilson, B.J. and Wilson, C.M. (1961) Oxalate formation in moldy feedstuffs as a possible livestock toxic disease. *Journal of Veterinary Research* 22, 961-969.

Chapter 37

Protective Effects of Cyclodextrins on Tunicaminyluracil Toxicity

P.L. Stewart, C. May and J.A. Edgar
CSIRO Australian Animal Health Laboratory, Post Office Box 24, Geelong, Victoria 3220, Australia

Introduction

The tunicaminyluracil antibiotics include tunicamycin (TCM) and corynetoxins. These toxins specifically and strongly inhibit N-acetylglucosamine-1-phosphate transferase (GlcNAc-1-P transferase), an enzyme that catalyzes the initial step in oligosaccharide assembly during the synthesis of N-linked glycoproteins (Lehle and Tanner, 1976). The membrane-associated enzyme occurs in the endoplasmic reticulum in particulate fractions of homogenates prepared from cells and tissues of various eukaryotes (Parodi and Leloir, 1979). Inhibition of GlcNAc-1-P transferase activity by TCM has been demonstrated in rat liver rough microsomes *in vitro* and *in vivo* (Eggens and Dallner, 1982; Jago *et al.*, 1983). Tunicamycin is a nucleoside antibiotic first isolated from *Streptomyces lysosuperificus* (Takatsuki *et al.*, 1971) and has been widely used experimentally in biochemical studies on glycosylation. Corynetoxins are the causative agents of annual ryegrass toxicity (ARGT), a commonly fatal neurological disease of livestock in South (SA) and Western Australia (WA) (Berry and Wise, 1975). The disease occurs as the result of the consumption of annual ryegrass seed heads containing the toxic chemical corynetoxin, produced by the bacterium *Clavibacter toxicus* (Riley and Ophel, 1992). Corynetoxin differs from TCM only in the nature of the fatty acid side chain (Edgar *et al.*, 1982), while the chemical and biological activities of the two toxins are essentially identical (Vogel *et al.*, 1982; Jago *et al.*, 1983). The pathological changes in the central nervous system seen in ARGT have been produced experimentally in sheep with TCM (Finnie and Jago, 1985).

Low molecular weight toxicants are capable of binding to some materials and are the principal causes of some poisoning diseases of livestock. Therefore, cyclodextrin was evaluated for the ability to ameliorate the overt clinical condition and death associated with tunicaminyluracil poisoning. The cyclodextrins are a class of cyclic glucose molecules with three members, alpha, beta and gamma, comprising six, seven and eight glucose units, respectively (Clarke *et al.*, 1988), which are water-soluble cavity molecules with a hydrophobic internal environment. Molecules that are either relatively hydrophobic or have a hydrophobic structural feature, i.e. a fatty acid side chain, will partition into the cavity of the cyclodextrins forming a molecular complex, changing the physical and biochemical properties of the 'guest' molecule. Complexation with cyclodextrins increases the water solubility of some molecules, a feature which has been used experimentally to rescue rats from hypervitaminosis A; vitamin A and other retinoids are known to form complexes with cyclodextrins (Pitha and Szente, 1983).

This laboratory observed previously that the TCMs partition into both alpha- and beta-cyclodextrin (α- and β-CD) with a relatively strong binding interaction. This complexed form of TCM has a far greater water solubility than TCM alone (May, unpublished). Cyclodextrins may have the ability to modify the inhibition of GlcNAc-1-P transferase activity by TCM. If so, cyclodextrins may have application as a therapeutic agent against the toxic effects of this class of toxin, since the consequences of the inhibition of glycosylation are believed to contribute to the toxicity. The typical course of ARGT gives the opportunity for a treatment program (antidote) to be administered during the transfer of animals from toxic to non-toxic areas. Experimental treatments for use in this type of scenario have been previously tested in both pen and field trial situations (Norris *et al.*, 1981; Richards, 1982) but, to date, none has proved sufficiently successful to warrant its use on a flock basis. Their low toxicity and increasing ease of availability make the cyclodextrins potentially useful in the treatment of animals affected by ARGT. The cyclodextrin derivatives with high water solubility were considered most suitable for this application. The derivative of choice was therefore hydroxypropyl beta-cyclodextrin (HPβ-CD).

The commercial availability of a pure form of TCM makes it desirable in preliminary work on ARGT. This chapter describes the cyclodextrins' effects on the inhibitory effect of TCM on GlcNAc-1-P transferase activity in rat and sheep liver rough microsomes *in vitro*, and on the body weight loss in TCM-poisoned rats. Additionally the therapeutic effects of cyclodextrin derivatives on the survival rate of sheep given TCM parenterally, and of sheep grazing toxic annual ryegrass were investigated.

Effect of Cyclodextrins on GlcNAc-1-P Transferase *in vitro*

Rat liver microsomes

Liver rough microsomes were prepared from adult male rats, following the method of Tetas *et al.* (1970) and Tkacz *et al.* (1974). The GlcNAc-1-P transferase activity was determined by measuring the radioactivity uptake from UDP[^{14}C]GlcNAc into dolichol-linked products (Tkacz and Lampen, 1975). The dose response of tunicamycin inhibition was measured, with and without alpha-cyclodextrin (α-CD) added to the assay mixture. The α-CD partially reversed the inhibitory effects of TCM on transferase activity, likely by binding sufficient toxin to prevent it from binding to the enzyme active site (Fig. 37.1). Thus the enzyme transfer process would be partially maintained even in the presence of toxin.

Sheep liver microsomes

Liver rough microsomes were prepared from a merino wether. Microsomal preparation and GlcNAc-1-P transferase assay were as above for rat liver. Partial protection from the inhibitory effects of TCM on transferase activity was achieved (Fig. 37.2).

Parenteral Protection using Hydroxypropyl Beta-cyclodextrin

Protective effects in rats

Body weight loss in rats due to inhibition of appetite is a sensitive measure of the toxicity of tunicaminyluracil antibiotics. Two groups of five adult male rats were

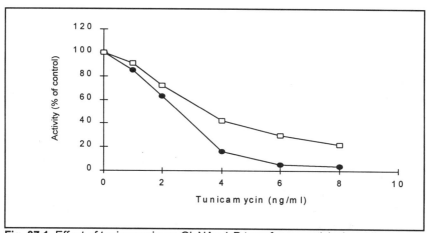

Fig. 37.1. Effect of tunicamycin on GlcNAc-1-P transferase activity in rat liver rough microsomes *in vitro*. Varying amounts of tunicamycin were preincubated with microsomes at 30°C for 5min before the addition of radiolabeled UDP-GlcNAc, with (□) and without (●) alpha-cyclodextrin (α-CD) in the assay mixture. Incubations were done at 30°C for 10min. Results are the average of triplicate determinations.

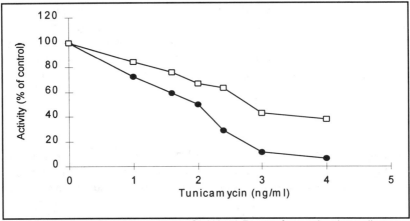

Fig. 37.2. Effect of tunicamycin on GlcNAc-1-P transferase in sheep liver rough microsomes *in vitro*, with (□) and without (•) alpha-cyclodextrins (α -CD) in the assay mixture.

given daily doses of TCM at 20µg/kg by subcutaneous (s.c.) injection for 4d for each of 3wks. One group was dosed s.c. twice daily with HPβ-CD at 100mg/kg for 5d. After 1wk, the weight loss in both groups was similar, but by the end of the third week the HPβ-CD-treated rats had lost substantially less weight than the untreated controls (Fig. 37.3).

Protective effects in sheep

Since work in rats had shown evidence of some protection against TCM toxicity by HPβ-CD, a trial was conducted in sheep. Two groups of merino wethers were injected s.c. with 14µg/kg of TCM on days one and two. On day three, one group was dosed twice daily with HPβ-CD intraperitoneally (i.p.) at 100mg/kg. Magnesium sulfate ($MgSO_4$) was administered intramuscularly (i.m.), once on day three (400mg/kg), and twice on days four and five (200 and 400mg/kg). Magnesium sulfate was used in the antidote treatment since it can delay the onset of convulsions (Richards, 1982). Survival rates were significantly different, with eight survivors in the HPβ-CD-treated and none in the untreated group.

While this treatment regime was successful in protecting sheep from toxicity, it would be impractical in a field situation, where two administrations of a treatment would be required. To prolong absorption and maintain a sufficient circulating level, the cyclodextrin was produced as a magnesium gluconate gel formulation, which could be given i.p. Pen trials in sheep demonstrated that significant protection against TCM poisoning could be achieved using HPβ-CD/magnesium gluconate/$MgSO_4$ formulation for 3 i.p. injections/d (May, unpublished).

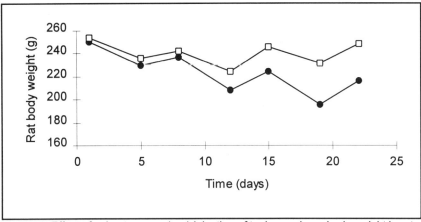

Fig. 37.3. Effect of subcutaneous (s.c.) injection of tunicamycin on body weight in rats, with (□) or without (●) s.c. injection of HPβ-CD.

Field trials of the experimental antidote were carried out in the summer of 1994-95 in ARGT endemic areas of both WA and SA. Testing was done on natural outbreaks of the disease as they occurred. Farmers in areas where ARGT was previously reported were made aware of the experimental antidote and the conditions under which trials were to be conducted. Trials were carried out by CSIRO and independently by both the WA and SA Departments of Agriculture with material supplied by CSIRO. In the nine field trials performed, seven substantially increased the survival rate of affected sheep and demonstrated the benefit of using the antidote treatment (May, unpublished). The antidote appeared to be effective even if given on the second or third day of an outbreak; however, the earlier it was used the greater the potential benefit.

Conclusion

Cyclodextrins are partially able to reverse the inhibition of GlcNAc-1-P transferase *in vitro* and show protective effects against the tunicaminyluracil toxins *in vivo*. Further testing of the ARGT antidote, including the use of alternative cyclodextrin derivatives, continues.

References

Berry, P.H. and Wise, J.L. (1975) Wimmera ryegrass toxicity in Western Australia. *Australian Veterinary Journal* 51, 525-530.

Clarke, R.J., Coates, J.H. and Lincoln, S.F. (1988) Inclusion complexes of the cyclomalto-oligosaccharides (cyclodextrins). *Advances in Carbohydrate Chemistry and Biochemistry* 46, 205-249.

Edgar, J.A., Frahn, J.L., Cockrum, P.A., Anderton, N., Jago, M.V., Culvenor, C.C.J., Jones, A.J., Murray, K. and Shaw, K.J. (1982) Corynetoxins, causative agents of annual ryegrass toxicity; their identification as tunicamycin group antibiotics. *Journal of The Chemistry Society Chemical Communication* 4, 222-224.

Eggens, I. and Dallner, G. (1982) Intramembranous arrangement of the glycosylating systems in rough and smooth microsomes from rat liver. *Biochimica Biophysica Acta* 686, 77-93.

Finnie, J.W. and Jago, M.V. (1985) Experimental production of annual ryegrass toxicity with tunicamycin. *Australian Veterinary Journal* 62, 248.

Jago, M.V., Payne, A.L., Peterson, J.E. and Bagust, T.J. (1983) Inhibition of glycosylation by corynetoxin, the causative agent of annual ryegrass toxicity: a comparison with tunicamycin. *Chemico-Biological Interactions* 45, 223-234.

Lehle, L. and Tanner, W. (1976) The specific site of tunicamycin inhibition in the formation of dolichol-bound *N*-acetylglucosamine derivatives. *FEBS Letters* 71, 167-170.

Norris, R.T., Richards, I.S. and Petterson, D.S. (1981) Treatment of ovine annual ryegrass toxicity with chlordiazepoxide: a field evaluation. *Australian Veterinary Journal* 57, 302-303.

Parodi, A.J. and Leloir, L.F. (1979) The role of lipid intermediates in the glycosylation of proteins in the eukaryotic cell. *Biochimica Biophysica Acta* 559, 1-37.

Pitha, J. and Szente, L. (1983) Rescue from hypervitaminosis A or potentiation of retinoid toxicity by different modes of cyclodextrin administration. *Life Science* 32, 719-723.

Richards, I.S. (1982) The effect of magnesium sulphate on convulsions induced by annual ryegrass toxicity. *Australian Veterinary Journal* 58, 115-117.

Riley, I.T. and Ophel, K.M. (1992) *Clavibacter toxicus* sp. nov., the bacterium responsible for annual ryegrass toxicity in Australia. *International Journal of Systemic Bacteriology* 42, 64-68.

Takatsuki, A., Arima, K. and Tamura, G. (1971) Tunicamycin, a new antibiotic. 1. Isolation and characterization of tunicamycin. *Journal of Antibiotics* 24, 215-223.

Tetas, M., Chao, H. and Molnar, J. (1970) Incorporation of carbohydrates into endogenous acceptors of liver microsomal fractions. *Archives of Biochemistry and Biophysics* 138, 135-146.

Tkacz, J.S. and Lampen, J.O. (1975) Tunicamycin inhibition of polyisoprenol N-acetylglucosaminyl pyrophosphate formation in calf-liver microsomes. *Biochemical and Biophysical Research Communications* 65, 248-257.

Tkacz, J.S., Herscovics, A., Warren, C.D. and Jeanloz, R.W. (1974) Mannosyltransferase activity in calf pancreas microsomes. *Journal of Biological Chemistry* 249, 6372-6381.

Vogel, P., Stynes, B.A., Coackley, W., Yeoh, G.T. and Petterson, D.S. (1982) Glycolipid toxins from parasitised annual ryegrass: a comparison with tunicamycin. *Biochemical and Biophysical Research Communications* 105, 835-840.

Chapter 38

Pyrrolizidine Alkaloid Detoxification by an Ovine Ruminal Consortium and its Use as a Ruminal Supplement in Cattle

W.H. Johnston, A.M. Craig, L.L. Blythe, J.T. Hovermale and K. Walker

College of Veterinary Medicine, Oregon State University, Corvallis, Oregon 97331, USA

Introduction

Tansy ragwort (*Senecio jacobaea*) is a major problem in the Pacific Northwest of the US. The toxic molecules of tansy plants are pyrrolizidine alkaloids. Pyrrolizidine alkaloid (PA) toxicosis has economic importance in the US, New Zealand, Australia, the UK, Russia, Argentina and Chile. In the US, the National Academy of Science has estimated that 9.7% of all animal deaths are due to plant toxicoses. In the past few years, a new solution to some of the plant toxicity problems has been found: ruminal microbes that can degrade the toxic principles from the poisonous plants. These microbes have been recently reviewed (Craig and Blythe, 1994; Craig, 1995). This chapter is one example of ruminal microbe protection against noxious weeds.

Tansy ragwort contains six different PAs. There are about 350 different PAs in over 6,000 plant species. The two principle molecules associated with tansy ragwort toxicosis are seneciphylline and jacobine, which are esters of necine base compounds and necic acids. They have been shown to cause irreversible liver disease in cattle (Craig *et al.*, 1991a) and horses (Craig *et al.*, 1991b), but not in sheep, which can ingest up to 300% of their body weight with no deleterious effects (Craig *et al.*, 1986). Studies incubating whole rumen fluid with PAs extracted from tansy ragwort *in vitro* gave evidence that the ruminal micoflora from sheep could rapidly biodegrade the toxins (Craig *et al.*, 1986). Further *in vitro* studies indicated that the small bacteria were responsible (Craig *et al.*, 1992).

Isolating and transferring the protective factor as a probiotic from sheep to cattle has been a long-term goal. An experiment was designed to demonstrate the efficacy of transferring ovine rumen fluid into the rumen of cattle to protect against PAs in a feeding trial with 5% tansy ragwort.

Materials and Methods

Animals and sampling methods

Eight 135-185kg beef calves had fistulas surgically placed within the rumen and were housed in a confinement facility at Oregon State University. After a 4wk period of adjustment and collection of baseline blood values and determination of ruminal PA detoxification rates, 1.5L of sheep whole rumen fluid was transferred to five of eight cattle (Group 1) *via* the ruminal fistula while three calves remained separated and non-inoculated (Group 2). The rumen fluid source was sheep that had been on a diet of 5% tansy ragwort for 3wks and had microbes with *in vitro* PA degradation times of 2-4hrs (Figs. 38.1 and 38.2). Following ruminal microbe transfer, all animals were converted to a PA-contaminated feed designed to consist of 5% tansy ragwort (dwt), 45% alfalfa hay, 47.5% grass hay and 2.5% molasses. The latter was added to increase palatability. The initial 5% tansy ragwort concentration had been selected since it approximates that which is common for a heavily contaminated tansy ragwort pasture (Craig *et al.*, 1986).

Upon analysis, it was found that the average PA concentration in the feed was considerably higher than expected, with 365µg PA/g as determined by soxhlet extraction and GC/MS. The diet actually delivered approximately 9% tansy ragwort, possibly due to the high concentration of dry tansy ragwort flowers used.

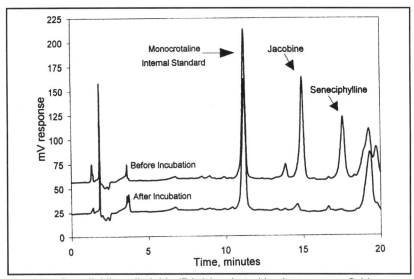

Fig. 38.1. Pyrrolizidine alkaloids (PAs) incubated in sheep rumen fluid.

Blood samples to monitor gamma glutamyl transferase (GGT) were taken weekly and analyzed using commercial reagents and instrumentation. One of the best early indicators of liver damage caused by tansy ragwort is GGT, since this enzyme elevates with the biliary hyperplasia characteristic of tansy ragwort toxicosis (Craig *et al.*, 1991a,b; Blythe and Craig, 1994). Percutaneous liver biopsies were taken with a Trucut biopsy needle as previously described (Pearson and Craig, 1980) every 2-3wks to histologically monitor hepatic pathology.

Analytical assays

Thin layer chromatography (TLC) was used to screen rumen digesta samples for PA biodegradation times (Craig *et al.*, 1992). The biodegradation times were used as a indirect indicator of microbial activity in the rumen. Rumen digesta samples were taken weekly during the course of the experiment and blended anaerobically with McDougall's buffer (McDougall, 1948). They were spiked with PA to give a final concentration of 50µg/ml. Samples were taken, made alkaline, extracted with dichloromethane and run on Whatman HPKF silica gel high performanceTLC plates. Pyrrolizidine alkaloids were visualized by spraying with Dragendorffs solution followed by 5% sodium nitrite.

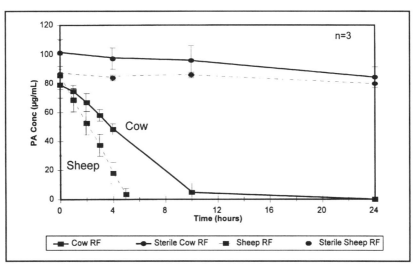

Fig. 38.2. Comparison of pyrrolizide alkaloid in biotransformation in bovine and ovine rumen fluid. RF=rumen fluid.

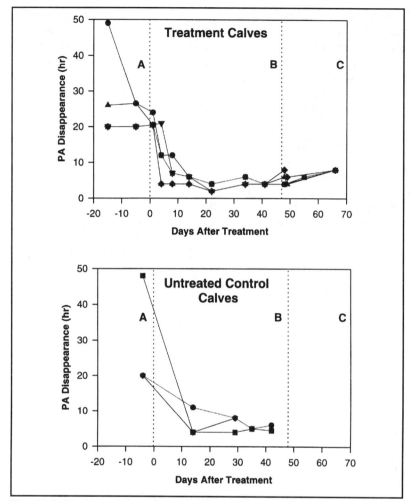

Fig. 38.3. Biodegradation of pyrrolizidine alkaloids in selected cattle rumen *vs.* days after treatment with sheep rumen fluid.

Results

The eight calves had initial PA biodegradation times of 20-26hrs. One calf in each of the two groups had an initial PA biodegradation time of 48hrs or more (Fig. 38.3). After supplementation, four of the five calves in Group 1 (treated) had biodegradation times reduced within 4d to 4-12hrs. The control calves once on the tansy ragwort feed did have an induction of PA-degrading ruminal microbes, but there was a lag of 14d compared to the Group 1 calves. One supplemented calf had a PA detoxification rate that remained unaffected at 21hrs. This calf was given a

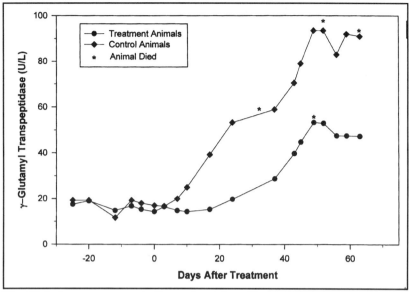

Fig. 38.4. Average concentrations of γ-glutamyl transpeptidase activities.

second supplement of sheep ruminal fluid on day seven, but it took another 8d for the biodegradation time to decrease to 7hrs. This calf and all of the controls developed clinical, histological and enzymological signs of tansy ragwort toxicosis and were euthanized and necropsied between day 32 and 60 of the experiment. Classic liver pathology was found in all euthanized calves (Craig *et al.*, 1991a).

Elevation of GGT activities in the blood were good indicators of hepatic dysfunction and paralleled the histopathological changes seen in the liver samples (Fig. 38.4). Bile duct hyperplasia and increased fibrosis were constant changes with some samples having hepatocellular megalocytosis and necrosis. This was identical to what has been seen in previous studies (Craig *et al.*, 1991a), but happened at a faster rate due to the higher concentration of PA in the feed. The calf in the treated Group 1 that had to be treated twice had increases in GGT similar to the untreated Group 2 calves. It is postulated that the failure of the sheep ruminal microbes to colonize in this calf in the early stages of the intoxication allowed enough initial damage that was irreversible and resulted in the classic liver toxicosis.

Conclusions

Whole ruminal fluid from sheep eating tansy ragwort was able to protect supplemented calves when challenged with tansy ragwort contaminated feed. This study inadvertently gave a higher dose of PA than would be encountered in a field situation, but the Group 1 treated calves were still protected, except for the one in which the ruminal transfer was not effective in increasing the rate of biodegradation of the toxins.

References

Blythe, L.L. and Craig, A.M. (1994) Clinical and preclinical diagnostic aids to hepatic plant toxicosis in horses, sheep and cattle. In: Colegate, S.M. and Dorling, P.R. (eds), *Plant-Associated Toxins: Agricultural, Phytochemical and Ecological Aspects*. CAB International, Wallingford, Oxon, pp. 313-318.

Craig, A.M. (1995) Detoxification of plant and fungal toxins by ruminant microbiota. In: Engelhardt, W.V., Leonhard-Marek, S., Breves, G. and Eisecke, D. (eds), *VIII International Symposium on Ruminant Physiology (VIII ISRP)*. Ferdinand Enke Verlag, Stuttgart, pp. 271-288.

Craig, A.M. and Blythe, L.L. (1994) Review of ruminal microbes relative to detoxification of plant toxins and environmental pollutants. In: Colegate, S.M. and Dorling, P.R. (eds), *Plant-Associated Toxins: Agricultural, Phytochemical and Ecological Aspects*. CAB International, Wallingford, Oxon, pp. 462-467.

Craig, A.M., Blythe, L.L., Lassen, E.D. and Slizeski, M.L. (1986) Ovine resistance to pyrrolizidine alkaloids. *Israel Journal of Veterinary Medicine* 42, 376-384.

Craig, A.M., Pearson, E.G., Meyer, C. and Schmitz, J.A. (1991a) Serum liver enzyme and histopathologic changes in calves with chronic and chronic-delayed *Senecio jacobaea* toxicosis. *American Journal Veterinary Research* 52, 1969-1978.

Craig, A.M., Pearson, E.G., Meyer, C. and Schmitz, J.A. (1991b) Clinicopathologic studies of tansy ragwort toxicosis in ponies: Sequential serum and histopathological changes. *Journal of Equine Veterinary Science* 11, 261-271.

Craig, A.M., Latham, C.J., Blythe, L.L., Schmotzer, W.B. and O'Conner, O.A. (1992) Metabolism of toxic pyrrolizidine alkaloids from tansy ragwort, *Senecio jacobaea*, in ovine ruminal fluid under anaerobic conditions. *Applied and Environmental Microbiology* 58, 2730-2736.

Pearson, E.G. and Craig, A.M. (1980) The diagnosis of liver disease in equine and food animals. Part I. *Modern Veterinary Practice* 61, 233-237.

McDougall, E.I. (1948) Studies on ruminant saliva. I. The composition and output of sheep's saliva. *Biochemistry Journal* 43, 99-109.

Chapter 39

A Comparison of a Nursling Rat Bioassay and an ELISA to Determine the Amount of Phomopsins in Lupin Stubbles

J.G. Allen[1], K.A. Than[2], E.J. Speijers[1], Z. Ellis[1], K.P. Croker[1], C.L. McDonald[1] and J.A. Edgar[2]

[1]*Agriculture Western Australia, 3 Baron-Hay Court, South Perth, Western Australia 6151, Australia;* [2]*CSIRO Division of Animal Health, Australian Animal Health Laboratory, Private Bag 24, Geelong, Victoria 3213, Australia*

Introduction

Lupinosis is a mycotoxicosis of livestock (van Warmelo *et al.*, 1970) caused by the consumption of phomopsins (Culvenor *et al.*, 1977) produced by *Diaporthe toxica* (anamorph *Phomopsis* spp.) that infects and colonizes lupins. Until 1990, the assays most often used to measure the toxicity and/or concentration of phomopsins in lupin samples were the nursling rat bioassay (NR; Peterson, 1978; Petterson *et al.*, 1985), the sheep bioassay (Allen, 1989) and an HPLC assay (Hancock *et al.*, 1987). In 1989, Than *et al.* (1992) developed an enzyme-linked immunoassay (ELISA) specific for the phomopsins, which has subsequently been used to assess phomopsins amounts in lupin samples (Iyer *et al.*, 1994; Than *et al.*, 1994). This is a comparison of the phomopsins concentrations in lupin stubbles determined using the nursling rat bioassay and the ELISA.

Materials and Methods

Fifty-nine lupin stubbles (*Lupinus angustifolius*, low alkaloid cultivars) were sampled (200-500g) during the summers of 1990-91 (17 samples), 1991-92 (24) and 1992-93 (18) from throughout the agricultural area of Western Australia. All but nine came from crops of *Phomopsis*-resistant cultivars. All samples collected in a summer/autumn period were tested at one time. Each sample was milled to a maximum particle length of 2mm, thoroughly mixed and two subsamples of 40g taken. For each stubble, one subsample was tested using the nursling rat bioassay (Peterson, 1978; Petterson *et al.*, 1985), while the other was tested using the ELISA (Than *et al.*, 1992; Gallagher *et al.*, 1994).

For the nursling rat bioassay, 300ml of methanol (MeOH):water (9:1) was added to 40g of ground stubble and shaken in a water bath (50°C) for 1hr before filtering (Whatman #1). This was repeated twice and the filtrates combined. The MeOH fraction was evaporated under reduced pressure at 50°C, and the resultant aqueous concentrate diluted to 200ml with water. After adjusting the pH to 8-9 with 1M NH_4OH, and then immediately to three with 1M HCl, the material was centrifuged and the supernatant discarded. This was washed with 2x75ml ethyl acetate and then extracted with 3x50ml of *n*-butanol. The *n*-butanol extract was adjusted to pH 6.5-7 with 1M NH_4Cl and evaporated to dryness under reduced pressure at 50°C. The residue was taken up in 10ml of distilled water and stored at -20°C until required.

For the ELISA, 12.5g of ground stubble was shaken with 250ml of reagent grade MeOH:water (4:1) overnight at room temperature. The extracts were filtered through GF/A glass microfiber paper (Whatman) and stored at -18°C until assayed. Just before analysis, aliquots of filtered extracts were diluted to a minimum of 1:10 with assay buffer and used (100µl/well) in the assay.

To examine the influence of the different extraction procedures related to the assays, surplus extracts prepared for the nursling rat bioassay in 1990-91 were also sent to Victoria to be tested using the ELISA. The association between the results (phomopsins concentrations in mg of equivalent phomopsin A bioactivity/kg) given by the two assays was tested by linear regression analysis, after $\log_{10}$ (1+value) transformation of the data.

Results

The phomopsins concentrations in the 59 stubble samples as determined by the two assays are shown in Fig. 39.1. The associations between the transformed results are indicated by the following equations (where E=ELISA result 0.05, NR=nursling rat result 0.05, r^2_a=proportion of variance accounted for by the regression, and the standard error (SE) of the slope and intercept are shown within brackets):

1990-91: $\log E = 1.004$ (0.128) $\log NR$ 0.514 (0.107), $r^2_a = 79\%$, $P < 0.001$.
1991-92: $\log E = 0.843$ (0.122) $\log NR$ 0.658 (0.080), $r^2_a = 67\%$, $P < 0.001$.
1992-93: $\log E = 0.827$ (0.119) $\log NR$ 0.234 (0.099), $r^2_a = 73\%$, $P < 0.001$.
All years: $\log E = 0.822$ (0.079) $\log NR$ 0.518 (0.061), $r^2_a = 65\%$, $P < 0.001$.

There was no significant difference between the slopes of the equations for the individual years, but the intercept for 1992-93 was significantly ($P<0.001$) lower than that for the other 2yrs, for which the intercepts were not different ($P>0.05$). For each year the slope of the line was not significantly different ($P>0.05$) from one, but the

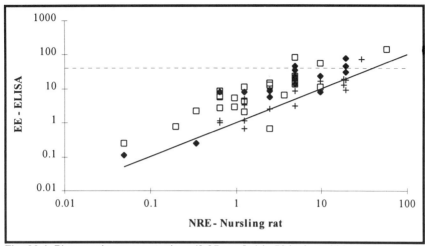

Fig. 39.1. Phomopsins concentrations (0.05; mg/kg) in 59 lupin stubbles measured by a nursling rat bioassay extraction procedure (NRE) and an ELISA extraction procedure (EE) in the summers of 1990-91 (♦) 1991-92 (□) and 1992-93 (+). The plotted full line represents points at which the ELISA and nursling rat results are equal and the dotted line indicates an ELISA result of 40mg/kg.

intercept was significantly ($P<0.05$) greater than zero.

In examining the influence of extraction methods on the 1990-91 stubbles, the association between the nursling rat bioassay and ELISA results for extracts prepared by the nursling rat extraction procedure was represented by the equation:

$$\log E = 0.959\ (0.167)\ \log NR\ 0.037\ (0.128),\ r^2_a = 67\%,\ P< 0.001$$

There was clearly significant difference between the results given by the two assays (the slope of the equation was not different from one and the intercept not different from zero, Fig. 39.2(a)). However, the association between the ELISA results for extracts from the same stubbles prepared by the nursling rat and the ELISA extraction procedures was given by the equation:

$$\log E\ \text{(nursling rat extraction)} = 0.480\ (0.134)\ \log E\ \text{(ELISA extraction)}$$
$$0.065\ (0.170),\ r^2_a = 43\%,\ P = 0.003$$

From the association depicted in Fig. 39.2(b), it is obvious that results for the extracts prepared using the ELISA extraction procedure were significantly higher (slope of the equation different from one).

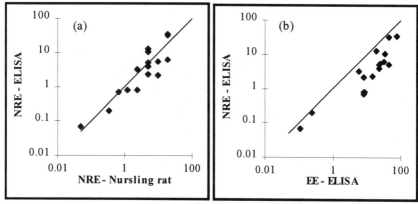

Fig. 39.2. Phomopsins concentrations (0.05; mg/kg) in 1990-91 lupin stubbles when (a) extracts prepared using the nursling rat bioassay extraction procedure (NRE) were tested using the nursling rat bioassay and the ELISA, and (b) extracts prepared using the NRE and ELISA extraction (EE) procedures were tested using the ELISA. The plotted line represents points at which the ELISA and nursling rat results are equal.

Discussion

There was a reasonable association between the concentrations of phomopsins in lupin stubbles determined using the nursling rat bioassay and the ELISA (65-79% of variance accounted for by the regression). However, the results provided by the nursling rat bioassay were generally lower (Fig. 39.1), particularly in the first 2yrs (the intercepts for their equations were significantly greater than that for the third year). There are several possible explanations for this difference, but extraction efficiency appears to be the major contributor.

Extracts prepared using the nursling rat bioassay extraction procedure contained similar concentrations of phomopsins when tested using the nursling rat bioassay and the ELISA (Fig. 39.2a), but extracts from the same stubbles prepared using the ELISA extraction procedure and tested using the ELISA had significantly greater concentrations of phomopsins (Fig. 39.2b). This clearly indicated a superior extraction efficiency for the ELISA method. The relatively complex extraction and concentration procedures employed in the nursling rat bioassay provide greater opportunity for physical losses and chemical breakdown than does the simple extraction method used for the ELISA.

The ELISA offers a rapid, inexpensive and sensitive means of measuring phomopsins concentrations in lupin stubbles (Than, 1994), and should replace the nursling rat bioassay for this purpose. Use of the ELISA avoids the need to use animals to obtain a test result.

References

Allen, J.G. (1989) Studies of the pathogenesis, toxicology and pathology of lupinosis and associated conditions. PhD Dissertation, Murdoch University, Perth.

Culvenor, C.C.J., Beck, A.B., Clarke, M., Cockrum, P.A., Edgar, J.A., Frahn, J.L., Jago, M.V., Lanigan, G.W., Payne, A.L., Peterson, J.E., Petterson, D.S., Smith, L.W. and White, R.R. (1977) Isolation of toxic metabolites of *Phomopsis leptostromiformis* responsible for lupinosis. *Australian Journal of Biological Sciences* 30, 269-277.

Gallagher, P.F., Walker, B. and Edgar, J.A. (1994) Quantitative analysis of phomopsins in *Phomopsis*-resistant and *Phomopsis*-susceptible lupin stubbles. In: Colegate, S.M. and Dorling, P.R. (eds), *Plant-Associated Toxins: Agricultural, Phytochemical and Ecological Aspects*. CAB International, Wallingford, Oxon, pp. 155-160.

Hancock, G.R., Vogel, P. and Petterson, D.S. (1987) A high performance liquid chromatographic assay for the mycotoxin phomopsin A in lupin stubble. *Australian Journal of Experimental Agriculture* 27, 73-76.

Iyer, L., Jackson, K. and Leclerc, J. (1994) Commercial measurement of phomopsin A in lupin seeds and stubble. In: Dracup, M. and Palta, J. (eds), *Proceedings of the First Australian Lupin Technical Symposium*. Western Australian Department of Agriculture, South Perth, pp. 274-277.

Peterson, J.E. (1978) *Phomopsis leptostromiformis* toxicity (lupinosis) in nursling rats. *Journal of Comparative Pathology* 88, 191-203.

Petterson, D.S., Peterson, J.E., Smith, L.W., Wood, P.McR. and Culvenor, C.C.J. (1985) Bioassay of the contamination of lupin seed by the mycotoxin phomopsin. *Australian Journal of Experimental Agriculture* 25, 434-439.

Than, K.A. (1994) Applications of phomopsin ELISA in lupin research. In: Dracup, M. and Palta, J. (eds), *Proceedings of the First Australian Lupin Technical Symposium*. Western Australian Department of Agriculture, South Perth, p. 318.

Than, K.A., Payne, A.L. and Edgar, J.A. (1992) Development of an enzyme immunoassay for the phomopsin mycotoxins. In: James, L.F., Keeler, R.F., Bailey, E.M. Jr, Cheeke, P.R. and Hegarty, M.P. (eds), *Poisonous Plants, Proceedings of the Third International Symposium*. Iowa State University Press, Ames, IA, pp. 259-263.

Than, K.A., Tan, R.A., Petterson, D.S. and Edgar, J.A. (1994) Phomopsin content of commercial lupin seed from Western Australia in 1991/2. In: Colegate, S.M. and Dorling, P.R. (eds), *Plant-Associated Toxins: Agricultural, Phytochemical and Ecological Aspects*. CAB International, Wallingford, Oxon, pp. 62-65.

van Warmelo, K.T., Marasas, W.F.O., Adelaar, T.F., Kellerman, T.S., van Rensburg, I.B.J. and Minne, J.A. (1970) Experimental evidence that lupinosis of sheep is a mycotoxicosis caused by the fungus *Phomopsis leptostromiformis* (Kühn) Bubák. *Journal of the South African Veterinary Medical Association* 41, 235-247.

Chapter 40

Towards a Commercial Vaccine Against Lupinosis

J.A. Edgar[1], K.A. Than[1], A.L. Payne[1], N. Anderton[1], J. Baell[1], Y. Cao[1], P.A. Cockrum[1], A. Michalewicz[1], P.L. Stewart[1] and J.G. Allen[2]
[1]*CSIRO Division of Animal Health, Australian Animal Health Laboratory, Private Bag 24, Geelong, Victoria 3220, Australia;* [2]*Agriculture Western Australia, South Perth, Western Australia 6151, Australia*

Introduction

Lupinosis is caused by phomopsin mycotoxins produced by a fungus, *Diaporthe toxica* (anomorph *Phomopsis* spp.) that infects and colonizes lupins (Williamson *et al.*, 1994). In 1987, a pen trial was conducted, which demonstrated that an immunogen made by conjugating phomopsin A to keyhole limpet hemocyanin was an effective vaccine against the livestock poisoning disease lupinosis (Payne *et al.*, 1992). A considerable amount of work has been conducted since then, aimed at converting the experimental vaccine into a commercial product (Than *et al.*, 1994; Allen *et al.*, 1994). Some issues that have been addressed are summarized here.

Production and Conjugation of Phomopsins

A low-cost high-yielding means of producing phomopsins is an essential requirement for vaccine manufacture. Phomopsins can be produced by growing *D. toxica* in liquid culture or using solid-phase culture on lupin seeds or other grains. Yields of phomopsins were consistently higher on solid-phase cultures. A large number of field isolates of the fungus, and mutants produced by UV irradiation, were examined before a high-yielding strain was selected for production of phomopsin A. A mixture of phomopsins, rich in phomopsin A, is produced after about 6wks incubation. The phomopsins are extracted into aqueous methanol and concentrated by adsorption onto nonionic polymeric adsorbent (XAD) resin and further purified by crystallization.

Phomopsin A has two free carboxyl groups, a phenolic hydroxyl group and a secondary amino group available for direct coupling to amino and carboxyl groups on carrier proteins. Many conjugation methods were examined before stirring a crystalline phomopsin-A-rich preparation (>75% phomopsin A) in aqueous solution with 1-ethyl-3-(3-dimethylaminopropyl) carbodiimide hydrochloride. An aqueous solution of the conjugate is isolated by filtration of the reaction mixture through a 10,000 molecular weight cutoff filter to remove low molecular weight matter. Usually 75% of the added phomopsin A attaches to the protein.

Carrier proteins investigated included keyhole limpet hemocyanin, tetanus toxoid, diptheria toxoid, ovalbumin, bovine serum albumin, bovine gamma globulin and bovine fetuin. Fetuin and gamma globulin conjugates were among the most effective in inducing anti-phomopsin IgG antibodies (Fig. 40.1).

Minimum Effective Dose of Phomopsin A per Injection

A minimum dose level of conjugated phomopsin A to produce protective levels of antibodies was determined by measuring antibody titers induced by conjugates carrying 50, 100, 200 and 300µg of phomopsin A/dose (Fig. 40.2). The carrier protein affected the amount of conjugated phomopsin A needed to induce desirable levels of anti-phomopsin IgG antibodies. The highest titers with bovine fetuin conjugates were obtained at the lowest level of hapten (50µg/dose). With gamma globulin as carrier, the highest titers occurred at 300µg phomopsin A/dose.

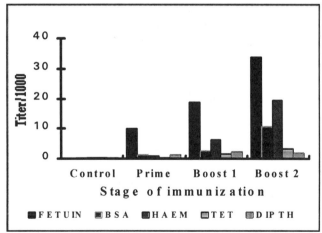

Fig. 40.1. Mean antibody titers before and after primary and booster injections. Groups of five sheep were vaccinated subcu-taneously with phomopsin A conjugated to bovine fetuin, bovine serum albumin, keyhole limpet hemocyanin, tetanus toxoid or diptheria toxoid as 1:1 emulsions with Freund's complete adjuvant, with boosters in Freund's incomplete adjuvant.

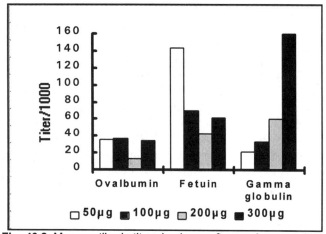

Fig. 40.2. Mean antibody titers in sheep after a primary and two booster injections of 50, 100, 200 or 300µg/dose of phomopsin A conjugated to ovalbumin, bovine fetuin, or bovine gamma globulin (n=10). Vaccines were injected as 1:1 emulsions with Freund's incomplete adjuvant.

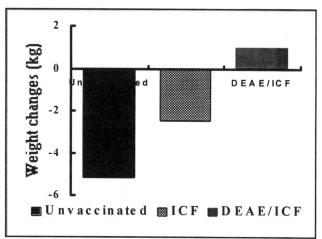

Fig. 40.3. Mean weight changes in unvaccinated sheep, sheep vaccinated with a phomopsin A/bovine fetuin conjugate in ICF; sheep vaccinated with a phomopsin A/ bovine fetuin conjugate in ICF containing 15% DEAE dextran (n=15). In 7wks exposure to lupin stubble, five unvaccinated sheep died from lupinosis and all showed clinical signs of lupinosis. Two sheep vaccinated using ICF showed clinical signs of lupinosis. No sheep vaccinated using 15% DEAE dextran/ICF showed signs of lupinosis.

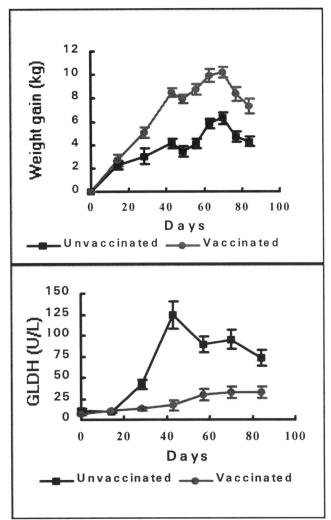

Fig. 40.4. a. Mean weight changes in unvaccinated sheep and sheep vaccinated with a commercially viable formulation of the lupinosis vaccine during exposure of the sheep to lupin stubble. **b.** Mean plasma GLDH changes for the same sheep.

Adjuvants investigated included Freund's incomplete (ICF), alum, Montanide oils, Quil A saponin and 15% diethyl aminoethyl (DEAE) dextran in ICF. The 15% DEAE dextran in ICF produced the highest and longest-lasting anti-phomopsin antibody titers. The high cost of DEAE dextran was overcome when it was demonstrated that 0.5% was as effective as 15% DEAE dextran. In field trials, the DEAE/ICF formulation out-performed ICF alone (Fig. 40.3).

Vaccination Protocol and Protection Studies

For a new vaccine to be accepted, it must impose a minimum additional burden on farmers. Lambs are typically "marked" and receive clostridial vaccine at 6-8wks of age, and this time was most suitable for the primary lupinosis vaccine injection.

A booster injection of the lupinosis vaccine must be given 2wks prior to exposure to poisonous lupin stubble to produce protective levels of antibodies in the first year. A 4mo delay between the primary and first booster injections significantly increased response to the vaccine when compared to a booster given 4wks after the primary injection, and raised antibody titers to well above protective levels. A single booster injection in subsequent years, given before exposure to poisonous stubbles, re-established titers.

During a field trial of the best formulations, even though little overt lupinosis was induced in the control sheep during 3mos subclinical exposure to phomopsins, significantly improved weight gains were seen with vaccinated compared to weight-matched, unvaccinated control animals (Fig. 40.4a). Liver damage, as measured by plasma enzyme levels, also demonstrated the effectiveness of the vaccine (Fig. 40.4b). Other trials have demonstrated the effectiveness of the vaccine against a more acute challenge (Fig. 40.3).

We believe the vaccine now meets the requirements of a commercial product.

References

Allen, J.G., Than, K.A., Edgar, J.A., Doncon, G.H., Dragicevic, G. and Kosmac, V.H. (1994) Field evaluation of vaccines against lupinosis. In: Colegate, S.M. and Dorling, P.R. (eds), *Plant-Associated Toxins: Agricultural, Phytochemical and Ecological Aspects*. CAB International, Wallingford, Oxon, pp. 427-432.

Payne, A.L., Than, K.A., Stewart, P.L. and Edgar, J.A. (1992) Vaccination against lupinosis. In: James, L.F., Keeler, R.F., Bailey, E.M. Jr, Cheeke, P.R. and Hegarty, M.P. (eds), *Poisonous Plants, Proceedings of the Third International Symposium*. Iowa State University Press, Ames, IA, pp. 234-238.

Than, K.A., Anderton, N., Cockrum, P.A., Payne, A.L., Stewart, P.L. and Edgar, J.A. (1994) Lupinosis vaccine: positive relationship between anti-phomopsin IgG concentration and protection in Victorian field trials. In: Colegate, S.M. and Dorling, P.R. (eds), *Plant-Associated Toxins: Agricultural, Phytochemical and Ecological Aspects*. CAB International, Wallingford, Oxon, pp. 433-438.

Williamson, P.M., Highet, A.S., Gams, W., Sivasithamparam, K. and Cowling, W.A. (1994) *Diaporthe toxica* sp. nov., the cause of lupinosis in sheep. *Mycological Research* 98, 1364-1368.

Chapter 41

Long-term Effects of Acute or Chronic Poisoning by Tunicaminyluracil Toxins

P.L. Stewart

CSIRO Australian Animal Health Laboratory, Post Office Box 24, Geelong, Victoria 3220, Australia

Introduction

Corynetoxins, the causative agents of annual ryegrass toxicity (ARGT), are toxic chemicals produced by the bacterium *Clavibacter toxicus* (Riley and Ophel, 1992), which parasitizes annual ryegrass seed heads. The disease is characterized by an acute neurological condition in livestock, with high morbidity and mortality rates (Berry and Wise, 1975). Corynetoxins and tunicamycins are members of the tunicaminyluracil antibiotic group and are specific inhibitors of the enzyme uridine diphospho *N*-acetylglucosamine:dolichyl phosphate *N*-acetylglucosamine-1-phosphate transferase (GlcNAc-1-P transferase). This enzyme catalyzes the initial step in the synthesis of *N*-linked glycoproteins (Lehle and Tanner, 1976) and its inhibition is a principal biochemical lesion in ARGT (Jago *et al.*, 1983; Culvenor and Jago, 1985). Tunicamycin is a nucleoside antibiotic first isolated from *Streptomyces lysosuperificus* (Takatsuki *et al.*, 1971) and tunicamycin has been widely used experimentally in biochemical studies on glycosylation. Inhibition of GlcNAc-1-P transferase activity by tunicamycin has been demonstrated in rat liver rough microsomes *in vitro* and *in vivo* (Eggens and Dallner, 1982; Jago *et al.*, 1983). Tunicamycin is structurally similar to corynetoxin, differing only in the nature of the fatty acid side chains (Edgar *et al.*, 1982), and the chemical and biological activities of the two toxins are essentially identical (Jago *et al.*, 1983; Vogel *et al.*, 1982). Experimental ARGT has been produced experimentally in sheep by administration of tunicamycin (Finnie and Jago, 1985).

Field cases of ARGT often present as acute poisoning, within 5-7d after introduction of sheep or cattle to an infected ryegrass paddock. However, there is also ample evidence of long-term effects of tunicaminyluracil toxicity following both acute or chronic toxin exposure, evidence that could assist in providing answers in the area of toxin distribution and clearance. There is a known variation in sensitivity of sheep treated experimentally with the tunicaminyluracil toxins (Jago and Culvenor, 1987), as well as a spread of clinical signs and mortality in field cases of ARGT. The implications are that there could be many survivors of ARGT outbreaks affected by subclinical toxicity. This report summarizes the known long-term adverse effects produced in animals by acute or chronic tunicamycin or corynetoxin poisoning.

Acute Poisoning

Long-term biochemical and pathological changes have been observed in animals after a single injection of tunicamycin. The most sensitive indicator of tunicaminyluracil poisoning known at present is the activity of liver rough microsomal GlcNAc-1-P transferase activity. In sheep given a single subcutaneous (s.c.) injection of tunicamycin at two dose rates (10 and 20µg/kg), transferase activity was still below 20% of normal activity after 14d (Stewart, unpublished). While enzyme activity recovered slowly over the next few months, it did not reach control levels until some time between 16-22wks. In rats injected s.c. with tunicamycin at 100 and 200µg/kg, recovery of transferase activity was more rapid, but still required 10-12wks to return to normal levels. The results suggest that inhibitory levels of toxin remain in the liver tissue for a considerable time.

Similar results were obtained when rats were given a single intraneural injection into the left sciatic nerve (Smith *et al.*, 1985). The GlcNAc-1-P transferase activity in peripheral nerve homogenates was inhibited for as long as 24d, when it was still depressed as low as 75% of control nerve activity.

Long-term reproductive effects have been demonstrated in tunicamycin-treated male rats. When administered as a single s.c. dose of 200µg/kg, tunicamycin caused permanent destruction of seminiferous tubules in adult rats, due to degenerative changes in the germinal epithelium (Peterson *et al.*, 1996). There was a 20% fall in body weight over the first week and the mean weight did not return to its pre-injection value until the eighth week.

Cumulative toxicity of tunicamycin has been demonstrated in two sheep receiving two s.c. injections of 25µg/kg up to 9wks apart (Jago and Culvenor, 1987). Although the sheep survived the initial dose, both died after receiving the second dose 9wks later, indicating that some residual toxic effects were present at the time of the second dose.

Chronic Poisoning

Chronic poisoning has been demonstrated by daily administration of very low doses of tunicamycin (1µg/kg/d), with the death of three sheep out of four within 29-48d (Jago and Culvenor, 1987). The daily dose rate was around 3% of a single lethal dose.

Progressive liver disease was shown by increasing levels of serum aspartate aminotransferase and bile acids.

Significant reduction in wool growth has been demonstrated in ewes given oral doses of corynetoxins twice weekly for 11wks (Davies *et al.*, 1996). While the dose of toxin was substantially less than that required to produce clinical signs, a 22% reduction in wool growth was observed.

Conclusions

The long-term effects of acute and chronic poisoning by the tunicaminyluracil toxins in rats and sheep are dependent on many factors, including size and frequency of dosing, and variation in sensitivity between animals. While the cumulative nature of the toxins has been proven experimentally and in the field, there are still many unanswered questions. It is not known why, for instance, not all sheep in a flock involved in an outbreak of ARGT show clinical signs or die, and why most deaths occur early in the season. Because of the cumulative toxicity, there should be greater sheep losses later in season, especially as sheep are often moved to another toxic paddock, rather than a 'safe' paddock. More work on distribution and clearance studies of the toxins is needed for a better understanding of the mechanism of toxicity. This work could also determine whether there are potential or real problems of tissue residues of toxins from livestock from ARGT-affected areas. The cumulative toxicity, the long-term biological effects and the extended depression of GlcNAc-1-P transferase activity suggest that corynetoxins may remain in liver, and possibly other tissues, for a considerable time.

References

Berry, P.H. and Wise, J.L. (1975) Wimmera ryegrass toxicity in Western Australia. *Australian Veterinary Journal* 51, 525-530.

Culvenor, C.C.J. and Jago, M.V. (1985) Annual ryegrass toxicity. In: Lacey, J. (ed), *Trichothecenes and Other Mycotoxins*. John Wiley and Sons, Brisbane, pp. 159-168.

Davies, S.C., White, C.L., Williams, I.H., Allen, J.G. and Croker, K.P. (1996) Sublethal exposure to corynetoxins affects production of grazing sheep. *Australian Journal of Experimental Agriculture* 36, 649-655.

Edgar, J.A., Frahn, J.L., Cockrum, P.A., Anderton, N., Jago, M.V., Culvenor, C.C.J., Jones, A.J., Murray, K. and Shaw, K.J. (1982) Corynetoxins, causative agents of annual ryegrass toxicity; their identification as tunicamycin group antibiotics. *Journal of the Chemistry Society Chemical Communications* 4, 222-224.

Eggens, I. and Dallner, G. (1982) Intramembranous arrangement of the glycosylating systems in rough and smooth microsomes from rat liver. *Biochimica et Biophysica Acta* 686, 77-93.

Finnie, J.W. and Jago, M.V. (1985) Experimental production of annual ryegrass toxicity with tunicamycin. *Australian Veterinary Journal* 62, 248.

Jago, M.V. and Culvenor, C.C.J. (1987) Tunicamycin and corynetoxin poisoning in sheep. *Australian Veterinary Journal* 64, 232-235.

Jago, M.V., Payne, A.L., Peterson, J.E. and Bagust, T.J. (1983) Inhibition of glycosylation by corynetoxin, the causative agent of annual ryegrass toxicity: a comparison with tunicamycin. *Chemico-Biological Interaction* 45, 223-234.

Lehle, L. and Tanner, W. (1976) The specific site of tunicamycin inhibition in the formation of dolichol-bound *N*-acetylglucosamine derivatives. *FEBS Letters* 71, 167-170.

Peterson, J.E., Jago, M.V. and Stewart, P.L. (1996) Permanent testicular damage induced in rats by a single dose of tunicamycin. *Reproductive Toxicology* 10, 61-69.

Riley, I.T. and Ophel, K.M. (1992) *Clavibacter toxicus* sp.nov., the bacterium responsible for annual ryegrass toxicity in Australia. *International Journal of Systematic Bacteriology* 42, 64-68.

Smith, M.E., Somera, F.P. and Sims, T.J. (1985) Enzymatic regulation of glycoprotein synthesis in peripheral nervous system myelin. *Journal of Neurochemistry* 45, 1205-1212.

Takatsuki, A., Arima, K. and Tamura, G. (1971) Tunicamycin, a new antibiotic. 1. Isolation and characterization of tunicamycin. *Journal of Antibiotics* 24, 215-223 .

Vogel, P., Stynes, B.A., Coackley, W., Yeoh, G.T. and Petterson, D.S. (1982) Glycolipid toxins from parasitised annual ryegrass: a comparison with tunicamycin. *Biochemical and Biophysical Research Communications* 105, 835-840.

Chapter 42

Experimental Modification of Larkspur (*Delphinium* spp.) Toxicity

B.L. Stegelmeier, K.E. Panter, J.A. Pfister, L.F. James, G.D. Manners, D.R. Gardner, M.H. Ralphs, and J.D. Olsen
USDA-ARS, Poisonous Plant Research Laboratory, 1150 East 1400 North, Logan, Utah 84341, USA

Introduction

Larkspur poisoning sporadically kills 5-15% of the cattle on North American mountain rangelands. *Delphinium* spp. contain more than 40 different diterpenoid alkaloids with variable prevalence and concentration. Several of these alkaloids, including highly toxic methyllycaconitine (MLA), reversibly bind and block nicotinic acetylcholine receptors (AchR). This study was designed to characterize *Delphinium* alkaloid toxicity and AchR binding affinity, and to further develop and examine potential therapies for intoxication.

A competitive binding assay with I^{125}-labeled bungarotoxin was used to determine the binding affinity of the larkspur alkaloids to rat and cow brain and skeletal muscle AchR (Macallan *et al.*, 1988; Rapier *et al.*, 1985). Comparing these binding affinities with affinities reported in the literature (Table 42.1) suggests that species variation in susceptibility to larkspur poisoning is related to receptor binding affinity to the different alkaloids, which may also explain the variation in clinical signs of poisoning seen between species. For example, cattle appear to be less susceptible to poisoning than rodents, and poisoned cattle have marked skeletal muscle weakness as opposed to the spastic convulsions observed in rodents (Nation *et al.*, 1982; Olsen and Sisson, 1991; Manners *et al.*, 1995). These differences are reflected in the binding affinities, as rat CNS AchR had nearly 100x higher affinity to MLA than cattle receptors. The toxic alkaloids such as MLA had higher binding affinities than the less toxic alkaloids tested (Table 42.2; Dobelis *et al.*, 1993; Kukel and Jennings, 1994). As binding affinity may be useful in determining animal susceptibility to larkspur poisoning or selecting resistant animals, further work is needed to better develop this technology.

Table 42.1. Methyllycaconitine (MLA) binding affinity of acetylcholine receptor (AchR) from the brain and skeletal muscle of rats, cows and humans. Affinities are expressed as the concentration of MLA required to inhibit 50% binding of I^{125}- labeled α-bungarotoxin (Ward *et al.*, 1990).

AchR	Binding	Reference
Rat Brain (P2)	1.1×10^{-9}	Current Report
Rat Muscle	2.0×10^{-7}	Current Report
Cow Brain (P2)	1.0×10^{-7}	Current Report
Rat Brain (P2)	1.4×10^{-9}	Ward, 1990
Neuroblastoma	6.3×10^{-5}	Ward, 1990
Human Muscle	7.8×10^{-6}	Ward, 1990

Table 42.2. Binding affinities of various larkspur alkaloids with different acetylcholine receptor preparations.

Diterpenoid Alkaloid	Occurrence[a]	LD_{50} mg/kg[b]	KI_{50} nM[c]	IC_{50} μM[d]
Lycoctonine	0.05	444	2,800	--
Deltaline	0.90	201	--	159
Dictyocarpine	0.09	283	--	--
N-Desethyldeltaline	--	210	--	--
Methyllycaconitine	0.20	4.0	1.7	1.50
Nudicauline	--	2.7	--	0.33
14-Deacetylnudicauline	0.03	4.0	100,000	0.65
Barbinine	0.01	57	--	11.7
Elatine	--	9.2	4,300	--
N-Desethylmethyllycaconitine	--	100	--	--
Zaliline	--	>230	--	--
Anthranoyllycoctonine	--	20.8	--	--
Delavaine	--	3.3	--	--

[a] Relative abundance (% dwt) of alkaloids identified in extracts from *Delphinium barbeyi* (Olsen *et al.*, 1990); [b] Estimated LD_{50} for mice given alkaloid intravenously (Manners *et al.*, 1995); [c] Concentration of larkspur alkaloid required to inhibit 50% of the binding of I^{125}-labeled α-bungarotoxin to rat brain AchR (Kukel and Jennings, 1994); [d] Concentration of larkspur alkaloid required to inhibit 50% of lizard muscle response (Dobelis *et al.*, 1993).

Modification of AchR Expression

As larkspur alkaloids bind and block the AchR, modification of AchR expression may alter animal response to poisoning. Previous studies have shown that chronic nicotine treatment upregulates AchR expression (Rogers and Wonnacott, 1995). To upregulate AchR expression, rats were treated with subcutaneous nicotine injections of 2 or 4mg/kg for 14d. Some have speculated that cattle can become tolerant to larkspur poisoning. Animals chronically ingesting larkspur could have altered AchR expression and susceptibility to larkspur poisoning. To test these hypotheses, two groups of animals were fed larkspur diets containing 10 and 40mg/kg MLA (1.1 and 4.4% larkspur). Northern blots demonstrated that nicotine treatment increased AchR mRNA expression 4-5x. No changes were detected in larkspur-treated rats (blots not shown). Rats with upregulated AchR did not have significant changes in susceptibility to MLA toxicity. The larkspur-treated rats tended to be more susceptible to MLA toxicity (Fig. 42.1). Small (4-5x) AchR upregulation or chronic larkspur ingestion did not change animal response to intoxication, and these treatments are unlikely to be useful as antidotes.

Metabolism Modifications

To determine if modifications of hepatic metabolism would alter animal responses to MLA toxicity, groups of 20 rats were treated with 3-methylcholanthrene (20mg/kg for 2d), SKF 525-A (75mg/kg 1hr before dosing), and phenobarbital (1mg/ml for 7d in the drinking water). These animals were then tested with MLA. Control animals received no treatment. Initial results were highly variable, as animals dosed with phenobarbital were occasionally resistant to MLA's effects (Fig. 42.2). Animals that were clinically depressed seemed most resistant. To test this hypothesis, two additional groups were treated with phenobarbital (1 mg/ml in the water, but with a 24hr withdrawal prior to dosing), and valium (10mg/kg, 15min prior to dosing). The valium treatment decreased toxicity, but again the response was variable (Fig. 42.2), and again the clinically sedate and depressed animals seemed to be most resistant to toxicity. Animals treated with phenobarbital with a withdrawal time prior to MLA testing responded similarly to controls. On the range, larkspur-poisoned cattle that remain calm and are not stressed, exercised or fatigued are less likely to die from larkspur poisoning. The response in rats treated with valium or phenobarbital was so variable that a confidence interval for the LD_{50} could not be determined using this modified high/low technique (Bruce, 1985, 1987). Though valium or phenobarbital treatment is not likely to be useful in cattle, it does allude to the mechanism of toxicity in rodents. As none of the P_{450} oxidase inducers or inhibitors altered animal response to MLA toxicity, it appears that hepatic metabolism is probably not important in larkspur toxicity.

Cholinesterase Inhibitors

Nearly every cowboy has a favorite remedy for larkspur poisoning, including cutting tails, bleeding, tobacco, rumen stimulants, mineral supplements and strychnine cocktails. Unfortunately, only the treatments including cholinesterase-inhibitors have been shown to have some clinical effect. Common cholinesterase-inhibitors include neostigmine or physostigmine. Although widely used, none of these has proven

Stegelmeier et al.

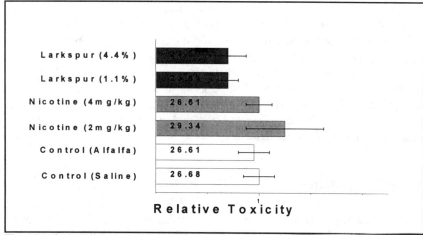

Fig. 42.1. Relative methyllycaconitine toxicity of rats chronically treated with nicotine and larkspur. Responses are LD_{50}s with 95% confidence intervals (CI).

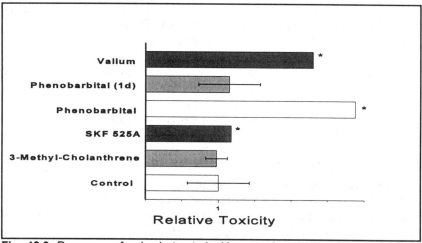

Fig. 42.2. Response of animals treated with several promoters and inhibitors of hepatic cytochrome P_{450} enzyme systems. Values (means ± 95% CI) marked with * had variable responses and animal numbers were too low to calculate a confidence interval (CI).

definitively to rescue fatally poisoned animals. As larkspur-induced neuromuscular blockage may be exacerbated by the stress of treatment or temporary increased physical activity, studies are needed to determine if these therapies affect the final outcome of poisoning. This study evaluated cholinesterase inhibition to see if it shifted the LD_{50} in a small animal model. Groups of 20 rats were treated with

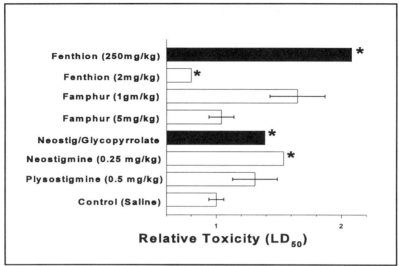

Fig. 42.3. Response of animals treated with several cholinesterase-inhibitors. Values (means ± 95% CI) marked with * had variable responses and animal numbers were too low to calculate a confidence interval (CI).

physostigmine (0.08mg/kg intraperitoneally (i.p.), 20min before dosing), neostigmine (0.08mg/kg i.p., 20min before dosing), neostigmine/glycopyrrolate (as above with 7μg/kg glycopyrrolate), famphur (5mg/kg transdermal (t.d.) 1x/d for 2d before dosing), fenthion (2mg/kg t.d., 1hr before dosing). These doses are similar to approved doses in cattle. The rats were then injected with MLA and a lethal dose was estimated (Bruce, 1985). There was no difference between the response of any of the groups and the control. Additional groups were treated with toxic doses of physostigmine (0.5mg/kg i.p., 20min before dosing), neostigmine (0.25mg/kg i.p., 20min before dosing), famphur (1g/kg t.d., 8hrs before dosing), fenthion (250mg/kg t.d., 1hr before dosing). All animals showed clinical signs of cholinergic intoxication. As seen in Fig. 42.3, these groups all had significantly lower responses to MLA toxicity (higher LD_{50}s). This suggests that, at or near toxic doses, cholinesterase-inhibitors can shift the toxicity of MLA and possibly all the larkspur alkaloids. These doses are considerably higher than the current doses recommended as larkspur antidotes (Pfister *et al.*, 1994). Work is needed to determine the appropriate dose and efficacy of treatment in livestock.

Conclusions

Binding affinity of AchR to the larkspur alkaloid correlates with species susceptibility to poisoning, and individual larkspur alkaloids bind AchR with affinities that correlate with their toxicity. Treatments to alter AchR expression or hepatic metabolism did not alter animal response to larkspur alkaloid (MLA) toxicity. The toxicity of MLA shifted in rats treated with toxic doses of cholinesterase-inhibiting drugs, which may be useful as larkspur antidotes. As cattle are very susceptible to anticholinesterase drugs, further research is needed to evaluate whether potential benefits outweigh the exacerbation of toxicity induced by stress of treatment.

References

Bruce, R.D. (1985) An up-and-down procedure for acute toxicity testing. *Fundamental and Applied Toxicology* 5, 151-157.

Bruce, R.D. (1987) A confirmatory study of the up-and-down method for acute oral toxicity testing. *Fundamental and Applied Toxicology* 8, 97-100.

Dobelis, P., Madl, J.E., Manners, G.D., Pfister, J.A. and Wolrond, J.P. (1993) Antagonism of nicotinic receptors by *Delphinium* alkaloids. *Neuroscience Abstracts* 631, 12.

Kukel, C.F. and Jennings, K.R. (1994) *Delphinium* alkaloids as inhibitors of alpha-bungarotoxin binding to rat and insect neural membranes. *Canadian Journal of Physiological Pharmacology* 72, 104-107.

Macallan, D.R., Lunt, G.G., Wonnacott, S., Swanson, K.L., Rapoport, H. and Albuquerque, E.X. (1988) Methyllycaconitine and (+)-anatoxin-a differentiate between nicotinic receptors in vertebrate and invertebrate nervous systems. *FEBS Letters* 226, 357-363.

Manners, G.D., Panter, K.E. and Pelletier, S.W. (1995) Structure-activity relationships of norditerpenoid alkaloids occurring in toxic larkspur (*Delphinium*) species. *Journal of Natural Products* 58, 863-869.

Nation, P.N., Benn, M.H., Roth, S.H. and Wilkens, J.L. (1982) Clinical signs and studies of the site of action of purified larkspur alkaloid, methyllycaconitine, administered parenterally to calves. *Canadian Veterinary Journal* 23, 264-266.

Olsen, J.D. and Sisson, D.V. (1991) Toxicity of extracts of tall larkspur (*Delphinium barbeyi*) in mice, hamsters, rats and sheep. *Toxicology Letters* 56, 33-41.

Olsen, J.D., Manners, G.D. and Pelletier, S.W. (1990) Poisonous properties of larkspur (*Delphinium* spp.). *Collections Botanica (Barcelona)* 19, 141-151.

Pfister, J.A., Panter, K.E., Manners, G.D. and Cheney, C.D. (1994) Reversal of tall larkspur (*Delphinium barbeyi*) poisoning in cattle with physostigmine. *Veterinary and Human Toxicology* 36, 511-514.

Rapier, C., Harrison, R., Lunt, G.G. and Wonnacott, S. (1985) Neosurugatoxin blocks nicotinic acetylcholine receptors in the brain. *Neurochemistry International* 7, 389-396.

Rogers, A.T. and Wonnacott, S. (1995) Nicotine-induced upregulation of alpha bungarotoxin (alpha bgt) binding sites in cultured rat hippocampal neurons. *Biochemistry Society Transactions* 23, S48.

Ward, J.M., Cockcroft, V.B., Lunt, G.G., Smillie, F.S. and Wonnacott, S. (1990) Methyllycaconitine: a selective probe for neuronal alpha-bungarotoxin binding sites. *FEBS Letters* 270, 45-48.

Chapter 43

Treatment and Prevention of Livestock Poisoning: Where to from Here?

J.A. Edgar

CSIRO Division of Animal Health, Australian Animal Health Laboratory, Private Bag 24, Geelong, Victoria 3220, Australia

Introduction

One of the primary objectives in studying intoxications of livestock is to develop rational prevention strategies or treatments that reduce or prevent livestock production losses. Unlike infectious agents, where vaccines and antibiotics provide methods for preventing and ameliorating disease, there are no well-developed generic approaches available for treating and preventing poisoning caused by natural chemical toxicants. The first step taken in studying livestock poisonings by natural toxicants is to identify the source of the poison. This provides an immediate opportunity to prevent poisonings by eliminating the plant or microbe from the field or preventing access to it as a feed source. Another chance for preventing poisoning is when the offending agent(s) is (are) identified and the timing of its formation studied so that a window of safe feeding opportunity may be exploited. Yet, traditional management and avoidance of poisoning have been unsuccessful with corynetoxin poisoning, phomopsin poisoning and pyrrolizidine alkaloidosis, and poisonings still occur with unsatisfactory frequency. Recent work at CSIRO in possible "generic" methods for treating and preventing livestock poisoning builds on some well-known ways of ameliorating livestock poisoning.

Three main approaches have been investigated: toxin binding agents, vaccination, and microbial detoxication. The focus has been on methods applicable to free-ranging livestock, since the most economically important poisoning diseases in Australia are primarily field diseases of livestock.

Treatments must be cost-effective, practical and in harmony with normal farm management practices if they are to be adopted by, and be useful to, livestock producers. Preventing poisoning through management avoids overt clinical signs and significant losses. Desirable treatments are those that combat poisoning and prevent absorption of chemicals into, or lead to their rapid elimination from, animal tissues/products.

Toxin Binding Agents

Activated charcoal has a considerable capacity to adsorb substances in the gastrointestinal tract, and has been used as an oral treatment for poisoning. Use in the field is hindered by its non-specific adsorption properties and the amount that must be administered to save poisoned stock. Hydrated sodium calcium aluminosilicates possess chemical adsorption properties and have been included in feed, where they irreversibly bind to aflatoxins, reducing their bioavailability.

Modern molecular technology may enable high specificity biocompatible molecular receptors to be designed for poisonous substances, allowing smaller quantities than charcoal or aluminosilicates in adsorbing natural toxicants. Initial work to design and synthesize *de novo* a specific receptor for the corynetoxins failed to produce binding in preliminary investigations. The multi-step, synthetic pathways to putative receptors were likely to be costly, and if binding agents produced were effective, they might not be economically feasible for livestock.

Host molecules, such as cyclodextrins (CDs) and callixarenes, that molecularly encapsulate guest molecules were appraised. Inclusion complexes of pyrrolizidine alkaloids with CDs produced weakly measurable binding interactions, likely insufficient to have any protective effect *in vivo* (Anderton *et al.*, 1994). However, the binding of corynetoxins to CDs was much more effective, and looks promising in preventing and treating corynetoxin poisoning (May *et al.*, 1995a; Chapter 37).

The CDs have potential for treating and preventing livestock poisonings. Since they are increasingly used in the pharmaceutical and food industries, they should be accepted for use in food-producing livestock. Small molecules with appropriate structural features can be entrained in the cylindrical hydrophobic CD cavity and have strong binding interactions. In corynetoxins, the fatty acid side chain is thought to be encapsulated by the CD. This converts the amphipathic corynetoxins into water-soluble corynetoxin-CD complexes (May *et al.*, 1995a), allowing altered tissue distribution and greater excretion. The CDs remove lipophilic molecules such as cholesterol from membranes. Corynetoxins and phorbol esters are distributed into membranes *in vivo* and interact with specific membrane-bound enzymes, which may be stripped from membranes by CDs. The increased water solubility of the toxicant-CD complex provides for excretion of cumulative, membrane-soluble toxicants. Besides the value in treating poisoned animals, if the CDs are in a controlled-release formulation they may be useful in preventing poisoning.

Vaccination

Vaccination against toxicants is well recognized, but attempts to produce useful vaccines against low molecular weight natural toxicants have been unsuccessful (Edgar, 1994). Molecules that induce their harmful effects by interaction with a natural receptor may be more amenable to a vaccination approach. Anti-toxicant antibodies and the CDs compete with the natural receptors for the toxicant, lowering the effective level of toxicant at its active site *in vivo*. Corynetoxins and the phomopsin mycotoxins were prime candidates for a vaccine approach, and effective vaccines have been developed in both cases (Chapters 34 and 40). Poisoning by the pyrrolizidine alkaloids and aflatoxins, whose hepatic metabolites alkylate nucleophilic sites on DNA and proteins, is resistant to mechanisms of the immune system and is not prevented by vaccination (Culvenor, 1978).

Microbial Detoxication in the Rumen

Detoxication of natural toxicants in the rumen is a mechanism of natural resistance to poisoning (Allison, 1978). The benefit was seen in cattle feeding on *Leucaena leucocephala* when naturally occurring mimosine-degrading microbes were added into their rumens (Jones, 1985) and a long-term solution was achieved.

Recombinant DNA technology offers the potential of introducing toxicant-degrading capacity to otherwise vulnerable ruminants (Gregg *et al.*, 1996). A few bacteria/microbes are capable of degrading corynetoxins (Payne *et al.*, 1994) but the work of transferring this activity to rumen bacteria has not been continued. Research in Australia has isolated a gene encoding fluoroacetate dehalogenase and transferred this gene into a rumen bacterium, *Butyrivibrio fibrisolvens* (Gregg *et al.*, 1994). Fluoroacetate, a natural constituent of a many indigenous Australian plants, causes significant livestock poisoning. Field testing of the recombinant organism in livestock has been halted by the Australian committee that oversees testing of genetically manipulated organisms because it has not been established that this capacity would not be transferred to other species, thereby endangering the survival of indigenous fluroacetate-producing plant species.

Another approach to using rumen bacteria to prevent poisoning is to modify the rumen environment to favor detoxicating bacteria. A bacterium was isolated that was capable of hydrogenolyzing and converting pyrrolizidine alkaloids into non-toxic, 1-methylene products *(Peptococcus helio-trinereducans*; now *Peptostreptococcus)*. Lanigan (1976) investigated this possibility by inhibiting methanogenic bacteria that competed with *P. heliotrinreducans* for free hydrogen. This failed because the antimethanogen employed, iodoform, was toxic to the sheep. Yet, a new class of antimethanogens has been developed in which otherwise volatile substances like bromochloromethane are encapsulated into a CD cavity to give products more amenable to delivery into the rumen (May *et al.*, 1995b). A bromochloromethane-α-

CD inclusion complex shows antimethanogenic activity in the rumen of cattle (McCrabb *et al.*, 1997) and sheep, and may prevent pyrrolizidine alkaloidosis by the Lanigan *et al.* (1978) method.

References

Allison, M.J. (1978) The role of ruminal microbes in the metabolism of toxic constituents from plants. In: Keeler, R.F., van Kampen K.R. and James L.F. (eds), *Effects of Poisonous Plants on Livestock*. Academic Press, New York, NY, pp. 101-118.

Anderton, N., Gosper, J.J. and May, C. (1994) The inclusion of pyrrolizidine alkaloids by á- and â-cyclodextrins. *Australian Journal of Chemistry* 47, 853-857.

Culvenor, C.C.J. (1978) Prevention of pyrrolizidine alkaloid poisoning - animal adaptation or plant control? In: Keeler, R.F., van Kampen, K.R. and James, L.F. (eds), *Effects of Poisonous Plants on Livestock*. Academic Press, New York, NY, pp. 189-200.

Edgar, J.A. (1994) Vaccination against low molecular weight natural toxicants. In: Wood, P.R., Willadsen, P., Vercoe, J.E., Hoskinson, R.M. and Demeyer, D. (eds), *Vaccines in Agriculture*. CSIRO Publications, Melbourne, pp. 149-153.

Gregg, K., Cooper, C.L., Schafer, D.J., Sharpe, H., Beard, C.E., Allen, G. and Xu, J. (1994) Detoxification of the plant toxin fluoroacetate by a genetically modified rumen bacterium. *BioTechnology* 12, 1361-1365.

Gregg, K., Allen, G. and Beard, C. (1996) Genetic manipulation of rumen bacteria: from potential to reality. *Australian Journal of Agricultural Research* 47, 247-56.

Jones, R.J. (1985) *Leucaena* toxicity and the ruminal degradation of mimosine. In: Seawright, A.A., Hegarty, M.P., James, L.F. and Keeler, R.F. (eds), *Plant Toxicology*. Queensland Poisonous Plants Committee, Yeerongpilly, QLD, pp. 111-119.

Lanigan, G.W. (1976) *Peptococcus heliotrinreducans*, sp. nov., a cytochrome-producing anaerobe which metabolises pyrrolizidine alkaloids. *Journal of Microbiology* 94, 1-10.

Lanigan, G.W., Payne, A.L. and Peterson, J.E. (1978) Antimethanogenic drugs and *Heliotropium europaeum* poisoning in penned sheep. *Australian Journal of Agricultural Research* 29, 1281-1292.

May, C., Pope, M., Stewart, P.L. and Edgar, J.A. (1995a) Methods for treating animals. Australian Patent Application No. 36628/95.

May, C., Payne, A.L., Stewart, P.L. and Edgar, J.A. (1995b). A delivery system for agents and composition. International Patent Application No. PCT/AU95/00733.

McCrabb, G.J., Berger, K.T., Manger, T., May, C. and Hunter, R.A. (1997) Inhibiting methane production in Brahman cattle by dietary supplementation with a novel compound and its effects on growth. *Australian Journal of Agricultural Research* 48, 323-329.

Payne, A.L., Cockrum, P.A. and Edgar, J.A. (1994) Metabolic transformation of corynetoxin and tunicamycin by *Alternaria alternata*. In: Colegate, S.M. and Dorling, P.R. (eds), *Plant-Associated Toxins: Agricultural, Phytochemical and Ecological Aspects*, CAB International, Wallingford, Oxon, pp. 445-449.

Chapter 44

Toxicity and Diagnosis of Oleander (*Nerium oleander*) Poisoning in Livestock

F.D. Galey[1], D.M. Holstege[1], B.J. Johnson[1] and L. Siemens[2]
[1]*California Veterinary Diagnostic Laboratory System;* [2]*Cardiology Department, University of California, Davis, California 95616, USA*

Introduction

Oleander (*Nerium oleander*) is an ornamental, evergreen shrub with leathery, dark green leaves. The leaves have a prominent midrib with secondary veins that are parallel to each other. The plant is found commonly in the Southern US, including most of California, as well as in many other parts of the world (Kingsbury, 1964; Everist, 1981). All parts of the plant are toxic, and ingestion of clippings from oleander is a common cause of poisoning in animals.

The toxicity of oleander is from several cardiac glycosides; the most prominent are oleandrin (aglycone is oleandrigenin) and neriine (Everist, 1981). Cardiac glycosides inhibit Na^+/K^+ ATPase, and as little as 0.005% of an animal's weight in dry leaves may be lethal (10-20 leaves for an adult horse; Kingsbury, 1964).

Animals exposed to oleander are often found dead or present with rapidly developing, nonspecific signs that may resemble colic (Everist, 1981; Galey *et al.*, 1996). Clinical signs, if observed, develop 2-4hrs after exposure and may include abdominal pain, weakness, rumen atony, and excessive salivation. Cardiac changes, including a variety of tachy and brady arrhythmias, often lead to death within 12-36hrs of the onset of signs.

Diagnosis in fatal cases may be facilitated by finding leaves in the ingesta or in the environment. However, leaves may be macerated beyond identification or passed into the posterior gastrointestinal tract. This chapter describes development of an assay for oleandrin in ingesta and body fluids. The toxicity and effects of oleander were determined in horses and cattle dosed orally with plant material. The impact of new testing methodology, along with improved recognition of the pathologic features of oleander toxicosis on diagnosis of oleander, is described.

Analytical Method for Oleandrin

Samples of plant, ingesta, urine, or other fluids (10g) are homogenized with 100ml dichloromethane (DCM). Aliquots are evaporated to dryness, reconstituted in 150µl methanol (MeOH) followed by 1ml DCM and loaded onto a C-18 solid phase extraction (SPE) column (Mycochar®, Romer Laboratories, Columbia, MO) and eluted with DCM. The eluate is evaporated and re-dissolved in 50µl MeOH. An aliquot of extract (20µl ingesta, 50µl urine) is spotted on a 10x10cm silica gel plate with oleandrin standard (0.25µg). The plate is developed in one direction with DCM followed by ethyl acetate. It is then turned 90°, oleandrin (0.25µg) spotted, developed 1cm with MeOH in the new direction and allowed to dry. Final development is with toluene:acetonitrile (ACN):acetic acid (50:45:5). Dry plates are sprayed with 20% $AlCl_3$ in H_2O:ethanol (1:1) and heated 5min at 110°C. Oleandrin appears as a blue spot under long wavelength UV (366nm) (Galey *et al.*, 1996).

The extract can also be derivatized with 1-naphthoyl chloride and quantified by HPLC with fluorescence detection (λ_{ex} 220nm; λ_{em} 345nm). The derivatization is followed by clean-up using isooctane (4ml) in distilled H_2O (2ml). The isooctane layer (2ml) is then evaporated, reconstituted in DCM, and loaded on a silica SPE column (Waters Corp, Milford, MA). Oleandrin is eluted with 6ml of 10% MeOH/DCM (v/v). The eluate is evaporated to dryness, redissolved in mobile phase of 35% H_2O/ACN (v/v) and analyzed (Tor *et al.*, 1996).

Chromatography of plant material and ingesta fortified with plant revealed a consistent pattern of other spots in addition to oleandrin not found in blank or samples spiked with oleandrin alone. Fortification of diagnostic samples of stomach and rumen contents at levels ranging from 0.05-5ppm all resulted in detection of oleandrin by thin layer chromatography (TLC). Additionally, oleandrin spikes at 0.02 and 0.1ppm were identified in both rumen contents and urine. No spikes at these levels failed; thus, the two-dimensional TLC method has a 0.02ppm limit of detection for urine and rumen content (Galey *et al.*, 1996). The limit of detection for oleandrin by HPLC was 0.05ppm. Replicate fortifications of stomach contents (n=6) revealed a mean recovery of 85% (4.6% coefficient of variation (CV)) (Tor *et al.*, 1996).

Toxicity of Oleander in Horses and Cattle

Light-breed mares (n=6; 2-24yrs) and crossbred beef heifers (n=7, 130kg) were dosed *via* nasogastric tube with dried, ground oleander according to a modified up-and-down statistical design (Bruce, 1985; Sananthanan *et al.*, 1987). Clinical signs, electrocardiographic, cardiophysiologic, and pathologic effects were recorded for horses. Clinical and pathological signs were recorded for cattle. Samples of urine and ingesta were obtained for analysis. The ground, mixed oleander (*N. oleander*) had approximately 1,000ppm of oleandrin by HPLC.

Horses given 20mg of plant/kg (total dose of approximately six medium-sized leaves) developed no, or mild, signs. All three horses given 40mg/kg developed signs of toxicosis within 8hrs of dosing. Initial signs included lethargy and mild colic, then diarrhea and weakness. By 24hrs after dosing, a wide variety of cardiac arrhythmias, including various atrioventricular conduction blocks and ventricular arrhythmias were present. Monitoring of cardiac physiology revealed no primary abnormalities until arrhythmias led to a drop in function. Clinical pathology parameters had changes that were typical of severe colic in the horse. Two of the horses required euthanasia at 36hrs after dosing. Pathologic lesions included mild pulmonary edema, large amounts of fluid in the pericardium and body cavities, and histologic evidence of multifocal myocardial degeneration and necrosis. Other cardiac changes included mural thrombi and hemorrhage under the epicardial surface. The median toxic dose of oleander in the horse was estimated to be 26mg/kg.

Two heifers dosed with 25mg of plant/kg and one given 50mg/kg were unaffected (n=3 total). Three of the heifers given 50mg/kg and one given 100mg/kg developed clinical toxicosis (n=4 total). The median toxic dose of oleander for cattle was estimated to be 45mg of plant/kg, appoximately equivalent to 14 average-sized leaves. Affected cattle developed signs of weakness and ataxia by 12hrs after dosing. By 24hrs after dosing at the higher level, diarrhea, weakness, and lethargy had developed. Most animals had audible dropped heartbeats, pulse deficits, and/or tachycardia by 24hrs after dosing. One heifer was skipping a beat every third heartbeat at 12hrs after dosing with 50mg/kg. All of the animals that developed signs died or required euthanasia by 48hrs after dosing. Affected heifers had evidence of fluid in the bowel and varying degrees of hemorrhage or thrombosis in the heart. Several animals had mild renal tubular lesions. Histologically, there was multifocal myocardial inflammation, degeneration, and necrosis. One heifer dosed with 1g/kg died within 45min with no pathologic lesions.

Diagnostic Cases of Oleander Poisoning

During the past 8yrs, 73 cases of oleander poisoning have been confirmed at the California Veterinary Diagnostic Laboratory System. Cases were selected for the study based on the confirmation of exposure to oleander by analytical testing or observation of ingested leaves or from the history. Other causes of weakness and/or sudden death in livestock were ruled out.

The 73 cases resulted in 262 dead, 658 sick, and 7901 animals at risk. At risk was defined as having come from the same enclosure as affected animals. Morbidity and mortality varied 1-100% (Galey *et al.*, 1996). Cases occurred throughout the year, and came from ranches located as far north as Yolo and Napa counties of Central California, south to the US/Mexico border and as far east as parts of Arizona. Cattle were most often affected (37 cases with 222 dead from 599 sick and out of 7758 animals at risk). Horses, llamas, sheep, ratites, and a dog were also affected (Table 44.1). All cases with adequate history and investigations revealed that oleander clippings and/or dried leaves were the source of the toxin.

For all species, sudden death was the most common presentation. Clinical signs reported included colic, diarrhea, weakness, ataxia, lethargy anorexia, and sometimes abnormal cardiac rhythm (tachycardia or blocks, one ECG). Some peracute cases had no lesions whatsoever. Gross lesions, if present, included endocardial hemorrhage, fluid in the pericardium and body cavities, and in cattle very fluid colon and cecal contents. Histologically, the most common lesions were observed in the sub-endocardial region of the left ventricle, but all parts of the heart could be affected, including the auricles. Several animals had pulmonary edema, and others had mild lesions in the liver or renal tubules.

Diagnoses in 59 of 73 cases were confirmed by testing for oleandrin. Of these, 33 cases had no history of oleander exposure or evidence of leaves in ingesta at the time of examination. Oleander was identified in the environment in most of those 33 cases, however, after the presence of the toxin was revealed analytically.

The method was not sensitive enough to detect oleandrin in urine consistently. One Holstein dairy cow was found to have detectable (by TLC) oleandrin in milk after exposure to oleander leaves from hay. Repeat testing of milk from previously affected survivors and the bulk tank 5d after exposure revealed no oleandrin. The toxin was not always found in the upper gastrointestinal tract, but rather in the cecum or fecal material. Testing of feces allowed diagnosis in two horses on an antemortem basis.

Conclusions

Some radioimmunoassays for digoxin may cross-react with oleandrin (Osterloh *et al.*, 1982), although samples spiked in this laboratory at 0.01ppm did not (Galey *et al.*, 1996). Therefore, development of a sensitive chemical assay for oleandrin was necessary. The 2-dimensional TLC method was repeatable and relatively straight-forward. The initial development in DCM resolves oleandrin from residual lipids. Routine use of the method allows diagnosis of oleander in ingesta and urine at 0.02ppm, which was adequate in all cases for which exposure was suspected, and finding the characteristic pattern of spots for oleander plant adds to method confidence. The HPLC method, though quantitative, was slightly less sensitive than TLC (0.05ppm in ingesta), besides being unacceptably labor-intensive for most diagnostic uses. Further investigation of HPLC methods is ongoing.

Controlled dosing of animals with oleander facilitated development of the analyses and allowed description of the physiology of oleander poisoning in the horse, suggesting that ECG changes were very diverse and severe. Both controlled and epidemiologic studies highlighted the potential for finding heart lesions in animals that die several hours after exposure. The clinical signs reported in all case animals were consistent with literature reports (Kingsbury, 1964; Everist, 1981).

All cases for which evidence was available suggested that clippings and/or dried leaves were the source of poisoning, which is consistent with reports that fresh oleander on the bush is apparently not palatable to animals (Everist, 1981).

Table 44.1. Distribution of oleander poisoning cases diagnosed at the California Veterinary Diagnostic Laboratory System from (1989-1997).

Species	Cases	At risk	Clinical signs	Dead
Cattle	37	7,758	599	222
Horses	17	46	21	19
Llamas	13	88	30	15
Big horn sheep	4	4	4	4
Ratites	1	4	3	1
Dog	1	1	1	1
Total	73	7,901	658	262

The toxicity of dried plant material is also supported. Although these findings suggest increased incidence of oleander toxicosis, it is more likely that diagnosis has improved with development of the test for oleandrin (Galey *et al.*, 1996).

Acknowledgments

These studies were partly supported by the California Center for Equine Health, The California Livestock Disease Research Laboratory, and USDA Formula Funding.

References

Bruce, R.D. (1985) An up-and-down procedure for acute toxicity testing. *Fundamental and Applied Toxicology* 5, 151-157.

Everist, S.L. (1981) Apocyanaceae. In: *Poisonous Plants of Australia*, 2nd ed. Angus and Robertson, London, pp. 77-89.

Galey, F.D., Holstege, D.M., Plumlee, K.H., Tor, E., Johnson, W., Anderson, M.L., Blanchard, P.C. and Brown, F. (1996) Diagnosis of oleander poisoning in livestock. *Journal of Veterinary Diagnostic Investigation* 8, 358-364.

Kingsbury, J.M. (1964) *Poisonous Plants of the United States and Canada*. Prentice-Hall, Englewood Cliffs, NJ, pp. 264-267.

Osterloh J., Herold S. and Pond, S. (1982) Oleander interference in the digoxin radioimmunoassay in a fatal ingestion. *Journal of the American Veterinary Medical Association* 247, 1596-1597.

Sananthanan, L.P., Gade, E.T. and Shipkowitz, N.L. (1987) Trimmed logit method for estimating the ED50 in quantal bioassay. *Biometrics* 43, 825-832.

Tor, E.R., Holstege, D.M. and Galey, F.D. (1996) Determination of oleander glycosides in biological matrices by high performance liquid chromatography. *Journal of Agricultural and Food Chemistry* 44, 2716-2719.

Chapter 45

Are there Toxic and Non-toxic Varieties of *Eupatorium rugosum* Houttuyn?

R.A. Smith and D.G. Lang
Livestock Disease Diagnostic Center, Veterinary Science Department, University of Kentucky, 1429 Newtown Pike, Lexington, Kentucky 40511, USA

Introduction

Eupatorium rugosum Houttuyn (white snakeroot) was responsible for many human deaths (milk sickness) in the US in the 60yrs prior to the Civil War. The disease, which also killed livestock, occurred in epidemic proportions in Kentucky, Ohio, Indiana, Illinois, Michigan and parts of adjacent states. The plant has been studied to determine the cause of its sporadic toxicity. Recently (Beier *et al.*, 1987; Beier *et al.*, 1993) microsomal activation was proposed to account for its unpredictable toxicity. *Eupatorium rugosum* continues to cause animal losses (Olson *et al.*, 1984), and reports of actual animal poisonings following a field investigation contain line drawings showing plants with abundant inflorescence confined to the top of the plant with very small stipules at the bases of the remaining petioles (type A morphology; Kingsbury, 1964). More recently the plant has been depicted as bearing fertile branches with less abundant inflorescence at the base and having stipules much larger than any other leaflet on the plant (type B morphology; Herron and LaBore, 1972).

Tremetone had been proposed as the toxic principle of *E. rugosum* (Couch, 1929). Later work suggested that there were three toxic ketones present in the plant (Bonner and DeGraw, 1962), although, since all oxygen atoms present in the precocenes were present as ether groups, they would not give a positive chemical test for ketones.

Several *E. rugosum* plants with A and B morphologies growing side by side were gathered in Cincinnati, Ohio. The plants were taken to Dr John Thieret, a taxonomic botanist at Northern Kentucky University, and all were identified as *E. rugosum*. Analysis of plants with types A and B morphology were carried out by GC/MS (Chapter 46). Plants with type A morphology had more tremetone, while plants with

220

type B morphology contained very little tremetone, but more of the precocenes 1 and 2. This plant is difficult even for botanists to identify as to species (Thieret, 1996), and a chemotaxonomical approach may be key for identification. This investigation indicated that there were two varieties of *E. rugosum*: that with type A morphology, the authors proposed to call var. *toxicarium*, and type B morphology, var. *innocuosus*.

Discussion

Two plants that illustrate the power and limitations of chemotaxonomical data in botanic speciation are *Corydalis aurea* and *Conium maculatum*. Speciation of *C. aurea* in a pre-flowering stage has been done by chemotaxonomy alone (Smith and Lewis, 1990). Based on the literature, capaurine and sendaverine were found together only in this plant (Manske, 1970). At of the other extreme, in *C. maculatum* the sprectrum of alkaloids changes on an hourly basis (Leete and Olson, 1972; Roberts, 1975). The plant can be identified as to species by the presence of an array of minor alkaloids and one or more of the major alkaloids; γ-coniceine, coniine, or ψ-conhydrine.

The molecular structures of these three major alkaloids and those of tremetone and the precocenes are given in Fig. 45.1. Biochemical alteration in *C. maculatum*

Fig. 45.1. The molecular structures of the major alkaloids of *Conium maculatum*, and of tremetone and precocenes 1 and 2.

in either direction from coniine can occur. *Eupatorium rugosum* is different; although the precocenes could easily be biochemically converted to each other, interconversion of either with tremetone is unlikely.

Historically some confusion has existed over the proper scientific name for the plant. Many scientific names have been proposed (Moseley, 1941), implying that in this genus, botanists themselves are delineating subspecies. A better understanding of the exact mechanism of toxicity and the identification difficulties of species in the genus *Eupatorium* by conventional taxonomic botany is needed.

References

Beier, R.C., Norman, J.O., Irvin, R., Witzel, D.A. (1987) Microsomal activation of constituents of white snakeroot (*Eupatorium rugosum* Houtt) to form toxic products. *American Journal of Veterinary Research* 48, 583-585.

Beier, R.C., Norman, J.O., Reagor, J.C., Rees, M.S. and Mundy, B.P. (1993) Isolation of the major component in white snakeroot that is toxic after microsomal activation: Possible explanation of sporadic toxicity of white snakeroot plants and extracts. *Natural Toxins* 1, 286-293.

Bonner, W.A. and DeGraw, J.I. (1962) Ketones from "white snakeroot," *Eupatorium urticaefolium. Tetrahedron* 18, 1295-1309.

Couch, J.F. (1929) Tremetol, the toxic component that produced trembles. *Journal of the American Chemical Society* 51, 3617-3619.

Herron, J.W. and LaBore, D.E. (1972) *Some Plants of Kentucky Poisonous to Livestock.* University of Kentucky College of Agriculture Cooperative Extension Service, p. 46.

Kingsbury, J.M. (1964) *Poisonous Plants of the United States and Canada.* Prentice-Hall, Englewood Cliffs, NJ, p. 398.

Leete, E. and Olson, J.O. (1972) Biosynthesis and metabolism of the hemlock alkaloids. *Journal of the American Chemical Society* 94, 5472.

Manske, R.H.F. (1970) *The Alkaloids: Chemistry and Physiology, Vol XII.* Academic Press, New York, NY, p. 335.

Moseley, E.L. (1941) *Milk Sickness caused by White Snakeroot.* The Ohio Academy of Science, Bowling Green, OH, frontispiece.

Olson, C.T., Keller, W.C., Gerken, D.F. and Reed, S.M. (1984) Suspected tremetol poisoning in horses. *Journal of the American Veterinary Medical Association*, 185, 1001-1003.

Roberts, M.F. (1975) Gamma-coniceine reductase in *Conium maculatum. Phytochemistry* 14, 2395.

Smith, R.A. and Lewis, D.L. (1990) Apparent *Corydalis aurea* intoxication of cattle. *Veterinary and Human Toxicology* 32, 63-64.

Thieret, J. (1996) personal communication.

Chapter 46

The Chemical Identification of Plant Toxins in Ingesta and Forage

D.G. Lang and R.A. Smith

Livestock Disease Diagnostic Center, Veterinary Science Department, University of Kentucky, 1429 Newtown Pike, Lexington, Kentucky 40511, USA

Introduction

The correct identification of plants and the assessment of their potential toxicity can require considerable knowledge of botany. The staff of veterinary toxicology laboratories are frequently presented with masticated and partially digested plant fragments, which complicates classical botanical methods of identification, leaving chemotaxonomy as a very viable route of identification. This chapter presents case histories involving cattle deaths following ingestion of *Conium maculatum*, *Nicotiana tabacum*, *Zygadenus venenosus*, *Taxus baccata* and *Kalmia latifolia*.

The Detection of *Conium maculatum* in Rumen Content

Poison hemlock, *C. maculatum*, is toxic and teratogenic. The chemistry and biosynthesis of the conium alkaloids were recently reviewed (Panter and Keeler, 1989). The plant is highly dangerous to livestock, and the alkaloids are reported to vary hour by hour in the living plant (Kingsbury, 1964).

A 6yr-old bull was found down and dehydrated, with a temperature of 35.5°C. It died shortly, and at necropsy the rumen content had a strong odor of hemlock. A portion of the rumen content and a plant specimen were analyzed for conium alkaloids. Each sample was blended with distilled water, made basic with NaOH and filtered through copper tinsel. The filtrate was extracted with dichloromethane (DCM), which was dried with Na_2SO_4, then extracted with 1N H_2SO_4. The acidified aqueous layer was then made basic and extracted with fresh DCM. The pooled DCM was dried, acetylated by adding a few drops of acetic anhydride, evaporated under nitrogen (N_2) and reconstituted in methanol (MeOH). The extract was injected onto a Finnigan-MAT Incos 50 GC/MS with a 12m x 0.2mm x 0.33μm film Hewlett-Packard Ultra 2 (crosslinked 5% phenyl-methyl silicon) column.

The rumen content extract showed two large peaks early in the chromatogram, the first and largest of which was acetyl coniine, and the second, acetyl conhydrinone. Acetylation allowed these highly volatile compounds to be separated easily with better peak shape, and the added weight also kept the compounds from eluting with the solvent. In the plant, coniceine was the predominant alkaloid.

Nicotiana tabacum Poisoning in Cattle

Tobacco (*N. tabacum*) is a commonly cultivated plant in Kentucky, and livestock occasionally eat large amounts of it after breaking into a field. A year-old steer was found dead in a tobacco field, and the rumen contained large leaf fragments resembling tobacco. The rumen content was extracted and the extract acetylated as above. An array of alkaloids was detected, in order of elution: nicotine, methylanabasine, acetylnornicotine, acetylanatabine and acetylanabasine. These compounds were unaltered from the forms found in the tobacco leaves.

Herd Poisoning Involving *Zygadenus venenosus*

Death camas (*Z. venenosus*) is common in the western US, and is not consumed if other forage is present. A herd of nearly 350 cows and calves in northwest Nebraska suffered severe losses in late April, 1994 (Collett *et al.*, 1996). Seventeen cows were found dead and a number of cattle showed severe clinical signs: rapid heart rate, open mouth breathing, prostration, hyperexcitability, excessive salivation and muscle tremors. Alkaloids were detected with thin layer chromatography (Majak *et al.*, 1992).

A portion of the plant was extracted with DCM, and the dried extract reconstituted in MeOH before injection onto the GC/MS. The 112 ion, which was the main ion found in zygadenine, was found in the array of compounds detected in the chromatogram (Fig. 46.1). The chromatogram showed that there were many more toxins in this plant than had been formerly identified.

Detecting *Taxus* Poisoning Using GC/MS

Recently, several cases of *Taxus* spp. poisoning in cattle have occurred in Kentucky. In two of the cases, the poisoning was due to the dumping of hedge clippings over the fence. The cattle showed no prior signs of illness, but five head were found dead next to the clippings. Another case involved a 5mo-old calf that was found on the ground, kicking. The illness lasted about 5min before the calf died. There was also a case of *Taxus* spp. poisoning in a horse found dead in the field with no signs of struggle or discharge of bodily fluids (Lang *et al.*, 1997).

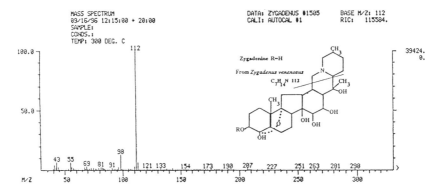

Fig. 46.1. The mass spectrum of *Zygadenus venenosus* extract.

Taxus poisoning in cattle is generally identified by examining rumen content at necropsy. Identification in non-ruminants is impossible since the plant is exposed to acid in the stomach.

Smith (1989) acetylated *Taxus* extracts and using a direct insertion probe to analyze for the major fragment of taxine (β-methylamino-β-phenyl-α-hydroxy-propionic acid). Lang *et al.* (1997) have demonstrated the presence of taxines in equine stomach content. The sample was extracted as above, with the exception that acetylation was not necessary. Taxine is too large a molecule to chromatograph intact on the GC/MS, but taxine fragments in the injection port to form a rapidly eluting moiety that produces a large sharp peak at 152°C. The fragmented product of taxine was distinguished by a large 134 ion, which is caused by the breaking off of the side chain between the carbon attached to the imido and phenyl groups and the hydroxylated carbon (Budavari *et al.*, 1996). The mass spectrum of taxine fragments listed by mass (% abundance) is: 134(100), 42(12), 91(10), 72(6), 118(5), 207(3), and 192(2).

Kalmia latifolia Poisoning

Mountain laurel (*K. latifolia*) is an evergreen bush that grows on the tops of high hills and along ridges in eastern Kentucky, and animals will eat it if other forage is scarce (Kingsbury, 1964). Two cows were found dead on a hillside near Berea, Kentucky. A DCM extraction was performed on the rumen contents of both cows and on a sample of the plant. The extracts were not acetylated.

The distinctive arrangement of chemical compounds in the chromatogram of mountain laurel was also seen in the rumen samples (Fig. 46.2). One of these is β-amyrin, which is seen in many plants (Budavari *et al.*, 1996). The other compounds have yet to be identified, but the arrangement of peaks is characteristic.

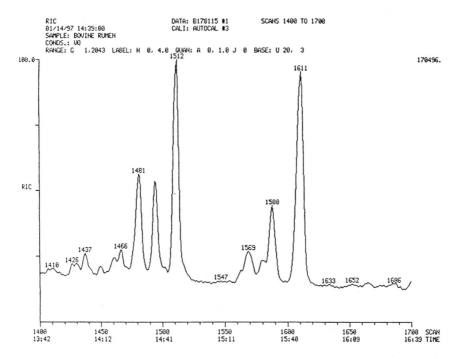

```
RIC                          DATA: B178115 #1        SCANS 1400 TO 1700
01/14/97 14:39:00            CALI: AUTOCAL #3
SAMPLE: BOVINE RUMEN
CONDS.: VO
RANGE: C   1,2043  LABEL: N  0, 4.0  QUAN: A  0, 1.0 J  0  BASE: U 20,  3
```

Fig. 46.2. Total ion chromatograph showing the characteristic array of peaks from *Kalmia latifolia* extracts.

References

Budavari, S., O'Neil, M.J., Smith, A., Heckelman, P.E. and Kinneary, J.F. (eds), (1996) *The Merck Index*, 12th ed. Merck and Company, Inc., Whitehouse Station, NJ.

Collett, S., Grotelueschen, D., Smith, R.A. and Wilson, R. (1996) Deaths of 23 adult cows attributed to intoxication by the alkaloids of *Zygadenus venenosus* (Meadow Death Camas). *Agri-Practice* 17, 5-9.

Kingsbury, J.M. (1964) *Poisonous Plants of the United States and Canada*. Prentice-Hall, Englewood Cliffs, NJ, pp. 251-260, 379-383.

Lang, D.G., Smith, R.A. and Miller, R.E. (1997) Detecting taxus poisoning using GC/MS. *Veterinary and Human Toxicology* 39, 314.

Majak, W., McDiarmid, R.E., Critofoli, W., Sun, F. and Benn, M. (1992) Content of *Zygadenus venenosus* at different stages of growth. *Phytochemistry* 31, 3417-3418.

Panter, K.E. and Keeler, R.F. (1989) Piperidine alkaloids of poison hemlock (*Conium maculatum*). In: Cheeke, P.R. (ed), *Toxicants of Plant Origin, Vol 1: Alkaloids*. CRC Press, Boca Raton, FL, pp. 109-132.

Smith, R.A. (1989) Comments on diagnosis of intoxication due to *Taxus. Veterinary and Human Toxicology* 31, 177.

Chapter 47

Conditioned Food Aversion: A Management Tool to Prevent Livestock Poisoning

M.H. Ralphs and J.D. Olsen
USDA-ARS, Poisonous Plant Research Laboratory, Logan, Utah 84341, USA

Introduction

Conditioned food aversion is a prominent field of research in the behavioral sciences (Braveman and Bronstein, 1985). It has been used to prevent predation on livestock and depredation of crops (Gustavson and Gustavson, 1985). Zahorik and Houpt (1977) first demonstrated that livestock could be averted to specific foods, and Laycock (1978) suggested aversions may have potential to prevent livestock from eating poisonous plants. Provenza and Balph (1988) proposed this type of diet training may enable managers to shape grazing behavior of animals to meet management goals, to condition livestock to avoid eating poisonous plants, and thus avoid poisoning.

Principles of Conditioning

A series of experiments was conducted to determine if cattle could be aversely conditioned to avoid eating tall larkspur (*Delphinium barbeyi* L. Huth), an important poisonous plant on mountain rangelands. Olsen and Ralphs (1986) confirmed that cattle have the neurophysiologic mechanisms to associate the taste of a common food (alfalfa pellets) to an illness induced by either lithium chloride (LiCl) or larkspur extract. Lane *et al.* (1990) fed heifers fresh larkspur in a pen, then infused LiCl through a rumen catheter (80mg/kg). The heifers associated the induced illness with the taste of larkspur and avoided eating larkspur when it was offered in the pen. When released in larkspur-infested pastures, the heifers abstained from eating larkspur for 2yrs. However, when the averted heifers were placed with non-averted cohorts that were readily grazing larkspur, they started sampling larkspur and the aversion was extinguished. Social facilitation is a strong force compelling animals to sample foods and plants they see others eating (Galef, 1986). Therefore, if animals sample plants without adverse consequences, they will continue to eat them and eventually extinguish the aversion.

Several experiments were conducted to try to overcome the influence of social facilitation and maintain the aversion under field grazing conditions when averted and non-averted cattle grazed together. First, heifers were dosed with LiCl to reinforce the aversion whenever they consumed larkspur in a group with non-averted cohorts that were readily eating larkspur (Ralphs and Olsen, 1990). Next, using native cattle that were familiar with the plant community was tried (Ralphs, unpublished data). Finally, larkspur alkaloid extract was used as an internal feed-back to reinforce aversions if cattle started to eat larkspur in the field (Ralphs and Olsen, 1992). None of these procedures was successful and it was concluded that averted cattle must be grazed separately to maintain the aversion.

Mature animals may retain aversions better than younger animals. Livestock learn to forage most efficiently at approximately the time of weaning (Provenza and Balph, 1988), but the inquisitive character of young animals in sampling feeds may be a liability in maintaining an aversion. Lambs extinguished aversions to pelleted feed more rapidly than did their mothers (Thorhallsdottir *et al.*, 1990). Mature cows required a lower dose of LiCl (200mg/kg) to maintain aversions to beet pulp, compared to yearling heifers (300mg/kg) (Ralphs and Cheney, 1993). However, the 300mg/kg dose creates severe illness for up to 5d. It was concluded that the 200mg/kg dose was optimum for cattle. The optimum LiCl dose for sheep was 150mg/kg (du Toit *et al.*, 1991).

Novelty of taste is important in forming strong aversions because the first exposure presents the orienting response to the new taste (Nachmann *et al.*, 1977; Burritt and Provenza, 1996). Cattle maintained stronger aversions to novel beet pulp than to familiar alfalfa pellets (Ralphs and Cheney, 1993). Steers that were readily grazing locoweed had difficulty forming aversions to it, compared to naive steers (Ralphs *et al.*, 1997). However, aversion to locoweed was eventually created by repeatedly allowing steers to graze locoweed in the pasture, then returning them to the pen and dosing with LiCl (100mg/kg). There was success in averting native cattle that have grazed on larkspur ranges all their lives (Ralphs, 1997).

There are some limitations using LiCl as an emetic. Lithium is retained at significant levels in the body for up to 96hrs (Johnson *et al.*, 1980; Ralphs, 1998). Treated cattle are most severely ill the second day after dosing, requiring a recovery period of at least 3d. Because of its caustic nature, the relatively large quantities of LiCl required to create aversions in livestock (80-200mg/kg) must be administered into the rumen either in solution by gavage or in boluses, allowing dilution by the rumen fluid. Lithium residue in milk was a potential concern, but there was not a sufficient amount to adversely affect nursing calves (Ralphs, 1998). Apomorphine, an intense but short-lived emetic used in large animals, was tested as an alternative to LiCl (given intramuscularly at 0.1 or 0.2mg/kg), but it did not create total aversions to flavored alfalfa pellets, and the partial aversions were extinguished rapidly (Ralphs and Stegelmeier, 1998).

Field Verification of Aversions

Three grazing studies were conducted on tall larkspur to implement the knowledge gained to create and maintain long-lasting aversions under field grazing conditions (Ralphs, 1997). The grazing studies were conducted 16km west of Yampa in western Colorado, on a Forest Service grazing allotment that had a history of serious losses to larkspur.

Naive cows, 1993-95

In the first study, 12 mature cows naive towards larkspur were used. Lithium chloride was given by gavage at a dose of 200mg/kg, and the averted cows were grazed separately from non-averted cohorts. The evening before aversion conditioning, regular feed was withheld and the cows were offered 9kg fresh larkspur. Some were reluctant to eat it on the first exposure. They were offered 15kg of larkspur the next morning and all of them readily ate it. Cows in the averted group (n=6) were then restrained in a chute and gavaged with LiCl *via* a stomach tube. Their regular ration of hay was withheld until evening so there was no interference between the induced illness and the taste of the novel larkspur. The averted cows were allowed to recover for 3d. The aversion was then tested by offering larkspur in group feedings with the non-averted control group (n=6), and reinforced by dosing with LiCl (100mg/kg) if any of the averted cows sampled it.

The two groups were placed in separate larkspur-infested pastures to avoid the influence of social facilitation. The groups switched pastures half-way through the trials to avoid pasture bias. Diets were estimated by bite counts, where each cow was observed for 5min periods and the number of bites of larkspur and other forages were counted. The cows were retained for 3yrs and the aversions were not reinforced in 1994 or 1995. Following conditioning with LiCl in 1993, the averted cows abstained from grazing larkspur for 3yrs while they grazed separately. The optimal conditions for maintaining aversions (mature animals, averted to a novel food, using a high dose of LiCl to induce intense gastrointestinal illness, and groups grazed separately to avoid social facilitation) combined to promote a long-lasting aversion to tall larkspur.

The control cows consumed larkspur from 11-20% of their bites. In 1993, three of the control cows died from larkspur poisoning. Symptoms of poisoning (muscular tremors and collapse) were observed in two other cows, and one cow showed symptoms of poisoning in 1994.

The averted and control groups were placed together at the end of the trial in 1995, to determine if the long-term aversion to larkspur would hold under the influence of social facilitation. The averted group did not consume larkspur for 7d after the groups were placed together, but then started to sample larkspur and gradually extinguished the aversion. In other larkspur-aversion grazing trials, the averted groups started to eat larkspur within 3d after being placed together (Lane *et al.*, 1990; Ralphs and Olsen, 1990, 1992). The optimal conditions described above apparently created a stronger aversion that lasted a few days longer than in previous trials, but the aversion was eventually extinguished.

Native cows, 1994-95

The objective of this trial was to determine if native cows that were familiar with larkspur could be averted and would abstain from eating it. Five cows, obtained from a local rancher in 1994, had grazed the allotment for 2-5yrs and been exposed to larkspur. They were successfully averted to larkspur as described in the first trial, then allowed to graze larkspur-infested pasture. Three of the five cows were available in 1995 to repeat the trial, and aversions were not reinforced in 1995.

The native cows generally abstained from eating larkspur while grazing separately in both 1994 and 1995. Control cows consumed larkspur in up to 10% of bites. The averted cows were also placed with the control cows of the previous study at the end of the 1995 study to evaluate the influence of social facilitation. These native cows extinguished the aversion and started eating larkspur after 2d, but mean consumption was less than half that of the controls (6 *vs.* 13% of bites).

Native cows, allotment scale trial, 1996

The objective of this trial was to test the aversion on native cattle that were grazing freely on a mountain allotment (2,000ha). A local rancher holding the grazing permit to the allotment provided the cows. Larkspur was in the early flower stage at the beginning of the trial, and the cattle had started to eat it before they were averted. Eleven cows were averted to larkspur as described above, then released onto the allotment and allowed to roam freely throughout the larkspur area. There were no other non-averted cows on the allotment. A rider located the cows daily and took informal bite counts to verify they were not eating larkspur.

The native cattle completely abstained from eating larkspur. They were herded to a limited degree to keep them in the larkspur areas so they would have an opportunity to select larkspur. Two of these cows were carry-overs from the previous aversion trial in which they extinguished the aversion at the end of the 1995 trial. However, they were successfully re-averted and abstained from eating larkspur in 1996.

Conclusion

Conditioned food aversion is a powerful experimental tool to modify animal diets (Galef, 1985). We have shown that it is a potential management tool to prevent livestock from grazing larkspur (Ralphs, 1997), locoweed (Ralphs *et al.*, 1997) and ponderosa pine (Pfister, unpublished data). The following principles will increase the strength and longevity of aversions: mature animals retain aversions better than young animals; novelty of the plant is important, although aversions have been created to familiar plants; LiCl is the most effective emetic and the optimum dose for cattle is 200mg/kg; and averted animals should be grazed separately to avoid the influence of social facilitation which will extinguish the aversion.

Social facilitation is the most important factor preventing widespread application of aversion conditioning. Averted animals, seeing others eating the target food, will sample it. If there is no adverse reaction, they will continue sampling and eventually extinguish the aversion. However, if averted animals can be grazed separately, aversion conditioning may provide an effective management tool to prevent animals from eating selected poisonous plants.

Aversion conditioning is an intensive management tool. It requires confining animals, forcing them to consume the target plant, dosing with an emetic, then testing the aversion. However, ranches that persistently lose animals to poisonous plants can afford to invest time and effort in management practices that prevent losses.

References

Braveman, N.S. and Bronstein, P. (eds) (1985) *Experimental Assessments and Clinical Applications of Conditioned Food Aversion.* Annals of the New York Academy of Science, p. 443.

Burritt, E.A and Provenza, F.D. (1996) Amount of experience and prior illness affect the acquisition and persistence of conditioned food aversions in lambs. *Applied Animal Behavior Science* 48, 73-80.

du Toit, J.T., Provenza, F.D. and Nastis, A. (1991) Conditioned taste aversions: how sick must a ruminant get before it learns about toxicity in foods? *Applied Animal Behavior Science* 30, 35-40.

Galef, B.G. Jr (1985) Socially induced diet preference can partially reverse a LiCl-induced diet aversion. *Animal Learning and Behavior* 13, 415-418.

Galef, B.G. Jr (1986) Social interaction modifies learned aversions, sodium appetite, and both palatability and handling-time induced dietary preference in rats. *Journal of Comparative Psychology* 100, 432-439.

Gustavson, C.R. and Gustavson, J.D. (1985) Predation control using conditioned food aversion methodology: theory, practice, and implications. In: Braveman, N.S. and Bronstein, P. (eds), *Experimental Assessments and Clinical Applications of Conditioned Food Aversion.* Annals New York Academy of Science 443, pp. 348-356.

Johnson, J.H., Crookshank, H.R. and Smolley, H.E. (1980) Lithium toxicity in cattle. *Veterinary and Human Toxicology* 22, 248-251.

Lane, M.A., Ralphs M.H., Olsen J.D., Provenza, F.D. and Pfister, J.A. (1990) Conditioned taste aversion: potential for reducing cattle loss to larkspur. *Journal of Range Management* 43, 127-131.

Laycock, W.A. (1978) Coevolution of poisonous plants and large herbivores on rangelands. *Journal of Range Management* 31, 335-342.

Nachmann, M., Rauschenberger, J. and Ashe, J.H. (1977) Stimulus characteristics in food aversion learning. In: Milgram, N.W., Krames, L. and Alloway, T.M. (eds), *Food Aversion Learning.* Plenum Press, New York, NY, pp. 105-131.

Olsen, J.D. and Ralphs, M.H. (1986) Feed aversion induced by intraruminal infusion with larkspur extract in cattle. *American Journal of Veterinary Research* 47, 1829-1833.

Provenza, F.D. and Balph, D.F. (1988) Development of dietary choice in livestock on rangelands and its implications for management. *Journal of Animal Science* 66, 2356-2368.

Ralphs, M.H. (1997) Persistence of aversions to larkspur in naive and native cattle. *Journal of Range Management* 50, 367-370.

Ralphs, M.H. (1998) Lithium residue in milk from doses used to create food aversions; effect on nursing calves. *Journal of Animal Science* (in review).

Ralphs, M.H. and Cheney, C.D. (1993) Influence of cattle age, lithium chloride dose level, and food type in the retention of food aversions. *Journal of Animal Science* 71, 373-379.

Ralphs, M.H. and Olsen, J.D. (1990) Adverse influence of social facilitation and learning context in training cattle to avoid eating larkspur. *Journal of Animal Science* 68, 1944-1952.

Ralphs, M.H. and Olsen, J.D. (1992) Comparison of larkspur alkaloid extract and lithium chloride in maintaining cattle aversion to larkspur in the field. *Journal of Animal Science* 70, 1116-1120.

Ralphs, M.H. and Stegelmeier, B.E. (1998) Ability of apomorphine and lithium chloride in creating food aversions in cattle. *Applied Animal Behavior Science* 56, 129-137.

Ralphs, M.H., Graham D., Galyean, M.L. and James, L.F. (1997) Creating aversions to locoweed in naive and familiar cattle. *Journal of Range Management* 50, 361-363.

Thorhallsdottir, A.G., Provenza, F.D. and Balph, D.F. (1990) Social influences on conditioned food aversions in sheep. *Applied Animal Behavior Science* 25, 45-50.

Zahorik, D.M. and Houpt, K.A. (1977) The concept of nutritional wisdom: applicability of laboratory learning models to large herbivores. In: Barker, L.M., Best, M.R. and Domjan, M. (eds), *Learning Mechanisms in Food Selection.* Baylor University Press, Waco, TX, pp. 45-67.

Chapter 48

Development and Validation of a Multi-residue Alkaloid Screen

D.M. Holstege, F.D. Galey and M.C. Booth
California Veterinary Diagnostic Laboratory System, University of California, Davis, California 95616, USA

Introduction

Alkaloids are naturally occurring plant toxicants containing a heterocyclic nitrogen; they are basic in character and are usually toxic. Sudden death or nonspecific illness in animals can be caused by ingestion of many alkaloid-containing plants (Kingsbury, 1964; Cheeke and Shull, 1985), and exposure to these plants must always be considered in diagnosing sudden death in animals. Plants of interest in the western US include *Conium maculatum* (poison hemlock), *Nicotiana glauca* (tree tobacco), *Delphinium barbeyi* (tall larkspur), *Datura wrightii* (also *Datura metaloides*, jimsonweed) and *Taxus baccata* (English yew).

A multiresidue analytical screen for determining alkaloid exposure in livestock has been developed as a tool for use in a veterinary diagnostic laboratory (Holstege *et al.*, 1995). Key features include its applicability to this chemically diverse class of compounds in a wide variety of sample matrices, including animal tissue, plant material, biological fluids and ingesta. Sample preparation avoids evaporative steps due to the volatility of some alkaloids. Extreme conditions that could hydrolyze esters are also avoided. The method is rapid, providing a result within 1d. Samples are sufficiently purified to provide diagnostically relevant detection limits qualitatively, with good quantitative precision and accuracy. Analytical standards for many alkaloids of interest are unavailable, so the multiresidue method (MRM) uses gas chromatographic (GC) and thin layer chromatographic (TLC) patterns, along with mass spectral (MS) information from plant samples to aid in the determination of alkaloid exposure. The method was validated using fortification and animal dosing trials.

233

Multiresidue Method for Alkaloids

The analysis followed that of Holstege *et al.* (1995). Chopped samples (5g) were extracted with 100ml 5% ethanol in ethyl acetate (EtOAc, v/v) after the addition of 1ml 10N NaOH and 50g Na_2SO_4. Hexane (100ml) was added to a 40ml aliquot, which was extracted with 10ml 0.5N HCl (x2), followed by 5ml 0.5N HCl. The combined aqueous extract was sparged with N_2 and the pH increased to greater than pH10 with 10N NaOH. The extract was pulled through a polymeric C-18 solid phase extraction (SPE) column (Interaction Chemicals Inc., Mountain View, California) under vacuum. The alkaloids were eluted with 2ml EtOAc, which was dried over 0.5g of Na_2SO_4 for 1-2min, and an aliquot taken for GC/nitrogen phosphorus detector (NPD) analyses. The extract was analyzed quantitatively using GC/NPD (Autosystem, Perkin Elmer) with a 5m x 0.53mm x 1.0μm DB-5 capillary column (J & W Scientific). The thermal gradient program was 60°C for 0.5min, 10°C/min to 120°C, 20°C/min to 280°C, hold for 5.5min, with a helium carrier flow rate of 12ml/min. A second aliquot (0.5ml) of EtOAc extract (1g/ml) was evaporated to dryness at 40°C and re-dissolved in 50μl EtOAc for semi-quantitative analysis by GC/MS (Model HP 5890 with HP 5970 MSD, Hewlett-Packard) using a 12m x 0.2mm x 0.33μm HP-1 capillary column (Hewlett-Packard). This temperature program was 50°C for 0.5 min, 10°C/min to 200°C, 30°C/min to 290°C, hold for 6.5 min, with a helium head pressure of 10psi. A 1ml aliquot of extract (1g/ml) was evaporated to dryness and re-dissolved in 25μl EtOAc. The extract was spotted on a blank Toxi-Disc® A, and qualitative analysis was performed by a modified commercial TLC system (Toxi-Lab® A, Toxi-Lab Division of Marion Laboratories, Laguna Hills, CA) developed with EtOAc/diethyl amine (95:5, v/v) and visualized with potassium iodide/iodine/bismuth subnitrate solution (Toxi-Dip® A-3 reagent, Dragendorff's reagent), then 5% aqueous sodium nitrite (w/v).

Method Validation

Alfalfa hay, bovine rumen content, urine, serum and liver were fortified with each of six alkaloids at one and 10ppm (Holstege *et al.*, 1995). The alkaloids were coniine, nicotine, atropine, retrorsine, solanidine and strychnine. Four replicate fortifications for each matrix type at each spike level were prepared.

All alkaloids were recovered at 73-113%, with the coefficients of variation (CVs) 1-14%. Retrorsine and solanidine gave the lowest recoveries and higher percent CVs in liver and serum, especially at the 1μg/g level, but both were well detected by both GC/MS and TLC at this low level. Library-searchable mass spectra were provided by GC/MS, with detection limits ranging from 0.25-1ppm. Detection limits using TLC were 0.25-1ppm in the matrices tested.

Nine female, adult, mixed breed goats were given sub-lethal doses of plant material in two stages (Holstege *et al.*, 1996). One goat was used as a control. After withholding food for 36hrs, plant material was diluted with water and pumped into the rumen of the each animal. The first dose was a low level to permit collection of serum samples at 15 and 30min, and at 1, 2, 4, 8, 12, 24 and 48hrs after dosing. The animals received a higher dose 1wk later. One goat each was dosed with *D. wrightii* (10g/kg), *N. glauca* (15g/kg), *D. barbeyi* (10g/kg) and *T. baccata* (10g/kg), and three goats were dosed with *C. maculatum* (20g/kg). After 3-7hrs, the animals were euthanized, necropsied and typical samples collected for analytical toxicology testing.

The total concentration of alkaloids found in each sample is presented in Table 48.1. Samples of *C. maculatum* contained primarily γ-coniceine, N-methylconiine, coniine and a urinary metabolite, 1'-oxo-γ-coniceine. The major alkaloids in the *D. wrightii* samples were scopolamine, noratropine and hyoscyamine, while samples from animals dosed with *N. glauca* contained mostly anabasine. Samples of *D. barbeyi* contained numerous alkaloids, especially deltaline, lycoctonine and dictyocarpine. The method hydrolyzed methyllycaconitine to lycoctonine. Four principle alkaloids were seen by TLC from *T. baccata* samples, but were not detected by GC/MS, and so were not quantitated. Sera contained no alkaloid at over 0.5ppm.

Analyses of rumen contents, abomasal contents and urine showed that a 1ppm method detection limit (MDL) was sufficient to detect ingestion of these alkaloid-containing plants in goats dosed once at sub-lethal amounts 3-7hrs prior to death. This MDL was readily obtained with the TLC and GC/MS screening methods. Kidney and liver offered some utility for testing, while serum was a poor sample.

Forty alkaloid-containing plants were analyzed using the MRM. Mass spectra from GC/MS analysis of the primary GC peaks were added to a spectral library. TLC plates were photographed to record patterns of spots. Pooled bovine rumen content was fortified at 0.6%, 0.25% and 1% with the six alkaloid-containing plants *C. maculatum, D. wrightii, N. glauca, L. sericeus, D. barbeyi* and *T. baccata* to validate the ability of the MRM to identify alkaloid presence by automatic GC/MS spectral search and TLC pattern (Table 48.2). The MRM was able to detect automatically 0.06% plant material in rumen contents by GC/MS. Figure 48.1 shows a typical

Table 48.1. Summary of total alkaloids found (in ppm) of plant material and rumen contents, abomasal contents, kidney, liver and urine samples from goats dosed with alkaloid-containing plants.

Plant species	Plant*	Rum.*	Abo.*	Kid.*	Liver*	Urine*
Conium maculatum	1,600	80	35	7.9	2.5	101
Datura wrightii	630	41	6.0	0.1	0.2	40
Nicotiana glauca	440	20	10	2.0	4.3	8.3
Delphinium barbeyi	20,000	780	410	1.7	2.1	8.6
*Taxus baccata***	22,670	++	+	(+)	-	-

* Average of three replicates ** TLC results only ++ Very positive + Positive (+) Slightly positive - Negative

analysis of rumen contents fortified with *D. wrightii*. Peaks for scopolamine, atropine and noratropine provided automatic library search matches greater than 90%. The sensitivity for *C. maculatum* was lower than for the other plants because the primary alkaloids have low masses with low abundance molecular ions, reducing the capabilities of the library search program. A severe disadvantage to the GC/MS method was the inability to detect *Taxus* alkaloids. Analysis by TLC detected alkaloids at 0.06% plant material, but patterns of spots were not readily interpreted at this level. Samples could be screened for the presence of alkaloids by TLC, but could not definitively identify the plants found.

GC/MS Analysis of *Taxus* Alkaloids

The method was modified to allow qualitative analysis for *Taxus* alkaloids. An aliquot of the final extract (0.5g/ml) in EtOAc was evaporated to dryness and re-dissolved in 35µl acetonitrile and 15µl bis(trimethylsilyl)trifluoracetamide (BSTFA Pierce Chemical Co., Rockford, IL). This was injected onto the GC/MS using the MRM parameters. Extract of *T. baccata* gave over ten characteristic late-eluting peaks, probably trimethylsilyl derivatives of *Taxus* alkaloid thermal breakdown products. Fortifications of rumen contents with 1% *T. baccata* gave many of the same peaks, while negative control rumen contents gave no peaks in this region, indicating that these peaks were diagnostic for exposure to *T. baccata*. Sensitivity was enhanced by generating a chromatogram of the primary ions of the characteristic peaks. Figure 48.2 shows a extracted ion chromatogram of a *Taxus* extract, using the sum of ions 472, 508, 418, 244, 420, 492, 536, 361, 400 and 458 m/z. The method found *Taxus* alkaloids in bovine urine by matching several peaks to those of *Taxus* extract (Fig. 48.3), and negative control urine had no matching peaks.

Table 48.2. Summary of analyses of bovine rumen content fortified with plant material.

Plant species	Fortification level					
	GC/MS analysis			TLC analysis		
	1%	0.25%	0.06%	1%	0.25%	0.06%
Conium maculatum	++	++	+	++	++	-
Nicotiana glauca	++	++	++	++	++	+
Lupinus sericeus	++	++	++	++	++	+
Delphinium barbeyi	++	++	++	++	++	++
Datura wrightii	++	++	++	++	++	++
Taxus baccata	-	-	-	++	++	+

++ Plant identified   + Sample positive, plant not identified   - Negative

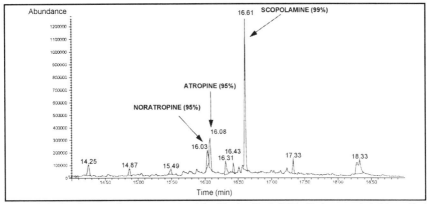

Fig. 48.1. Analyses of rumen content fortified with 0.25% *Datura meteloides*, (2µl of 10g/ml extract) by GC/MS. Numbers in parentheses are the automatic library search match quality.

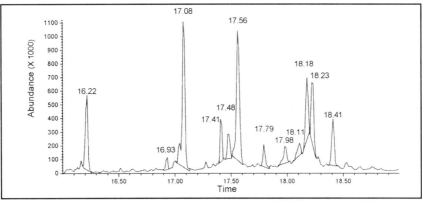

Fig. 48.2. Summed extracted ion chromatogram of GC/MS analysis of *Taxus baccata* extract derivatized with BSTFA (2µl of a 10g/ml extract).

Conclusions

The alkaloid MRM gave good precision and accuracy for biological samples fortified with six model alkaloids. The method performed well with alkaloids of high volatility (coniine), high polarity (retrorsine, atropine), poor chromatographic behavior (strychnine, solanidine) and low polarity (nicotine). Exposure to alkaloid-containing plants could be determined in biological samples typically submitted to a veterinary diagnostic laboratory. Sucessful automatic spectral searches were made by GC/MS at less than 0.1% plant material in rumen contents. With added analysis of BSFTA derivatized sample, determination of exposure to *Taxus* alkaloids was made.

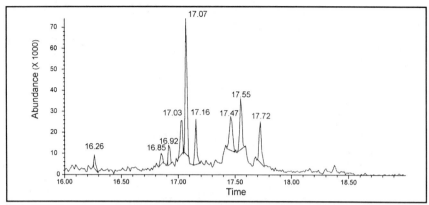

Fig. 48.3. Summed extracted ion chromatogram of GC/MS analysis of urine sample positive for *Taxus baccata* alkaloids, derivatized with BSTFA (2μl of a 10g/ml extract).

Acknowledgments

James Seiber, University of Nevada, Reno and Bill Johnson, Frank Brown, Jeanine Swensson, John Tahara and Gene Whitehead, California Veterinary Diagnostic Laboratory System, Davis, California. Analytical standards were provided by Gary Manners, USDA Albany, California, Kip Panter, USDA Logan, Utah, and Hank Segall, University of California, Davis, California. These studies were supported in part by the Livestock Disease Research Laboratory, University of California, Davis, California.

References

Cheeke, P.R. and Shull, L.R. (1985) *Natural Toxicants in Feeds and Poisonous Plants.* AVI Publishing Company, Westport, CT, pp. 1-8, 92-172.

Holstege, D.M., Seiber, J.N. and Galey, F.D. (1995) Rapid multiresidue screen for alkaloids in plant material and biological samples. *Journal of Agricultural and Food Chemistry* 43, 691-699.

Holstege, D.M., Galey, F.D. and Seiber, J.N. (1996) Determination of alkaloid exposure in a model ruminant (goat) using a multiresidue screening method. *Journal of Agricultural and Food Chemistry* 44, 2310-2315.

Kingsbury, J.M. (1964) *Poisonous Plants of the United States and Canada.* Prentice-Hall, Englewood Cliffs, NJ, pp. 264-267.

Chapter 49

Tolerance in Cattle to Timber Milkvetch (*Astragalus miser* var. *serotinus*) due to Changes in Rumen Microbial Populations

W. Majak, C. Hunter and L. Stroesser

Agriculture and Agri-Food Canada, Range Station, 3015 Ord Road, Kamloops, British Columbia V2B 8A9, Canada

Introduction

The Interior Douglas Fir (*Pseudotsuga menziesii*) zone is the most important forest zone for cattle grazing in southern British Columbia (Wikeem *et al.*, 1993). Timber milkvetch (TMV), also known as Columbia milkvetch (*Astragalus miser* var. *serotinus*), is a legume found mainly in this zone and in the associated rough fescue (*Festuca scabrella*) upper grassland communities. The crude protein content of TMV is relatively high (Majak *et al.*, 1996), but TMV also synthesizes large quantities of miserotoxin (Quinton *et al.*, 1989), a glycoside of 3-nitropropanol (NPOH) that causes acute and chronic poisoning in ruminants. The glycoside is rapidly hydrolyzed by rumen bacteria to the aglycone, NPOH, which is then absorbed and oxidized to 3-nitropropionic acid, a potent inhibitor of mitochondrial enzymes essential to respiration (Majak and Pass, 1989). However, rumen bacteria also have the capacity to degrade and detoxify NPOH, and the rate of degradation can be enhanced by increasing the amount of protein in the diet (Majak, 1992; Anderson *et al.*, 1996). Nitroethane (NE), a relatively innocuous analogue of NPOH, can be used as a feed additive at low concentrations to expose and adapt rumen bacteria to the aliphatic nitro group, which is rare in nature (Majak, 1992).

The purpose of this study was to determine microbial rates of NPOH detoxification in four rumen-fistulated Jersey steers grazing a grassland site during 1990-91 and a forest site during 1992-93. Two of the steers were given, free-choice, molasses blocks containing 25% crude protein and the other two were given blocks containing protein plus 0.5% NE. Each pair of steers was accompanied by ten Hereford cows (3-11yrs) with calves to test the efficacy of the treatments in the prevention of TMV poisoning. The grazing system, the description of the pastures and the TMV densities, which were similar at the two sites, were earlier reported (Majak *et al.*, 1996) as were the details of the anaerobic *in vitro* procedures and the spectrophotometric methods for determining rates of NPOH disappearance in rumen fluid (Majak, 1992). The cattle were exposed to the molasses blocks in the field for at least 2wks before *in vitro* assays were conducted. In field conditions, samples of rumen fluid were incubated with NPOH (2mM), then subsamples were pretreated with the protein precipitant and cooled prior to transportation and centrifugation. Rates of NPOH disappearance in rumen fluid were determined 4x at weekly intervals for each treatment in each year. It was confirmed by TLC and HPLC that NPOH disappearance resulted in the appearance of aminopropanol, the detoxified form of NPOH (Anderson *et al.*, 1993).

Field Trials on Grassland Range in 1990-91

Forty cow/calf pairs grazed the upper grassland range during 1990-91. There was no evidence of TMV poisoning, nor were there signs of poisoning among the rumen-fistulated steers. The cattle avoided TMV on the grassland range, selecting more palatable forages first and consuming increasing amounts of TMV as other forage species became depleted (Majak *et al.*, 1996). All of the animals had previously been exposed to TMV and all had access to protein supplements in the form of molasses blocks with an average consumption rate of 0.82 kg/hd/d.

In 1990, the average rate of NPOH disappearance in the NE group was almost twice the rate for the control group, but in 1991 differences between treatments were not detected (Table 49.1). A third pair of steers, without protein supplements in 1991, had much lower rates of NPOH detoxification ($195\pm27\mu M/L/hr$, n=3). A linear increase in the rate of NPOH detoxification was observed during the 5wks of the 1991 grazing trial (Fig. 49.1). These results suggested that the steers that had also been used in 1990 were adapting to their TMV diet without ill effects.

Field Trials on Forest Range in 1992-93

Twenty cow/calf pairs grazed the forest range during 1992, but again there were no clinical signs of TMV poisoning. In contrast to the grassland site, TMV was a preferred forage at the forest site probably because it was associated with pinegrass (*Calamagrostis rubescens*), which is less palatable than TMV because of its high silica and low protein content (Majak *et al.*, 1996). In 1992, the average rate of NPOH disappearance was higher in the NE group than in the controls, but significant differences between treatments were not detected in 1993 (Table 49.1). The absence of treatment effects in 1991 and 1993 might be attributed to the adaptation of rumen bacteria to TMV itself, without the requirement for synthetic NE. Earlier it was shown that TMV supplements can enhance NPOH detoxification (Majak, 1992). A

Table 49.1. Mean rates of nitropropanol detoxification by ruminal bacteria (µM/L/hr) in cattle grazing TMV and treated with supplementary nitroethane and/or protein.

Site	Year	Supplement		SE*	P
		Protein	Protein + NE		
Grassland	1990	507	889	41	<0.01
	1991	501	616	40	n.s.
	1992	382	568	24	<0.01
Forest	1993	899	865	82	n.s.

*SE=standard error of the mean.

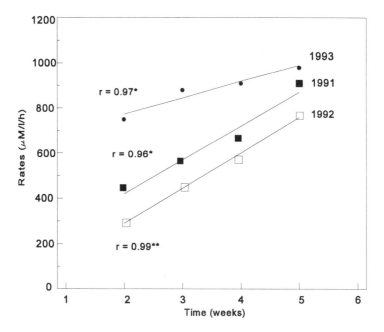

Fig. 49.1. Relationship between rates of NPOH detoxification and grazing time during 1991-93. Rates have been averaged over treatments.

significant linear increase in the rate of NPOH detoxification was evident during the grazing periods of 1992 and 1993, again indicating an adaptation to TMV and a resultant protection from poisoning (Fig. 49.1).

Two-year-old first-calf heifers were used for the first time in TMV grazing trials in 1993. Also, a third group of 10 cow/calf pairs concurrently grazed the TMV pastures but without the provision of free-choice protein supplements. After 3wks of grazing, one of five heifers in this third group showed early signs of poisoning, and then acute signs (Maricle *et al.*, 1996) 2d later. The animal was taken off test, as were two other heifers in the same group that developed signs. The older cows in all the groups and the heifers in the groups receiving protein supplements completed the grazing trial without showing significant signs of TMV poisoning. Thus the age of the cow may be important in determining susceptibility to TMV poisoning. First-calf heifers were most susceptible, especially in the absence of adequate protein supplements. The poisoned heifers had unknown history of TMV exposure at purchase. Exposure to TMV at a young age may be a factor in determining resistance to TMV and this possibility is being explored.

Acknowledgments

The authors would like to thank Ruth McDiarmid and Monica Durigon for assistance with *in vitro* assays in field conditions; Greg Syme for livestock care; the BC Ministry of Forests, 100 Mile District for their support; and Uniblok Canada, Rockyford AB for supplying molasses blocks. The study was partially funded by the Agri-Food Regional Development Subsidiary Agreement (project #11010) and by the BC Ministry of Agriculture, Fisheries and Food (DATE project #337).

References

Anderson, R.C., Rasmussen, M.A. and Allison, J.M. (1993) Metabolism of the plant toxins nitropropionic acid and nitropropanol by ruminal microorganisms. *Applied and Environmental Microbiology* 59, 3056-3061.

Anderson, R.C., Rasmussen, M.A. and Allison, M.J. (1996) Enrichment and isolation of a nitropropanol-metabolizing bacterium from the rumen. *Applied and Environmental Microbiology* 62, 3885-3886.

Majak, W. (1992) Further enhancement of 3-nitropropanol detoxification by ruminal bacteria in cattle. *Canadian Journal of Animal Science* 72, 863-870.

Majak, W. and Pass, M.A. (1989) Aliphatic nitrocompounds. In: Cheeke, P.R. (ed), *Toxicants of Plant Origin*. CRC Press, Boca Raton, FL, pp. 143-159.

Majak, W., Stroesser, L., Hall, J.W., Quinton, D.A. and Douwes, H.E. (1996) Seasonal grazing of Columbia milkvetch by cattle on rangelands in British Columbia. *Journal of Range Management* 49, 223-227.

Maricle, B., Tobey, J., Majak, W. and Hall, J.W. (1996) Evaluation of clinicopathological parameters in cattle grazing timber milkvetch. *Canadian Veterinary Journal* 37, 153-156.

Quinton, D.A., Majak, W. and Hall, J.W. (1989) The effect of cattle grazing on the growth and miserotoxin content of Columbia milkvetch. *Journal of Range Management* 42, 368-371.

Wikeem, B.M., McLean, A., Bawtree, A. and Quinton, D. (1993) An overview of the forage resource and beef production on Crown land in British Columbia. *Canadian Journal of Animal Science* 73, 779-794.

Chapter 50

Fourier Transform Infrared Spectroscopy for Rapid Analysis of Toxic Compounds in Larkspurs and Milkvetches

T.K. Schoch, D.R. Gardner, G.D. Manners, J.A. Pfister and M.H. Ralphs

USDA-ARS, Poisonous Plant Research Laboratory, 1150 East 1400 North, Logan, Utah 84321, USA

Introduction

Detection of the toxic compounds in larkspurs (*Delphinium* spp.) and milkvetches (*Astragalus* spp.) has been accomplished by many different methods. Larkspurs contain several norditerpenoid alkaloids, including toxic *N*-(methylsuccinimido)-anthranoyllycoctonine (MSAL) alkaloids (Fig. 50.1). These toxins have been measured by titrimetric (Williams and Cronin, 1963), gravimetric (Pelletier *et al.*, 1981; Pfister *et al.*, 1988), fluorometric (Luo *et al.*, 1983), near-infrared (Clark *et al.*, 1987), HPLC (Manners and Pfister, 1993), GC (Manners and Ralphs, 1989), and Fourier transform infrared spectroscopic (FT-IR; Gardner *et al.*, 1997) methods.

In addition to the MSAL alkaloids, some species of *Astragalus* (milkvetch) contain glycosides of 3-nitropropanol (3-NPOH) and 3-nitropropionic acid (3-NPA) (Fig. 50.2). Methods for these nitrotoxins include gravimetric (Stermitz *et al.*, 1972), colorometric (Majak and Bose, 1974), and HPLC (Muir and Majak, 1984) procedures. These methods, although effective and accurate, suffer from a number of flaws. Fourier transform infrared spectroscopy provides a fast, sensitive, and simple method for quantitative analysis of specific compounds in complex mixtures. An FT-IR technique for quantifying alkaloids in tall larkspur species (*Delphinium barbeyi*, *D. glaucescens*, *D. glaucum* and *D. occidentale*) and nitrotoxins in milkvetches (*Astragalus miser*, *A. cibarius*, *A. canadensis*, *A. pterocarpus* and others) is reported in this chapter.

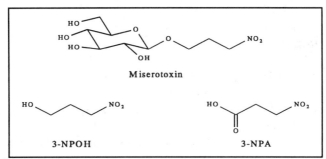

Fig. 50.1. *N*-(methylsuccinimido)anthranoyllycoctonine (MSAL) alkaloids from tall larkspur species.

Fig. 50.2. Structure of aglycones from *Astragalus* toxins, and miserotoxin, the glycoside isolated from *A. miser*.

Experimental

Plant material

Larkspur samples were collected at several range study sites throughout the western US (Ralphs *et al.*, 1998). Plants were dried and ground to pass a 1mm screen. Twenty samples of *D. barbeyi* covering the typical range of alkaloid levels found in the plant populations were selected to provide standards for the calibration of the FT-IR method, levels of MSAL alkaloids in these standards were measured in triplicate by the HPLC method of Manners and Pfister (1993), and total alkaloid levels were measured by gravimetric analysis after crude base isolation.

Astragalus samples were collected from various sites in the western hemisphere.

Astragalus miser var. *oblongifolius* and *A. cibarius* were collected in northern Utah. *Astragalus bakaliensis* and *A. pterocarpus* were collected by M.C. Williams in *c.*1975 from unknown locations. *Astragalus miser* var. *serotinus*, *A. canadensis*, and *A. collinus* were obtained from W. Majak. A sample of GS-571 was collected in South America and obtained from Edith Gomez-Sosa through R.J. Molyneux. For the standard curves, a species of *Astragalus* without nitrotoxins *(A. tenellus)* was spiked with varying amounts of 3-NPA or acetylated miserotoxin. The plant material was dried, then ground or cut finely prior to analysis. Small samples (<50mg) were allowed to air dry and were then ground.

Extraction procedures

A detailed description of the extraction procedure for larkspur has already been published (Gardner *et al.*, 1997). The method involves acid/base partitioning followed by evaporation of the chloroform alkaloid-containing fraction to dryness.

For the *Astragalus* species, 100mg of plant material was weighed into 7ml vials. A solution of potassium thiocyanate (KSCN, internal standard) in acetone (400mg/L) was prepared, and 2ml of this solution was added to each vial. The samples were extracted by gentle mechanical shaking for 1hr for qualitative, or overnight for quantitative, analysis. Each vial was centrifuged and the supernatent liquid was used for analysis.

FT-IR spectroscopic analysis

Infrared spectra were collected using a Nicolet Magna 550 FT-IR spectrometer and Nicolet Omnic software packages (Nicolet Instrument Corp., Madison, WI). Larkspur alkaloid spectra were recorded by dissolving the alkaloid fraction in 2.0ml of carbon disulfide and transferring an aliquot to a 5mm path length liquid cell. The spectra were recorded after eight scans at 4cm^{-1} resolution. The 20 standard samples of *D. barbeyi* were analyzed using FT-IR, and calibration curves for MSAL alkaloids and total alkaloids were generated. The method was validated using a matrix of standard alkaloid solutions. A detailed description of the method has been published (Gardner *et al.*, 1997).

For nitrotoxin analysis, a diffuse reflectance (DRIFTS) micro-sample cup was filled with KBr, and 5µl of the acetone extract was added and allowed to dry. The sample cup was placed in a DRIFTS apparatus (Spectra-Tech, Stamford, CT) and the spectra recorded after 16 scans at 4cm^{-1} resolution. Calibration curves for both 3-NPOH and 3-NPA were generated. For 3-NPOH, samples of *A. tenellus* were spiked with varying amounts of acetylated miserotoxin (Baer *et al.*, 1973) and analyzed as above. Potassium thiocyanate was added as an internal standard, since the DRIFTS method results in variable cell path lengths and a direct use of Beer's law is difficult using diffuse reflectance. For the 3-NPA analysis, a calibration curve was generated by spiking samples of *A. tenellus* with pure 3-NPA (Aldrich Chemical Company, St Louis, MO) in a similar manner.

Results and Discussion

FT-IR analysis of larkspur alkaloids

Many of the alkaloids contained in tall larkspur have been isolated and tested for
toxicity (Olsen *et al.*, 1990; Manners and Pfister, 1993; Manners *et al.*, 1995). The
MSAL alkaloids have been found to be more toxic than the other types by one to two
orders of magnitude, and therefore the quantitation of MSAL alkaloids has been
crucial in estimations of larkspur toxicity. After examining the FT-IR spectra of
isolated alkaloids and plant extracts, two spectral regions were selected for measuring
the levels of MSAL and total alkaloids. The three carbonyl groups of the MSAL
alkaloids exhibit a C=O stretch at 1723cm^{-1}, while all the norditerpenoid alkaloids
exhibit a strong C-O absorbance at 1091cm^{-1}.

Using these spectral regions, the FT-IR results for the *D. barbeyi* standard
samples were calibrated against HPLC results for MSAL alkaloids and gravimetric
results for total alkaloids. The average relative errors of the predicted values versus
the known HPLC and gravimetric values were 7.15% for MSAL and 0.8% for total
alkaloids. The estimated detection limits are 0.5mg/g and 1.0mg/g for MSAL and
total alkaloids, respectively.

Although the FT-IR method does not provide concentrations of individual
alkaloids, it is a fast and accurate method for determining alkaloid levels in tall
larkspur. The rapid analysis time has enabled studies on palatability, environmental
factors, and seasonal changes to proceed more quickly than previously possible. At
present, the method is in use only for the tall larkspurs, but work is in progress to
extend its capabilities to other larkspur species.

FT-IR analysis of nitrotoxins

Nitrotoxin levels obtained by FT-IR analysis of eight *Astragalus* species are reported
in Table 50.1. Nitrotoxins found in these species fall into two types: glycosides of 3-
NPOH, such as miserotoxin, and glycosides of 3-NPA. Derivatives of 3-NPOH tend
to be more toxic due to the higher toxicity of 3-NPOH as compared to 3-NPA
(Stermitz *et al.*, 1972). The characteristic spectral feature of the nitro group is an
absorbance at 1550-1560cm^{-1}, and the two nitrotoxins in *Astragalus* can be
differentiated by their slightly different absorbances. Species that contain 3-NPOH
exhibit a peak at 1552 cm^{-1}, while 3-NPA derivatives appear at 1557cm^{-1} with a
strong C=O absorbance at 1737 cm^{-1} (Fig. 50.3). When screening for nitrotoxins, a
quick extraction with ethanol or acetone and subsequent FT-IR analysis can
determine if 3-NPOH or 3-NPA is present. A longer extraction time is necessary for
quantitative measurements, as was use of the KSCN internal standard.

These results have not yet been validated by other analytical methods. In spike-
recovery experiments, the 3-NPA method performed well, accurately predicting the
final 3-NPA concentration in a sample of *A. cibarius* spiked with a known amount of
pure 3-NPA. The same experiment using the 3-NPOH method and *A. miser* was

Table 50.1. Nitrotoxin concentrations in various *Astragalus* species.

Species	3-NPOH (mg NO₂/g)	3-NPA (mg NO₂/g)
A. miser var. *oblongifolius*	8.6	-
A. miser var. *serotinus*	1.6	-
A. pterocarpus	4.7	-
A. bakaliensis	-	2.2
A. canadensis	-	8.6
A. cibarius	-	5.8
A. collinus	-	2.1
GS-571	-	4.7

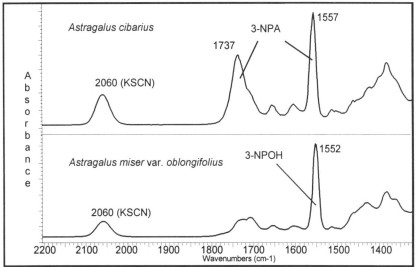

Fig. 50.3. Acetone plant extract diffuse reflectance Fourier transform infared spectra (potassium thiocyanate internal standard).

unsuccessful. This could be due to use of impure material in the spike or decomposition of 3-NPOH during the extraction procedure. In future work, both methods will be validated by NMR or other techniques, and the amount of plant material extracted will be decreased. Preliminary results indicate that qualitative results can be obtained on samples as small as 10mg.

References

Baer, H.H., Chiu, S.L. and Shields, D.C. (1973) The synthesis of miserotoxin and structural analogs. *Canadian Journal of Chemistry* 51, 2828-2835.

Clark, D.H., Ralphs, M.H. and Lamb, R.C. (1987) Total alkaloid determination in larkspur and lupine with near infrared reflectance spectroscopy. *Agronomy Journal* 79, 481-485.

Gardner, D.R., Manners, G.D., Ralphs, M.H. and Pfister, J.A. (1997) Quantitative analysis of norditerpenoid alkaloids in larkspur (*Delphinium* spp.) by Fourier transform infrared spectroscopy. *Phytochemical Analysis* 8, 55-62.

Luo, S., Wu, S. and Tang, J. (1983) A simple fluorochromatographic method for the identification of total alkaloids in *Delphinium*. *Zhongcaoyao* 14, 399-400.

Majak, W. and Bose, R.J. (1974) Chromatographic methods for the isolation of miserotoxin and detection of aliphatic nitro compounds. *Phytochemistry* 13, 1005-1010.

Manners, G.D. and Pfister, J.A. (1993) Normal phase liquid chromatographic analysis of toxic norditerpenoid alkaloids. *Phytochemical Analysis* 4, 14-18.

Manners, G.D. and Ralphs, M.H. (1989) Capillary gas chromatography of *Delphinium* diterpenoid alkaloids. *Journal of Chromatography* 466, 427-432.

Manners, G.D., Panter, K.E. and Pelletier, S.W. (1995) Structure-activity relationships of norditerpenoid alkaloids occurring in toxic larkspur (*Delphinium* species). *Journal of Natural Products* 58, 863-869.

Muir, A.D. and Majak, W. (1984) Quantitative determination of 3-nitropropionic acid and 3-nitropropanol in plasma by HPLC. *Toxicology Letters* 20, 133-136.

Olsen, J.D., Manners, G.D. and Pelletier, S.W. (1990) Poisonous properties of larkspurs (*Delphinium* spp.). *Collectanea Botanica* (Barcelona) 19, 141-151.

Pelletier, S.W., Daily, O.D. Jr, Mody, N.V. and Olsen, J.D. (1981) Isolation and structure elucidation of the alkaloids of *Delphinium glaucescens*. *Journal of Organic Chemistry* 46, 3284-3293.

Pfister, J.A., Ralphs, M.H. and Manners, G.D. (1988) Cattle grazing tall larkspur on Utah mountain rangeland. *Journal of Range Management* 41, 118-122.

Ralphs, M.H., Manners, G.D. and Gardner, D.G. (1998) Influence of light and photosynthesis on alkaloid concentration in larkspur. *Journal of Chemical Ecology* 24, 167-182.

Stermitz, F.R., Lowry, W.T., Norris, F.A., Buckeridge, F.A. and Williams, M.C. (1972) Aliphatic nitro compounds from *Astragalus* species. *Phytochemistry* 11, 1117-1124.

Williams, M.C. and Cronin, E.H. (1963) Effect of silvex and 2,4,5-T on alkaloid content of tall larkspur. *Weeds* 11, 317-319.

Chapter 51

DNA Alkylation Properties of Dehydromonocrotaline: Sequence Selective N7 Guanine Alkylation and Heat and Alkali-resistant Multiple Crosslinks

T.N. Pereira[1, 2], A.A. Seawright[1], P.E.B. Reilly[2] and A.S. Prakash[1]

[1]*National Research Centre for Environmental Toxicology, 39 Kessels Road, Coopers Plains, Queensland 4108, Australia;* [2]*Department of Biochemistry, The University of Queensland, St Lucia, Queensland 4072, Australia*

Introduction

Pyrrolizidine alkaloids (PAs) are a group of plant toxins that have been responsible for several episodes of fatal human poisoning (Bras *et al.*, 1954; Mohabbat *et al.*, 1976) and large-scale losses of livestock (Hooper and Scanlan, 1977; Johnson and Molyneux, 1984; Gaul *et al.*, 1994). Hepatic venoocclusive disease (VOD), pulmonary hypertension and right ventricular hypertrophy are the main features of the toxic sequelae. In addition, studies have demonstrated the carcinogenicity of these compounds upon chronic administration to rodents (Hirono *et al.*, 1983; Chan *et al.*, 1994).

The liver forms pyrrolic metabolites (Fig. 51.1), which are strong electrophiles that can bind to cellular nucleophiles such as proteins and DNA. Robertson (1982) has shown that dehydroretronecine (DR) (Fig. 51.1), a bifunctional metabolite of the PA monocrotaline, binds to nucleophiles *via* the C7 and C9 positions, which does not involve the pyrrole ring. Bifunctional pyrroles can alkylate DNA and then form interstrand DNA/DNA crosslinks and DNA/protein crosslinks (Petry *et al.*, 1984; Hincks *et al.*, 1991). A correlation exists between the cytotoxicity and crosslinking abilities of these compounds. Crosslinking is not crucial for toxicity, as shown by the cytotoxic effect of the synthetic monofunctional pyrrole 3-hydroxymethyl-1,2-dimethylpyrrole in cell culture (Mattocks *et al.*, 1980).

In this work the DNA alkylation properties of activated monocrotaline were examined using gel electrophoresis and electron microscopy. The results show that dehydromonocrotaline (DHM) reacts sequence selectively with DNA in the major groove. Multiple crosslinks were also observed with electron microscopy.

Materials and Methods

Monocrotaline was converted to DHM by the method of Mattocks *et al.* (1989). The purity of the resulting pyrrole was determined by NMR. It was stored at -80°C and, since it is highly unstable, was made in small quantities as required.

Alkylation of DNA was carried out by the method of Prakash *et al.* (1990). Briefly, a 375-base-pair *Bam*H1/*Eco*R1 fragment of pBR322 was 3'-end-labeled at the *Eco*R1 site by [$\propto^{32}$-P] dATP with Klenow and the fragment was isolated on a 4% nondenaturing polyacrylamide gel.

Labeled DNA (at 10,000cpm) was preincubated with DHM at 37°C in 10mM tris-EDTA buffer (pH 7.6) for various times (15, 30, 60, 120min). The reaction was stopped by precipitating in ethanol. The modified DNA was treated with 1M piperidine at 90°C for 10min, causing cleavage of DNA where alkylation has occurred at the N-7 of guanine residues (Prakash *et al.*, 1990). Solvent was removed by lyophilization before samples were dissolved in a formamide dye. The samples and sequencing lanes corresponding to guanine and purine were electro-phoresed on a 6% denaturing polyacrylamide gel at 50°C until the xylene cyanol dye front had migrated 25cm. The gel was dried and applied to X-ray film, which was developed.

Plasmid (pBR322) DNA was linearized with *Eco*R1. The enzyme was removed by phenol/chloroform/isoamyl alcohol (25:24:1) extraction and the DNA was precipitated with ethanol. The reaction of DNA with the DHM at a 1:5 molar ratio took place at 37°C for 60min in 50µl of 10mM tris-EDTA buffer (pH 7.6). A control

Fig. 51.1. Chemical structure of monocrotaline and its metabolite, dehydromonocrotaline and dehydroretronecine.

sample of DNA without DHM was included. The hyperphase consisted of 25µl of 2M ammonium acetate/4mM EDTA (pH 7.5), 25µl cytochrome *c* at 0.4mg/ml and 50µl of the test sample solution. The hypophase was 0.25M ammonium acetate (pH 7.5). The hyperphase was spread as described by Coggins (1987) and was adsorbed onto carbon-coated colloidion grids, stained with 50mM uranyl acetate/50mM HCl and rotary shadowed with platinum/carbon. Grids were examined with a JEOL 1010 electron microscope operating at 80kV.

Results

Figure 51.2 shows electron micrographs of linearized pBR322 plasmid DNA and DHM-induced multiply-crosslinked plasmid DNA. Multiple crosslinks appear to be joined at a focal point, seen as a dark spot. These spots are independent of the DNA (Fig. 51.2b), showing that DHM in the absence of DNA can produce these structures.

Figure 51.3 shows the autoradiogram of the strand cleavage pattern produced by DHM-treated DNA. Dehydromonocrotaline produces piperidine-labile N-7G alkylation with a slight preference for 5'-GG and 5'-GA sequences (arrows). The band intensity is maximal at 15min (lane 2) and decreases with time (lanes 3, 4 and 5). The intensities of the bands corresponding to the unmodified DNA (lanes 3, 4 and 5) do not recover to control levels. Instead, piperidine-resistant bands of lower mobility are seen near the top of the gel (arrow). The reduction in the band intensity of piperidine-labile DNA, with the concomitant increase in intensity of piperidine-resistant DNA with time, suggests crosslinking of the DNA carrying the N-7G alkylation.

Discussion

Monocrotaline toxicity has been attributed to its reactivity with cellular DNA after activation by P_{450}-3A4 enzyme. The exact mechanism is still unclear, although DNA/DNA interstrand crosslinks and DNA/protein crosslinks have been implicated by various studies (Hincks *et al.*, 1991). The reaction with DNA has been reported to occur at the exocyclic amino group of guanine in the minor groove involving the C7 and C9 of DHM (Robertson, 1982). There has been significant variation between the toxicity values of different PAs, which moderately correlate with *in vitro* DNA crosslinking levels (Hincks *et al.*, 1991). This study found that DHM alkylates N7 of guanines in the major groove, an effect that has not been reported elsewhere. The reaction reaches a maximum within 15min with a preference for guanines in 5'-GG and 5'-GA sequences (Fig. 51.3). In addition to these lesions, piperidine-resistant, low-mobility bands were observed. These bands appeared at a slower rate and seemed to reach a maximum by 1hr. These lesions were probably due to crosslinked DNA fragments arising from non-N7-G lesions. In addition, material was found that failed to migrate from the wells, suggesting that its structures were bigger than the single-crosslinked product.

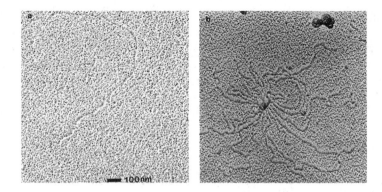

Fig. 51.2. Electron micrograph of linearized plasmid (pBR322) DNA (a) control untreated DNA (b) exposed to 10mM dehydromonocrotaline for 60min.

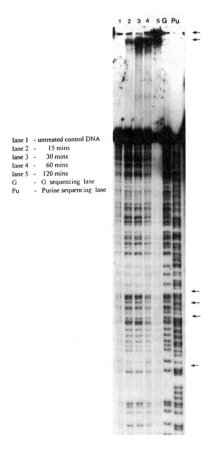

lane 1 - untreated control DNA
lane 2 - 15 mins
lane 3 - 30 mins
lane 4 - 60 mins
lane 5 - 120 mins
G - G sequencing lane
Pu - Purine sequencing lane

Fig. 51.3. Autoradiogram of PAGE showing guanine N7 alkylation of the ^{32}P 3'-end-labeled *Eco*R1/*Bam*HI fragment of pBR322 DNA exposed to dehydromonocrotaline.

To confirm the existence of such structures, electron microscopic analysis was carried out on these samples. The micrographs (Fig. 51.2) clearly show the presence of multiple-crosslinked DNA. A similar finding has been reported for DR (Reed *et al.*, 1988). Such multiple crosslinks have not been observed for any other crosslinking agents. It is difficult to speculate on the biological significance of such lesions until they can be shown to occur in cells.

Acknowledgments

We thank Dr B. Cribb and R.I. Webb (Centre for Electron Microscopy and Microanalysis, University of Queensland) for assistance with electron microscopy. T. Pereira gratefully acknowledges a travel grant to attend ISOPP5 from the Queensland Cancer Fund. National Research Centre for Environmental Toxicity is funded by the National Health and Medical Research Council, Queensland Health, The University of Queensland and Griffith University.

References

Bras, G., Jelliffe, D.B. and Stuart, K.L. (1954) Veno-occlusive disease of liver with nonportal type of cirrhosis, occurring in Jamaica. *Archives of Pathology* 57, 285-300.

Chan, P.C., Mahler, J., Bucher, J.R., Travlos, G.S. and Reid, J.B. (1994) Toxicity and carcinogenicity of riddelliine following 13 weeks of treatment to rats and mice. *Toxicon* 32, 891-913.

Coggins, L.W. (1987) Preparation of nucleic acids for electron microscopy. In: Sommerville, J. and Scheer, U. (eds), *Electron Microscopy in Molecular Biology*. IRL Press, Oxford, pp. 1-29.

Gaul, K.L., Gallagher, P.F., Reyes, D., Stasi, S. and Edgar, J. (1994) Poisoning of pigs and poultry by stock feed contaminated with heliotrope. In: Colegate, S.M. and Dorling, P.R. (eds), *Plant-Associated Toxins: Agricultural, Phytological and Economic Aspects.* CAB International, Wallingford, Oxon, pp. 137-142.

Hincks, J.R., Kim, H.Y., Segall, H.J., Molyneux, R.J., Stermitz, F.R. and Coulombe, R.A. Jr (1991) DNA cross-linking in mammalian cells by pyrrolizidine alkaloids: structure-activity relationships. *Toxicology and Applied Pharmacology* 111, 90-98.

Hirono, I., Ueno, I., Aiso, S., Yamaji, T. and Haga, M. (1983) Carcinogenic activity of *Farfugium japonicum* and *Senecio cannabifolius. Cancer Letters* 20, 191-198.

Hooper, P.T. and Scanlan, W.A. (1977) *Crotalaria retusa* poisoning of pigs and poultry. *Australian Veterinary Journal* 53, 109-114.

Johnson, A.E. and Molyneux, R.J. (1984) Toxicity of threadleaf groundsel (*Senecio douglasii* var *longilobus*) to cattle. *American Journal of Veterinary Research* 45, 26-31.

Mattocks, A.R. and Legg, R.F. (1980) Anti-mitotic activity of dehydroretronecine, a pyrrolizidine alkaloid metabolite, and some analogous compounds, in a rat liver parenchymal cell line. *Chemico-Biological Interactions* 30, 325.

Mattocks, A.R., Jukes, R. and Brown, J. (1989) Simple procedures for preparing putative toxic metabolites of pyrrolizidine alkaloids. *Toxicon* 27, 561-567.

Mohabbat, O., Younos, M.S., Merzad, A.A., Srivastava, R.N., Sediq, G.G. and Aram, G.N. (1976) An outbreak of hepatic veno-occlusive disease in north-western Afghanistan. *Lancet* 2, 269.

Petry, T.W., Bowden, G.T., Huxtable, R.J. and Sipes, I.G. (1984) Characterization of hepatic DNA damage induced in rats by the pyrrolizidine alkaloid monocrotaline. *Cancer Research* 44, 1505-1509.

Prakash, A.S., Deny, W.A., Gourde, T.A., Val, K.K., Wedged, P.D. and Wakelin, L.P. (1990) DNA-directed alkylating ligands as potential anti-tumor agents: sequence specificity of alkylation by intercalating aniline mustards. *Biochemistry* 29, 9799-9807.

Reed, R.L., Ahern, K.G., Pearson, G.D. and Buhler, D.R. (1988) Crosslinking of DNA by dehydroretronecine, a metabolite of pyrrolizidine alkaloids. *Carcinogenesis* 9, 1355-1361.

Robertson, K.A. (1982) Alkylation of N2 in deoxyguanosine by dehydroretronecine, a carcinogenic metabolite of the pyrrolizidine alkaloid monocrotaline. *Cancer Research* 42, 8-14.

Chapter 52

Acute and Chronic Toxicity Induced by Activated Ptaquiloside in Rats: Comparison of Pathogenesis due to Oral and Intravenous Administrations

M. Shahin[1], M.R. Moore[1], B.L. Smith[2], A.A. Seawright[1] and A.S. Prakash[1]

[1]National Research Centre for Environmental Toxicology, The University of Queensland, 39 Kessels Road, Coopers Plains, Queensland 4108, Australia; [2]AgResearch, Ruakura Agricultural Research Centre, East Street, Private Bag Hamilton 3123, New Zealand

Introduction

Bracken fern (*Pteridum* spp.) produces a number of toxic effects in grazing animals including carcinoma of the upper alimentary tract and urinary bladder (Evans and Mason, 1965; Pamucku and Price, 1969; Smith, 1990). The toxic signs induced by ingestion of a large quantity of bracken are different in each species (Evans *et al.*, 1982). The thiaminase in bracken fern triggers anorexia and ataxia in horses, while bracken toxicity in sheep generally leads to bright blindness because of retinal neuroepithelium degeneration. In cattle, bracken poisoning causes depression of bone marrow associated with leukopenia, thrombocytopenia and hematuria.

The carcinogenicity of bracken in experimental animals has been shown. Rats fed a diet containing bracken developed multiple ileal, urinary bladder and mammary gland adenocarcinomas (Evans and Mason, 1965; Pamucku and Price, 1969; Hirono *et al.*, 1983). The principle carcinogen in bracken is ptaquiloside (PT), which when activated in alkaline conditions to activated ptaquiloside (APT) (Fig. 52.1), alkylates adenines and guanines in DNA in a sequence-selective fashion (Smith *et al.*, 1994). The role of APT in the initiation of carcinogenesis is yet to be demonstrated. Recently, it was shown that bracken-fed calves harbored PT/DNA adducts arising from N3-adenine alkylation, which resulted in H-*ras* activation *via* mutation in codon

Fig. 52.1. Chemical structure of PT and formation of DNA adducts *via* its intermediates.

61 (Prakash *et al.*, 1996). A pilot study using Sprague-Dawley rats showed that intravenous (i.v.) exposure to APT gave rise to PT/DNA adducts and subsequently to mammary gland adenocarcinoma (Shahin *et al.*, 1995; Shahin *et al.*, 1996). In that study a significant increase in monocyte levels was observed in treated animals during the dosing period. Peripheral blood monocytes and macrophages exhibited tumoricidal activity both *in vitro* and *in vivo* (Fidler and Raz, 1981). They expressed maximal levels of several potent cytokines including tumor necrosis factor-alpha (TNFα) when exposed to specific signals in their micro-environment; the cytokines then selectively bound and destroyed neoplastic cells (Fidler and Schroit, 1988; Chensue *et al.*, 1988).

Rodents are sensitive to oral (p.o.) and subcutaneous (s.c.) delivery of PT. Depending on quantities and time of exposure, multiple ileal, esophageal and pharyngeal neoplasias and bladder and mammary gland carcinomas have been seen in rats (Pamucku and Price, 1969; Hirono *et al.*, 1983; Hirono *et al.*, 1984). These routes of toxin administration have some drawbacks: 1) they require large amounts of toxin; 2) the toxin travels to organs not known to be associated with PT toxicity; and 3) inactivation of the toxin in the acid environment of the stomach can occur.

This work was carried out using APT to evaluate the usefulness of i.v. dosing in rats for studying bracken toxicity. The results suggest that the route of administration may determine the nature of disease from the toxin exposure.

Materials and Methods

Young bracken fronds were collected from Southeast Queensland, Australia, freeze dried, milled into a fine powder and stored at -80°C. Extraction of PT was carried out using CHCl₃ followed by MeOH-CHCl₃ and recovered by adsorption on nonionic polymeric (XAD-2) resin (Oelrichs *et al.*, 1995). The PT was eluted in MeOH and stored at -80°C. Electrospray ionization mass spectrometry (ESIMS) and HPLC methods were used to confirm the purity of the compound. Activation of PT was done with 1hr incubation in 10mM NaOH at 37°C, and conversion was confirmed by ESIMS.

One group (n=5) of female Sprague-Dawely rats were dosed i.v. weekly with 3mg APT for 10wks. Two other groups received 6mg (acute dose) and 3mg (chronic, 10wks of dosing) APT, respectively, intragastrically (p.o.). The animals receiving chronic doses were euthanized 30wks after the final dose. The acutely dosed (6mg, p.o.) animals required termination after 3wks due to weakness, anorexia and hemorrhages.

Fortnightly, blood samples were collected from the tail vein for hematology including total white cell count (TWBC), monocyte, differential leukocyte and platelet numbers. Plasma samples collected from individual rats at the end of experiment were stored at -80°C for TNFα analysis. Gross pathology (liver, kidney, ileum, urinary bladder, lung, lymph nodes and mammary gland) was noted. Tissue samples were preserved in 10% buffered neutral formalin and were prepared for routine histopathological examination.

Results

Histopathology studies showed profound apoptotic bodies in the hepatocytes (Fig. 52.2), coagulation necrosis in the kidney and bone marrow hypoplasia in animals in the acute study (6mg APT, p.o.). The chronically dosed animals (3mg APT/d, p.o.) had a monocytosis associated with TNFα elevation in the plasma (Fig. 52.3), with only renal tubular necrosis histopathologically.

The animals dosed i.v. with 3mg APT/wk for 10wks showed formation of mammary gland and papillary adenocarcinoma (40%, $P<0.05$). Furthermore, there was type II pneumocyte proliferation and focal renal tubular necrosis. Hematology results indicated monocytosis associated with plasma TNFα elevation.

Discussion

A parallel study (Chapter 65) demonstrated that activation of PT is required for PT carcinogenicity. In this study, the i.v. route of administration compares well with the intragastric route for producing PT-related toxicity. The total amount of toxin required was 10x less than what was used for oral dosing in a prior study (Hirono *et al.*, 1984).

The finding that the two different routes of PT administration lead to increased monocyte and TNFα levels while control animals show normal levels suggests that

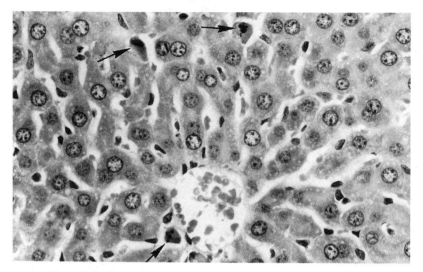

Fig. 52.2. Acute intragastric toxicity: apoptotic bodies in the hepatocytes (arrows).

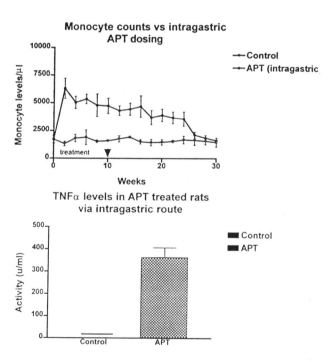

Fig. 52.3. Monocytosis associated with TNFα elevation.

may provide a useful means of monitoring bracken carcinogen exposure.

The differences observed in gross and histopathological appearance in the animals dosed *via* the two routes indicate that the route of the toxin administration determines the organs of target and the nature of disease.

References

Chensue, S.W., Remick, D.G., Shymr-Forsch, C., Beals, T.F. and Kunkel, S.L. (1988) Immunohistochemical demonstration of cytoplasmic and membrane-associated tumor necrosis factor in murine macrophages. *American Journal of Pathology* 133, 564-572.

Evans, I.A. and Mason, J. (1965) Carcinogenic activity of bracken. *Nature,* 208, 913.

Evans, W.C., Patel, M.C. and Koohy, Y. (1982) Acute bracken poisoning in monogastric and ruminant animals. *Proceedings - Royal Society of Edinburgh* 81, 29.

Fidler, I.J. and Raz, A. (1981) *Lymphokines, Vol 3.* Academic Press, New York, NY, pp. 345-363.

Fidler, I.J. and Schroit, A.J. (1988) Recognition and destruction of neoplastic cells by activated macrophages: discrimination of altered self. *Biochemica et Biophysica Acta* 948, 151-173.

Hirono. I., Aiso, S., Yamaji, T., and Haga, M. (1983) Induction of mammary cancer in CD rats fed bracken fern. *Carcinogenesis* 4, 885-887.

Hirono, I., Aiso, S., Yamaji, T., Mori, H., Yamada, K., Niwa, H., Ojika, M., Wukamatsu, K., Kigoshi, H., Niiymama, K. and Uosaki, Y. (1984) Carcinogenicity in rats of ptaquiloside isolated from bracken. *Gann* 75, 833.

Oelrichs, P.B., Ng, J. and Bartely, J. (1995) Purification of ptaquiloside, a carcinogen from *Pteridium aquilinum. Phytochemistry* 40, 53-56.

Pamucku, A.M. and Price, J.M. (1969) Induction of intestinal and urinary bladder cancer in rats by feeding bracken fern (*Pteris aquilina*). *Journal of the National Cancer Institute* 43, 275-281.

Prakash, A.S., Pereira, T., Smith, B.L., Shaw, G. and Seawright, A.A. (1996) Mechanism of bracken fern carcinogenesis: Evidence for H-*ras* activation *via* initial adenine alkylation by ptaquiloside. *Natural Toxins* 4, 221-227.

Shahin, M., Smith, B.L., Moore, M.R. and Prakash, A.S. (1995) Development of a rat cancer model for ptaquiloside, a bracken fern carcinogen. *Proceedings of the Australian Society for Clinical and Experimental Pharmacology and Toxicology* 2, 174.

Shahin, M., Smith, B.L., Moore, M.R., Seawright, A.A. and Prakash, A.S. (1996) Induction of monocytosis, DNA damage and mammary gland carcinoma by ptaquiloside. *Proceedings of the Australian Society for Clinical and Experimental Pharmacology and Toxicology* 3, 21.

Smith, B.L. (1990) Bracken and animal health in Australia and New Zealand. In: Thomson, J.A. and Smith, R.T. (eds), *Bracken Biology and Management.* The Australian Institute of Agriculture Science, Wahroonga, NSW, pp. 227-232.

Smith, B.L., Shaw, G., Prakash, A.S. and Seawright, A.A. (1994) Studies on DNA adduct formation by ptaquiloside, the carcinogen of bracken fern (*Pteridium* spp.). In: Colgate, S.M. and Dorling, P.R. (eds), *Plant-Associated Toxins: Agricultural, Phytochemical and Ecological Aspects.* CAB International, Wallingford, Oxon, pp. 167-172.

Chapter 53

Immunotoxic Effects of Selenium in Mammals

M.F. Raisbeck, R.A. Schamber and E.L. Belden

Department of Veterinary Sciences, University of Wyoming, Laramie, Wyoming 82070, USA

Introduction

Although naturally occurring selenosis was recorded by workers at the Agricultural Experiment Stations of South Dakota and Wyoming nearly 70yrs ago (Franke *et al.*, 1934), the recent discovery that human activities such as irrigation or strip mining may mobilize sufficient selenium (Se) to cause toxicity in water-fowl has rekindled interest in spontaneous selenosis. There are very few reports attesting to the immunotoxic effects of Se in mammals (Koller *et al.*, 1986; Larsen, 1988), and none in cattle. There was circumstantial evidence of a likely association between excess dietary Se and immunosuppression in cattle in field studies between 1988-1991. This chapter summarizes a series of trials to explore this association.

In vitro Immunocyte Response to Selenium

Peripheral blood lymphocytes (PBL) from healthy cattle were the basis of a series of experiments utilizing *in vitro* exposure of immunocytes to various forms of Se. Mononuclear cells were harvested and cultured as described by Belden *et al.* (1981). Final Se concentrations in the media were Na_2SeO_3: 0.007, 0.015, 0.031, 0.062, 0.125, 0.25, 0.5, 1, 3, or 5ppm Se; selenocystine: 0.015, 0.031, 0.062, 0.125, 0.25, 0.5, 1, 3, or 5ppm Se; and L-selenomethionine (SEMET): 0.125, 0.25, 0.5, 0.75, 1, 3, or 5ppm Se. Cells in untreated culture media served as controls for each trial. Selenium inhibition thresholds were defined as the concentration of added Se that resulted in greater than 50% inhibition of the function being measured.

Blastogenesis assays were performed as reported (Belden and Strelkauskaus, 1981; Schamber, 1994). Results were expressed as the difference between mitogen stimulated and unstimulated counts. Production and bioassay of interleukin 2 (IL-2) by PBL were as previously reported (Zelarney and Belden, 1988).

B-cell function was evaluated by measuring the ability of pokeweed mitogen (PWM) stimulated lymphocytes to produce antibody with an enzyme-linked anti-globulin (ELISPOT) assay. Cells were cultured with PWM for 96hrs and plated on antiglobulin treated microtiter plates. After antiglobulin addition and substrate development, the ratio of chromogen positive spots in stimulated and unstimulated wells was expressed as a stimulation index. Interleukin 4-like (IL-4) activity in the supernatant of Concanavalin A (ConA) and Phytohemaglutinin (PHA) stimulated PBL's was also evaluated using the ELISPOT assay.

All three forms of Se produced an obvious dose-related inhibition of both ConA and PHA-induced lymphocyte blastogenesis (Fig. 53.1). The median inhibitory concentrations for Na_2SeO_3 and selenocystine (SECYS) were less than blood or plasma concentrations commonly associated with bovine selenosis; however, that for SEMET was somewhat greater. There was no difference in inhibition between forms of Se at the highest and lowest concentrations. At intermediate concentrations, SEMET was less inhibitory than either Na_2SeO_3 or SECYS, which were approximately equipotent (Table 53.1). This ranking is consistent with the cytotoxic potency observed in other *in vitro* experimental systems (Spallholz, 1994) and probably reflects nonspecific cell death rather than a specific immunotoxic response. These results contrast with reports indicating that SEMET is more immunotoxic than Na_2SeO_3 *in vivo* (Larsen, 1988; Fairbrother and Fowles, 1990). This may be due to the tendency of SEMET to accumulate to higher tissue levels *in vivo*,

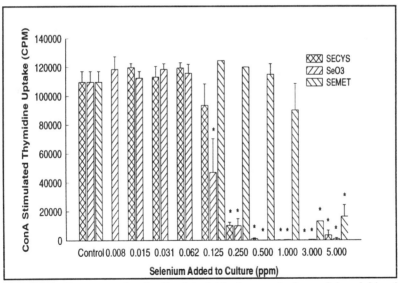

Fig. 53.1. Concanavalin A stimulated blastogenesis in peripheral blood lymphocytes cultured in selenium. Each bar=increased ³H-thymidine uptake as a result of mitogen stimulation. *=significant decrease from control values.

Table 53.1. Selenium inhibition thresholds (ppm, greater than 50% inhibition) in various *in vitro* bioassays.

Form	Blastogenesis	Elispot	Cytokine production
Selenomethionine	3.0	3.0	None
Selenocystine	0.25	0.125	0.5
Selenite	0.125	0.25	0.25

resulting in a larger pool of Se for local conversion to some as yet unidentified proximate toxicant.

In vivo Studies

Mice

Sixteen 5wk-old BALB/C mice were assigned to one of four treatment groups (control, SEMET, SECYS or Na_2SeO_3) and housed by treatment group in plastic cages with rodent chow and deionized water *ad libitum*. Experimental diets were rodent chow plus water to which sufficient of each of the aforementioned forms of Se was added to achieve 7ppm Se. After 14d on Se-treated water each mouse received a single subcutaneous (s.c.) injection of 2mg hen egg albumin (OVA).

After 47d, serum for antibody analysis was collected from each mouse *via* the infraorbital sinus, after which all were euthanized by cervical dislocation. After the spleens were removed aseptically, splenocytes were harvested by mechanical teasing and centrifugation in Hank's balanced salt solution and diluted to 2×10^6 viable cells/ml in Roswell Park Memorial Institute media (RPMI-1640) +10% fetal bovine serum. Blastogenesis assays were performed as described above for bovine PBLs using ConA, PHA and lipopolysaccharide (LPS) as mitogens. A separate plate was prepared for blastogenesis using OVA as an antigen. B-cell function was assessed in splenocytes using the ELISPOT assay described above. Total Ig concentrations were determined in sera with an antigen trap ELISA, and were calculated by comparison with standards. An indirect ELISA was used to determine OVA specific antibody.

There were few statistically significant differences in mitogen-stimulated blastogenesis between treatment group means. Proliferative response to ConA reflected a slight but non-significant decline from controls for each of the three forms of Se. Immunotoxic potency was $SeO_3^=$ >SECYS >SEMET. B-cell function, as measured by ELISPOT, was significantly decreased from controls by SEMET and SECYS (Fig. 53.2). Total Ig concentration did not vary between treatment groups. Interestingly, there were no differences between Se-treated and control groups in

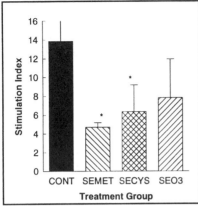

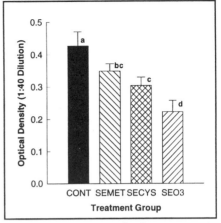

Fig. 53.2. B-cell function in splenocytes from mice fed three forms of selenium for 47d. ✳ denotes less than control (*P*<0.05). Each bar depicts mean±sd ratio of PWM-stimulated to unstimulated splenocytes.

Fig. 53.3. Primary antibody response to OVA in mice treated with 7ppm Se in drinking water for 47d. Superscripts denote statistical similarity (ANOVA, *P*<0.05). Each bar represents mean±sd.

OVA-stimulated blastogenesis, but all Se-treated groups had significantly lower OVA-specific antibody concentrations than did controls (Fig. 53.3).

Cattle

Twenty yearling Hereford-cross steers were housed in concrete-floored pens equipped with heated waterers and electronically gated feed bunks keyed to individual steers. After a 45d acclimation period, steers were divided into one control and six experimental groups. Each experimental steer received Se at 0.15, 0.28 or 0.8 Se mg/kg as either Na_2SeO_3 or SEMET daily. Doses were prepared each day by absorbing aqueous Se solution into 500g ground corn cob and carefully hand mixing with the amount of chopped hay each steer would consume in 18hrs.

After 42d on Se, each steer received a single subcutaneous (s.c.) 40mg dose of OVA. Blood was collected at 21, 42, and 63d post-immunization. Lymphocyte blastogenesis, B-cell function, total Ig and OVA-specific antibody were evaluated as above with proper species-specific reagents. In an effort to duplicate Se content of the cellular microenvironment *in vitro*, the Se (10% of *in vivo* blood levels) form being fed to each animal was added to an another set of cultured cells at 120d.

No significant differences existed between responses to T-cell mitogens, B-cell stimulation, IL-2 or IL4-like activity at any time (data not shown). Addition of SEMET to *in vitro* PBL cultures resulted in no change from the control group or the cultures without added Se. Addition of Na_2SeO_3 to PBL cultures from Na_2SeO_3-treated steers inhibited mitogen-stimulated blastogenesis (Fig. 53.4). Addition of

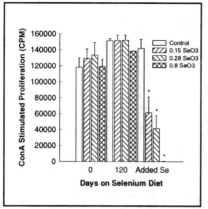

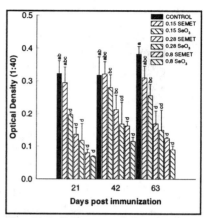

Fig. 53.4. ConA stimulated blastogenesis in bovine PBL from steers on
seleniferous diets. PBLs collected at
120d were assayed with and without
added Se equivalent to 10% blood Se
concentration. * denotes less than
untreated (*P*<0.05).

Fig. 53.5. Primary antibody response
to OVA in steers fed seleniferous diets.
Similar superscripts denote statistical
similarity (ANOVA, *P*<0.05). Each bar
depicts group mean±sd.

0.27ppm Se (high dose group) resulted in cytotoxicity like that seen *in vitro* (Table
53.1). In the groups that received Na₂SeO₃ at 0.15 and 0.28mg/kg much less Se than
the *in vitro* toxic threshold, proliferation was inhibited (Fig. 53.4).

Mean total Ig concentrations did not vary between treatment groups (data not
shown). Primary antibody response to OVA exhibited a dose-related inhibition, with
larger Se doses resulting in lower OVA specific antibody concentrations (Fig. 53.5).
Greater inhibition of antibody formation resulted from Na₂SeO₃ than SEMET.

Antelope

Eight captive-raised pronghorn antelope (*Antelocapra americana*), randomly assigned
to two groups, were housed in concrete-floored pens with heated waterers. The control
group received a ration (0.3-0.5ppm Se) of native grass hay and alfalfa. The
experimental group received a ration (13-16ppm Se) prepared similarly from
seleniferous grass (Raisbeck *et al.*, 1996). Both diets were fed free choice for 167d.

After 28d on Se, each antelope received a s.c. injection of 40mg OVA. A battery
of tests similar to that used previously with the steers was applied to blood taken at
3wk intervals after immunization. There were no significant differences between treatment groups in PBL response to T-cell mitogens, B-cell function, or total Ig
concentration. Primary antibody response to OVA, was less in the Se-fed group than
controls at all sampling intervals after immunization (data not shown).

Summary

Exposure to dietary Se at concentrations that did not result in clinical selenosis compromised immune function in rodents, cattle and antelope. This effect was manifested as decreased primary antibody response *in vivo* to antigen administered while the animal was on a seleniferous diet. With the possible exception of ConA-stimulated blastogenesis in Na_2SeO_3-treated mice, there were no significant Se-related effects in any assay that relied on *in vitro* manipulation of immunocytes. These results are in contrast to Larsen (1988), who reported that lymphocytes from sheep fed 1.3ppm SEMET exhibited decreased ConA-stimulated blastogenesis, but lymphocytes from sheep fed Na_2SeO_3 did not.

Inhibition of various immunocyte functions by Se compounds *in vitro* corresponded with overt cytotoxicity and cell death. Cell death does not explain the depressed *in vivo* primary antibody response in these experiments, as the latter occurred in Se-treated animals despite normal circulating lymphocyte counts. Interestingly, Se concentrations that inhibited various functions in Na_2SeO_3- or SECYS-treated PBL cultures were within the range of plasma concentrations commonly accepted as "adequate," but the SEMET concentration required was an order of magnitude higher. This difference in potency is consistent with the relative cytotoxic potential of these compounds using other *in vitro* endpoints such as chemiluminescence (Spallholz, 1994).

These observations re-emphasize the artificial nature of *in vitro* bioassays. It is unlikely that the majority of blood Se exists in either of these forms *in vivo*. In all probability the proximate toxicants in selenosis are highly reactive moieties with relatively short half-lives (Burk, 1991). Without better data on Se toxicokinetics and metabolism at toxic doses it is impossible to model Se immunotoxicity realistically *in vitro*.

In vitro tests on PBLs from Se-exposed animals in these trials further support this conclusion. These assays were essentially negative despite the fact that PBLs being assayed came from animals in which primary antibody response was inhibited. Finch and Turner (1989) reported that lymphocytes from Se-deficient lambs regained normal function during the course of a typical blastogenesis assay if exposed to adequate Se in culture media. Cells from Se-intoxicated animals should also regain normal function by equilibrating with normal-Se media.

In the intact animal, Se somehow interferes with the cellular events responsible for an immune response. Elevated Se has been shown to promote peroxidative damage in *in vitro* and *in vivo* systems (Spallholz, 1994; Bjornstedt *et al.*, 1996). Lymphocyte cell membranes are especially susceptible to free radical damage by virtue of their relatively high unsaturated fatty acid content. Selenium-induced free radicals could also increase the oxidation potential (pE) of the immediate cellular microenvironment or the pE of the cytosol, inhibiting these functions by a similar mechanism. Removal to a normal-Se environment, i.e. a typical *in vitro* bioassay, would decrease oxidant stress and thus permit recovery of apparently normal lymphocyte function.

Obviously, none of these questions will be answered until there is an understanding of the forms of Se in the cellular microenvironment and how the *in vivo* redox environment influences lymphocyte function and immunity. This will necessarily require additional *in vivo* and *in vitro* examination of Se immunotoxicity.

References

Belden, E.L. and Strelkauskas, A.J. (1981) Mitogen and mixed lymphocyte culture responses of isolated bovine lymphocyte subpopulations. *American Journal of Veterinary Research* 42, 934-937.

Belden, E.L., McCrosky, J.K. and Strelkauskas, A.J. (1981) Subpopulations of bovine lymphocytes separated by rosetting techniques. *Veterinary Immunology and Immunopathology* 2, 267-270.

Bjornstedt, M., Odlander, B., Kuprin, S., Claesson, H. and Holmgren, A. (1996) Selenite incubated with NADPH and mammalian thioredoxin reductase yields selenide, which inhibits lipoxygenase and changes the electron spin resonance spectrum of the a active site iron. *Biochemistry* 35, 8511-8517.

Burk, R.F. (1991) Molecular biology of selenium with implications for its metabolism. *FASEB Journal* 5, 2274-2279.

Fairbrother, A. and Fowles, J. (1990) Subchronic effects of sodium selenite and selenomethionine on several immune functions in mallards. *Archives of Environmental Contamination and Toxicology* 19, 836-844.

Finch, J.M and Turner, R.J. (1989) Enhancement of ovine lymphocyte responses: A comparison of selenium and vitamin E supplementation. *Veterinary Immunology and Immunopathology* 23, 245-256.

Franke, K.W., Rice, T.D., Johnson, A.G. and Schoening, H.W. (1934) *Report on a Preliminary Field Survey of the So-Called "Alkali Disease" of Lvestock*. US Department of Agriculture Circular No. 320, pp. 1-10.

Koller, L.D., Exon, J.H., Talcott, P.T., Osborne, C.A. and Henningsen, G.M. (1986) Immune responses in rats supplemented with selenium. *Clinical and Experimental Immunology* 63, 570-576.

Larsen, H.J. (1988) Effect of selenium on sheep lymphocyte responses to mitogens. *Research in Veterinary Science* 45, 11-15.

Raisbeck, M.F., O'Toole, D., Schamber, R.A., Belden, E.L. and Robinson, L.J. (1996) Toxicologic effects of a high-selenium hay diet in captive adult and yearling pronghorn antelope (*Antilocapra americana*). *Journal of Wildlife Diseases* 32, 9-16.

Schamber, R.A. (1994) *Selenium Immunotoxicity*. MS Thesis, University of Wyoming.

Spallholz, J.E. (1994) On the nature of selenium toxicity and carcinostatic activity. *Free Radical Biology and Medicine* 17, 45-64.

Zelarney, P.T. and Belden, E.L. (1988) Bovine interleukin 2: Production by an E-rosette-defined lymphocyte subpopulation. *Veterinary Immunology and Immunopathology* 18, 297-305.

Chapter 54

In vivo and *in vitro* Effects of L-Canavanine and L-Canaline

L. Ogden[1], S. Gorham[2], T. Hanner[1], D. Norford[1] and S. Hurley[1]

[1]*Animal Science, North Carolina Agricultural and Technical State University, Greensboro, North Carolina 27411, USA;* [2]*Department of Pathology, School of Veterinary Medicine, Tuskegee University, Tuskegee, Alabama 36117, USA*

Introduction

Canavanine is a non-protein amino acid from the large West Indian jack bean, *Canavalia ensiliformis*. It is grown in the southern US and imported for livestock feed (Kingsbury, 1964).

L-canavanine, the guanidinooxy structural analogue of L-arginine, is synthesized by alfalfa and other leguminous plants, some of which constitute a significant proportion of the diets of livestock and humans. Canavanine makes up a considerable portion of some legumes (Bell, 1958), approximately 1.5% of the dry weight of alfalfa sprouts (Bell *et al.*, 1978) and up to 13% of the dry weight of alfalfa and other legumes (Cheeke and Shull, 1987). The exact physiological role of canavanine in plants is not well known, but it is believed to function in maintaining nitrogen requirements of developing plants and to contribute significantly to plant chemical defense.

The antineoplastic effects of canavanine became evident by the increased life span of mice bearing L1210 leukemic cells (Green *et al.*, 1980) and Fisher rats that were given 2mg/kg canavanine by subcutaneous (s.c.) administration (Keeler *et al.*, 1978). Antineoplastic potential effects have also been reported on human pancreatic adenocarcinoma cells (MIA PaCa-2)(Swaffer *et al.*, 1994).

Some scientists attribute the toxicity of canavanine in insects to its structural similarity to arginine, a conditionally indispensable amino acid (Rosenthal, 1977). The arginine-like structure enables canavanine to bind many enzymes that usually interact with arginine, and it is incorporated into polypeptide chains, resulting in structurally aberrant canavanine-containing proteins. Canavanine is subject to hydrolytic cleavage by arginase, yielding urea and L-canaline (Rosenthal, 1977).

Studies show that canavanine induces many harmful effects in insects and prokaryotes, including infertility and reduced fecundity. Chronic exposure to canavanine caused reduced egg fertility, ovarian mass and fecundity during the last larval stadium of the adult tobacco hornworm, *Manduca sexa* (Dahlman and Berge, 1986; Rosenthal *et al.*, 1995).

Consumption of jack bean seeds at a rate exceeding 4% of the body weight is toxic to cattle, causing diarrhea, weakness, enteritis and nephritis (Kingsbury, 1964). Canavanine was only slightly toxic to adult and neonatal rats following single s.c. injections, but multiple injections caused growth inhibition, alopecia, appetite depression and weight loss. The pancreas was affected more than other organs that were evaluated (Thomas and Rosenthal, 1987a, b).

Monkeys fed alfalfa sprouts developed a systemic lupus erythematosus-like syndrome, which was attributed to canavanine toxicity (Montanaro and Bardana, 1991). Human consumption of alfalfa seeds gives an initial reduction in serum cholesterol levels, but prolonged ingestion of alfalfa seeds has been associated with pancytopenia, anemia, leukopenia, and with the development of antinuclear antibodies (Montanaro and Bardana, 1991). Canavanine can alter reproductive and immune functions and digestive processes, possibly presenting chronic insidious effects in both livestock and humans. This study was designed to investigate the toxic potential of L-canavanine and L-canaline.

Materials and Methods

L-canavanine and L-canaline were obtained from Sigma Chemical Company. Alfalfa (crude ground) dietary supplement was obtained from a local health food store. Cell line AR42J was obtained from ATCC. Dulbecco's Modified Eagle's Medium (DMEM), L-glutamine, penicillin-streptomycin (100IU/ml), trypsin-EDTA, and William E media were obtained from Fisher Scientific, Bio-Rad or Gibco BRL.

Dietary exposure of L-canavanine to rats

Weanling (6wks) male Sprague-Dawley rats (90±10g) were allowed access *ad libitum* to municipal water and a commercial rat chow throughout the 3wk study. Almost 0.75ml blood samples were obtained from anesthetized rats *via* the orbital sinus, alternating sampling sites. The following serum parameters were evaluated: lipase, amylase, glucose, creatinine, blood urea nitrogen, alkaline phosphatase, creatine, transaminases, albumin, total protein, cholesterol, sodium, potassium, chloride, calcium, and phosphorus. Hematological indices (CBC, differential, RBC counts, MCV, MCHC, HCT) were determined on whole blood.

Body and organ weights and organ/body weight ratios were measured on all rats. Liver, pancreas, spleen, and kidney were evaluated for morphological effects. Each pancreas was examined by light and electron microscopy.

Parenteral administration of L-canavanine and L-canaline in mice

Mice (BALB/C, 18±1g) were randomly assigned to three groups each containing five mice of each sex. They were given 100mg/g of L-canavanine, L-canaline or saline (0.5ml) intraperitoneally (i.p.) and observed for 48hrs. Mice were then anesthetized, intracardiac blood samples were drawn, and animals were euthanized with CO_2. Parameters used as indicators of canavanine and canaline toxicity included blood urea nitrogen, amylase, lipase, alkaline phosphatase, clinical signs, and morphological changes in liver, spleen, kidney, and pancreatic tissues.

Mice treated with L-canaline appeared depressed and less active during the first 2hrs after treatment. Mice treated with L-canavanine or saline appeared normal and were active throughout the study. There were no significant changes in serum chemistries or abnormalities in tissue morphologies. Based on this comparative biochemical phase, there were no significant differences in biochemical parameters between the L-canavanine and L-canaline treated groups and the control group.

Effects of L-canavanine and L-canaline on normal and pancreatic cancer cells

Rat pancreatic cancer cells (AR42J) were grown in DMEM supplemented with fetal bovine serum (10%), penicillin/streptomycin, and L-glutamine in 60mm culture dishes. Cells were treated for 24hrs at concentrations of 5, 10, and 50μM of either L-canavanine, L-canaline or phosphate buffered saline (PBS). Pancreatic acinar cells were isolated from clinically healthy Sprague-Dawley rats, exposed to 10μM concentrations of L-canavanine, L-canaline or to PBS for 24hrs. Treated cultures were counted by automated Coulter Counter.

L-canavanine and L-canaline inhibited growth of pancreatic cancer cells at all concentrations evaluated, compared to controls. Growth of normal rat exocrine pancreatic cells was not inhibited significantly by L-canavanine or L-canaline.

Statistics were done on Sigma Stat (Jandel), with significance at $P=0.05$. Analysis of variance and Student's *t* test were used for comparisons of data groups.

Discussion

Dietary canavanine significantly affected serum amylase and lipase in rats at 1g/kg, but there were no significant changes in the morphology of the pancreas by light or electron microscopy. The elevated levels returned to normal ranges 24-48hrs after canavanine exposure ceased. Morphological changes were not induced by canavanine or canaline in these studies. Clinical chemistries, hematology, and tissue morphology in treated Sprague-Dawley rats and BALB/C mice were normal.

According to the *in vitro* studies, L-canavanine and L-canaline inhibited the growth of cancer cells, but not normal cells. Further studies involving cell morphology, molecular biology and receptor pharmacology of normal cells treated with L-canavanine and L-canaline are currently in progress.

References

Bell, E.A. (1958) Canavanine and related compounds in Leguminosae. *Biochemical Journal* 70, 617-619.

Bell, E.A., Lakey, J.A. and Polhill, R.M. (1978) Systemic significance of canavanine in the Papilionoidean (Faboideae). *Biochemical and Systemic Ecology* 6, 201-212.

Cheeke, P.R. and Shull, L.R. (1987) *Natural Toxicants in Feeds and Poisonous Plants*. AVI Publishing Company, Westport, CT, pp. 275-276.

Dahlman, D. and Berge, M. (1986) Possible mechanisms for adverse effects of L-canavanine on insects. In: Green, M.B. and Hedin, P.A. (eds), *Natural Resistance of Plants to Pests: Role of Alleochemicals*. American Chemical Society Symposium, Series No. 296, American Chemical Society, Washington, DC, pp. 118-129.

Green, M.H., Brooks, T.L., Mendelson, J. and Howell, S.B. (1980) Antitumor activity of L-canavanine against L1210 murine leukemia. *Cancer Research* 40, 535-537.

Keeler, R.F., Van Kampen, R.F. and James, L. (1978) *Toxic Amino Acids*. Academic Press, New York, NY, p. 583.

Kingsbury, J.M. (1964) *Poisonous Plants of the United States and Canada*. Prentice-Hall, Englewood Cliffs, NJ, pp. 313-314.

Montanaro, A. and Bardana, E.J. (1991) Dietary amino acid-induced systemic lupus erythematosus. *Nutrition and Rheumatic Diseases of North America* 17, 323-332.

Rosenthal, G.A. (1977) The biological effects and mode of action of L-canavanine, a structural analog of L-arginine. *Quarterly Review of Biology* 52, 155-178.

Rosenthal, G.A., Dahlman, D.L., Crooks, P.A., Phuket, S.N. and Trifonov, L.S. (1995) Insecticidal properties of some derivatives of L-canavanine. *Journal of Food Chemistry* 43, 2728-2734.

Swaffer, D.S., Ang, C.Y., Desai, P.B. and Rosenthal, G.A. (1994) Inhibition of the growth of human pancreatic cancer cells by the arginine antimetabolite L-canavanine. *Cancer Research* 54, 6045-6048.

Thomas, D.A. and Rosenthal, G.A. (1987a) Toxicity and pharmacokinetics of the nonprotein amino acid L-canavanine in the rat. *Toxicology and Applied Pharmacology* 91, 395-405.

Thomas, D.A. and Rosenthal, G.A. (1987b) Metabolism of L-[guanidinooxy-^{14}C] canavanine in the rat. *Toxicology and Applied Pharmacology* 91, 406-414.

Chapter 55

The Isolation of a Hepatotoxic Compound from *Eupatorium adenophorum* (Spreng) and Some Preliminary Studies on its Mode of Action

P.B. Oelrichs[1], A.A. Seawright[1], J.K. MacLeod[2] and J.C. Ng[1]
[1]The *National Research Centre for Environmental Toxicology, 39 Kessels Road, Coopers Plains, Queensland 4108, Australia;* [2]*The Research School of Chemistry, Australian National University of Canberra, Australia*

Introduction

Eupatorium adenophorum (Spreng) or Crofton weed is native to southern Mexico and Costa Rica, but was introduced as a weed to Hawaii, the Philippines, Thailand, New Zealand, Australia, India, Nepal, Sikkim and California (Fuller, 1981). It was naturalized in Australia, where it has spread from near Sydney up the eastern coast to north of Brisbane (Everist, 1981).

The plant (Fig. 55.1) was suspected of causing a chronic equine respiratory disease in the 1920s in Hawaii, where it was called "blowing" disease. In the early 1940s this respiratory disease was recognized in parts of the hinterland of south-east Queensland and northern New South Wales. The disease became known as Numinbah horse sickness or Tallebudgera horse disease after the river and valleys where it occurred. Horses are the only animals affected by the disease, and then only after being stressed. Cattle grazing where the plant grows profusely do not eat it, but horses apparently eat the plant readily.

Numinbah horse sickness was first reproduced by feeding a 300kg horse 2-3kg of fresh *E. adenophorum* in lucerne chaff daily for 8mos (O'Sullivan, 1979). Typical lung lesions were seen after only 50d of feeding the plant while in flower, implying that this was the most toxic stage. Similar lung lesions were seen after feeding horses *E. riparium* at 1.5kg/d for 12-22mos (Gibson and O'Sullivan, 1984).

As it was not practical to use horses as test animals, laboratory animals were used in attempts to isolate the toxins involved. When male Quackenbush mice (50g) were given 100mg or more of freeze-dried powdered *E. adenophorum* in normal saline by gavage, liver lesions were produced. These were characterized by multiple areas of focal necrosis of the parenchyma associated with degeneration and loss of the epithelium lining of the small bile ducts, usually with periductal edema and fibrosis (Sani *et al.*, 1992). No changes were observed in the livers of the mice dosed orally with a 50mg dose of ground freeze-dried plant, and no lesions were seen in other organs.

The present study was directed only at isolating the toxin(s) that produced these specific liver lesions in mice. The compound 9-oxo-10,11-dehydroagerophorone (the toxin) (Fig. 55.2) was isolated using the mouse test, and dosing mice with the purified compound produced identical lesions to the ground freeze-dried plant. Although this compound had been previously isolated from *E. adenophorum* (Bohlmann and Gupta, 1981) there is no report of its mammalian toxicity. In a study of the mechanism of the toxic effect of the compound (Fig 55.2) Scawright *et al.* (1996) showed that its hepatotoxicity could be modified by altering glutathione levels in the test animal.

Methods and Materials

To test the initial crude extracts of *E. adenophorum*, the various fractions from the plant were added to ground mouse pellets and the mixture dried under reduced pressure before feeding. At the latter stages of the purification procedure, 1% Tween 80 suspensions of the toxin were used for dosing by oral gavage. Two animals were used for each test, and the bioassay for hepatotoxicity was conducted at all stages of the purification procedure.

Fig. 55.1. Flowering plant of *Eupatorium adenophorum*.

The milled, freeze-dried leaf (100g) was extracted by stirring with methanol (MeOH) (3x700ml) at room temperature, and the extracts concentrated to dryness under reduced pressure at 40°C (Oelrichs *et al.*, 1995). Water (100ml) was added and the mixture extracted with ethyl acetate (EtOAc) (5x100ml). The organic extract was washed with water (100ml) and then dried under reduced pressure. The residue was redissolved in MeOH (50ml) and absorbed on High Flow Super Cell (HFSC), an inert material, and dried under reduced pressure. The HFSC was added to the top of a silica-gel (0.07-0.15mm) dry packed column (50g), washed with dichloromethane (DCM), and eluted with 2% EtOAc/DCM, which was then evaporated to dryness. To remove colored impurities, the dried residue was dissolved in MeOH/H_2O/$CHCl_3$ (chloroform) (85:15:5) (20ml) added to an XAD-2 reverse phase column (2x10cm) and the column washed with the same solvent (200ml). The eluate was evaporated to dryness, dissolved in 2% EtOAc/DCM (10ml), added to a Kieselgel 60 ART 7734 dry packed column (50g). Elution with this solvent yielded a pure toxin.

Results and Discussion

The yield of the pure toxin (730mg) was about 0.7% of the original weight of freeze-dried plant. From the analysis of its NMR and MS spectra, the toxin was identified as 9-oxo-10,11-dehydroagerophorone (Fig. 55.2).

This cadenine sesquiterpene was first isolated by Bohlmann and Gupta (1981) from *E. adenophorum* and its structure assigned on the basis of its ^{1}H NMR and MS. The NMR and MS of the obtained toxin agree with those published by Bohlmann and Gupta (1981). The structure of the toxin was confirmed by 2-D NMR techniques including short- and long-range HECTOR and HMBC experiments, as well as the complete ^{13}C assignments for the toxin (Fig.55.2), which had not been previously reported.

The highest non-fatal oral dose of the toxin for male mice was 350-400mg/kg. Mice affected with the toxin were jaundiced within 2d of dosing, and the jaundice persisted for 3wks. Histologically, there was degeneration and loss of the epithelium of most of the intrahepatic bile ducts. The duct walls were thickened from edema and fibrosis, and the small portal tracts were more prominent than usual due to an increase of connective tissue and histiocytes. The lesions caused by the toxin were identical to those caused by the freeze-dried plant (Fig. 55.3).

Focal hepatic necrosis has also been demonstrated in rats dosed with α-naphthyl isothiocyanate (ANIT), and has also been shown to occur following obstruction of the bile duct in normal non-intoxicated rats due to regional loss of integrity of the bile duct wall (allowing escape of concentrated bile components toxic to hepatocytes) (Seawright *et al.*, 1996). In a study of the mechanism of action of the toxic principle Seawright *et al.* (1996) showed that the biliary tract lesion caused in the mouse by a single oral dose of 300mg/kg of the toxin (Fig. 55.2) was identical to that caused by ANIT.

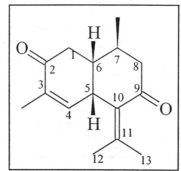

Fig. 55.2. The chemical structure of 9-oxo-10,11-dehydroagerophorone.

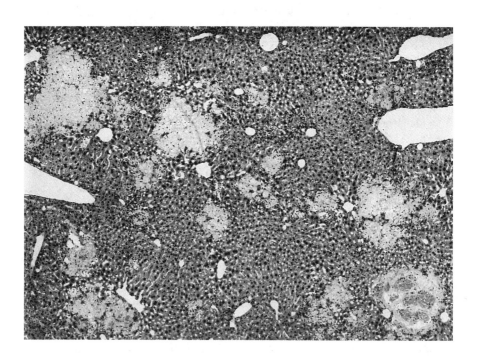

Fig. 55.3. The liver of a mouse dosed with the toxin from *E. adenophorum.*

When mice were treated with oltipraz, a dithiolthione known to increase both GSH and GSH-S transferase levels in the liver, and dosed 48hrs later with 300mg/kg of the toxin the necrogenic effect on the intrahepatic biliary tract 24hrs later was increased. Pretreating mice with butathionine sulfoximine and diethylmaleate to suppress GSH levels, followed 2hrs later by 300mg/kg of the toxin resulted in periacinar hepatocellular necrosis but no cholestasis or biliary tract injury.

It is of interest that a given dose of the toxin appears to cause two completely different liver lesions depending on the level of GSH metabolism. Further work is being done and the results will be published in due course. The relevance of the toxin for lung damage in horses ingesting *E. adenophorum* remains undetermined.

References

Bohlmann, F. and Gupta, R.K. (1981) Six Cadenine Derivatives from *Ageratina adenophora*. *Phytochemistry* 20, 1432-1433.

Everist, S.L. (1981) *Poisonous Plants of Australia*. Angus and Robertson, Sydney, pp. 167-169.

Fuller, T.C. (1981) Introduction and spread of *Eupatorium adenophorum* in California. In: *Proceedings of the 8th Asian-Pacific Weed Science Conference*. Asian-Pacific Weed Science Society, Bangalore , India, pp. 277-280.

Gibson, J.A. and O'Sullivan, B.M. (1984) Lung lesions in horses fed mist flower *Eupatorium riparium. Australian Veterinary Journal* 61, 271.

Oelrichs, P.B., Calanasan, C.A., MacLeod, J.K., Seawright, A.A. and Ng, J.C. (1995) Isolation of a compound from *Eupatorium adenophorum* (Spreng) and *Ageratina adenophora* (Spreng) causing hepatotoxicity in mice. *Natural Toxins* 3, 350-354.

O'Sullivan, B.M. (1979) Crofton weed toxicity in horses. *Australian Veterinary Journal* 55, 19-21.

Sani, Y., Harper, P.A.W., Cook, R.L., Seawright, A.A. and Ng, J.C. (1992) The toxicity of *Eupatorium adenophorum* for the liver of the mouse. In: James, L.R., Keeler, R.F., Bailey, E.M. Jr, Cheeke, P.R. and Hegarty, M.P. (eds), *Poisonous Plants*. Iowa State University Press, Ames, IA, pp. 626-629.

Seawright, A.A., Nolan, C.C., Oelrichs, P.B. and Ng, J.C. (1996) Increasing or decreasing hepatic glutathione levels in mice modifies the hepatoxicity of *Eupatorium* toxin (9-oxo-10,11-dihydroageraphorone). *Human and Experimental Toxicology* 15, 674.

Chapter 56

Glycosidase Inhibitors in British Plants as Causes of Livestock Disorders

R.J. Nash[1], A.A. Watson[1], A.L. Winters[1], G.W.J. Fleet[2], M.R. Wormald[3], S. Dealler[4], E. Lees[4], N. Asano[5] and R.J. Molyneux[6]

[1]Institute of Grassland and Environmental Research, Aberystwyth SY23 3EB, UK; [2]Dyson Perrins Laboratory, South Parks Road, Oxford OX1 3QY, UK; [3]Oxford Glycobiology Institute, South Parks Road, Oxford OX1 3QU, UK; [4]Burnley General Hospital, Casterton Avenue, Burnley BB10 2PQ, UK; [5]Faculty of Pharmaceutical Sciences, Hokuriku University, Kanazawa, Japan; [6]USDA, Western Regional Research Center, Albany, California 94710, USA

Introduction

Inhibition of mammalian glycosidases and glycosyltransferases by ingestion of plant and microbial compounds has been shown to cause many livestock disorders around the world. The mannosidase-inhibiting alkaloid swainsonine, found in *Swainsona*, *Astragalus*, *Oxytropis* and *Ipomoea* species (Colegate *et al.*, 1979), causes "locoism" and "peastruck" in livestock. Clinical signs include depression, tremors, emaciation and gastrointestinal disorders with a toxic dose as low as 0.001% in the diet (Molyneux *et al.*, 1994). Castanospermine, the related alkaloid from the Australian tree *Castanospermum australe*, causes gastrointestinal distress in horses, pigs and cattle and degenerative vacuolation of hepatocytes and skeletal monocytes by inhibition of glucosidases (Stegelmeier *et al.*, 1995). Tunicaminyl-uracils are antibiotic complexes produced by *Clavibacter* and *Streptomyces* species that are specific inhibitors of glycosylation of *N*-linked glycoproteins. These cause cerebellum dysfunction by disturbance of brain vasculature resulting in incoordination, head swaying and cerebral convulsions (e.g. in annual ryegrass toxicity) (Bourke, 1994). Polyhydroxylated *nor*tropane alkaloids (calystegines), potent inhibitors of mammalian glucosidases and galactosidases, have been implicated in neurological disorders associated with *Solanum* species: for example, *S. dimidiatum* in Texas, which causes "Crazy Cow Syndrome," and *S. kwebense* in South Africa, which causes "Maldronksiekte" (Menzies *et al.*, 1979). Calystegines were discovered as major alkaloids in the skin of potatoes and other solanaceous human foods in 1993. An investigation of British plants for glycosidase-inhibiting compounds was initiated to help explain a variety of livestock problems in the UK.

Detection of Glycosidase Inhibitors in British Species

Extracts of aerial parts of over 250 herbaceous plant species were tested for inhibition of glycosidase activities on crude homogenates of bovine brain material collected from abattoirs. *P*-nitrophenyl- (α-glucosidase) and methyl-umbelliferyl substrates were used to screen for α-glucosidase, β-glucosidase, β-galactosidase, β-*N*-acetylglucosaminidase and β-*N*-acetylgalactosaminidase activities. Extracts prepared in 70% ethanol were subject to cation exchange chromatography. Bound material displaced with 2M ammonia was dried and standardized to concentrations equivalent to 10mg fwt/ml. Over 20% of the randomly selected plants inhibited glycosidases (usually specific activities), and examples of the results obtained are shown in Tables 56.1 and 56.2. Assay details will be published later.

Extracts of plants showing strong inhibition of glycosidases were derivatized with trimethylsilyl-ether (prepared using Sigma-Sil-A) and analyzed by GC/MS. In addition, HPLC with a Pulsed Electrochemical Detector (PED) (Dionex DX 500) was found useful in detecting alkaloid glycosidase inhibitors. The mobile phase was 100mM HCl, with post column addition of 300mM NaOH. Sensitivity was in the nanogram range for glycosides of alkaloids and the method was able to separate most previously known plant glycosidase-inhibiting alkaloids and glycosides.

Calystegines in the Solanaceae and Convolvulaceae

Table 56.1 illustrates the results of analysis from four plant species containing calystegine. Besides the reported inhibition of β-glucosidases and galactosidases by calystegines (Fig. 56.1), there was strong inhibition of the α-glucosidase activity, which is not explained by compounds currently described from these plants. Calystegines in classes A, B and C were detected in all the samples of these plants but at very low concentrations in *Calystegia sylvatica*. However, this plant was mature and the concentrations of calystegines decrease in vegetative material as it ages. Besides free calystegines, *Atropa belladonna* contained several glycosides including 3-O-β-D-glucopyranosyl-calystegine B_1 and 4-O-α-D-galactopyranosyl-calystegine B_2. The concentration of the galactoside of calystegine B_2 increased markedly in potatoes stored at 4°C. The glycosides can inhibit glycosidases, but their specificities differ from those of the aglycones. For example, the 3-O-β-D-glucopyranosyl-calystegine B_1 inhibited rice α-glucosidase (Asano *et al.*, 1997). Activities of the glycosides and of the calystegines will be reported later.

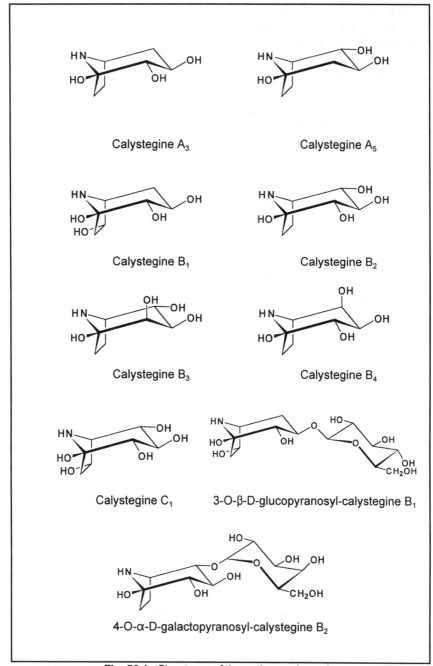

Fig. 56.1. Structures of the various calystegines.

Table 56.1. Examples of inhibition of bovine brain glycosidase activities by British Solanaceae and Convolvulaceae.

Plant species[1]	Inhibition of enzyme activity (%)				
	α-glucoside-ase	β-glucoside-ase	β-galactosid-ase	β-N-acetyl-galactos-aminidase	β-N-acetyl-glucos-aminidase
Convol-vulus avensis	98	78	50	5	11
Caly-stegia sylvatica	28	50	0	30	25
Atropa bella-donna	90	94	40	20	15
Solanum dulcam-ara	70	65	25	20	10

[1]70% EtOH extracts were tested after cation exchange chromatography (Dowex 50 H+ form) and made up to the equivalent of 10mg fwt/ml.

New Sources of Glycosidase Inhibitors

Table 56.2 shows the inhibition profiles obtained from some common wild plant species in the UK. While many of the plants are still undergoing investigations and many of the compounds are still being evaluated for biological activity, it is clear that polyhydroxylated alkaloids of various structural types are common in British plants. Some of the interesting new structures discovered in the plant families Hyacinthaceae and Campanulaceae will be considered here.

The analysis of the Bluebell (*Hyacinthoides non-scripta*) yielded several alkaloids, potent inhibitors of glycosidases, found for the first time in this family. The major alkaloid in leaves, fruits and bulbs is 2,5-dihydroxymethyl-3,4-dihydroxy-pyrrolidine (DMDP, up to 1% dwt), reportedly a strong inhibitor of α- and β-glucosidases. Another glucosidase-inhibiting alkaloid, 1,4-dideoxy-1,4-imino-D-arabinitol (D-AB1), was detected in the leaf extract. Other novel alkaloids and glycosides including 2,5-dideoxy-2,5-imino-DL-*glycero*-D-*manno*-heptitol (homo-DMDP) and homo-DMDP-7-*O*-apioside have been identified (Watson *et al.*, 1997) (Fig. 56.2). Novel poly-hydroxylated pyrrolizidine alkaloids have been identified in the Hyacinthaceae such as *Scilla* species and in Bluebells. *Hyacinthus orientalis* is cultivated in the UK and the glucosidase inhibiting piperidine alkaloid α-homo-

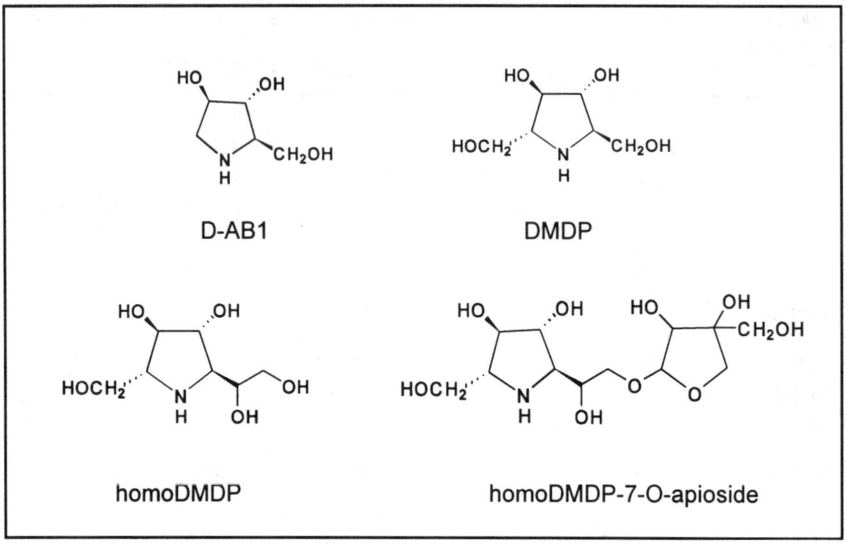

Fig. 56.2. Structures of other glucosidase-inhibiting alkaloids.

nojirimycin is a major alkaloid in both leaves and bulbs. HomoDMDP is a more potent inhibitor ($K_i=1.5\mu M$) of the almond β-glucosidase than DMDP ($K_i=10\mu M$) or than D-AB1 ($K_i=280\mu M$). Many of the pyrrolizidine alkaloids isolated from the Hyacinthaceae are not potent inhibitors of standard commercial glycosidases, and some appear to be completely inactive.

Although rare, there are reports of livestock that have been poisoned by grazing *H. non-scripta*. The signs in the horse are choking, cold and clammy skin, reduced pulse rate, abdominal pain and dysentery. Cows suffer from lethargy and dullness, with heads hanging low and ears drooping, slow and deliberate movements and staggering when driven (Thursby-Pelham, 1967). The inhibition of glycosidase activities by the polyhydroxylated pyrrolidine alkaloids purified from *H. non-scripta* suggests that these compounds could be responsible for the toxicity of this plant to mammals. Other polyhydroxylated piperidine, pyrrolizidine alkaloids and glycosides identified as major alkaloids in the Hyacinthaceae may contribute to toxic effects of other members of this family, such as *Hyacinthus* itself (Shaw and Williams, 1986).

The potent inhibition of glucosidase activity by *Campanula rotundifolia* Harebell (Table 56.2) was due to 1% dwt of DMDP, the major alkaloid during the spring and summer. Minor related alkaloids exist, and analysis of other *Campanula* species had high concentrations of these, permitting the isolation of a phenolic derivative of DMDP (compound designated PLB-1 (Fig. 56.3) or its enantiomer). There is a structural resemblance between PLB-1 and the antibiotic anisomycin. The standard commercial glycosidases tested are not strongly inhibited by PLB-1.

Table 56.2. Examples of the selective inhibition of bovine brain glycosidase activities by common British wild flower species.

Plant species[1]	α-gluco-sidase	β-gluco-sidase	β-galacto-sidase	β-N-acetyl-galactos-aminidase	β-N-acetyl-glucos-aminidase
	Inhibition of enzyme activity (%)				
Hyacinth-oides non-scripta	100	70	70	33	26
Petasites hybridus	70	0	0	0	0
Meland-rium rubrum	20	40	0	70	93
Geranium dissectum	10	60	0	83	9
Primula vulgaris	85	20	10	45	50
Campan-ula rotundi-folia	80	85	44	59	65
Vicia cracca	20	90	0	0	0
Lotus cornicul-atus	25	50	0	91	88
Taraxa-cum officinale	95	30	14	1	20
Cirsium arvense	20	96	0	4	0
Urtica dioica	0	10	0	0	0

[1]70% EtOH extracts were tested after cation exchange chromatography (Dowex 50H+ form) and made up to the equivalent of 10mg fwt/ml.

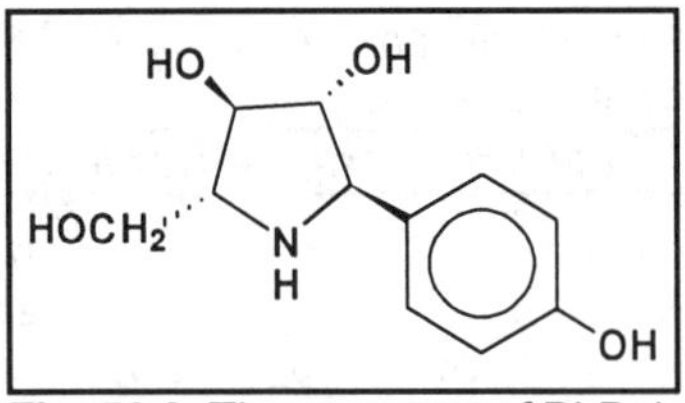

Fig. 56.3. The structure of PLB-1.

The pyrrolidine alkaloid DMDP, initially reported from tropical legume genera *Lonchocarpus* and *Derris*, appears to be a common secondary metabolite in temperate species. These produced a wider range of derivatives of DMDP than those reported from tropical species. Low concentrations of DMDP were detected in lush growth from ryegrass (*Lolium perenne*) surrounding cow dung. However, a *Streptomyces* spp. has been reported to produce an alkaloid that could have been taken up from the soil. Several plant parasitic nematode species are inhibited by DMDP, and several glycosidase inhibitors have been reported to be antifeedants to insects (Nash *et al.*, 1996). They may therefore act as defense to a number of classes of potential predators.

Other studies are investigating the inhibition of cellulase and β-glucosidase activities by certain varieties of the grasses (*Festuca arundinacea* and *L. perenne*), which is seasonal with the inhibition peaking in May of 1994 and 1995 in Wales. The timing corresponds with the peak number of cases of Equine Grass Sickness and the related disorder Leporine Dysautonomia. Both diseases have a neurological and gastrointestinal component (Griffiths and Whitwell, 1993). Since cellulase inhibitors have been reported from grass species, e.g. *Dactylis glomerata* (Sidhu *et al.*, 1967), the possible relationship to endophytes is being investigated.

Significance to Animal Health in the UK

A number of the plants investigated have been reported to be toxic, but the identification of glycosidase-inhibiting alkaloids may require a fresh look at the suspected mode of toxicity. Every year in the UK there are cases of suspected poisoning for which no clear etiology can be discerned. For example, in 1992, a dairy herd in Norfolk had an outbreak of a disorder similar to bovine spongiform encephalopathy (BSE) affecting 50% of the herd with signs of apprehension, salivation, tremors, abnormal head carriage and loss of condition (Pritchard and Bowman, 1993). The clinical signs were virtually identical to those listed for BSE (Wilesmith *et al.*, 1992). This dairy had already had 21 confirmed cases of BSE. The findings were inconclusive, but it is now significant with the knowledge of calystegines that the herd was fed potatoes, though the quantities were not reported. Young green shoots of potatoes contain high concentrations of calystegines (exceeding 0.1% fwt). Since neurological disorders caused by swainsonine (histopathologically characterized by neuronal vacuolation and axonal dystrophy) and the presence of many structural classes of glycosidase-inhibiting alkaloids in UK wild plants and crops, these classes of compound clearly deserve consideration in the differential diagnosis of BSE. At present, about 15% of suspected BSE cases cannot be confirmed by histopathological studies. "Cold cow syndrome" has symptoms of staggering and muscle tremor, and has no known etiology. These inhibitors may be involved in other disorders, involving neurolgical or gastrointestinal and wasting disorders such as Equine Grass Sickness and Leprine Dysautonomia.

It may be that these inhibitors have sub-clinical effects which may be noticed only occasionally. They may be involved in poor weight gain and have synergistic effects with other disorders (or toxins). For example, with *Vicia cracca* the potent inhibition of β-glucosidase may serve to retain its defensive secondary compounds vicine and convicine, which are hydrolyzed by almond β-glucosidase activity (Meijer and Muuse, 1988). The ability of alkaloid glycosidase inhibitors to inhibit synthesis and degrade oligosaccharides on glycoproteins might mean that they could be factors associated with the glycosylation of the prion proteins of BSE and scrapie. Studies are beginning on the inhibition of glycosyl-transferases by the natural inhibitors from British plant species.

Acknowledgments

In the UK we thank the British and Biological Sciences Research Council and Ministry of Agriculture, Fisheries and Food for partly funding this work.

References

Asano, N., Kato, A., Kizu, H., Matsui, K., Griffiths, R.C., Jones, M.G., Watson, A.A. and Nash, R.J. (1997) Enzymatic synthesis of glycosides of calystegines B$_1$ and B$_2$ and their glycosidase inhibitory activities. *Carbohydrate Research* 304, 173-178.

Bourke, C.A. (1994) Tunicaminyluracil toxicity, an emerging problem in livestock fed grass or cereal products. In: Colegate, S.M. and Dorling, P.R. (eds), *Plant-Associated Toxins: Agricultural, Phytochemical and Ecological Aspects*. CAB International, Wallingford, Oxon, pp. 399-404.

Colegate, S.M., Dorling, P.R. and Huxtable, C.R. (1979) A spectroscopic investigation of swainsonine: an α-mannosidase inhibitor isolated from *Swainsona canescens*. *Australian Journal of Chemistry* 32, 2257-2264.

Griffiths, I.R. and Whitwell, K. (1993) Leporine dysautonomia: further evidence that hares suffer from grass sickness. *Veterinary Record* 132, 376-377.

Meijer, M.M.T. and Muuse, B.G. (1988) Optimalisation of dehulling technique and enzyme hydrolysis of vicine/convicine to eliminate ANFs of faba beans. In: Huisman, J., Poel, T.F.B. and Liener, I.E.J. (eds), *Recent Advances of Research in Antinutritional Factors in Legume Seeds*. Centre for Agricultural Publishing and Documentation, Wageningen, pp. 268-271.

Menzies, J.S., Bridges, C.H. and Bailey, E.M. Jr (1979) A neurological disease associated with *Solanum dimidiatum*. *The Southwestern Veterinarian* 32, 45-49.

Molyneux, R.J., James, L.F., Ralphs, M.H., Pfister, J.A., Panter, K.P. and Nash, R.J. (1994) Polyhydroxy alkaloid glycosidase inhibitors from poisonous plants of global distribution: analysis and identification. In: Colegate, S.M. and Dorling, P.R. (eds), *Plant-Associated Toxins: Agricultural, Phytochemical and Ecological Aspects*. CAB International, Wallingford, Oxon, pp. 107-112.

Nash, R.J., Watson, A.A. and Asano, N. (1996) Polyhydroxylated alkaloids that inhibit glycosidases. In: Pelletier, S.W. (ed), *Alkaloids: Chemical and Biological Perspectives, Vol 11*. Elsevier Science, Oxford, pp. 344-376.

Pritchard, G. and Bowman, A. (1993) To BSE or not to BSE. *State Veterinary Journal* 2, 4.

Rice, D., McMurray, C. and McFarland, P. (1983) Cold cow syndrome. *Veterinary Record* 112, 531.

Shaw, R.J. and Williams, M.C. (1986) Consider the lilies of the field. *Utah Science* 47, 30-35.

Sidhu, K.S., Hargus, W.A. and Pfander, W.H. (1967) Metabolic inhibitor(s) in fractions of orchardgrass (*Dactylis glomerata* L.) by *in vitro* rumen fermentation technique. *Proceedings of the Society for Experimental Biology and Medicine* 124, 1038-1041.

Stegelmeier, B.L., Molyneux, R.J., Elbein, A.D. and James, L.F. (1995) The lesions of locoweed (*Astragalus mollissimus*), swainsonine, and castanospermine in rats. *Veterinary Pathology* 32, 289-298.

Thursby-Pelham, R.H.C. (1967) Suspected *Scilla non-scripta* (bluebell) poisoning in cattle. *Veterinary Record* 80, 709-710.

Watson, A.A., Nash, R.J., Wormald, M.R., Harvey, D.J., Dealler, S., Lees, E., Asano, A., Kizu, H., Kato, A., Griffiths, R.C., Cairns, A. and Fleet, G.W.J. (1997) Glycosidase inhibiting pyrrolidine alkaloids from *Hyacinthoides non-scripta*. *Phytochemistry* 46, 255-259.

Wilesmith, J.W., Hoinville, L.J., Ryan, J.B.M. and Sayers, A.R. (1992) Bovine spongiform encephalopathy: aspects of the clinical picture and analyses of possible changes 1986-1990. *Veterinary Record* 130, 197-201.

Chapter 57

The Immunologic and Toxic Effects of Chronic Locoweed (*Astragalus lentiginosus*) Intoxication in Cattle

B.L. Stegelmeier, P.W. Snyder, L.F. James, K.E. Panter, R.J. Molyneux, D.R. Gardner, M.H. Ralphs and J.A. Pfister

USDA-ARS, Poisonous Plant Research Laboratory, 1150 East 1400 North, Logan, Utah 84341, USA

Introduction

Swainsonine, the toxin that causes locoism or locoweed poisoning, inhibits several mannosidases, causing abnormal lysosomal catabolism and glycoprotein metabolism (Molyneux and James, 1982). Locoweed poisoning in livestock has an insidious onset, with clinical signs not apparent until the animal has grazed locoweed for several weeks. Swainsonine also inhibits Golgi mannosidase II, a key enzyme in the glycosylation of many glycoproteins including enzymes, receptors, signal transducers, hormones and cytokines (James and Panter, 1989; Stegelmeier *et al.*, 1995a). Swainsonine is a potential chemotherapeutic agent, as glycoproteins are important in viral infections, carcinogenesis, and the metastasis of neoplasms (Dennis *et al.*, 1990; Mohla *et al.*, 1990; Olden *et al.*, 1991). Anecdotal reports suggest that locoweed-poisoned animals are more susceptible to common respiratory diseases. *In vitro*, swainsonine enhanced concanavalin A-induced blastogenesis, lymphokine-activated killer cell and null killer cell activity, cytotoxicity of large granular lymphocytes and activation of macrophages and other antigen-presenting cells (Bowlin *et al.*, 1989; Yagita and Saksela, 1990; Bowlin *et al.*, 1991; Fujieda *et al.*, 1994; Galustian *et al.*, 1994; Das *et al.*, 1995). In other studies, swainsonine inhibited pokeweed mitogen-induced and phytohemagglutinin-induced lymphocyte blastogenesis, lymphocyte response to soluble antigens and lysosomal function (Sharma *et al.*, 1984; Wall *et al.*, 1988; Marijanovic *et al.*, 1990; Karasuno *et al.*, 1992; Tulsiani and Touster, 1992). The long-term or residual effects of locoweed poisoning on the immune system *in vivo* are unclear.

Material and Methods

Four groups of three mixed-breed beef heifers were gavaged with ground *Astragalus lentiginosus* to obtain swainsonine doses of 0.0, 0.25, 0.75 and 2.25 mg/kg/d for 45d, after which they were allowed to recover for 45d. During dosing and recovery, blood samples and liver and lymph node biopsies were collected. On dosing days 25 and 39 all heifers were immunized with keyhole limpet hemocyanin (KLH) in complete Freund's adjuvant (first injection) and incomplete Freund's adjuvant (booster). On recovery days 25 and 39 similar injections were made with ovalbumin (OVA). On dosing day 45 and recovery day 43 the animals were intradermally injected with 0.1ml tuberculin, and responses were measured using a caliper 72hrs later. After 45d of recovery, the heifers were slaughtered and tissues were collected for immunology analyses and microscopic evaluation.

Hematology, serum biochemistry and serum swainsonine analyses were done using reported techniques (Stegelmeier *et al.*, 1995b). Immunoglobulins were quantitated using radial immunodiffusion (VMRD, Pullman, WA). Lymphocyte populations were compared using normal hemograms and by flow cytometry with specific lymphocyte markers (FAB, WCI, CD2, CD3, CD4, CD8, and TCR; VMRD, Pullman, WA). Lymphocytes were isolated, cultured and incubated with KLH, OVA, or mitogens [concanavalin A (ConA) or pokeweed mitogen (PWM)]. After pulsing with tritiated thymidine, cells were harvested, washed, counted and a proliferation index calculated. Responses to KLH and OVA were compared using the proliferation index, serum titers and a solid phase immuno-enzymatic technique to identify antibody-secreting cells (Sedgwick and Holt, 1983). Data were analyzed by analysis of variance with a generalized linear model for a repeated measures design. Mean separations were done using Duncan's method at $P < 0.05$ after a significant R test at $\alpha < 0.05$ (SAS Institute, Cary, NC).

Results and Discussion

All heifers dosed with locoweed started showing signs of locoweed poisoning after 3wks of dosing. The high-dose animals were most severely affected as they ate less, stood around with low head carriage and were slow to respond to stimuli. Within 30d of locoweed dosing, all treated animals had significantly lower hematocrits than the controls (Fig. 57.1), and these did not return to normal until 14d after dosing was discontinued. Hemogram analysis showed that the erythrocytes were morphologically normal. Swainsonine reportedly stimulates hematopoietic precursors and protects them from the effects of cancer chemotherapeutics including methotrexate, fluorouracil, cyclophosphamide and doxorubicin (Olden *et al.*, 1991; Oredipe *et al.*, 1991; White *et al.*, 1991). It may be that chronic swainsonine intoxication disrupts or inhibits hematopoiesis, but it does not appear to affect the leukocyte numbers or populations.

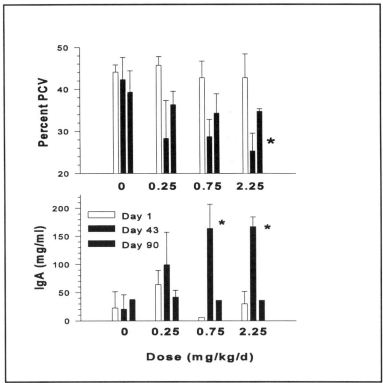

Fig. 57.1. Histogram of hematocrit and serum IgA concentrations of cattle dosed with locoweed at 0.0, 0.25, 0.75 and 2.25mg swainsonine/kg/d for 45d followed by a recovery time of 45d. * =significantly different (*P*<0.05).

Vacuolation of monocytes and large lymphocytes in the peripheral blood was first observed after 6d of dosing. Later, nearly all large lymphocytes and monocytes were severely vacuolated. Small lymphocytes or granulated cells were not affected. Lymphocyte vacuolation persisted for 12d into recovery. No significant differences were detected in peripheral blood lymphocyte numbers or populations.

Significant biochemical changes included increased aspartate aminotransferase and alkaline phosphatase with decreases in serum iron, transferrin and thyroid hormones. Serum swainsonine concentrations were similar to previous reports at 0, 2.4±7.0, 180±43 and 510±102ng/ml for the controls, low, medium and high dose groups (Stegelmeier *et al.*, 1995b). Histologically, the Kupffer cells of the liver and lymph node macrophages were severely vacuolated in medium and high dose animals, and this persisted for 2wks into recovery. Tissues were histologically normal with rare axonal spheroids in the white tracts of the cerebellar crus and medulla. No significant lesions were identified in the controls or low-dose animals.

Stegelmeier et al.

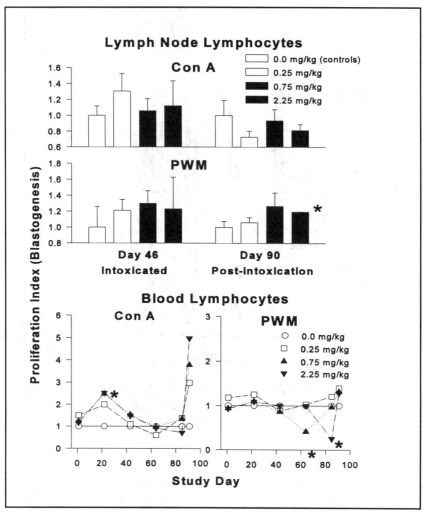

Fig. 57.2. Histogram of lymph node and blood lymphocyte response (blastogenesis) to concanavalin A (ConA) and pokeweed mitogen (PWM). Lymphocytes were isolated from cattle dosed with locoweed at 0.0, 0.25, 0.75 and 2.25mg swainsonine/kg/d for 45d followed by a recovery period of 45d. * = significantly different (*P*<0.05).

Locoweed, especially at 0.25 and 0.75mg swainsonine/kg/d doses, promoted proliferation of lymph node and peripheral lymphocytes in response to ConA and PWM (Fig. 57.2), with a maximum promotion on day 20 in response to ConA. The promotion was brief, as locoweed-treated animals had lower lymphocyte responses after weeks of treatment. All animals responded similarly to tuberculin injections.

Locoweed-treated animals also had enhanced responses to injected antigens, especially in lymphocytes isolated from lymph nodes. Some responses were nonspecific, with increased titers even to antigens to which the animals had not yet been exposed. No differences were detected in serum titers or specific circulating lymphocytes. Mouse lymphocytes *in vitro* had similar increased responses to novel protein antigens (OVA) when pretreated with swainsonine or isolated from swainsonine-treated mice (Marijanovic *et al.*, 1990). The stimulative effect was quickly reversed in the recovery period, when no groups had increased responses.

Locoweed-treated animals had significantly higher serum IgA concentrations (Fig. 57.1). Other natural toxins including trichothecene and vomitoxin enhance secretion of several interleukins (Azcona-Olivera *et al.*, 1995). Vomitoxin has been shown to dysregulate IgA, and this may contribute to IgA-induced nephropathy (Greene *et al.*, 1994; Ouyang *et al.*, 1995). As vomitoxin also inhibits protein synthesis, swainsonine-induced stimulation may occur *via* the same mechanism. Additional work is needed to identify the mechanism and determine the point at which the effects of locoweed intoxication become detrimental.

References

Azcona-Olivera, J.I., Ouyang, Y.L., Warner, R.L. and Linz, J.E. (1995) Effects of vomitoxin (deoxynivalenol) and cycloheximide on IL-2, 4, 5 and 6 secretion and mRNA levels in murine CD4+ cells. *Food and Chemical Toxicology* 33, 433-441.

Bowlin, T.L., McKown, B.J., Kang, M.S. and Sunkara, P.S. (1989) Potentiation of human lymphokine-activated killer cell activity by swainsonine, an inhibitor of glycoprotein processing. *Cancer Research* 49, 4109-4113.

Bowlin, T.L., Schroeder, K.K. and Fanger, B.O. (1991) Swainsonine, an inhibitor of mannosidase II during glycoprotein processing, enhances concanavalin A-induced T cell proliferation and interleukin 2 receptor expression exclusively *via* the T cell receptor complex. *Cellular Immunology* 137, 111-117.

Das, P.C., Roberts, J.D., White, S.L. and Olden, K. (1995) Activation of resident tissue specific macrophages by swainsonine. *Oncology Research* 7, 425-433.

Dennis, J.W., Koch, K., Yousefi, S. and van der Elst, I. (1990) Growth inhibition of human melanoma tumor xenografts in athymic nude mice by swainsonine. *Cancer Research* 50, 1867-1872.

Fujieda, S., Noda, I., Saito, H., Hoshino, T. and Yagita, M. (1994) Swainsonine augments the cytotoxicity of human lymphokine-activated killer cells against autologous thyroid cancer cells. *Archives of Otolaryngology Head and Neck Surgery* 120, 389-394.

Galustian, C., Foulds, S., Dye, J.F. and Guillou, P.J. (1994) Swainsonine, a glycosylation inhibitor, enhances both lymphocyte efficacy and tumour susceptibility in LAK and NK cytotoxicity. *Immunopharmacology* 27, 165-172.

Greene, D.M., Bondy, G.S., Azcona-Olivera, J.I. and Pestka, J.J. (1994) Role of gender and strain in vomitoxic-induced dysregulation of IgA production and IgA nephropathy in the mouse. *Journal of Toxicology and Environmental Health* 43, 37-50.

James, L.F. and Panter, K.E. (1989) Locoweed poisoning in livestock. In: James, L.F., Elbein, A.D., Molyneux, R.J. and Warren, C.D. (eds), *Swainsonine and Related Glycoside Inhibitors*. Iowa State University Press, Ames, IA, pp. 23-38.

Karasuno, T., Kanayama, Y., Nishiura, T., Nakao, H., Yonezawa, T. and Tarui, S. (1992) Glycosidase inhibitors (castanospermine and swainsonine) and neuraminidase inhibit pokeweed mitogen-induced B cell maturation. *European Journal of Immunology* 22, 2003-2008.

Marijanovic, D., Norred, W.P. and Riley, R.T. (1990) Effect of swainsonine on antigen specific immune response in mice. *Veterinary and Human Toxicology* 32, 111.

Mohla, S., White, S., Grzegorzewski, K., Nielsen, D., Dunston, G., Dickson, L., Cha, J.K., Asseffa, A. and Olden, K. (1990) Inhibition of growth of subcutaneous xenografts and metastasis of human breast carcinoma by swainsonine: modulation of tumor cell HLA class I antigens and host immune effector mechanisms. *Anticancer Research* 10, 1515-1522.

Molyneux, R.J. and James, L.F. (1982) Loco intoxication: Indolizidine alkaloids of spotted locoweed (*Astragalus lentiginosus*). *Science* 216, 190-191.

Olden, K., Breton, P., Grzegorzewski, K., Yasuda, Y., Gause, B.L., Oredipe, O.A., Newton, S.A. and White, S.L. (1991) The potential importance of swainsonine in therapy for cancers and immunology. *Pharmacology Therapy* 50, 285-290.

Oredipe, O.A., White, S.L., Grzegorzewski, K., Gause, B.L., Cha, J.K., Miles, V.A. and Olden, K. (1991) Protective effects of swainsonine on murine survival and bone marrow proliferation during cytotoxic chemotherapy. *Journal of the National Cancer Institute* 83, 1149-1156.

Ouyang, Y.L., Azcona-Olivera, J.I. and Pestka, J.J. (1995) Effects of trichothecene structure on cytokine secretion and gene expression in murine CD4+ T-cells. *Toxicology* 104, 187-202.

Sedgwick, J.D. and Holt, P.G. (1983) A solid phase immunoenzymatic technique for the enumeration of specific antibody-secreting cells. *Journal of Immunology Methods* 57, 301-309.

Sharma, R.P., James, L.F. and Molyneux, R.J. (1984) Effect of repeated locoweed feeding on peripheral lymphocytic function and plasma proteins in sheep. *American Journal of Veterinary Research* 45, 2090-2093.

Stegelmeier, B.L., Molyneux, R.J., Elbein, A.D. and James, L.F. (1995a) The comparative pathology of locoweed, swainsonine, and castanospermine in rats. *Veterinary Pathology* 32, 289-298.

Stegelmeier, B.L., James, L.F., Panter, K.E. and Molyneux, R.J. (1995b) Serum swainsonine concentration and alpha-mannosidase activity in cattle and sheep ingesting *Oxytropis sericea* and *Astragalus lentiginosus* (locoweeds). *American Journal of Veterinary Research* 56, 149-154.

Tulsiani, D.R. and Touster, O. (1992) Evidence that swainsonine pretreatment of rats leads to the formation of autophagic vacuoles and endosomes with decreased capacity to mature to, or fuse with, active lysosomes. *Archives of Biochemistry and Biophysics* 296, 556-561.

Wall, K.A., Pierce, J.D. and Elbein, A.D. (1988) Inhibitors of glycoprotein processing alter T-cell proliferative responses to antigen and to interleukin 2. *Proceedings of the National Academy of Sciences USA* 85, 5644-5648.

White, S.L., Nagai, T., Akiyama, S.K., Reeves, E.J., Grzegorzewski, K. and Olden, K. (1991) Swainsonine stimulation of the proliferation and colony forming activity of murine bone marrow. *Cancer Communications* 3, 83-91.

Yagita, M. and Saksela, E. (1990) Swainsonine, an inhibitor of glycoprotein processing, enhances cytotoxicity of large granular lymphocytes. *Scandinavian Journal of Immunology* 31, 275-282.

Chapter 58

Sawfly (*Perreyia flavipes*) Larval Poisoning of Cattle, Sheep and Swine[1]

F. Riet-Correa[1], F. Dutra[2], M.P. Soares[1] and M.C. Méndez[1]

[1]*Laboratório Regional de Diagnóstico, Faculdade de Veterinária, Pelotas University, 96010-900, Pelotas RS, Brazil;* [2]*DILAVE Miguel C Rubino, Rincón 203, Treinta y Tres, Uruguay*

Introduction

Perreyia flavipes (Konow, *c*. 1899; Perreyiinae, Pergidae: Hymenoptera) has been reported in Argentina (provinces of Buenos Aires and Entre Rios) and Brazil (states of Espirito Santo, Rio Grande do Sul, Santa Catarina and Distrito Federal) (Smith, 1990), and is also widespread in Uruguay (Dutra *et al.*, 1997). The larvae, which were previously unknown, are glossy black and up to 2.5cm long. From June to September, masses of larvae (40-80g) are found crawling over the grass in orderly columns approximately 15cm long and 8cm wide. The masses contain an average of approximately 100 larvae, and vary from 6-200. Some days, great numbers of these masses are seen, while on other days very few or no groups are found. This behavior is probably used to search for food.

There are no previous studies on the biology of *P. flavipes*, and its feeding habits are unknown. Rodrigues Camargo (1955) discussed the biology and the habits of *Perreyia lepida* Brullé (syn: *Paraperreyia dorsuaria*), a species found in Brazil that is similar to *P. flavipes*. The larvae of *P. lepida* are gregarious, and farmers from the states of Santa Catarina and Rio Grande do Sul observed deaths in pigs after casual ingestion, which inspired the name "mata porco" ("pig killer") (de Costa Lima, 1941; Rodrigues Camargo, 1955). The larvae of *P. lepida* have been seen feeding on grasses and small shrub plants like *Eryngium* spp., as well as on cattle feces (Rodrigues Camargo, 1955). Pupation occurs in the soil, at 5mm below the surface,

---

1. Portions of this chapter are previously published as an article and reprinted here with permission from *Veterinary and Human Toxicology*.

where the prepupae construct cocoons, gumming the earth particles together to form a compact, black shell. Adults emerge in about 6mos and may be seen from November to autumn flying over small shrub plants such as *Senecio* spp., *Baccharis* spp. and *Eryngium* spp. (Rodrigues Camargo, 1955).

This chapter reviews recent reports on poisonings by the larvae of *P. flavipes* in sheep and cattle in Uruguay. Experimental intoxication in sheep, cattle and swine, insect identification, and some aspects of its biology are also reported.

Spontaneous Intoxication in Cattle and Sheep

Intoxication by *P. flavipes* larvae was observed in Uruguay. From June to early October of 1993-95, at least 40 outbreaks of a highly lethal disease occurred in cattle and sheep in central Uruguay. During 1995, total losses of cattle exceeded 1,000hd. The disease affected weaner and yearling beef cattle of both sexes more frequently than older stock. Mortalities were 1.6, 7.0 and 1.3% for calves, yearlings and adults, respectively, but mortalities up to 28% occurred on some farms. All cattle breeds reared in the area (mainly Hereford, Aberdeen, Angus and crosses) were affected. Sheep were less often affected than cattle. A remarkable feature of the outbreaks was the great loss of animals in a short time (4-8d). On most farms, the disease was seen in only one paddock, and on some it reoccurred 1-2wks after the initial incident. Movement of remaining animals to another pasture effectively stopped some outbreaks (Dutra *et al.*, 1997).

Most animals were found dead. Cattle with clinical signs showed weakness, depression and stupor. Others became highly excited, with fine trembling of the neck and head, and were belligerent. Most affected cattle died within 2d, but some survived for as long as a week. Jaundice and mild photosensitization were seen in some of these survivors (Dutra *et al.*, 1997).

At necropsy the liver was consistently affected. In some animals it was pale, the cut surface presenting a prominent acinar pattern, which took the form of a fine, regular, pallid network of parenchyma standing above red, hemorrhagic, depressed areas. In others, the organ was deep reddish purple and obviously swollen with rounded edges; on the cut surface it presented a mosaic appearance of few and scattered bulging gray or yellow islands of tissue intermingled with extensive, collapsed, dark red areas. Where both patterns were seen together, the acinar pattern prevailed in the left lobe while the dark mosaic pattern was more common in the right lobe (Dutra *et al.*, 1997). In most cases there was prominent edema of the gallbladder wall and its attachments, extending to the mesenteric border of the small intestine.

Sawfly larval body fragments and heads were found in the rumen and omasum of all ten animals necropsied. Gross lesions in other organs were prominent when present. Widespread ecchymoses and petechiae were observed on serous membranes, especially the epi- and endocardium. The spleen was hemorrhagic and slightly enlarged. Peritoneal vessels were injected, and diffuse hemorrhages were seen on the small intestine, mesentery and abomasum. The edematous mucosa had lines of congestion and hemorrhages along the rugae. Dried content was observed in the omasum, and the colon content was scant and drier than normal, containing blood and mucus. A moderate clear, yellow ascites was observed in some cases (Dutra *et al.*, 1997).

Histologically, there was centrilobular to extensive hemorrhagic necrosis in the liver. In less severe cases, corresponding to the acinar pattern described above, the surviving periportal hepatocytes were swollen with clear, vacuolated cytoplasm, and the centrilobular areas of necrosis were eccentric about hepatic venules. In other cases necrotic, hemorrhagic areas coalesced, isolating the portal units. Massive necrosis was depicted by the destruction of every cell of the affected lobule except some few degenerated hepatocytes surrounding the portal spaces, giving the mosaic appearance seen grossly. Hemorrhage and necrotic debris replaced the dead parenchyma. Moderate bile duct proliferation was seen in the portal units, even in peracute cases. The splenic white pulp of two animals was depleted, and the lymphatic nodules were nearly absent. The periarteriolar sheaths contained no germinal centers. The colonic lamina propria was congested and edematous, with exfoliating foci of epithelium. Mucus and erythrocytes were seen in the lumen. No significant lesions were found in other organs (Dutra *et al.*, 1997).

Experimental Intoxication in Cattle and Sheep

Two calves were dosed with nine and 40g/kg of whole fresh larvae, respectively. The calf given 9g/kg became anorexic 4-5d after dosing but recovered transiently on day six. By day nine the calf again became depressed, slightly ataxic, anorexic and had a respiratory rate of 60/min. During days five to nine the animal passed a mildly dark diarrhea. It recovered clinically 10d after dosing. Liver biopsy of that calf showed swollen, hydropic hepatocytes in the periportal area. The laminae of the periacinar region were dissociated and replaced by cells with small to medium-sized basophilic nuclei. Apoptotic cells, pyknotic fragments and some hemorrhages were also present in these regenerating foci (Dutra *et al.*, 1997). The calf given 40g/kg had a profuse and fluid diarrhea 9hrs after the ingestion, and within 14hrs it became anorexic and depressed. After 18hrs, the animal was unwilling to move and was mildly ataxic. Menace and sound responses were diminished, and the calf soon became recumbent with a respiratory rate of 24/min. By 21hrs the feces it passed were drier, and almost normal. The calf developed some degree of muscle fasciculation and was reluctant to rise. After 24hrs, the animal was severely affected, almost comatose, and was euthanized. At necropsy the liver was enlarged and mottled with rounded edges. The right lobe was swollen, deep reddish purple and the cut surface had prominent, bulging islands of pale tissue. The left lobe was pale and the cut surface was finely mottled red and yellow, giving a nutmeg appearance. Serous edema of the gallbladder that extended to the duodenum and surrounded the pancreas was a striking finding. Edema was also present between the liver and the right kidney. The gallbladder was mildly distended with bile. Some subepicardial and subendocardial petechiae were also seen and the upper small intestinal mucosa was mildly edematous. Histopathology revealed centrilobular to diffuse hepatocyte necrosis, which was most severe in the right lobe. Some necrotic areas extended up to the portal triads, where a few degenerated periportal hepatocytes remained. Hemorrhage replaced necrotic centrilobular cells, and Kupffers cells were prominent. In the left lobe, the zones of necrosis were more restricted to the hepatocytes surrounding the hepatic venules, with the surviving cells having varying degrees of degeneration. Necrotic lymphocytes were seen in the splenic white pulp and the mesenteric lymph nodes.

Three sheep were dosed with whole, fresh larvae at rates of 10, 20 and 40g/kg, respectively. All sheep were severely affected. Clinical signs were characterized by depression, anorexia, forced respiration, fine muscular tremors, dry and firm feces covered with streaks of blood and mucus, and recumbency with paddling movements, opisthotonos and intermittent seizures before dying. They died 14-68hrs after the ingestion of larvae. At necropsy the liver was enlarged, deep reddish purple or pale, with rounded borders and a nutmeg appearance on the cut surface. Edema of the gallbladder and around the pancreas and duodenum, and a scant yellowish ascites were also observed. Microscopically, all three sheep had severe hemorrhagic centrilobular necrosis. The animal given 10g/kg had more normal periportal hepatocytes than the others two animals; most of those of the sheep given 20 and 40g/kg were pyknotic and had an eosinophilic, homogeneous cytoplasm. No germinal centers were present in the cortex of mesenteric lymph nodes of sheep given 40g/kg. The paracortical area and the medullary cords were thin and depleted of lymphocytes. Many macrophages with hemosiderin were found in the medullary sinuses (Dutra *et al.*, 1997).

Experimental Intoxication in Pigs

Fresh *P. flavipes* larvae were fed to two pigs at five and 10g/kg, respectively. The pig dosed at 5g/kg became anorexic and depressed, but recovered in 5d. Serum values of aspartate amine transferase (AST) from both pigs are shown in Fig. 58.1 (Soares, 1997). The pig given 10g/kg became anorexic and depressed 48hrs after the ingestion. It was very depressed on the next day, with loss of equilibrium, muscular tremors and a slight jaundice. It died 4d after the ingestion of larvae. Necropsy findings included a yellow mottled liver, yellow fluid in the abdominal and pericardial cavities, mild jaundice, and edema of the stomach wall. Microscopic lesions of the liver were characterized by centrilobular necrosis and vacuolated hepatocytes in the periportal and midzonal regions.

In another study, larvae were dried at 100°C for 24hrs. The dry matter represented 17% of the fresh larval weight. Two pigs were given 0.87 and 1.7g/kg representing fresh larvae doses of five and 10g/kg, respectively. On the second day after ingestion, both pigs became anorexic and depressed, but recovered two or 3d later. Serum values of AST from both pigs are shown in Fig. 58.1 (Soares, 1997).

These results suggest that the toxicity of *P. flavipes* larvae is similar for cattle, sheep and pigs. The clinical signs and the rise in the serum AST in the pigs dosed with dry larvae demonstrated the toxicity of the larvae after being dried at 100°C for 24hrs.

Identification of the Adult Insect and Some Biologic Aspects

During September, 1996, several groups of live larvae were collected in Treinta y Tres, Uruguay, near the area where sawfly larval poisoning had occurred in the previous years. One adult sheep died after being fed 30g/kg of this collection. Larvae were also collected during July and August, 1996, in Rio Grande do Sul, Brazil, and this batch was lethal to pigs at 10g/kg (Soares, 1997).

Larvae were reared to adults in both laboratories by the same method. Immediately on arrival the larvae were placed in boxes containing swards of native grasses and covered with gauze to prevent the escape of adults on emergence. In order to provide the larvae with abundant live and dead grass leaves, the sward was watered and trimmed regularly to maintain a sward surface height of 10-15cm. Larvae were seen feeding on young leaves and senescent vegetation from different grass species, and also on dry cattle feces. Pupation occurred in September, 3-10cm below the soil surface. The larvae constructed cocoons from the surrounding earth,

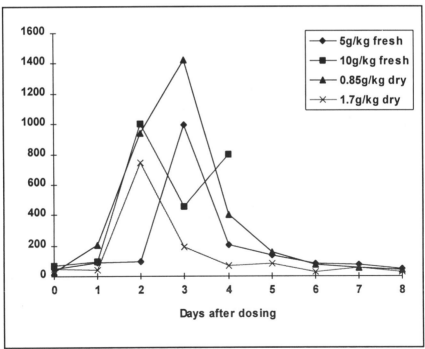

Fig. 58.1. Serum AST (IU/l) in pigs intoxicated by *P. flavipes.*

gumming the earth particles together to form compact, black rounded oblong shells approximately 1cm long by 0.5cm wide.

Adults emerged during February and March, 1997. The adults did not feed. The females died in approximately 24hrs, immediately after the deposition of the eggs, while the males lived for 36-72hrs. The females were very awkward fliers, showed little activity and flew only short distances, whereas the males were more active and flew longer distances. Copulation occurred on the top of grass leaves. The females deposited oblong, white-yellowish eggs under the vegetation or immediately below the soil surface in clusters of 100-400 each. The incubation period was 4-8wks.

Adult males and females were mounted and submitted for identification to Dr David R. Smith from the Systematic Entomology Laboratory, National Museum of Natural History, Smithsonian Institution, in Washington, DC. All specimens both from Uruguay and Brazil were identified as *P. flavipes*, a distinctive black species with contrasting bright orange legs, 14-15 segmented antenna and characteristic male genitalia. A description of the adults is given by Smith (1990).

In the outbreaks observed in Uruguay during 1995, the morbidity of the intoxication correlated with lower stocking rates (Dutra *et al.,* unpublished). Some of the biologic aspects observed in *P. flavipes* suggest that the accumulation of green and senescent grasses in the field in late autumn and winter, which gives protection and food for the larvae, enhances the frequency of intoxication. This accumulation occurs after abundant rainfalls in summer and early autumn and is also favored by low stocking rates.

References

de Costa Lima, A. (1941) Bichos "Mata Porcos". *Chácaras e Quintais* 63, 43.

Dutra, F., Riet-Correa, F., Méndez, M.C. and Paiva, N. (1997) Sawfly larval poisoning in cattle and sheep in Uruguay. *Veterinary and Human Toxicology* 39, 281-286.

Rodrigues Camargo, O. (1955) Contribuição ao estudo do Tenthredinideo "Mata Porcos," *Paraperreya dorsuaria* (Konow, 1899), no Rio Grande Do Sul. Thesis, Escola de Agronomia e Veterinária da Universidade do Rio Grande do Sul.

Smith, D.R. (1990) A synopsis of the sawflies (Hymenoptera, symphyta) of the America South of the United States: Pergidae. *Revista Brasileira de Entomologia* 34, 7-200.

Soares, M.P. (1997) Intoxicação por larvas de *Perreyia flavipes* em suínos. Thesis, Faculdade de Veterinária, Universidade Federal de Pelotas, Brazil.

Chapter 59

Comparison of the Reproductive Effects of *Baptisia australis*, *Iva annua* and *Sophora nuttalliana* in Rats

G.E. Burrows, R.P. Schwab, L.E. Stein, J.G. Kirkpatrick, C.W. Qualls and R.J. Tyrl
Oklahoma State University, Stillwater, Oklahoma 74078, USA

Introduction

Sophora nuttalliana Turner (*S. sericea*), silky sophora or white loco, is a rhizomatous, perennial herb forming colonies of plants from northern Mexico, Texas and New Mexico, northward to Wyoming and the Dakotas. It has long been suspected of being neurotoxic and was considered to be a locoweed by early ranchers (Chesnut, 1898; Marsh, 1909). This reputation has never been confirmed experimentally. The quinolizidine alkaloids represented in *Sophora* spp. are mainly tetracyclic matrine-types such as matrine, allomatrine and sophoridine, which have strong cardiac and neurologic effects (Kimura *et al.*, 1989; Tsai *et al.*, 1993). Sophocarpine (the major alkaloid of *S. nuttalliana*) (Khan *et al.*, 1992) and oxymatrine have cardiac effects but they are more noteworthy as potent neurodepressants and hypothermics (Yuan *et al.*, 1986; 1987).

Besides the neurotoxic compounds, the teratogenic quinolizidine alkaloids anagyrine and baptifoline are present in some *Sophora* spp., although they have not been specifically reported in *S. nuttalliana* (Aslanov *et al.*, 1987; Jin *et al.*, 1993). Preliminary studies on *S. stenophylla* and *S. arizonicus* indicated that both are teratogenic (Keeler, 1972). Field observations of tendon contractures in horses, similar to those reported for *Astragalus* (McIlwraith and James, 1982), may be associated with *S. nuttalliana* ingestion late in gestation. Therefore, these studies were to determine the effects of *S. nuttalliana* on development of the rat fetus.

Two other plants were evaluated. *Baptisia australis* (L.) Br., blue wild indigo or blue false indigo, was evaluated because of its anagyrine content (Wink *et al.*, 1983; Keeler, 1972; 1976). *Iva annua* L., marsh-elder or rough marsh-elder (*I. ciliata*), was included because of the abortifacient effects of a related species, *Iva angustifolia* (Murphy *et al.*, 1983).

Materials and Methods

Sixty virgin female Sprague-Dawley rats of breeding age were divided among 11 groups (Table 59.1). A female rat was placed with a male, and separated on the day that a vaginal plug was observed (day 0), and the female put on the treatment diet (plant material mixed with commercial rat chow). Feed for Groups 2 and 3 was restricted to match feed intake of *Sophora*-fed dams, which exhibited reduced feed intake and weight gain. The teratogens dexamethasone and phenytoin were administered as positive controls.

One dam from Group 9 and two from Group 11 died during the experiments. Animals were euthanized on day 20 and the fetuses removed (Barrow, 1990). Half the fetuses were placed in ethanol (EtOH) and cleared in KOH for staining of cartilage and bone with alcian blue and alizarin red S. The other fetuses were fixed in Bouin's solution, hand-sectioned, and examined for visceral abnormalities. Evaluations were based upon fetal measurements and parameters of the dams.

Table 59.1. Experimental groups established to evaluate the reproductive effects of *Sophora nuttalliana*, *Iva annua* and *Baptisia australis*.

Group	Number of dams	Treatment
Negative - controls		
1	6	Feed, full
2	5	Feed, partially restricted (PF)
3	6	Feed, restricted (R)
Sophora - fed		
4	8	15% dried leaves, day 1-20
5	4	15% dried leaves, day 1-13, then control diet to day 20
6	3	10-15% dried pods, day 1-8, then 15% dried leaves
7	6	25% dried leaves, day 1-20
Positive - controls		
8[a]	6	chow + 0.5mg dexamethasone s.c. b.i.d. from day 8-13
9[a]	5	chow + 150mg/kg or 200mg/kg phenytoin i.p. daily from day 8-14
Iva - fed		
10	8	25% dried leaves and stems, day 1-20
Baptisia - fed		
11	5	15% dried seeds/pods, day 1-20

[a] Groups 8 and 9 given full-feed rat chow (*ad libitum*) as per Group 1.
Subcutaneous=s.c., intraperitoneal=i.p., twice daily=b.i.d.

Results and Discussion

A total of 747 fetuses were obtained from 57 females. Resorption of all embryos was observed in all dams fed 25% *S. nuttalliana* (Group 7), 1/6 dexamethasone dams, 3/3 surviving dams fed 15% *B. australis,* and 2/6 dams on the control diet. Resorption of fetuses to a lesser extent was also observed in rats of other groups but there were no significant differences between groups.

Parameter evaluations for dams and fetuses may be seen in Table 59.2. Feed intake by dams fed 15% and 25% *Sophora* leaves (4, 7) was significantly decreased relative to the control group given unrestricted feed (1), and weight gain during gestation was disproportionately less for the *Sophora*-fed groups (4, 7) than for the feed-restricted groups (2, 3). This effect was especially severe in the group fed 25% *Sophora* leaves. When given 15% *Sophora* for only the first 13d of gestation (5), the decrease in dam weight gain was reversed by day 20, and final dam weights were not different from the full-feed control animals (1). Similarly, the decrease in fetal weight was significantly greater in the *Sophora* group (4) than for the matched partially restricted feed control group (2). These effects were not seen when *Sophora* was only fed for the first 13d of gestation (5).

Table 59.2. Parameter evaluations in rat dams and fetuses (means ±SD).

Group	Treatment	N*	Feed intake (g), 20d	Weight gain, (g) 20d	N	Fetal weight (g)	Placenta weight (g)
1	Control, full feed	6	546.7[a] ±22.4	169.2[a] ±25.0	131	4.13[a] ±0.54	5.58[a] ±0.59
2	Control, feed, PF	5	342.0[b,c] ±90.0	94.2[b] ±37.7	61	3.53[b,e] ±0.30	4.74[b] ±0.33
3	Control, feed, R	4	210.8[b] ±5.5	27.2[c,d] ±17.0	73	2.76[c] ±0.26	3.83[c] ±0.32
4	*Sophora,* 15% leaves	8	349.1[c] ±33.3	28.9[c,d] ±20.3	109	3.10[d] ±0.67	4.48[b] ±0.75
5	*Sophora,* 15% leaves day 1-13	4	503.0[a] ±72.1	109.2[a,b] ±40.4	62	4.12[a] ±0.86	5.53[a] ±0.72
6	*Sophora,* 25% leaves	6	296.8[b,c] ±103.1	72.3[e] ±11.4	----	NA	NA

[a] Values within columns with the same superscript are not significantly different ($P>0.05$) using Tukey's Highest Significant Difference Multiple Comparisons (Systat).
* Dams in which all fetuses were resorbed are not included.

Fetal ossification (Table 59.3) was slightly influenced by feed restriction, but was significantly decreased with 15% *Sophora* leaves (4) for all parameters measured, especially digit ossification and skull length. These effects were also seen when *Sophora* was fed for only the first 13d of gestation (5).

Changes in soft tissue and skeletal characters of fetuses were apparent in several groups. For the restricted-feed controls (3), the skulls were significantly shorter when compared to other negative controls, and there was enlargement of the subdural spaces and dimunition of the lateral ventricles in 10 of 19 fetuses from two of the dams. Among the dams given *Sophora*, the most distinct and consistent effects were seen in those given 15% in the feed for the entire 20d of gestation (4). Skull lengths were significantly shortened and were similar to those of the feed-restricted controls. However, there was also absence or marked reduction in ossification of the interparietal and supraoccipital bones of the skull in all fetuses of three of the eight dams. There was marked enlargement of the subdural space in all fetuses of three dams and in 50% of those from three other dams. This was accompanied by substantial diminution of the lateral ventricular dilation in the brain, an effect not prevented by feeding the *Sophora* for only the first 13d of gestation (5).

Table 59.3. Ossification scores and other parameter means ±SD in fetuses.

Group	N	Digits*	Vertebrae**	Skull+	Skull Length (mm)	N	Lateral Ventricular Dilation[++]
1	64	13.1[a]	15.8[a]	14.8[a]	14.4[a]	67	2.2[a]
		±2.2	±0.7	±1.0	±0.5		±0.6
		(12.0)	(16.0)	(15.0)	(14.5)		(2.0)
2	29	12.1[a,b]	16.0[a]	14.9[a]	14.3[a]	32	2.4[a,d]
		±1.2	±0.2	±0.3	0±.6		±0.6
		(12.0)	(16.0)	(15.0)	(14.0)		(2.0)
3	35	10.5[b]	15.9[a]	13.8[a,b]	13.4[b,c]	37	2.1[a,b]
		±0.6	±0.2	±1.0	±0.4		±0.6
		(10.5)	(16.0)	(14.0)	(13.5)		(2.0)
4	53	8.6[c]	15.1[b]	13.4[b]	12.9[c]	56	1.2[c]
		±3.7	±1.8	±3.0	±0.9		±1.1
		(9.0)	(16.0)	(15.0)	(13.0)		(1.0)
5	30	13.1[a,d]	15.8[a]	14.9[a]	14.3[a]	32	1.4[b,c]
		±3.6	±0.7	±0.4	±1.0		±0.8
		(11.25)	(16.0)	(14.0)	(14.0)		(1.0)

[a] Values within columns with the same superscript are not significantly different (*P*>0.05) using TUKEY HDS Multiple Comparisons (Systat); ( ) median value; Scoring System: *Digits - 1.5 points for each metacarpal and metatarsal and 1 point (pt) for each phalanx ossified on one side; ** Vertebrae - T_8-L_2- 0.5 pts, 1 pt, 1.5 pts, 2 pts, per vertebrae; + Skull - degree of ossification of the parietal, interparietal and supraoccipital bones on a rating of 0, 1.5, 3 or 5 pts; [++] Ventricular dilation scored from 0 for no space to a maximum of 6 for wide spaces.

Ossification of the vertebrae and digits was impaired in fetuses of four of the eight dams fed 15% *Sophora* (4). Similar but less distinctive changes were noted in the two groups given *Sophora* for less than the entire gestation (5, 6). For the rats given 10-15% pods (6), the changes were similar to those given 15% leaves (4) except for the lack of changes in ventricular dilation.

For the dams given dexamethasone (8), dam weight gain, placental weight, and ossification of digits and vertebrae were decreased. For phenytoin (9), fetal weights were decreased and lateral ventricular dilation scores were reduced.

Changes in rats fed *I. annua* consisted of decreased placental weights, increased lateral ventricular dilation and delayed digital ossification. Ingestion of *B. australis* produced such severe effects that either the dams died or all fetuses were resorbed.

Conclusions

The reproductive effects observed in rats ingesting *S. nuttalliana* were caused directly by the plant and indirectly by the plant's effect on animal nutrition. There were distinctive effects on ossification of both cartilaginous and membranous bone as shown by the decrease or delay in development of the vertebrae, digits, and bones of the skull. The most important change was the distortion of the skull and accompanying compression of the brain.

When *S. nuttalliana* is fed for only the first 13d of gestation, many of the plant's toxic effects are prevented or attenuated. However, effects on the brain still occurred. Lateral ventricular compression was noted whether the *Sophora* was fed for the entire gestation period or only the first 13d. The effects on the brain may be due to actual deformation rather than developmental immaturity.

The lack of changes in most parameters indicate that *I. annua* is of little consequence as a toxic plant when fed to rats for sustained periods. In contrast, *B. australis* caused severe effects including death of rat dams and/or their fetuses.

Acknowledgments

This research was supported by Agricultural Experiment Station and College of Veterinary Medicine, Oklahoma State University.

References

Aslanov, K.A., Kushmuradov, Y.K. and Sadykov, A.S. (1987) Lupine alkaloids. In: Brossi, A. (ed), *The Alkaloids, Vol 31*. Academic Press, San Diego, CA, pp. 117-192.

Barrow, P. (1990) *Technical Procedures in Reproductive Toxicology.* Royal Society of Medicine, London, pp. 26-40.

Chesnut, V.K. (1898) *Principal Poisonous Plants of the United States*. US Department of Agriculture, Division of Botany Bulletin 20.

Jin, L.X., Cui, Y.Y. and Zhang, G.D. (1993) HPLC analysis of alkaloids in *Sophora flavescens* Ait. *Yao Hsueh Hsueh Pao* 28, 136-139.

Keeler, R.F. (1972) Known and suspected teratogenic hazards in range plants. *Clinical Toxicology* 5, 529-565.

Keeler, R.F. (1976) Lupin alkaloids from teratogenic and nonteratogenic lupins. III. Identification of anagyrine as the probable teratogen by feeding trials. *Journal of Toxicology and Environmental Health* 1, 887-898.

Khan, M.A., Burrows, G.E. and Holt, E.M. (1992) (5α,6α,7α,11β)-Δ13,14-sophocarpine monohydrate. *Acta Crystallographica* C48, 2051-2053.

Kimura, M., Kimura, I., Li, X., Kong, X. and Cai, Y. (1989) Positive inotropic action and conformation of lupine alkaloids in isolated cardiac muscle of guinea pig and bullfrog. *Phytotherapia Research* 3, 101-105.

Marsh, C.D. (1909) *The Loco-weed Disease of the Plains*. US Department of Agriculture Bureau Animal Industries Bulletin 112, 130 pp.

McIlwraith, C.W. and James, L.F. (1982) Limb deformities in foals associated with ingestion of locoweed by mares. *Journal of the American Veterinary Medical Association* 181, 255-258.

Murphy, M.J., Reagor, J.C., Ray, A.C. and Rowe, L.D. (1983) Bovine abortion associated with ingestion of *Iva angustifolia* (narrowleaf sumpweed). *Proceedings of the American Association of Veterinary Laboratory Diagnostics* 26, 161-166.

Tsai, K.J., Lin, T.J., Lin, W.L. and Deng, J.F. (1993) The potential neurotoxicity of bitter tea drinking - a case report. *Veterinary and Human Toxicology* 35, 330.

Wink, M., Witte, L., Hartmann, T., Theuring, C. and Volz, V. (1983) Accumulation of quinolizidine alkaloids in plants and cell suspension cultures: genera *Lupinus, Cytisus, Baptisia, Genista, Laburnum*, and *Sophora*. *Planta Medica* 48, 253-257.

Yuan, H., Yin, Y., He, H. and Zhao, Y. (1986) Pharmacological studies on *Sophora alopecuroides* (II) neuropharmacological effects of oxymatrine. *Yaouri Fenxi Zazhi* 6, 349-352.

Yuan, H., He, H., Zhao, Y. and Wang, Z. (1987) Inhibitory action of sophocarpine on the central nervous system. *Zhongyao Tongbao* 12, 237-239.

Chapter 60

Livestock Poisoning by Teratogenic and Hepatotoxic Range Plants

D.R. Gardner, K.E. Panter, B.L. Stegelmeier, L.F. James, M.H. Ralphs, J.A. Pfister and T.K. Schoch

USDA-ARS, Poisonous Plant Research Laboratory, 1150 East 1400 North, Logan, Utah 84341, USA

Teratogenic Plants

Approximately 1% of all livestock conceived on US rangelands are born with birth defects from maternal ingestion of teratogenic plants. The identified teratogenic genera include *Veratrum*, *Lupinus*, *Conium*, *Nicotiana*, *Astragalus* and *Solanum*. Consumption by livestock during critical periods of gestation can result in severe congenital deformities or embryonic losses.

Veratrum californicum

Western false hellebore (*Veratrum californicum*) causes severe poisoning in sheep and can affect cattle and goats (Binns *et al.*, 1963). The toxic and teratogenic compounds are steroidal alkaloids that include cyclopamine, cycloposine, and jervine. The steroidal alkaloids interfere with the embryo during neural tube development (day 14 of gestation in sheep), causing gross craniofacial deformities (monkey-face) in lambs (Keeler, 1978). A high incidence of embryonic/fetal loss occurs when ewes ingest the plant during gestation days 19-21, and severe fetal tracheal stenosis and limb malformations may occur when ewes ingest *Veratrum* during gestation days 28-33 (Keeler *et al.*, 1986). Several species of *Solanum* also contain steroidal alkaloids and induce similar deformities in laboratory animals. Current management practices to reduce livestock losses include keeping sheep, goats, and cattle away from *Veratrum* during early gestation.

Lupinus, *Conium* and *Nicotiana* spp.

Certain *Lupinus*, *Conium* and *Nicotiana* species have induced birth defects, multiple congenital contractures (MCC) and cleft palate (Panter and Keeler, 1993). The teratogenic compounds are quinolizidine and piperidine alkaloids. Some *Lupinus* species contain the quinolizidine alkaloid anagyrine, which is believed to be responsible for birth defects only in cattle. Other *Lupinus* plants contain teratogenic piperidine alkaloids. The piperidine alkaloids ammodendrine and N-acetylhystrine are the probable teratogenic compounds in cattle, sheep, and goats. Coniine, γ-coniceine and N-methyl coniine in *Conium*, and anabasine in *Nicotiana*, cause similar birth defects in swine, sheep, goats and cattle.

The MCC defects include overextension or flexure of the limbs, spinal curvature, twisted neck, sway back, hump back and secondary rib cage anomalies. The critical gestational periods are days 30-60 for sheep, goats and pigs, and days 40-70 for cattle. The windows for cleft palate formation are much narrower: days 35-41 for sheep, goats and pigs, and days 40-50 for cattle (Panter and Keeler, 1992).

Management includes avoidance of these plants during the critical gestational periods or grazing when plants are least dangerous (while flowering or after seed has dropped). Alkaloid content is highly variable among the different *Lupinus* species. Chemical analysis of the plants can identify specific populations that may contain teratogenic alkaloids and others which can be safely grazed.

Locoweeds

The locoweeds (*Astragalus* and *Oxytropis* spp.) are found throughout the western US and cause neurological disturbances in sheep, cattle and horses. Some species also cause fetotoxicity, which may result in abortion and occasionally in skeletal deformities (James *et al.*, 1967; James *et al.*, 1969).

The indolizidine alkaloid swainsonine causes the neurological syndrome termed "locoism" (Molyneux and James, 1982). Swainsonine was first isolated from an Australian *Swainsona* species that induces an identical locoism syndrome (Colegate *et al.*, 1979). It is unknown if swainsonine causes abortions or birth defects.

Deformaties due to locoweed may occur during almost any period of gestation. Livestock owners can reduce losses by keeping pregnant animals off locoweed-infested ranges when it is relatively palatable (usually spring and fall).

Hepatotoxic Plants

Plants containing hepatotoxic pyrrolizidine alkaloids (PAs) are found worldwide, and often cause poisoning in livestock and man. These plants infest pastures, fields and rangelands and the entire plant or its seeds can contaminate stored forages and grain.

Important PA-containing plants in the US include *Senecio, Cynoglossum* and *Crotalaria* spp. *Senecio jacobaea* (tansy ragwort), *S. longilobus* (threadleaf groundsel) and *S. riddellii* (Riddell's groundsel) are responsible for most livestock poisonings in the western US (Johnson *et al.*, 1989). *Cynoglossum officinale* (houndstongue) is a common noxious weed that may contaminate some harvested forages (Stegelmeier *et al.*, 1996). *Crotalaria retusa* and *C. spectabilis* are common weeds of grain fields (Cheeke, 1988), and their seeds may contaminate grain often fed to poultry. The toxicity of some *Senecio* and *Cynoglossum* plants are listed in Table 60.1.

High doses of PA result in acute liver failure; however, in most cases PA-containing plants are unpalatable, and acute intoxication is rare. More common is chronic intoxication resulting from ingestion of low doses of PA through feed contamination. Chronically poisoned animals are often hepatic cripples that perform poorly and are culled as poor doers. Human poisonings can occur when PA-containing plants are used medicinally or when cereal crops are contaminated.

Poisoning from PA-containing plants is often difficult to diagnose, since the chronic disease may develop several months or years after exposure. Changes in serum biochemistry, including changes in serum enzymes and bile acids, are good diagnostic markers, but are nonspecific indicators of liver disease. Liver biopsy and histopathology are also good indicators of liver disease, but are equally nonspecific. Pyrrolizidine alkaloid metabolites can be extracted from the liver of poisoned animals and detected using GC/MS, which may be useful in diagnoses of acutely poisoned animals. However, they occur at extremely low levels in chronically poisoned animals, and current chemical methods lack the sensitivity needed to detect and quantify unstable pyrrolic metabolites. Ensuring availability of good quality forage, especially when PA content is high, will prevent livestock from consuming PA-containing plants.

Summary

Livestock poisoning by teratogenic and hepatotoxic range plants must be well understood, managed and not forgotten. Given the nature of many of these plants, a particular location may not have any serious problems for several years, and then unforeseen environmental conditions will result in exploding plant populations and an "outbreak" of livestock poisonings. Such poisoning incidents result in direct losses for the local producers and ultimately affect the economic viability of our rural communities. Currently there are no substitutes for good fundamental descriptions of rangeland poisonous plants, the conditions under which poisonings may occur and effective management tools.

Table 60.1. Toxicity of pyrrolizidine alkaloid-containing plants commonly found on ranges of the western United States.

Plant species	PA lethal dose mg/kg	PA level in plant %dw	Lethal amount of plant for a 220kg calf (g dwt/d)
Senecio jacobea	2.3	0.31	161
S. longilobus	10	2.19	118
S. riddellii	15	6.4	63
Cynoglossum officinale	<15	0.7	470

References

Binns, W., James, L.F., Shupe, J.L. and Everett, G. (1963) A congenital cyclopian type malformation in lambs induced by maternal ingestion of a range plant *Veratrum californicum*. *American Journal of Veterinary Medicine* 24, 1164-1175.

Cheeke, P.R. (1988) Toxicity and metabolism of pyrrolizidine alkaloids. *Journal of Animal Science* 66, 2343-2350.

Colegate, S.M., Dorling, P.R. and Huxtable, C.R. (1979) A spectroscopic investigation of swainsonine: an alpha-mannosidase inhibitor isolated from *Swainsona canescens*. *Australian Journal of Chemistry* 32, 2257-2264.

James, L.F., Shupe, J.L., Binns, W. and Keeler, R.F. (1967) Abortive and teratogenic effects of locoweed on sheep and cattle. *American Journal of Veterinary Research* 28, 1379-1388.

James, L.F., Keeler, R.F. and Binns, W. (1969) Sequence in the abortive and teratogenic effects of locoweed fed to sheep. *American Journal of Veterinary Research* 30, 377-380.

Johnson, A.E., Molyneux, R.J. and Ralphs, M.H. (1989) Senecio: a dangerous plant for man and beast. *Rangelands* 11, 261-264.

Keeler, R.F. (1978) Cyclopamine and related steroidal alkaloid teratogens: their occurrence, structural relationship and biological effects. *Lipids* 13, 708-715.

Keeler, R.F., Stuart, L.D. and Young, S.Y. (1986) When ewes ingest poisonous plants: the teratogenic effects. *Food Animal Practice, Veterinary Medicine* 449-454.

Molyneux, R.J. and James, J.L. (1982) Loco intoxication: indolizidine alkaloids of spotted locoweed (*Astragalus lentiginosus*). *Science* 216, 190-191.

Panter, K.E. and Keeler, R.F. (1992) Induction of cleft palate in goats by *Nicotiana glauca* during a narrow gestational period and the relation to reduction in fetal movement. *Journal of Natural Toxins* 1, 25-32.

Panter, K.E. and Keeler, R.F. (1993) Quinolizidine and piperidine alkaloid teratogens from poisonous plants and their mechanism of action in animals. *Veterinary Clinics of North America* 9, 33-40.

Stegelmeier, B.L., Gardner, D.R., James, L.F. and Molyneux, R.J. (1996) Pyrrole detection and the pathologic progression of *Cynoglossum officinale* (houndstongue) poisoning in horses. *Journal of Veterinary Diagnostic Investigation* 8, 81-90.

Chapter 61

Pine Needle (*Pinus ponderosa*) and Broom Snakeweed (*Gutierrezia* spp.) Abortion in Livestock

K.E. Panter[1], D.R. Gardner[1], L.F. James[1], B.L. Stegelmeier[1], J.A. Pfister[1], R.J. Molyneux[2], M.H. Ralphs[1] and J.N. Roitman[2]

[1]USDA-ARS, Poisonous Plant Research Laboratory, 1150 East 1400 North, Logan, Utah, USA; [2]USDA-ARS, Western Regional Research Center, 800 Buchanan St, Albany, California, USA

Introduction

Ponderosa pine and broom snakeweeds (BSW) cause serious losses to the livestock industry from abortions and toxicoses. Ponderosa pine needles (PN) induce abortion and occasional toxicosis in cattle (Stegelmeier *et al.*, 1996), while snakeweeds causes abortions and toxicoses in cattle, sheep and goats (Smith *et al.*, 1991). The toxic constituents in snakeweeds have not been characterized, but both plants contain labdane resin acids (Roitman *et al.*, 1994). The resin acids identified in the pines are known to have abortifacient properties (Gardner *et al.*, 1994; Gardner *et al.*, 1996).

Ponderosa Pine

Ponderosa pine (*Pinus ponderosa* Laws) forests are throughout North America from southern British Columbia into northern Mexico and from the Pacific coast to western Nebraska. The tree is a 2- or 3-needled pine that ranges from 25-180cm in diameter and from 12.5-61m tall. A common lumber species, it is an economic resource to communities throughout the West, providing around 35,000 jobs and over US$1b/yr in economic activity (Van Hooser and Keegan, 1988). Economically and aesthetically, it is not feasible to eliminate the Ponderosa pine to prevent poisoning; cattle management strategies must be devised to allow the valuable forest forages to be used while preventing PN-induced abortions and losses.

Ponderosa PN-induced abortion causes over US$20m losses to the cattle industry annually (Miner *et al.*, 1987). This reflects calf losses only but does not include costs associated with additional management, supplemental feeding, lost forage, added medical care, lengthened breeding intervals or increased culling rates.

Toxicology

Ponderosa pine causes abortion and associated complications in cows during the last trimester of gestation (James *et al.*, 1994). Onset of early parturition is characterized by depression, mucus discharge, weak uterine contractions, occasional incomplete cervical dilation, uterine hemorrhage and birth of a weak, but live, premature calf. Calf survival increases with decreasing prematurity. Maternal complications include retained placenta, septic metritis, agalactia, rumen stasis and death if prompt treatment is not provided. High doses of PN and new growth tips also cause toxicity in cows, resulting in depression, anorexia, rumen stasis, dyspnea, peripheral neuropathy, nephrosis and death (Panter *et al.*, 1990; Stegelmeier *et al.*, 1996).

The toxic and abortifacient components in PN have been identified (Gardner *et al.*, 1994; Gardner *et al.*, 1996; Stegelmeier *et al.*, 1996). They are the labdane resin acids: isocupressic acid (ICA), acetyl-ICA and succinyl-ICA. These resin acids have been isolated from PN, new growth tips and bark, and all have induced late-term abortions in cows (Gardner *et al.*, 1994). The rumen rapidly hydrolyzes acetyl- and succinyl-ICA to ICA, which is believed to be the abortifacient (Gardner *et al.*, 1996).

Other resin acids, such as the diterpene abietane-type abietic and dehydroabietic acids, were toxic, but not abortifacient, at high doses (Stegelmeier *et al.*, 1996). These compounds are found in high levels in rosin gum and new growth tips and in lesser amounts in needles (Gardner *et al.*, 1994). Biochemical manifestations include azotemia, hypercreatinemia, hyperphosphatemia, proteinuria, and elevation of serum enzymes. Histologically, intoxicated cows had nephrosis, vacuolation of basal ganglia neuropil with patchy perivascular and myelinic edema and skeletal myonecrosis (Stegelmeier *et al.*, 1996).

Grazing studies

Pine needle abortion usually occurs in the last trimester and most commonly in winter or early spring. Cattle will consume up to 30-40% of their diets as PN at certain times of year (Pfister *et al.*, 1992). Green needle consumption increases as ambient temperatures fall, and consumption of needle litter is related to temperature and snow depth (Pfister and Adams, 1993). High levels of PN will adversely affect cellulolytic bacteria and therefore nutrition. Consumption of PN and risk of abortion are greatly diminished during mild winter weather.

Management to prevent losses

Methods to control PN such as clear cutting, burning and spraying have been used to prevent young stands of trees from encroaching. Fencing pregnant cows away from pines in the last trimester and providing supplemental feed is the only practical strategy to reduce losses. Once cows have eaten pine needles there is no known treatment to prevent abortions. Supportive therapy for the cow and calf is critical once PN-induced parturition has occurred, and will enhance survival of both.

Broom Snakeweed

Broom snakeweeds (*Gutierrezia* spp.) cause abortion and poisoning of livestock on arid rangelands of the southwestern US. Snakeweeds are widely distributed throughout western North America. The plants are short-lived perennial half-shrubs that range from 15-60cm tall. Numerous stems originate from a woody base and die back when the plant enters dormancy. Snakeweeds have taproots and dense lateral roots, narrow leaves and clusters of thread-like yellow flowers. Snakeweed taxonomy has undergone numerous revisions (Lane, 1985), and the plant is not easily identified. Lack of reproducibility of animal feeding trials may thus be due to use of incorrectly identified plant material.

Broom snakeweeds cause significant economic losses to livestock production in the southwestern US. These include death losses, calf losses, lost forage and reduced grazing capacity. McGinty and Welch (1987) estimated annual losses from BSW of over US$34m in West Texas alone.

Toxicology

Snakeweeds begin to grow in late winter and early spring when more desirable forage is dormant. Most livestock intoxications and abortions are reported during these seasons. Broom snakeweed toxicity varies among phenological stages and soils. It is most toxic when rapidly growing, in the early stages of leaf growth. Although BSW grows abundantly on limestone soils, only sporadic losses have been reported. Experiments have shown that abortions are much more likely with BSW grown on sandy soils than on soils derived from limestone (Dollahite and Anthony, 1957). Poor nutrition may be a contributing factor in snakeweed toxicosis (Martinez *et al.*, 1993). In feeding trials, no abnormal effects on the reproductive cycle or pregnancy in cycling heifers and late pregnant ewes and heifers in good condition resulted when they were fed diets containing up to 30% BSW (Williams *et al.*, 1993; Martinez *et al.*, 1993).

The abortifacient and toxic constituents of BSW have not been identified. Experiments suggesting that steroidal saponins might be responsible for snakeweed abortions (Dollahite *et al.*, 1962; Shaver *et al.*, 1964) have not been confirmed. Roitman *et al.* (1994) isolated several diterpene acids from broom snakeweeds grown on both sandy and gravelly soil, and found marked differences in secondary metabolite patterns between the two soils. Some of these metabolites have structural characteristics similar to the labdane resin acids in PN.

Management and control

Snakeweeds are not very palatable and are unlikely to be grazed if good-quality feed is available. In spring, livestock will often eat the early-greening snakeweed while native grasses are still dormant. Poor nutrition and stress of pregnancy may contribute to an animal's propensity to graze snakeweeds and their susceptibility to the toxins.

Control of snakeweed populations may be accomplished by a variety of methods including herbicide application (application of picloram or metsulfuron in the fall) (McDaniel and Duncan, 1987), prescribed burning, biological control or a combination of these methods, followed by good range management practices (McDaniel *et al.*, 1982).

Conclusions

Three abortifacient labdane resin acids and a class of toxic compounds have been identified in Ponderosa pine needles. While there are related compounds in BSW, abortifacient or toxic components have not yet been specifically identified. Future research will concentrate on the absorption, metabolism and excretion of PN abortifacient compounds and isolation and identification of abortifacient or toxic components in BSW.

References

Dollahite, J.W. and Anthony, W.V. (1957) Poisoning with *Gutierrezia microcephala* a perennial broomweed. *Journal of the American Veterinary Medical Association* 130, 525-530.

Dollahite, J.W., Shaver, T. and Camp, B.J. (1962) Injected saponins as abortifacients. *American Journal of Veterinary Research* 23, 1261-1263.

Gardner, D.R., Molyneux, R.J., James, L.F., Panter, K.E. and Stegelmeier, B.L. (1994) Ponderosa pine needle-induced abortion in beef cattle: Identification of isocupressic acid as the principal active compound. *Journal of Agricultural and Food Chemistry* 42, 756-761.

Gardner, D.R., Panter, K.E., Molyneux, R.J., James, L.F. and Stegelmeier, B.L. (1996) Abortifacient activity in beef cattle of acetyl- and succinyl- isocupressic acid from ponderosa pine. *Journal of Agricultural and Food Chemistry* 44, 3257-3261.

James, L.F., Molyneux, R.J., Panter, K.E., Gardner, D.R. and Stegelmeier, B.L. (1994) Effect of feeding ponderosa pine needle extracts and their residues to pregnant cattle. *Cornell Veterinarian* 84, 33-39.

Lane, M. (1985) Taxonomy of *Gutierrezia* Laf. (Compositae: Asteraceae) in North America. *Systematic Botany* 10, 7-28.

Martinez, J.H., Ross, T.T., Becker, K.A., Williams, J.L., Campos, D. and Smith, G.S. (1993) Snakeweed toxicosis in late gestation ewes and heifers. In: Sterling, T.M. and Thompson, D.C. (eds), *Snakeweed Research; Updates and Highlights.* Agriculture Experiment Station Cooperative Extension Service, New Mexico State University, Las Cruces, NM, pp. 48-49.

McDaniel, K.C. and Duncan, K.W. (1987) Broom snakeweed control with picloram and metsulfuron. *Weed Science* 35, 837-841.

McDaniel, K.C., Peiper, R.D. and Donert, G.B. (1982) Grass response following thinning of broom snakeweed. *Journal of Range Management* 35, 219-222.

McGinty, A. and Welch, T.G. (1987) Perennial broomweed and Texas ranching. *Rangelands* 9, 246-249.

Miner, J.L., Bellows, R.A., Staigmiller, R.B., Peterson, M.K., Short, R.E. and James, L.F. (1987) Montana pine needles cause abortion in beef cattle. In: *Montana Agricultural Research* 4, Montana Agriculture Experiment Station, Montana State University, MT, pp. 6-9.

Panter, K.E., James, L.F., Short, R.E., Molyneux, R.J. and Sisson, D.V. (1990) Premature bovine parturition induced by ponderosa pine: Effects of pine needles, bark, and branch tips. *Cornell Veterinarian* 80, 329-333.

Pfister, J.A. and Adams, D.C. (1993) Factors influencing pine needle consumption by grazing cattle during winter. *Journal of Range Management* 46, 394-398.

Pfister, J.A., Adams, D.C., Wiedmeier, R.D. and Cates, R.G. (1992) Adverse effects of pine needles on aspects of digestive performance in cattle. *Journal of Range Management* 45, 528-533.

Roitman, J.N., James, L.F. and Panter, K.E. (1994) Constituents of broom snakeweed (*Gutierrezia sarothrae*), an abortifacient rangeland plant. In: Colegate, S.M. and Dorling, P.R. (eds), *Plant-Associated Toxins: Agricultural, Phytochemical and Ecological Aspects.* CAB International, Wallingford, Oxon, pp. 345-350.

Shaver, T.N., Camp, B.J. and Dollahite, J.W. (1964) The chemistry of a toxic constituent of *Xanthocephalum* species. *Annals of the New York Academy of Sciences* 111, 737-743.

Smith, G.S., Ross, T.T., Flores-Rodriguez, G.I., Oetting, B.C. and Edrington, T.S. (1991) Toxicology of snakeweeds, *Gutierrezia microcephala* and *G. sarothrae*. In: James, L.F., Evans, J.O., Ralphs, M.H. and Child, R.D. (eds), *Noxious Range Weeds*. Westview Press, Boulder, CO, pp. 236-246.

Stegelmeier, B.L., Gardner, D.R., James, L.F., Panter, K.E. and Molyneux, R.J. (1996) The toxic and abortifacient effects of ponderosa pine. *Veterinary Pathology* 33, 22-28.

Van Hooser, D.D. and Keegan, C.E.III (1988) Distribution and volumes of ponderosa pine forests. In: Baumgartner, D.M. and Lotan, J.E. (eds), *Ponderosa Pine: The Species and Its Management.* Office of Conferences and Institutes, Washington State University, WA, pp.1-6.

Williams, J.L., Campos, D., Toss, T.T., Smith, G.S., Martinez. J.M. and Becker, K.A. (1993) Heifer reproduction is not impaired by snakeweed consumption. In: Sterling, T.M. and Thompson, D.C. (eds), *Snakeweed Research; Updates and Highlights.* Agriculture Experiment Station Service, New Mexico State University, Las Cruces, NM, pp. 46-47.

Chapter 62

Comparative Study of Prenatal and Postnatal Monocrotaline Effects in Rats

R.M.T. Medeiros[1], S.L. Górniak[2] and J.L.Guerra[2]

[1]Faculty of Veterinary Medicine of the Federal University of Paraiba, Patos, Brazil; [2]Department of Pathology of the Veterinary Medicine and Zootechny of the University of São Paulo, São Paulo, Brazil

Introduction

Crotalaria spectabilis (Fabaceae) is a native poisonous plant that grows in almost all Brazilian territory, and is extensively used as a soil builder and green manure (Joly, 1977). *Crotalaria* spp. contain high concentrations of pyrrolizidine alkaloids (PAs), most of which accumulate in the seeds (Johnson *et al.*, 1985; Williams and Molyneux, 1987). Contamination of grain with *Crotalaria* seed has resulted in livestock and poultry poisonings in Australia (Hooper, 1978) and Brazil (Nobre *et al.*, 1994). In India in the early 1970s, contamination of millet grain with *Crotalaria* caused epidemic human mortalities (Mattocks, 1986).

Monocrotaline (MCT) is metabolized into a toxic pyrrole (MCTP) that produces lesions in lungs, liver and kidneys (Mattocks, 1986). Although studies have shown the toxicity of MCT in several animal species (Allen *et al.*, 1970; Schultze *et al.*, 1991), not much is known about its perinatal toxicity.

These alkaloids are secreted into the milk of cattle on feed or pasture infested with PA-containing plants (Shoental, 1959; Johnson, 1976), and Sundareson (1942) has shown that senecionine from *Senecio jacobea* crosses the placenta. Because the plants containing these alkaloids are globally distributed, PA poisoning is a public health problem in many areas of the world. This work was to verify possible toxic effects of the administration of MCT during the perinatal period.

Material and Methods

Twenty pregnant female Wistar rats were divided into four groups of five animals.

Rats from the two experimental groups were fed a commercial ration containing 0.012% of MCT (Sigma Chemical) during lactation (LE) or during the entire gestation and lactation periods (GLE). Animals from control groups (LC or GLC) were fed the commercial diet without MCT.

Soon after birth, four young female rats and four young male rats were randomly chosen to remain with their dams until weaning (21d). On day 13, a 5cm high wood frame was connected to the cage edges to deny the offspring access to contaminated ration.

The young rats of each litter were weaned and 10d later were anesthetized and blood samples from a hepatic vein were collected for serum enzyme analyses. The activity of alanine aminotransferase (ALT), aspartate aminotransferase (AST), γ-glu-tamyltransferase (GGT), lactate dehydrogenase (LDH), alkaline phosphatase (ALP), conjugated bilirubin and unconjugated bilirubin, glucose, urea and creatinine were determined using commercially available reagents (Reactoclin-CELM®). All samples were run in duplicate and the data were averaged.

Immediately after blood collection the animals were euthanized and lungs, liver and kidney tissues were collected, fixed in Bouin's solution, processed and stained with hematoxylin and eosin (HE) for histologic examination.

Results

The clinical pathological data on pups from LE dams treated with MCT showed significantly ($P<0.05$) increased levels of ALT and urea in the males and increased urea in the females. No significant differences were found between treated and control in the other parameters. Biochemical analyses of serum from pups in the MCT treated GLE group could not be performed since these animals died before blood sample collection at 21d.

Histopathological study demonstrated alterations in almost all offspring from dams that received ration containing MCT, although they were more pronounced in GLE offspring. The litters of both experimental groups showed alterations in the lungs (Fig. 62.1) and kidneys (Fig. 62.2). The lungs showed interstitial pneumonia, thickened alveolar septa and mixed cell infiltration. There was an accumulation of exudate in the perivascular-peribronchiolar interstitium and thickening of the arterial wall. Alveolar sacs contained an abundance of foamy cells. Lesions in the kidneys were characterized as mild, toxic, tubular nephrosis; however, alterations in the liver were found only in the GLE litter group, which presented a partially preserved lobular hepatic architecture. Lesions predominated in focal areas of hepatocellular hydropic degeneration and nuclear changes (karyorrhexis, pyknosis, and karyolysis). Periportal necrosis was observed, and there was mononuclear cell infiltration and hemorrhage. There was no fibrosis.

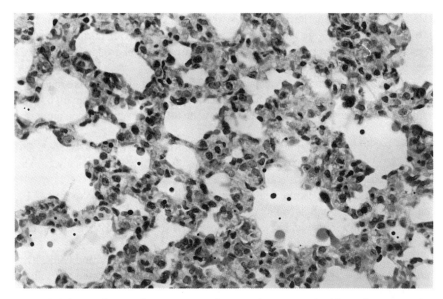

Fig. 62.1. Lung tissue of a young rat from a dam treated with a diet containing monocrotaline showing chronic interstitial pneumonia. Note enlarged alveolar septa due to intense mononuclear inflammatory infiltrate (HE). 160x.

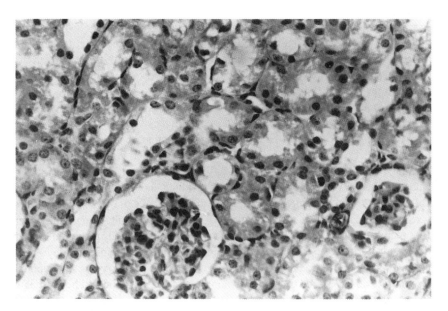

Fig. 62.2. Kidney of a young rat from a dam consuming a diet containing monocrotaline during lactation. Note the vascular degeneration of the renal tubular epithelium (HE). 400x.

Discussion

The PAs are metabolized in the liver into toxic pyrroles (Mattocks, 1986). These compounds react with nucleophiles, leading to edema, centrilobular necrosis, megalocytosis and loss of hepatic function (Cheeke and Shull, 1985). Because it is more acidic than plasma, basic compounds concentrate in milk. Elimination of toxicants *via* milk benefits the lactating animal, but it can poison the suckling animal because the toxin may be concentrated in milk and neonates are less able to eliminate or detoxify xenobiotics (Panter and James, 1990).

Sucklings of lactating females fed *S. jacobea* displayed hepatotoxicity (Shoental, 1959; Johnson, 1976; Goeger *et al*., 1979). *Crotalaria spectabilis* seeds fed to rats (Kay and Heath, 1966; Sriraman *et al*., 1987) and swine (Hooper and Sanlan, 1977) produced lesions in pulmonary parenchyma that were more severe than those in liver and kidneys. This study confirms these reports, since all groups had lesions in the pulmonary parenchyma as the main finding, showing that the pneumotoxicity of MCT is greater than its hepatotoxicity.

This study was conducted to determine if MCT could pass into milk or cross the placenta. In the LE group, no pup mortality was found, but alterations in serum biochemistry were observed. Elevated ALT levels indicated hepatopathy in the experimental animals. The high urea levels and microscopic finding of mild toxic tubular nephrosis indicated loss of renal function. Histopathologic study of GLE pups revealed severe lung lesions, characterized by interstitial pneumonia, alveolar septal thickening and mixed inflammatory cell infiltration. Kidney and liver alterations were milder, with hepatocellular degeneration and necrosis and toxic tubular nephrosis. Lesions in pups were more prominent than those observed in dams, which suggests that MCT or MCTP crosses the placenta and also enters the milk, and that pups are more sensitive to the toxic effects of the PAs than are dams. Although the placenta separates fetal and maternal compartments, it often does not protect the fetus against xenobiotics (Klassen, 1986). Pups from GLE groups showed high mortality levels, and some rats from this group did not give birth. Sundareson (1942) observed failure to implant and postnatal mortality after repeated administration of the PA senecionine to pregnant female rats.

This study confirms that the toxicity of MCT *via* milk, and its diffusion through the placenta may be important to the health of the fetus and neonate. Furthermore, MCT may have implications for human food safety.

References

Allen, J.R., Chesney, C.F. and Castens, L.A. (1970) Clinical signs and pathologic changes in *Crotalaria spectabilis*-intoxicated rats. *American Journal of Veterinary Research* 31, 1059-1070.

Cheeke, P.R. and Shull, L.R. (1985) *Natural Toxicants in Feeds and Poisonous Plants.* AVI Publishing, Westport, CT, pp.106-111.

Goeger, D.E., Cheeke, P.R. and Buhler, D.R. (1979) The effect of dietary tansy ragwort (*Senecio jacobea*) on dairy goats and toxicity on their milk. *Journal of Animal Science* 49, 370.

Hooper, P.T. (1978) Pyrrolizidine alkaloid poisoning with particular reference to differences in animal and plant species. In: Keeler, R.F., Vankampen, K.R. and James, L.F. (eds), *Effects of Poisonous Plants on Livestock*. Academic Press, New York, NY, pp.161-176.

Hooper, P.T. and Sanlan, W.A. (1977) *Crotalaria retusa* poisoning of pigs and poultry. *Australian Veterinary Journal* 53, 109-114.

Johnson, A.E. (1976) Changes in calves and rats consuming milk from cows fed chronic lethal doses of *Senecio jacobea* (tansy ragwort). *American Journal of Veterinary Research* 37, 107-110.

Johnson, A.E., Molyneux, R.J. and Merril, G.B. (1985) Chemistry of toxic range plants. Variation in pyrrolizidine alkaloid content of *Senecio*, *Amsinckia and Crotalaria* species. *Journal of Agricultural and Food Chemistry* 33, 50.

Joly, A.B. (1977) *Botânica: Introdução à Taxonomia Vegetal*, 4 ed. Biblioteca Universitária, Série 3, Ciências Puras, 4, Editora Nacional, São Paulo, pp. 381-382.

Kay, J.M. and Heath, D. (1966) Observation on pulmonary arteries and heart weight of rats fed on *Crotalaria spectabilis* seeds. *Journal Pathology and Bacteriology* 92, 385-394..

Klassen, C.D. (1986) Distribution, excretion, and absorption of toxicants. In: Klassen, C.D., Admur, M.O. and Doull, J. (eds) *Casarett and Doull's Toxicology: The Basic Science of Poisons*. Macmillan Publishing Co., New York, NY, pp. 43-44.

Mattocks, R.R. (1986) *Chemistry and Toxicology of Pyrrolizidine Alkaloids*. Academic Press, London, p. 393.

Nobre, D., Dagli, M.L.Z. and Haraguchi, M. (1994) *Crotalaria juncea* intoxication in horses. *Veterinary and Human Toxicology* 36, 445-447.

Panter, K.E. and James, L.F. (1990) Natural plant toxicants in milk: a review. *Journal of Animal Science* 68, 892-904.

Schultze, A.E., Wagner, G.J., White, S.M. and Roth, R.A. (1991) Early indications of monocrotaline pyrrole-induced lung injury in rats. *Toxicology and Applied Pharmacology* 109, 41-50.

Shoental, R. (1959) Liver lesions in young rats suckled by mothers treated with the pyrrolizidine (*Senecio*) alkaloids, lasiocarpine and retronecine. *Journal of Pathology and Bacteriology* 77, 485-495.

Sriraman, P.K., Gopal Naidu, N.R. and Rama Rao, P. (1987) Effects of monocrotaline in rats, an experimental study. *Indian Journal of Animal Sciences* 57, 1060-1087.

Sundareson, A.E. (1942) An experimental study on placental permeability to cirrhogenic poisons. *Journal of Pathology and Bacteriology* 49, 289-298.

Williams, M.C. and Molyneux, R.J. (1987) Occurrence, concentration and toxicity of pyrrolizidine alkaloids in *Crotalaria* seeds. *Weed Science* 35, 476-81.

Chapter 63

Embryotoxic Effect of *Plumeria rubra*

V.K. Gunawardana[1], M.M. Goonasekera[2], G.M.K.B. Gunaherath[3],
A.A.L. Gunatilaka[3] and K. Jayasena[2]

[1]*Department of Veterinary Preclinical Studies, Faculty of Veterinary Medicine and
Animal Science;* [2]*Department of Pharmacology, Faculty of Medicine;* [3]*Department
of Chemistry, Faculty of Science;* [4]*University of Peradeniya, Peradeniya, Sri Lanka*

Introduction

Many plants are known for their medicinal properties, and plant-based traditional medical practice forms a substantial part of the health care system in many developing countries, including Sri Lanka. *Plumeria rubra* L. (Apocynaceae) has been used as a febrifuge, a purgative and in the treatment of leprosy. It is also reported to have hypoglycemic properties (Dhar *et al.*, 1968). However, the plant appears to be used most widely for its effects on reproduction. Aqueous extracts of stem and root bark have been used orally as emmenagogues or menstrual inducers (Quisumbing, 1951; Petelot, 1954; Watt and Breyer-Brandwijk, 1962; Datta and Datta, 1975). The fruit, latex and other parts of the plant have been administered orally as abortifacients (Chopra *et al.*, 1965; Oliver-Bever, 1986), while the branches are reported to have been used intravaginally to produce abortion (Saha *et al.*, 1961). Despite the ethnomedical claims, there is little experimental information on the anti-fertility properties of this plant. The present study was undertaken to verify claims of anti-fertility activity in *P. rubra*, to investigate its action on pregnant rats and to identify active constituents.

Materials and Methods

Several collections of stem bark of *P. rubra* were made at different times of year. Identity was verified by Dr S. Balasubramaniam, Department of Botany, University of Peradeniya, Sri Lanka. Plant material was air dried, and voucher specimens were deposited in the herbarium at the University of Peradeniya.

The stem bark of *P. rubra* was exhaustively extracted with water or 95% ethanol (EtOH) by using a soxhlet apparatus. It was subsequently subjected to sequential extraction with hot petroleum ether (PET), hot dichloromethane (DCM) and hot methanol (MeOH) under reflux conditions. The extracts were reduced to thick syrups in a rotary evaporator and then vacuum evaporated to a constant weight and freeze dried. During the course of extraction with DCM, an off-white solid precipitated, while the concentrated extract was a dark green solid. Further extraction of this dark green material with fresh DCM yielded more of the precipitate.

For purposes of the bioassay, the solubility of freeze-dried extracts was increased by co-precipitation with polyvinylpyrrolidone (PVP). A known weight of extract was dissolved in a known volume of MeOH, and an equal weight of PVP was dissolved separately in an equal volume of MeOH. When these solutions were mixed and the solvent removed, water-soluble complexes of PVP:extract (1:1) were obtained. Solutions used to dose the animals were prepared fresh daily. Control groups were dosed with the same amount of PVP as the treatment group. Extracts subjected to bioassays are summarized in Table 63.1.

Proven fertile male and virgin female Sprague-Dawley rats were maintained under natural lighting of 12hrs of daylight and darkness at a temperature of about 21°C. The females (180-200g) were individually housed. Their vaginal smears were examined daily to determine cyclicity. Females showing regular estrus cycles were selected on the basis of two consecutive cycles. They were then paired with proven males and the males were kept with the female until mating, which was indicated by the presence of spermatozoa in the vaginal smear. The day of mating was considered day one of pregnancy, and animals were assigned to control or experimental groups randomly until each group consisted of ten animals. Dose was determined daily

Table 63.1. Details of *Plumeria rubra* extracts and fractions investigated.

Extract/fraction code	Collection number	Type of extract
E001	01	Hot aqueous
E002	01	Hot ethanol
E003	02	Hot ethanol
E004	03	Petroleum ether (sequential extraction)
E005	03	Dichloromethane (sequential extraction)
E006	03	Methanol (sequential extraction)
E007	04	Dichloromethane (sequential extraction)
E008	04	Methanol (sequential extraction)
K001	03	Plumieride

according to the recorded body weight and was administered by gavage. Rats were dosed for 10d and autopsied on day 16.

The number of implantation sites (IS), corpora lutea (CL), normal (NF) and degenerated (DF) fetuses, and the state of the other organs were noted and recorded for each animal.

The proportion of pregnant animals in the treated versus the control group was compared by using the Fisher/Irwin method on the exact test of fourfold tables (Armitage, 1977). The differences in the mean number of IS, NF and CL per pregnant rat between the treated and control group were evaluated by employing a Student's *t* test. For purposes of this study, an animal with at least one implantation site was considered pregnant, and in calculating means only pregnant animals were included.

Results

The results obtained with crude aqueous and EtOH extracts of *P. rubra* are given in Table 63.2. At a dose of 3.1g/kg, the aqueous extract (E001) exhibited toxicity, with all animals showing signs of diarrhea and four animals dying prematurely. No statistically significant anti-fertility activity was observed with this extract.

The hot EtOH extract (E002) was significantly active but toxic at a dose of 1.52g/kg. The animals in this group showed severe weight loss, with one animal dying on the fifth day of pregnancy. Some activity was observed but toxicity was less marked at lower doses (0.75, 1.35g/kg). An EtOH extract (E003) derived from a different collection of *P. rubra* gave equivocal results at 1.25g/kg. There was significant activity at 2.34 and 2.7g/kg, but four animals died in each group.

The results of the bioassays carried out on crude extracts obtained by sequential extraction are given in Table 63.3. The PET extract (E004) did not exhibit any anti-fertility activity at 1.0g/kg. However, the DCM extracts showed significant anti-fertility activity at 1.36g/kg (E005) and equivocal activity at 0.6g/kg (E007). Only one pregnant animal was found in the E005 group, but two animals died prematurely. The MeOH extracts (E006 and E008) also exhibited significant anti-fertility activity. Two of the pregnant rats in the group dosed with E008 had only pin-point implantation sites.

The white precipitate from the DCM extract consisted of one major compound with some minor impurities of varying polarity. Chemical and physical data for the major compound and its penta-acetate agreed well with those of plumieride, and comparison with an authentic sample confirmed its identity as plumieride.

The precipitate was found to cause statistically significant early post-implantation $(0.05 > P > 0.01)$ and fetal resorption $(0.01 > P > 0.001)$ effects at a dosage of 0.83g/kg. The anti-implantation effect was marginal when compared with the control group. Reduced feed intake and weight loss were observed during the period of dosing though the other toxic symptoms were not marked. Normal feed intake and weight gain were restored after the dosing was stopped.

Table 63.2. Results obtained with crude aqueous and ethanol extracts.

Extract code	Dose (g/kg)	Pregnant/ dosed (control)	IS mean±SD (Control)	NF mean±SD (Control)	CL mean±SD (Control)
E001	1.5	7/ 10 (9 / 10)	9.1 ± 4.3 (10.2 ± 2.2)	6.6 ± 4.2 (8.6 ± 2.8)	11.1 ± 3.2 (11.3 ± 1.8)
E001	2.0	7 / 10 (9 / 10)	10.1 ± 1.4 (10.4 ± 1.3)	10.0 ± 1.6 (9.7 ± 1.6)	10.9 ± 1.6 (10.9 ± 1.5)
E001	3.1	4 / 6 (10 / 10)	10.0 ± 0.8 (8.7 ± 2.6)	8.0 ± 0.8 (8.3 ± 2.6)	10.5 ± 0.6 (10.7 ± 1.7)
E002	0.75	6 / 10 (10 / 10) P=0.05	11.7 ± 1.9 (9.5 ± 3.3)	11.3 ± 1.9 (9.0 ± 0.1)	11.7 ± 1.9 (10.5 ± 1.5)
E002	1.35	6 / 10 (9 / 10)	12.5 ± 3.4 (13.1 ± 1.5)	9.7 ± 5.4 (12.3 ± 1.4)	13.2 ± 2.6 (13.7 ± 1.2)
E002	1.52	2 / 9 (9 / 10) P=0.01	10.0 ± 1.4 (10.0 ± 2.3)	10.0 ± 1.4 (9.3 ± 1.9)	10.5 ± 2.1 (10.3 ± 2.7)
E002	2.7	1 / 9 (10 / 10) P<0.005	14.0 ± 0 (12.7±3.8)	14.0 ± 0 (12.1 ± 3.6)	15.0 ± 0 (14.8 ± 2.3)
E003	1.25	7 / 10 (10 / 10)	13.4 ± 2.4 (11.1 ± 4.0)	13.0 ± 2.7 (10.8 ± 3.9)	13.9 ± 2.8 (12.3 ± 2.8)
E003	2.34	2 / 10 9 / 10 P<0.005	11.0 ± 4.2 (13.3 ± 1.1)	11.0 ± 4.2 (13.1 ± 1.1)	12.0 ± 4.2 (14.7 ± 1.7)

IS=implantation sites; NF=normal fetuses; CL=corpora lutea

Discussion and Conclusions

Pregnancy is a complex series of events starting with fertilization and terminating with the birth of live offspring. The bioassay used in this investigation was to evaluate the efficacy of the plant extracts in interfering with all events following fertilization. A proportion of pregnant animals detected extracts that exerted an effect prior to implantation. A reduction in the number of IS in the treated group indicated an early post-implantation effect. By comparing the number of NF in the two groups, any undesired abortifacient, teratogenic or fetal resorption activity was assessed. Luteolytic activity of the extract was evaluated by comparison of CL.

Table 63.3. Results obtained with the sequential extracts.

Extract code	Dose (g/kg)	Pregnant/ dosed (Control)	IS mean±SD (Control)	NF mean±SD (Control)	CL mean±SD (Control)
E004	1.0	10/10 (10/10)	13.5±2.0 (12.7±3.8)	12.7±1.8 (12.1±3.6)	15.3±2.7 (14.8±2.3)
E005	1.36	1/8 (10/10) $P<0.005$	14.0±0 (11.8±1.7)	14.0±0 (11.0±1.3)	14.0±0 (13.3±2.3)
E006	1.5	5/10 (10/10) $P=0.025$	11.6±2.7 (11.8±1.7)	9.4±5.3 (11.0±1.3)	13.0±2.7 (13.3±2.3)
E007	0.6	7/10 (10/10)	10.9±1.1 (10.4±1.1)	10.6±1.0 (9.8±1.1)	11.7±1.7 (12.4±2.0)
E008	1.6	4/10 (9/10) $P=0.05$	9.8±5.0 (12.1±1.7)	6.5±7.5 (11.4±2.1)	11.5±2.4 (12.7±1.6)
K001	0.83	6/10 (9/10)	7.7±3.1 (11.6±2.2) $0.05>P>0.01$	5.8±4.7 (11.2±2.3) $0.01>P>0.001$	9.7±1.2 (13.0±2.1)

IS=implantation sites; NF=normal fetuses; CL=corpora lutea

Due to the marginal activity of the aqueous extracts, studies were conducted on plant material extracted with different organic solvents. The hot EtOH extract of the stem bark consistently showed toxic effects (with animal deaths) at doses above 1.5g/kg. These extracts showed significant anti-implantation activity while being devoid of any post-implantation anti-fertility effects. The DCM and MeOH extracts showed the presence of active anti-fertility agents. Reproducible results were obtained with extracts prepared from different collections.

The off-white precipitate was embryotoxic, causing fetal resorption. Plumieride is an iridoid glycoside previously isolated from the leaf, root and stem bark of *P. rubra* (Mahran *et al.*, 1974), but data regarding its biological activities are lacking.

The weight loss observed with some extracts that showed significant anti-fertility activity varied considerably. Groups of animals in which a statistically significant anti-fertility activity was not observed still lost considerable weight. Treated animals gained weight after dosing was discontinued.

Further work should be done on *P. rubra* to determine whether the anti-fertility constituents can be separated from the toxic constituents. In addition, since plumieride seems to be an interesting lead compound, further investigations are merited in order to determine its role in fertility regulation. It is essential to test an authentic sample of plumieride for anti-fertility activity. Since plumieride is a glycoside it would be worthwhile to test the aglycone obtained after removing the sugar moiety.

References

Armitage, P. (1977) *Statistical Methods in Medical Research.* Blackwell Scientific Publications, Oxford.

Chopra, R.N., Badhwar, R.L. and Ghosh, S. (1965) *Poisonous Plants of India, Vol 1.* Indian Council of Agricultural Research, New Delhi.

Datta, S. and Datta, P.C. (1975) Bark drugs of *Plumeria. Quarterly Journal of Crude Drug Research* 14, 129-142.

Dhar, M.L., Dhar, M.M., Dhawan, B.N., Mehrotra, B.N. and Ray, C. (1968) Screening of Indian plants for biological activity: Part 1. *Indian Journal of Experimental Biology* 6, 232-247.

Mahran, G.H., Abdel-Wahab, S.M. and Salah Ahmed, M. (1974) Isolation and quantitative estimation of plumieride from the different organs of *Plumeria rubra* and *Plumeria rubra* Var. *alba. Planta Medica* 25, 226.

Oliver-Bever, B. (1986) *Medicinal Plants in Tropical West Africa.* Cambridge University Press, Cambridge.

Petelot, A. (1954) *Les plantes medicinales du Cambodge, du Laos et du Vietnam, Vols 1-4.* Archives des recherches agronomiques et pastorales au Vietnam No. 23.

Quisumbing, E. (1951) *Medicinal Plants of the Philippines.* Technical Bulletin 16, Department of Agriculture and Natural Resources, Manila.

Saha, J.C., Savini, E.C. and Kasinathan, S. (1961) Ecbolic properties of Indian medicinal plants. Part 1. *Indian Journal of Medical Research* 49, 130-151.

Watt, J.M. and Breyer-Brandwijk, M.G. (1962) *The Medicinal and Poisonous Plants of Southern and Eastern Africa,* 2nd ed. E&S Livingstone, London, pp. 94-95.

Chapter 64

Evaluation of the Toxicity of *Solanum malacoxylon* During the Perinatal Period

S.L. Górniak[1], M.L.Z. Dagli[1], N.V.M. Arruda[2] and P.C.F. Raspantini[1]
[1]University of São Paulo, Faculty of Veterinary Medicine and Zootechny, Department of Pathology, CEPTOX-São Paulo-Brazil; [2]EMPAER-Mato Grosso, Brazil

Introduction

Since the end of the last century, an endemic disease of cattle and other grazing animals called "espichamento" in Brazil, or "enteque seco" in Argentina, has been described. The "espichamento" is a pathological deposition of calcium phosphate in soft tissue, and it occurs when animals graze *Solanum malacoxylon* (Sm).

There are many studies showing physiological (Corradino and Wasserman, 1974; Uribe *et al.*, 1974; Napoli *et al.*, 1977) and chemical (Wasserman *et al.*, 1976) evidence of the presence of 1,25-dihydroxy vitamin D_3 (1,25-OHD$_3$) in *S. malacoxylon* leaves. Studies on the toxicity of the plant have been conducted on a variety of animal species including ruminants (Carrillo and Worker, 1967), pigs (Done *et al.*, 1976), rabbits (Mautalen, 1972), rats (Uribe *et al.*, 1974) and chickens (BaSudde and Humphreys, 1975). In all species, pathological features are similar to those of animals given massive doses of vitamin D. Despite studies on the toxic actions of the plant, ingestion during pregnancy has not been investigated. High doses of vitamin D in the gestational period can produce teratogenic effects such as aortic valve stenosis and impaired skeletal formation (Ornoy *et al.*, 1972; Zane, 1976). This study was to evaluate the toxic effects to young rats whose dams ingested *S. malacoxylon* during gestation and lactation.

Material and Methods

Leaves of *S. malacoxylon* were collected from the Pantanal region, Mato Grosso, Brazil. They were air-dried, powdered and homogenized with rat diet at 0.05% and

0.1% (0.05%Sm and 0.1%Sm). The Sm diets were pelleted and fed *ad libitum*.

Female Wistar rats (30) were paired overnight with proven sires. The morning that a sperm-positive vaginal smear was observed was counted as day 1 of gestation. Two trials were performed. In the first (GL), 12 females were divided at random into two equal groups that received 0.05%Sm and 0.1%Sm ration from the first gestational day to the last lactational day (42d of treatment). In the second (G), 12 females received 0.05%Sm and 0.1%Sm ration on days 6-21 of gestation. Control animals (6) were fed pelleted food containing no plant material throughout gestation and lactation. Ration consumption and body weights were recorded weekly. One day after weaning, the dams were anesthetized and blood samples were taken from the hepatic vein for evaluation of serum calcium (Ca), inorganic phosphorus (P), magnesium (Mg) and alkaline phosphatase. Dams were euthanized and lung, heart, kidney and aorta slices were taken, and routinely prepared histopathology slides were stained with hematoxylin and eosin (HE) or Von Kossa stain.

Young rats from each litter (8/dam) were weighed weekly from birth to weaning. Three days after weaning, all litters were anesthetized, euthanized, and sampled as described for the dams. Data were expressed as mean values and standard error of the mean and were evaluated by analysis of variance followed by Duncan's test. Results were considered significant when $P<0.05$.

Results

The feed intake of dams from the GL 0.05%Sm and 0.1%Sm groups significantly decreased during the 2nd week of gestation when compared to the control group. During the following week the food consumption of these groups was significantly higher than that of the control group. Thus, inclusion of *S. malacoxylon* in the diet produced no overall effect on feed intake. Body weights of G and GL dams and of their respective pups during lactation were not significantly different from controls.

The ranges of serum Ca, P, Mg and alkaline phosphatase of the dams are outlined in Fig. 64.1. Females from the GL 0.05%Sm, GL 0.1%Sm and G 0.1%Sm groups showed an increase in serum Ca levels, and inorganic P levels were significantly higher in dams from groups G 0.05%Sm and G 0.1%Sm. An increase in serum Mg was observed in females that received 0.1% *S.malacoxylon* in the ration. Serum alkaline phosphatase activity was significantly decreased in all treatment groups.

The biochemical results of the pups are summarized in Fig. 64.2. Increased serum Ca was seen in females from group G dams fed 0.05%Sm and 0.1%Sm and in males from group G mothers treated with 0.1%Sm. Serum P was higher in males from group G dams fed 0.5%Sm and 0.1%Sm and group GL dams fed 0.1%Sm. Serum Mg of both sexes from group G dams fed 0.05%Sm and 0.1%Sm differed significantly from controls. Alkaline phosphatase activity was significantly decreased in both sexes of pups from group GL mothers treated with 0.05%Sm.

Histopathological evaluation revealed alterations in almost all mothers that received ration containing *S. malacoxylon*. These lesions were more intense in

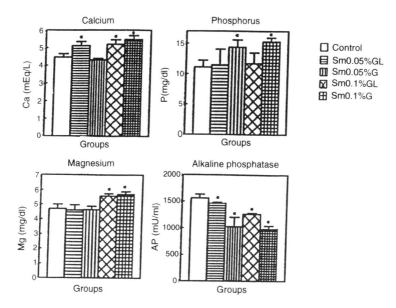

Fig. 64.1. Serum levels of calcium, phosphorus, magnesium and alkaline phosphatase in control and experimental dams treated during gestation and lactation (GL) and only during gestation (G) with rations containing 0.05% or 0.1% *S. malacoxylon* leaves; *P<0.05 compared to control.

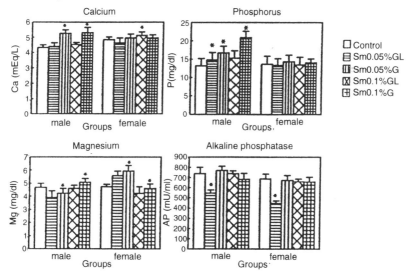

Fig. 64.2. Serum levels of calcium, phosphorus, magnesium and alkaline phosphatase in pups from control and experimental dams treated during gestation and lacatation (GL) and only during gestation (G) with rations containing 0.05% or 0.1% *S. malacoxylon* leaves; *P<0.05 compared to control.

animals that consumed a larger amount of Sm. Pups from treated mothers also presented several alterations. Lungs were the most affected organ, showing intense inflammatory infiltrates of mostly mononuclear cells around the bronchi, bronchioles and blood vessels. Peribronchial connective tissue was dissociated by edema. An important and characteristic finding was the presence of several mast cells accompanying the inflammatory infiltrate in lungs, as well as in other organs. A deposition of weak basophilic material in the most external layer of the blood vessels was also observed. Von Kossa staining was negative for Ca in these structures. Interstitial pneumonia was also observed in some rats. Some control animals also presented a discrete mononuclear inflammatory infiltrate around the bronchi and bronchioles, but this was considered very mild when compared to treated animals, and was not accompanied by mast cells. Hyperplasia of the ciliated pseudostratified columnar epithelium of bronchi was also observed. Other examined organs did not present morphological alterations.

Discussion

Enzootic calcinosis of domestic animals is a disease that causes heavy economic losses to the cattle industry in Brazil and Argentina (Morris, 1982; Boland, 1988). Decreased fertility has also been attributed to S. malacoxylon. Despite these losses, no studies on the effects of this plant on fetal development and the possible transmission of 1,25-OHD$_3$ through milk have been published. Several studies have shown transplacental (Keeler, 1984; Panter et al., 1992; James et al., 1994) and milk (Panter and James, 1990) transmission of different plant toxins.

Previous studies conducted on rabbits in this laboratory (unpublished data) clearly showed the passage of the calcinogenic glycoside through the placenta, producing increased calcemia, phosphatemia and extensive mineralization of soft tissue in litters from dams treated with S. malacoxylon. The present study supports early reports showing that toxic effects are produced by S. malacoxylon in rats not only during the gestational period but also during lactation. This is confirmed by the fact that the histopathological findings were more prominent in young rats from GL mothers. However, the absence of widespread mineralization shows that rats are less sensitive to the toxic compound than rabbits (Carrillo and Worker, 1967).

The possibility that the calcinogenic effects observed in the offspring of mothers treated with S. malacoxylon were induced by an excess of maternal Ca rather than by 1,25-OHD$_3$ is ruled out by the fact that maternal hypercalcemia is known to produce hypocalcemia in the offspring (Fairney and Weir, 1970). In the present study, young rats from mothers treated with S. malacoxylon had similar or increased serum calcium levels. In addition, increased P and Mg levels and decreased alkaline phosphatase activity were observed in these animals, suggesting a vitamin D effect.

The morphological alterations observed in the lungs of both dams and litters treated with S. malacoxylon raises questions about the pathogenesis of tissue calcification. It has been reported that the earliest changes are modifications of the elastic fibers. De Barros *et al.* (1981) reported that smooth muscle cells responsible for the synthesis of connective tissue components are the first to be morphologically modified. Alterations in peribronchial connective tissue, which will require further morphological and/or ultrastructural studies, were seen in this study.

Morris (1982) reported that deposition of minerals is favored by or dependent on damage to connective tissue elements; damaged elastin may act as a matrix in which Ca deposits can occur. Morris *et al.* (1979) also reported an increase in the number of mast cells in the lungs of animals fed *Solanum torvum*, and stated that this is an important finding since calcifications also occur around discharged mast cell granules. A possible explanation is that the glycosaminoglycans contained in mast cell granules may act as a matrix for Ca deposit. Another finding described by other authors in cases of Sm or vitamin D intoxication is the presence of mononuclear inflammatory infiltrates. This was also reported by Haschek *et al.* (1978), whereas Chimene *et al.* (1976) described a more intense granulomatous reaction. The toxic principle of *S. malacoxylon* could also be responsible for the epithelial proliferation observed in bronchi. In fact, Koh *et al.* (1988) showed that vitamin D produces proliferation of vascular smooth muscle cells. Thus, taken as a whole, the present data strongly support that the toxic principle of the plant can affect the fetus and neonate and may have important consequences for the livestock industry.

References

BaSudde, C.D.K. and Humphreys, D.J. (1975) The effect of the administration of *Solanum malacoxylon* on the chick. *Research in Veterinary Science* 18, 330-331.

Boland, R.L. (1988) *Solanum malacoxylon*: a toxic plant which affects animal calcium metabolism. *Biomedical and Environmental Science* 1, 414-423.

Carrillo, B.J. and Worker, N.A. (1967) Enteque seco: arteriosclerosis y calcificación metastática de origen tóxico en animales a pastoreo. *Revista de Investigaciónes Agropecuária Serie 4 Patologia de Animales Buenos Aires* 4, 9-30.

Chimene, C.N., Krook, L. and Pond, W.G. (1976) Bone pathology in hypervitaminosis D. An experimental study in young pigs. *Cornell Veterinarian* 66, 387-412.

Corradino, R.A. and Wasserman, R.H. (1974) 1,25-dihydroxycholecalciferol-like activity of *Solanum malacoxylon* extract on calcium transport. *Nature* 252, 716-718.

De Barros, S., Tabone, E., Dos Santos, M., Andujar, M. and Grimaud, J.A. (1981) Histopathological and ultrastructural alterations in the aorta in experimental *Solanum malacoxylon* poisoning. *Virchows Archives [Cell Pathology]* 35, 169-175.

Done, S.H., Tokarnia, C.H., Dämrich, K. and Dobereiner, J. (1976) *Solanum malacoxylon* poisoning in pigs. *Research in Veterinary Science* 20, 217-219.

Fairney, A. and Weir, A.A. (1970) The effect of abnormal maternal plasma calcium levels on the offspring of rats. *Journal of Endocrinology* 48, 337-345.

Haschek, W.M., Krook, L., Kallfelz, F.A. and Pond, W.G. (1978) Vitamin D toxicity. Initial site and mode of action. *Cornell Veterinarian* 68, 324-364.

James, L.F., Panter, K.E., Stegelmeier, B.L. and Molyneux, R.J. (1994) Effect of natural toxins on reproduction. *Veterinary Clinics of North America: Food Animal Practice* 10, 587-600.

Keeler, R.F. (1984) Teratogens in plants. *Journal of Animal Science* 58, 1029-1039.

Koh, E., Morimoto, S., Fuko, K., Itoh, K., Hironada, T., Shiraishi, T., Onishi, T. and Kumhara, Y. (1988) 1,25-dihydroxyvitamin D3 binds specifically to rat vascular smooth muscle cells and stimulates their proliferation *in vitro*. *Life Sciences* 42, 215-223.

Mautalen, C. (1972) Mechanism of action of *Solanum malacoxylon* upon calcium and phosphate metabolism in the rabbit. *Endocrinology* 90, 563-567.

Morris, K.M.L. (1982) Plant induced calcinosis: A review. *Veterinary and Human Toxicology* 24, 34-48.

Morris, K.M.L., Simonite, J.P., Pullen, L. and Simpson, J.A. (1979) *Solanum torvum* as a causative agent of enzootic calcinosis in Papua, New Guinea. *Research in Veterinary Science* 27, 264-266.

Napoli, J.L., Reeve, L.E., Eisman, J.A., Schones, H.K. and DeLuca, H.F. (1977) *Solanum glucophyllum* as a source of 1,25 dihidroxovitamin D3. *Journal of Biological Chemistry* 252, 2580-2583.

Ornoy, A., Horowitz, A., Kaspi, T. and Nebel, L. (1972) Anomalous fetal and neonatal bone development induced by administration of cortisone and vitamin D2 to pregnant rats. In: Klinberg, M.A., Abramovici, A. and Chemke, J. (eds), *Drugs: Fetal Development*. Plenum, New York, NY, pp. 219-226.

Panter, K.E. and James, L.F. (1990) Natural plant toxicants in milk: a review. *Journal of Animal Science* 68, 892-904.

Panter, K.E., Keeler, R.F., James, L.F. and Bunch, T.D. (1992) Impact of plant toxins on fetal and neonatal development: a review. *Journal of Range Management* 45, 52-57.

Uribe, A., Holick, M.F., Jorgensen, N.A. and DeLuca, H.F. (1974) Action of *Solanum malacoxylon* on calcium metabolism in the rat. *Biochemical and Biophysical Research Communications* 58, 257-262.

Wasserman, R.H., Henion, J.D., Haussler, M.R. and McCaine, T.A. (1976) Calcinogenic factor in *Solanum malacoxylon*: evidence that it is 1,25 dihydorxyvitamin D3-glycoside. *Science* 194, 853-854.

Zane, C.E. (1976) Assessment of hypervitaminosis D during the first trimester of pregnancy on the mouse embryo: preliminary report. *Arzneinmittel-Forschritt* 26, 1589-1590.

Chapter 65

Induction of Mammary Gland Carcinoma, Monocytosis and Type II Pneumonocyte Proliferation by Activated Ptaquiloside

M. Shahin[1], B.L. Smith[2], P.B. Oelrichs[1], M.R. Moore[1], S. Worral[3], A.A. Seawright[1] and A.S. Prakash[1]

[1]*National Research Centre for Environmental Toxicology, The University of Queensland, 39 Kessels Road, Coopers Plains, Queensland 4108, Australia;* [2]*Agriculture Research Centre, East Street, Private Bag 3123, Hamilton 2001, New Zealand;* [3]*Department of Biochemistry, The University of Queensland, Queensland 4067, Australia*

Introduction

Bracken fern (*Pteridum* spp.) produces toxic effects in grazing animals, including carcinoma of the upper alimentary tract and urinary bladder (Pamucku *et al.*, 1976; Smith, 1990). The syndromes induced by large quantities of bracken include anorexia, ataxia and incoordination in horses, bright blindness due to retinal neuroepithelial degeneration in sheep and depression of bone marrow associated with leukopenia, thrombocytopenia and hematuria in cattle (Evans *et al.*, 1982).

Rats fed a diet containing bracken developed multiple ileal, urinary bladder and mammary gland adenocarcinomas (Pamucku and Price, 1969; Hirono *et al.*, 1983). The principle carcinogen in bracken is ptaquilaside (PT), which when activated in alkaline states to activated ptaquilaside (APT) alkylates adenines and guanines in DNA in a sequence-selective fashion (Smith *et al.*, 1994). Bracken-fed calves harbor PT/DNA adducts, which results in H-*ras* activation *via* mutation in codon 61 (Prakash *et al.*, 1996). Initiation of carcinogenesis by APT has yet to be demonstrated.

Depending on quantities and time of exposure, bladder and mammary gland carcinomas and multiple ileal, esophageal and pharyngeal neoplasias have been seen in rats fed bracken (Pamucku and Price, 1969; Hirono *et al.*, 1983, Hirono *et al.*, 1984). This work uses PT and APT for rat tumor modeling, and is the first to describe the potential of activated PT as a carcinogen *in vivo*.

Materials and Methods

Bracken was collected from southeast Queensland and PT was isolated and purified (Oelrichs *et al.*, 1995). *Trans-anti*($\pm$)benzo[*a*]pyrenediolepoxide (BPDE) was obtained from Midwest Research Institute, US.

Two groups of ten female Sprague-Dawley rats (6wks, 145g) were dosed intravenously (i.v.) for 10wks with 3mg/wk of PT or APT using saline buffer as a vehicle. Activation of PT to APT was by incubation in 10mM NaOH for 1hr at 37°C. A control group received only the vehicle. The animals were kept alive for 30wks after the last dose to allow tumor formation. Animals were weighed weekly and the ethical requirements of the Australian National Health and Medical Research Council were followed. All three groups were euthanized after 280d.

Two more groups of five animals each were dosed i.v. with 3mg APT every week for 10wks. Since it is known from studies using other DNA-damaging agents that DNA lesions in tissues occur shortly after dosing and that both DNA damage and bone marrow degeneration (BMD) are repaired within 48-72hrs, the animals were euthanized 24hrs after the last dose for DNA damage and BMD studies. A group of five rats was included as control.

Peripheral blood was collected into EDTA every 2wks for hematology profiles (total white blood cells, differential leukocyte, platelet counts). Plasma was collected from each rat and kept at -80°C for α-tumor necrosis factor (TNFα) analysis. Gross and microscopic pathology were done on liver, kidney, ileum, urinary bladder, lung, lymph nodes and mammary gland. Tissue was preserved in 10% buffered formalin for routine histopathology.

DNA (20μg) isolated from each rat ileum was used in the ^{32}P-postlabeling assay (Prakash *et al.*,1996). Adducts were visualized using autoradiography. Assays were performed in duplicate and negative control (no DNA) and positive control (calf thymus DNA adducted with BPDE) were included in each assay.

Results

Both APT- and PT-treated animals had weight loss at the end of the trial, and an increase in monocytes, which appeared after the first treatment and remained elevated at the end of the trial (Fig. 65.1). Total leukocyte and polymorphonuclear leukocyte, lymphocyte and platlet counts did not show significant changes.

The level of TNFα at end of the trial was significantly higher in both PT- and APT-treated rats than in controls (Fig. 65.2). The difference in the TNFα levels between the two treated groups was not statistically significant ($P<0.05$).

Histologically, 40% of the animals that received APT produced mammary gland adenocarcinoma and papillary carcinoma ($P<0.05$). Also, one animal had ileal adenocarcinoma. No urinary bladder tumors were present. Focal type II pneumonocyte necrosis and focal renal tubular necrosis were seen in these animals.

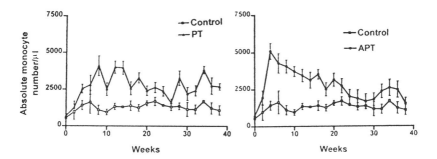

Fig. 65.1. Monocyte count *vs.* ptaquilaside and activated ptaquilaside dosing of rats.

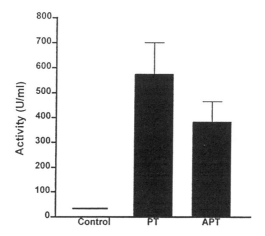

Fig. 65.2. Alpha tissue necrosis factor levels in ptaquilaside and activated ptaquilaside treated rats.

The rats that received PT did not develop any tumors, but they had focal renal tubular necrosis. Also, 30% of the animals from this group exhibited focal Type II pneumonocyte necrosis. Analysis for bone marrow degeneration in the animals from the BMD group showed no significant bone marrow degeneration. None of the control group developed tumors or other pathological changes.

Postlabeling showed the presence of DNA adduct formation in the ileum of the APT-treated animals (Fig. 65.3). The position of the adduct spot on thin-layer chromatography (TLC) is identical to that for PT-DNA adduct formed *in vitro* as reported (Shahin *et al.*, 1995). Mammary gland tissue samples could not be tested for DNA adducts because of insufficient quantities of tissue specimen.

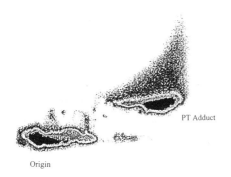

Fig. 65.3. Formation of DNA adducts in ileum of rats treated by APT.

Discussion

Administration of i.v. APT induces tumors in rats and hence is a good model for PT carcinogenesis. The total amount of toxin required was 10x less than needed for oral dosing (Hirono *et al.*, 1984).

An early event in the initiation of chemically induced carcinogenesis is DNA adduct formation. Most adducts disappear within a few days after exposure to DNA-damaging agents due to DNA repair mechanisms within cells. The post-labeling results showed that the ileum of APT-treated rats had DNA adducts that corresponded chromatographically to the PT-DNA adducts observed *in vitro* (Smith *et al.*, 1994; Prakash *et al.*, 1996). This provides confirmation of APT as the final metabolite involved in the DNA damage and initiation of PT carcinogenesis.

In contrast to the earlier studies with PT (Hirono *et al.*, 1984), tumors were not observed in rats administered PT. While C-D rats used in the previous study were genetically predisposed to mammary tumors, the S-D rats used here were not.

A surprising finding in this study was that treated animals had high monocyte and TNFα levels during dosing that persisted long after the last dose. Activated monocytes and macrophages were able to release soluble cytotoxic molecules including TNFα, capable of lysing cells *in vitro* (Cameron, 1982; Adams and Nathan, 1983). Necrosis and regression of some animal tumors have been caused by TNFα (Balkwill *et al.*, 1986). Peripheral blood monocytes and macrophages have tumoricidal activity both *in vitro* and *in vivo* (Fidler and Raz, 1981). The increase in serum TNFα was likely due to either a host response to the toxin or to tumorigenesis. The increased activity of the monocytes was possibly a consequence of cancer cell growth and was triggered as part of the defense against cancer. Further studies will be directed towards developing monocyte and TNFα as biomarkers of carcinogen exposure in both animals and humans.

References

Adams, D.O. and Nathan, C.F. (1983) Molecular mechanisms in tumor-cell killing by activated macrophages. *Immunology Today* 4, 166-170.

Balkwill, F.R., Lee, A. and Aldam, G. (1986) Human tumour xenografts treated with recombinant human tumour necrosis factor alone or in combination with interferons. *Cancer Research* 96, 3990-3993.

Cameron, D.J. (1982) *In vitro* killing of tumor cells by macrophage factors obtained from cancer patients and normal donors. *Journal of the Reticuloendothelial Society* 32, 247-256.

Evans, W.C., Patel, M.C. and Koohy, Y. (1982) Acute bracken poisining in monogastric and ruminant animals. *Proceedings of the Royal Society of Edinburgh* 81, 29.

Fidler, I.J. and Raz, A. (1981) The induction of tumoricidal capacities in mouse and rat macrophages by lymphokines. In: Pick, E. (ed), *Lymphokines, Vol 3*. Academic Press, New York, NY, pp. 345-363.

Hirono, I., Aiso, S., Yamaji, T. and Haga, M. (1983) Induction of mammary cancer in C-D rats fed bracken fern. *Carcinogenesis* 4, 885-887.

Hirono, I., Aiso, S., Yamaji, T., Mori, H., Yamada, K., Niwa, H., Ojika, M., Wukamatsu, K., Kigoshi, H., Niiymama, K. and Uosaki, Y. (1984) Carcinogenicity in rats of ptaquiloside isolated from bracken. *Gann* 75, 833.

Oelrichs, P.B., Ng., J.C. and Bartely, J. (1995) Purification of ptaquiloside, a carcinogen from *Pteridium aquilinum. Phytochemistry* 40, 53-56.

Pamucku, A.M. and Price, J.M. (1969) Induction of intestinal and urinary bladder cancer in rats by feeding bracken fern (*Pteris aquilina*). *Journal of the National Cancer Institute* 43, 275-281.

Pamucku, A.M., Goksoy, S.K. and Price, J.M. (1976) Urinary bladder neoplasm induced by feeding bracken fern (*Pteris aquilina*) to cows. *Cancer Research* 27, 917.

Prakash, A.S., Pereira, T., Smith, B.L., Shaw, G. and Seawright, A.A. (1996) Mechanism of bracken fern carcinogenesis: Evidence for H-*ras* activation *via* initial adenine alkylation by ptaquiloside. *Natural Toxins* 4, 221-227.

Shahin, M., Smith, B.L., Moore, M.R. and Prakash, A.S. (1995) Development of a rat cancer model for ptaquiloside, a bracken fern carcinogen. *Proceedings of the Australian Society for Clinical and Experimental Pharmacology and Toxicology* 3, 174.

Smith, B.L. (1990) Bracken and animal health in Australia and New Zealand. In: Thomson, J.A. and Smith, R.T. (eds), *Bracken Biology and Management*. The Australian Institute of Agriculture Science, Wahroonga, NSW, pp. 227-232.

Smith, B.L., Shaw, G., Prakash, A.S. and Seawright, A.A. (1994) Studies on DNA adduct formation by ptaquiloside, the carcinogen of bracken fern (*Pteridium* spp.). In: Colgate, S.M. and Dorling, P.R. (eds), *Plant-Associated Toxins: Phytochemical, Agricultural and Economic Aspects*. CAB International, Wallingford, pp.167-172.

Chapter 66

Ingestion of Nitrate-containing Plants as a Possible Risk Factor for Congenital Hypothyroidism in Foals

A.L. Allen[1], H.G.G. Townsend[2], C.E. Doige[1] and P.B. Fretz[3]

[1]Departments of Veterinary Pathology, [2]Veterinary Internal Medicine, and [3]Veterinary Anesthesiology, Radiology and Surgery, Western College of Veterinary Medicine, University of Saskatchewan, 52 Campus Drive, Saskatoon, Saskatchewan S7N 5B4, Canada

Introduction

In the late 1970s and early 1980s, veterinarians at the Western College of Veterinary Medicine conducted studies into angular limb deformities of foals (Fretz, 1980; McLaughlin *et al.*, 1981; Fretz *et al.*, 1984). Among the foals examined during this period were several that were atypical of the rest, but with similarities to each other. In 1981, McLaughlin and Doige described the lesions found post mortem in seven of these foals. Multiple musculoskeletal abnormalities were common, including mandibular prognathism, forelimb contracture, rupture of the common digital extensor tendons and severely retarded ossification of the carpal and tarsal bones. All of the foals also had thyroid glands that, despite a normal appearance, were grossly hyperplastic upon microscopic inspection, leading to speculation that these foals may have had abnormal thyroid function during the perinatal period. Surgical thyroidectomy of 1d-old foals resulted in a decrease in the rate of ossification of the carpal and tarsal bones (McLaughlin and Doige, 1982). Triiodothyronine and/or thyroxine levels were low in ten of 14 foals with various combinations of hyperplastic thyroid glands, angular limb deformities, forelimb contracture, ruptured common digital extensor tendons, mandibular prognathism and incompletely ossified carpal or tarsal bones. The response of two foals to the intramuscular administration of 15IU of thyroid-stimulating hormone was monitored and considered poor (McLaughlin *et al.*, 1986).

Thyroid-stimulating hormone response tests (Shaftoe *et al.*, 1988) on nine newborn foals suspected to be suffering with this disease, confirmed that affected foals were congenitally hypothyroid (Allen, 1996). It was unclear whether all the anomalies in affected foals were caused by the same agent, or if affected foals were primarily hypothyroid *in utero* and the developmental anomalies were the result of the hypothyroidism.

An experiment to determine the effects of fetal thyroidectomy on the growth and development of the equine fetus was performed (Allen *et al.*, 1997). Two sham-operated control foals and four partially thyroidectomized foals were carried to term following surgery at about 215d of gestation. Control foals were normally developed while partially thyroidectomized foals were hypothyroid and had similar histories, clinical signs, and lesions as the spontaneously occurring hypothyroid foals reported in western Canada.

Congenital hypothyroidism has been a major cause of reproductive loss and foal mortality in western Canada. The purpose of this investigation was to identify risk factors for the development of congenital hypothyroidism in foals.

Materials and Methods

A case-control study was conducted using privately owned foals born in the province of Alberta in 1993. A foal was classified as congenitally hypothyroid if it was found to have any two of the following musculoskeletal anomalies: 1) mandibular prognathism; 2) flexural deformities of the legs; 3) rupture of one or both common digital extensor tendons or 4) incomplete ossification of the carpal or tarsal bones. Normal foals from neighboring farms were used as controls.

A questionnaire was used to collect standard information during personal interviews with foal owners, farm managers and veterinary practitioners. The information gathered allowed for investigation of the potential transmission of infectious agents, the likelihood of exposure to a toxic substance and the possibility of a dietary deficiency. When possible, samples of the forage fed to pregnant mares were collected and stored. Following preliminary analysis of the data, the forage samples were analyzed for nitrate levels (Helrich, 1990).

Variables concerning mare and foal signalment, diet, dam management and farm environment were investigated for a possible association with the birth of congenitally hypothyroid foals (details in Allen *et al.*, 1996). Information regarding mares and farms producing congenitally hypothyroid foals was compared statistically to that of mares and farms producing normal foals. Stratified analyses were performed to examine the association of combinations of variables with the occurrence of disease (Mantel and Haenszel, 1959). The strength of association between a variable and the birth of a congenitally hypothyroid foal was expressed as the relative odds of disease.

The association of all variables, except the presence of nitrate in forage samples with the occurrence of disease, was examined using all affected foals and an equal number of randomly selected normal foals. The association between the presence of nitrate in forage samples and congenital hypothyroidism in foals was examined at the level of the farm and the individual animal. The latter analysis included all foals examined on farms from which forage was collected.

Results

Information concerning 39 congenitally hypothyroid foals from 26 farms, and 89 normal foals from 23 farms, was available for analysis. Of the 77 independent variables generated from the questionnaire, ten were associated with congenital hypothyroidism at $P \leq 0.15$. After stratified analyses and critical evaluation, only three variables were statistically associated with the birth of congenitally hypothyroid foals and were thought to have biological importance.

Gestation was significantly ($P<0.0001$) longer in congenitally hypothyroid (Allen *et al.*, 1994) foals (357.6d) compared to normal foals (338.9d). Pregnant mares not provided with supplemental mineral had 5.6x ($P=0.0472$) greater odds of producing a congenitally hypothyroid foal than mares that received mineral supplementation, and mares that were fed greenfeed had 13.1x ($P=0.0068$) greater odds of producing a congenitally hypothyroid foal than mares not fed greenfeed. The term greenfeed was used to refer to cereal crops, almost always oats, harvested prior to maturity, i.e., 'green,' and baled for use as a livestock feed. Green oats, green oat hay, oat hay, green oat forage, and oat straw are other terms used to describe similar types of forage. Since greenfeed grown in Alberta often contains high levels of nitrate (Smith and Suleiman, 1991), which is believed to impair thyroid gland function in a variety of animals (Wyngaarden *et al.*, 1953; Arora *et al.*, 1968; Jahreis *et al.*, 1987), the stored forage samples were analyzed for nitrate.

Samples of 20 different forages from 14/26 farms producing hypothyroid foals, and of ten different forages from 7/23 farms producing normal foals had been collected. Nitrate was present more often (8/14 farms) and at higher concentrations in those samples collected from farms producing congenitally hypothyroid foals compared to those farms producing normal foals (1/7). Further, the odds of at least one case of the congenital hypothyroidism occurring on farms feeding forage with at least a trace of nitrate was 8.0x greater ($P=0.0873$) than the odds of disease occurring on farms that fed forage free of nitrate. On an individual animal basis, the odds of a mare producing a congenitally hypothyroid foal when exposed to forage containing at least a trace of nitrate was 5.9x greater ($P=0.0007$) than for mares exposed to nitrate-free forage.

Discussion

This study found that the lack of mineral supplementation and the presence of greenfeed in the diet of pregnant mares significantly increased the odds of producing a congenitally hypothyroid foal. Failure to supplement pregnant mares with mineral may have been associated with an iodine deficiency, since western Canadian soils and the plants grown on these soils are believed to be very low in iodine (Allen *et al.*, 1994). Iodine is essential for normal thyroid function. As discussed above, the ingestion of nitrate has been associated with impaired thyroid function in a variety of animals. If the presence of nitrate in the diet of pregnant mares interferes with fetal thyroid function, it will be important to consider all sources of environmental nitrate to fully understand and prevent disease.

Unfortunately, the combined effects of the lack of supplemental mineral and the feeding of greenfeed on the occurrence of congenital hypothyroidism could not be pursued with this data set, as none of the farms in this study failed to provide supplemental mineral and fed greenfeed. There is evidence in rats (Lee *et al.*, 1970) and in pigs (Jahreis *et al.*, 1986) that increased levels of iodine in the diet can counteract, to some degree, the effects of nitrate on thyroid activity.

From this study, it appears that a deficiency of dietary iodine and the ingestion of nitrate are two underlying factors that may interfere with fetal thyroid function and may produce congenital hypothyroidism in foals.

References

Allen, A.L. (1996) Investigations into Congenital Hypothyroidism in Foals. PhD Thesis. University of Saskatchewan, Saskatoon, Saskatchewan.

Allen, A.L., Doige, C.E., Fretz, P.B. and Townsend, H.G.G. (1994) Hyperplasia of the thyroid gland and concurrent musculoskeletal deformities in western Canadian foals: Reexamination of a previously described syndrome. *Canadian Veterinary Journal* 35, 31-38.

Allen, A.L., Townsend, H.G.G., Doige, C.E. and Fretz, P.B. (1996) A case-control study of the congenital hypothyroidism and dysmaturity syndrome of foals. *Canadian Veterinary Journal* 37, 349-358.

Allen, A.L., Fretz, P.B., Card, C.E. and Doige, C.E. (1997) The effect of partial thyroidectomy on the development of the equine foetus. *Equine Veterinary Journal* 30, 53-59.

Arora, S.P., Hatfield, E.E., Garrigus, U.S., Romack, F.E. and Motyka, H. (1968) Effect of adaptation to dietary nitrate on thyroxine secretion rate and growth in lambs. *Journal of Animal Science* 27, 1445-1448.

Fretz, P.B. (1980) Angular limb deformities in foals. *Veterinary Clinics of North America: Large Animal Practice* 2, 125-150.

Fretz, P.B., Cymbaluk, N.F. and Pharr, J.W. (1984) Quantitative analysis of long-bone growth in the horse. *American Journal of Veterinary Research* 45, 1602-1609.

Helrich, K. (1990) Method #986.31 - Nitrate in forages. *Official Methods of Analysis, Vol 1.* 15th ed. Association of Official Analytical Chemists, Arlington, VA, pp. 357-358.

Jahreis, G., Hesse, V., Schone, F., Hennig, A. and Gruhn, K. (1986) Effect of chronic dietary nitrate and different iodine supply on porcine thyroid function, somatomedin-C level and growth. *Experimental and Clinical Endocrinology* 88, 242-248.

Jahreis, G., Schone, F., Ludke, H. and Hesse, V. (1987) Growth impairment caused by dietary nitrate intake regulated *via* hypothyroidism and decreased somatomedin. *Endocrinologia Experimentalis* 21, 171-180.

Lee, C., Weiss, R. and Horvath, D.J. (1970) Effects of nitrogen fertilization on the thyroid function of rats fed 40% orchard grass diets. *Journal of Nutrition* 100, 1121-1126.

Mantel, N. and Haenszel, W. (1959) Statistical aspects of the analysis of data from retrospective studies of disease. *Journal of the National Cancer Institute* 22, 719-748.

McLaughlin, B.G. and Doige, C.E. (1981) Congenital musculoskeletal lesions and hyperplastic goitre in foals. *Canadian Veterinary Journal* 22, 130-133.

McLaughlin, B.G. and Doige, C.E. (1982) A study of carpal and tarsal bones in normal and hypothyroid foals. *Canadian Veterinary Journal* 23, 164-168.

McLaughlin, B.G., Doige, C.E., Fretz, P.B. and Pharr, J.W. (1981) Carpal bone lesions associated with angular limb deformities in foals. *Journal of the American Medical Association* 178, 224-230.

McLaughlin, B.G., Doige, C.E. and McLaughlin, P.S. (1986) Thyroid hormone levels in foals with congenital musculoskeletal lesions. *Canadian Veterinary Journal* 27, 264-267.

Shaftoe, S., Pichler-Schick, M. and Chen, C.L. (1988) Thyroid stimulating hormone response tests in one-day-old foals. *Journal of Equine Veterinary Science* 8, 310-312.

Smith, R.A. and Suleiman, A. (1991) Nitrite intoxication from large round bales. *Veterinary and Human Toxicology* 33, 349-350.

Wyngaarden, J.B., Stanbury, J.B. and Rapp, B. (1953) The effects of iodine, perchlorate, thiocyanate, and nitrate administration upon the iodine concentrating mechanism of the rat thyroid. *Endocrinology* 52, 568-574.

Chapter 67

Diterpene Acid Chemistry of Ponderosa Pine and Implications for Late-term Induced Abortions in Cattle

D.R. Gardner, K.E. Panter, L.F. James, B.L. Stegelmeier and J.A. Pfister

USDA-ARS, Poisonous Plant Research Laboratory, 1150 East 1400 North, Logan, Utah 84341, USA

Introduction

Cattle eat relatively large amounts of ponderosa pine (*Pinus ponderosa*) needles (Pfister and Adams, 1993), grazing of fresh green needles, needles from windfalls and older dry needles from the trees or ground. Induced abortions in pregnant cows after consumption of ponderosa pine needles is documented (James *et al*., 1977; James *et al*., 1989). Generally, cattle in late gestation (third trimester) appear to be most susceptible. The abortions are characterized by weak uterine contractions, occasional incomplete cervical dilation, excessive mucus discharge, the birth of a small weak calf that soon dies and retained placenta (Stevenson *et al*., 1972; James *et al*., 1989). Direct losses associated with ponderosa pine needle abortion have been estimated to be as high as US$20m (Miner *et al*., 1987; Lacey *et al*., 1988).

After chemical extraction, isolation, and feeding trials in cattle, a diterpene resin acid, isocupressic acid, was identified as an abortifacient compound in pine needles (Gardner *et al*., 1994). This led to the need to further characterize the diterpene acid content of ponderosa pine, the bioactivity of isocupressic acid and structurally related compounds, and the biological distribution of such abortifacient compounds in other plants, especially those known to cause abortions in livestock.

Diterpene Acids of Ponderosa Pine Needles

The diterpene acid content of ponderosa pine has been extensively characterized and described (Fujii and Zinkel, 1984; Zinkel and Magee, 1991). As shown in Fig. 67.1, two broadly classified structural groups of compounds have been identified as either tricyclic (abietane, pimarane, isopimarane) or bicyclic (labdane) ring structures.

In feeding trials with pregnant cattle, the tricyclic compounds (1-6, Fig. 67.1) have shown no abortifacient activity (Gardner *et al.*, 1994). However, when cattle have been dosed with relatively high doses of the tricyclic compounds such as dehydroabietic acid, rosin gum (a commercially available mixture of the tricyclic diterpene acids) or new growth pine tips (high relative concentration of tricyclic compounds compared to the needles) they developed a toxicosis characterized by anorexia, mild rumen acidosis, dyspnea, paresis progressing to paralysis and death (Stegelmeier *et al.*, 1996). Histologically, all animals had nephrosis, vacuolation of the basal ganglia neuropil with patchy perivascular and myelinic edema, and skeletal myonecrosis. Intoxication from the tricyclic diterpene acids in cattle from grazing of ponderosa pine needles is probably rare based on the required dosage (100g/d), and the lower concentration of the tricyclic compounds in the needles as compared to new growth tips.

In contrast, the bicyclic labdane compound (7-11, Fig. 67.1), isocupressic acid (7), was found to induce abortions in cattle when fed in a purified form (Gardner *et al.*, 1994). In further feeding studies, the ester derivatives of isocupressic acid, acetylisocupressic acid (8) and succinylisocupressic acid (9) (which also occur naturally in pine needles), were also found to induce abortions in cattle. However, *in vitro* rumen metabolism studies demonstrated that the acetyl and succinyl compounds are hydrolyzed in the rumen to isocupressic acid (Gardner *et al.*, 1996).

Three other labdane compounds have been reported from ponderosa pine needles: imbricataloic (10), imbricatoloic (11) and dihydroagathic acid (12). The abortifacient activity of these compounds is unknown at this time. Based on analysis of ponderosa pine needles (Zinkel and Magee, 1991) and from analyses in this laboratory, imbricataloic acid occurs at concentrations comparable to that of isocupressic acid (10-30% relative concentration), and imbricatoloic and dihydroagathic acids usually occur at lower concentrations (1-5%). Imbricataloic acid was the major diterpene acid present in one of the fractions isolated from ponderosa pine needles during the original fractionation of abortifacient compounds from the pine needles but, when fed to a pregnant cows, no abortion was induced (Gardner *et al.*, 1994). The difficulty in isolating these additional labdane compounds in sufficient quantities and purity for feeding trials in cattle has prevented further studies. An alternative bioassay that is highly specific, sensitive and requires less material would greatly aid in the testing of additional compounds.

Metabolism

Acetyl and succinylisocupressic acid are metabolized (ester hydrolyzed) in the rumen to isocupressic acid with a half-life of 2hrs and 4hrs, respectively (Gardner *et al.*, 1996). Isocupressic acid was unchanged after 8hrs in the same *in vitro* rumen trials, and when dosed intravenously (i.v.) induced abortions in cattle (unpublished).

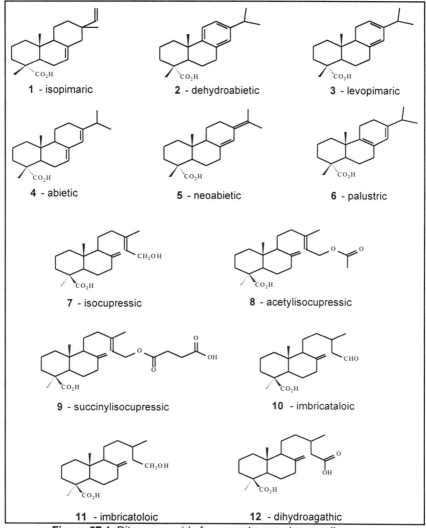

Figure 67.1. Diterpene acids from ponderosa pine needles.

Rumen metabolism of isocupressic acid is therefore not needed for abortifacient activity.

Isocupressic acid has been detected in serum samples taken from cattle treated by i.v. infusion; however, its presence is short-lived and is only readily detected in the serum for approximately 1hr after treatment. No isocupressic acid was detected in any of the serum samples taken over a 30hr period after treatment with a single oral dosage of isocupressic acid. Metabolism of isocupressic acid is suspected and several isocupressic acid metabolites have now been detected in serum samples of both orally and i.v.-treated animals. Identification of the isocupressic acid metabolites is underway at this time. The metabolism of isocupressic acid needs to be studied further, as an isocupressic acid metabolite may actually be bioactive and responsible for the induced abortions.

Mechanism of Action

The physiochemical mechanism by which the induced abortions occur is still not well defined. It has been proposed that the induced abortions occur after decreased uterine blood flow to the calf as a result of vasoconstriction of the caruncular arteries (Ford *et al.*, 1992; Christenson *et al.*, 1993). Isocupressic acid was tested in an *in vitro* placentome assay and found not to be vasoactive (Al-Mahmoud *et al.*, 1995). Based on the most recent data and analyses of serum samples, it is possible that isocupressic acid induces abortion when given orally or i.v. The biochemical response inducing abortion is a metabolite of isocupressic acid.

Occurrence of Isocupressic Acid in Other Plants

Diterpene acids are prominent plant secondary compounds in a number of species. Isocupressic acid was originally identified from *Cupressus sempervirens* (Mangoni and Belardini, 1964) and has since been identified in a variety of genera such as *Pinus*, *Juniperus*, *Araucaria* and a few others. Therefore, it is necessary to examine the occurrence of isocupressic acid in *Pinus* species and other related plant species for a better understanding of potential abortifacient plants. A survey is underway to test a variety of different coniferous trees from the western and southern US. Leaf collections are made from three to five locations per species in the general geographical areas where they occur. Samples are analyzed by GC for total available isocupressic acid. Acetyl- and succinylisocupressic acid, if present, are converted to isocupressic acid during the analytical extraction process.

Some preliminary results of the survey are listed in Table 67.1. Only a few genera have significantly high levels (comparable to ponderosa pine) of isocupressic acid. The highest levels of isocupressic acid were detected in common juniper (*Juniperus communis*). The authors are unaware of any reported abortions associated with this plant material. However, common juniper is typically found at the high elevations (>8,000ft) in the western US, where late-gestation cows would not have access during winter and spring calving months. Moderate levels of isocupressic acid were detected in Rocky Mountain juniper (*Juniperus scopulorum*). There are anecdotal reports of abortions occurring from consumption of this plant, but the abortions have not been reproduced experimentally. Lodgepole pine (*Pinus contorta*) also included significant levels of isocupressic acid. Again, there are several field cases of abortion on rangelands with stands of lodgepole pine. It would be beneficial to establish the

Table 67.1. Distribution of isocupressic acid (ICA) in *Pinus* and related genera as % dry weight of plant.

Moderate/high ICA levels detected >0.5%	Trace levels of ICA <0.2%	No ICA detected
Pinus ponderosa	*Pseudotsuga menziesii*	*Pinus montezumae*
Cupressus macrocarpa	*Pinus monophylla*	*Pinus arizonica*
Pinus jeffreyi	*Pinus flexilis*	*Pinus echinata*
Pinus contorta	*Pinus aristata*	*Pinus taeda*
Juniperus scopulorum	*Picea engelmannii*	*Pinus palustris*
Juniperus communis	*J. osteosperma*	*Pinus elliotii*
	J. monosperma	*Abies grandis*
	Abies lasiocarpa	*Abies concolor*

abortifacient potential of plants containing moderate or higher levels of isocupressic acid and to understand the potential grazing patterns of animals exposed to those plants.

Samples of Monterey cypress (*Cupressus macrocarpa*) were collected in New Zealand. This plant is indigenous to a small area on the central California coast (Monterey Peninsula); however, this tree was introduced into New Zealand and Southern Australia pastures for the purpose of providing shade and shelter belts.

There have been significant reports of cattle abortions associated with the consumption of Monterey cypress in New Zealand (MacDonald, 1956). Even though isocupressic acid was originally identified from *C. sempervirens* (European cypress), it was not until Parton *et al.* (1996) recognized the possible connection between cupressus-induced abortions in New Zealand and those from ponderosa pine needles in the US, that isocupressic acid was identified in the New Zealand plant material. Interestingly, a single sample of *C. macrocarpa* from a California arboretum was analyzed, and no isocupressic acid was detected.

References

Al-Mahmoud, M.S., Ford, S.P., Short, R.E., Farley, D.B., Christenson, L. and Rosazza, J.P.N. (1995) Isolation and characterization of vasoactive lipids from the needles of *Pinus ponderosa*. *Journal of Agricultural and Food Chemistry* 43, 2154-2161.
Christenson, L.K., Short, R.E., Farley, D.B. and Ford, S.P. (1993) Effects of ingestion of pine needles (*Pinus ponderosa*) by late-pregnant beef cows on potential sensitive Ca^{2+} channel activity of caruncular arteries. *Journal of Reproduction and Fertility* 98, 301-306.
Ford, S.P., Christenson, L.K., Rosazza, J.P. and Short, R.E. (1992) Effects of ponderosa pine needle ingestion on uterine vascular function in late-gestation beef cows. *Journal of Animal Science* 70, 1609-1614.

Fujii, R. and Zinkel, D.F. (1984) Minor components of ponderosa pine oleoresin. *Phytochemistry* 23, 875-878.

Gardner, D.R., Molyneux, R.J., James, L.F., Panter, K.E. and Stegelmeier, B.L. (1994) Ponderosa pine needle-induced abortion in beef cattle: identification of isocupressic acid as the principal active compound. *Journal of Agricultural and Food Chemistry* 42, 756-761.

Gardner, D.R., Panter, K.E., Molyneux, R.J., James, L.F. and Stegelmeier, B.L. (1996) Abortifacient activity in beef cattle of acetyl- and succinyl-isocupressic acid from ponderosa pine. *Journal of Agricultural and Food Chemistry* 44, 3257-3261.

James, L.F., Call, J.W. and Stevenson, A.H. (1977) Experimentally induced pine needle abortion in range cattle. *Cornell Veterinarian* 67, 294-299.

James, L.F., Short, R.E., Panter, K.E., Molyneux, R.J., Stuart, L.D. and Bellows, R.A. (1989) Pine needle abortion in cattle: a review and report of 1973-1984 research. *Cornell Veterinarian* 79, 39-52.

Lacey, J.R., James, L.F. and Short, R.E. (1988) Ponderosa pine: economic impact. In: James L.F., Ralphs, M.H. and Nielsen, D.B. (eds), *The Ecology and Economic Impact of Poisonous Plants on Livestock Production.* Westview Press, Boulder, CO, pp. 95-106.

MacDonald, J. (1956) Macrocarpa poisoning. *New Zealand Veterinarian Journal* 4, 30.

Mangoni, L. and Belardini, M. (1964) Sui componenti della resina di *Cupressus sempervirens.* Nota I. Acido communico, acido cupressico e acido isocupressico. *Gazzetta Chimica Italiana* 94, 1108-1121.

Miner, J.L., Bellows, R.A., Staigmiller, R.B., Peterson, M.K., Short, R.E. and James, L.F. (1987) Montana pine needles cause abortion in beef cattle. In: *Montana Agricultural Research.* Montana Agriculture Experimental Station, Montana State University, Bozeman, MT, 4, pp. 6-9.

Parton, K., Gardner, D. and Williamson, N.B. (1996) Isocupressic acid, an abortifacient component of *Cupressus macrocarpa. New Zealand Veterinarian Journal* 44, 109-111.

Pfister, J.A. and Adams, D.C. (1993) Factors influencing pine needle consumption by grazing cattle during winter. *Journal of Range Management* 46, 394-398.

Stegelmeier, B.L., Gardner, D.R., James, L.F., Panter, K.E. and Molyneux, R.J. (1996) The toxic and abortifacient effects of ponderosa pine. *Veterinary Pathology* 33, 22-28.

Stevenson, A.H., James, L.F. and Call, J.W. (1972) Pine needle (*Pinus ponderosa*)-induced abortions in range cattle. *Cornell Veterinarian* 62, 519-524.

Zinkel, D.F. and Magee, T.V. (1991) Resin acids of *Pinus ponderosa* needles. *Phytochemistry* 30, 845-848.

Chapter 68

Toxic and Teratogenic Piperidine Alkaloids from *Lupinus*, *Conium* and *Nicotiana* Species

K.E. Panter[1], D.R. Gardner [1], R.E. Shea[1], R.J. Molyneux [2] and L.F. James[1]

[1]USDA-ARS, Poisonous Plant Research Laboratory, 1150 East 1400 North, Logan, Utah 84341, USA; [2] USDA-ARS, Western Regional Research Center, 800 Buchanan St, Albany, California 94710, USA

Introduction

Piperidine alkaloids are widely distributed in nature; many are toxic and some are teratogenic in livestock. Plant genera containing piperidine alkaloids include *Lobelia*, *Pinus*, *Punica*, *Duboisia*, *Sedum*, *Withania*, *Carica*, *Hydrangea*, *Dichroa*, *Cassia*, *Prosopis*, *Genista*, *Ammodendron*, *Liparia*, *Collidium*, *Lupinus*, *Conium* and *Nicotiana*. Many of these contain piperidine alkaloids structurally similar to those known to be teratogenic in livestock. *Lupinus*, *Conium* and *Nicotiana* contain alkaloids known to be toxic and teratogenic in livestock (Panter *et al.*, 1990).

Through feeding trials with crude extracts, commercially obtained alkaloids or purified compounds, it has been demonstrated that coniine and γ-coniceine from *Conium*, ammodendrine from *Lupinus formosus* and *L. arbustus*, and anabasine from *Nicotiana* are teratogenic piperidine alkaloids. They all fit the structural requirements for teratogenicity suggested by Keeler and Balls (1978), including a piperidine ring with an attached structure of three carbons or larger located next to the ring. Coniine exemplifies the minimum structural requirement (Fig. 68.1). Gamma-coniceine is more toxic (Bowman and Sanghvi, 1963), and is believed to be more potent as a teratogen because of the double bond in the ring structure (Panter *et al.*, 1994). Anabasine is a known teratogen, and it is possible that the double bond derivative anabaseine would be more toxic and perhaps more potent as a teratogen (Panter *et al.*, 1994). Alkaloids in *L. formosus* and *L. arbustus* are thought to have a similar toxic/teratogenic structure activity relationship (Fig. 68.1).

The periods of gestation when the fetus is susceptible to these plant teratogens have been partially defined in cattle, sheep, swine and goats (Shupe *et al.*, 1967; Keeler and Crowe, 1983, 1984; Panter *et al.*, 1990). In swine, cleft palate only occurred when *Conium* was fed during days 30-41 of gestation (Panter *et al.*, 1985a). Skeletal defects, predominantly the forelimbs, spine and neck, without cleft palate were induced when dams were fed *Conium* during gestation days 40-53 (Panter *et al.*, 1985b). When fed on days 50-63, rear limbs were affected also. When the feeding period included days 30-60, all combinations of the defects described occurred. In sheep and goats, the teratogenic insult period is similar to pigs and includes days 30-60 (Keeler and Crowe, 1984; Panter *et al.*, 1990). In goats, a narrow period for cleft palate induction was defined to include days 35-41 (Panter and Keeler, 1992). In cattle, the susceptible period of gestation was thought to include days 40-70 (Shupe

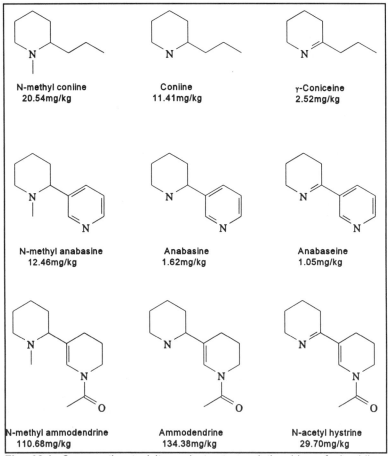

Fig. 68.1. Comparative toxicity and structure relationships of piperidine alkaloids from *Conium*, *Nicotiana* and *Lupinus* spp.

et al., 1967), but recent epidemiology data suggest that the insult period may extend to day 100 (Panter *et al.*, 1997). The severity and type of the malformations depend on the dosage of teratogenic alkaloids ingested, the length of time the plants are ingested and the stage of pregnancy when ingestion takes place. The mechanism of action appears to be inhibited fetal movement by putative alkaloids crossing the placental barrier (Panter *et al.*, 1990).

The objectives of this research were (1) to determine structure/activity relationships of known and suspected teratogenic piperidine alkaloids, (2) to narrowly define the cleft palate induction period in the cow and (3) to determine what contribution *N*-acetyl hystrine or *N*-methyl ammodendrine might have in the teratogenicity of *L. formosus* or *L. arbustus*.

Materials and Methods

Piperidine alkaloids were purchased commercially (coniine), extracted from plant material (anabasine, ammodendrine) or partially synthesized from parent compounds (γ-coniceine, *N*-methyl coniine, anabaseine, *N*-methyl anabasine, *N*-acetyl hystrine and *N*-methyl ammodendrine). Methods of partial synthesis and extraction will be published elsewhere. The toxicities of these compounds were tested in a mouse bioassay using 14-25g male white Swiss Webster mice. The compounds were solubilized in buffered saline and injected intravenously (i.v.). The LD_{50} was calculated using a modified up-and-down method (Bruce, 1985).

Seven crossbred 2-3yr-old first-calf heifers were time bred to a Hereford bull and after confirmation of pregnancy were divided into three groups. Group 1 contained three cows fed *L. formosus* (collections 87-3, 89-1), Group 2 contained two cows fed *L. arbustus* (collection 94-4) and Group 3 contained two cows that were water-gavaged controls. *Lupinus formosus* and *L. arbustus* were collected, dried, ground and stored at room temperature until feeding trials. Each cow was gavaged twice daily (AM and PM) through gestation days 40-49. *Lupinus formosus* contained high levels of ammodendrine and *N*-acetyl hystrine but very little *N*-methyl ammodendrine (Table 68.1), whereas *L. arbustus* contained ammodendrine and *N*-methyl ammodendrine but no *N*-acetyl hystrine. The plants were fed to each cow at equivalent daily ammodendrine dosages of about 27mg/kg.

Results and Discussion

Nine alkaloids from three poisonous plant species were tested in a mouse bioassay (Fig. 68.1). Toxicity (LD_{50}) of alkaloids from *N. glauca* was 1.6mg/kg for anabasine, 12.4mg/kg for *N*-methyl anabasine and 1.05mg/kg for anabaseine. Toxicity of alkaloids from *C. maculatum* was 2.5mg/kg for γ-coniceine, 11.4mg/kg for coniine and 20.5mg/kg for *N*-methyl coniine. Toxicity of alkaloids from *L. formosus* was 29.7mg/kg for *N*-acetyl hystrine, 110.7mg/kg for *N*-methyl ammodendrine and

134.4mg/kg for ammodendrine.

The most toxic of the alkaloids from within plant species in this study were those containing a double bond in the position next to the N in the piperidine ring. The *N*-methyl derivatives appear to be the least potent, with the exception of ammodendrine and *N*-methyl ammodendrine (Fig. 68.1).

Table 68.1. Plant analysis of *L. formosus* and *L. arbustus* (% alkaloid in dry plant).

	L. formosus		L. arbustus
	87-3[a]	89-1	94-4
Total alkaloid	1.69	1.09	1.40
Ammodendrine	0.34	0.36	0.95
N-methyl ammodendrine	0.12	0.15	tr
N-acetyl hystrine	0.40	0.23	tr

[a]Indicates year and collection number; tr=trace

Table 68.2. Alkaloid daily dosage (mg/kg) and outcome of pregnancy in cows fed *L. formosus* and *L. arbustus*.

Cow	Group	Treatment	T	A	B	C	Results
5991	1	87-3	122	25	8.6	28.8	Clinical signs mod-sev, calf normal
6038	1	87-3	133	27	10.0	31.6	Clinical signs mod, CP
6041	1	89-1	84	27	11.5	17.7	No effects, calf normal
6008	2	94-1	42	28	----	----	No effects, ED/R
6025	2	94-1	42	28	----	----	No effects, BK/CF
6030	3	ctr	----	---	----	----	Calf normal
6026	3	ctr	----	---	----	----	Calf normal

T=Total alkaloid daily dosage (mg/kg); A=Daily ammodendrine dosage; B=Daily *N*-methyl ammodendrine dosage; C=Daily *N*-acetyl hystrine dosage; CP=Cleft palate ED/R=Embryonic death/resorption; BK/CF=Bucked knees/carpal flexure; 87-3 and 89-1= *L. formosus*; 94-4= *L. arbustus*; ctr= water control

Lupinus formosus fed at the dosages used in the bovine study caused clinical toxicoses in 2/3 cows of Group 1 by day five after repeated AM and PM dosing of 3.5g/kg. Clinical signs included depression, loss of appetite, frothing, muscular weakness and ataxia. Signs were moderate to severe in one cow, but recovery was rapid once feeding of *L. formosus* ceased. Cow 5991 (Table 68.2) had moderate to severe clinical signs of poisoning, preventing one dose during the treatment period; thus Table 68.2 reflects a slightly lower average daily dosage. Clinical signs of toxicity can be related to total alkaloid in the plants fed. No clinical signs were observed in Group 2 cows or in cow 6041 fed *L. formosus* (collection 89-1), which were fed lower alkaloid dosages. This suggests that *N*-acetyl hystrine and *N*-methyl ammodendrine contributed to the toxicosis and teratogenicity of *L. formosus*.

One calf in Group 1 had a full bilateral cleft palate with no other deformities evident. The two other calves in Group 1 were normal. In Group 2, one calf was born with minor contractures of the carpal joints, which resolved spontaneously within 3wks. The other fetus in Group 2 died *in utero* (determined by ultrasound) on day five of treatment and was resorbed over the next 2wks.

Conclusions

While the mouse bioassay demonstrated relative toxicity of nine potential alkaloid teratogens, it did not reflect teratogenic potency. Our hypothesis, however, stated that the toxicity and teratogenicity are related. The bovine feeding trials indicate a narrow period for cleft palate induction (40-49d) from these plants. While this information is inconclusive as to which alkaloids in *L. formosus* and *L. arbustus* are teratogens, there is evidence that *N*-acetyl hystrine and *N*-methyl ammodendrine contribute to teratogenicity. Further studies will be conducted to confirm these observations.

References

Bowman, W.C. and Sanghvi, I.S. (1963) Pharmacological actions of hemlock (*Conium maculatum*) alkaloids. *Journal Pharmacy and Pharmacology* 15, 1-25.

Bruce, R.D. (1985) An up-and-down procedure for acute toxicity testing. *Fundamental and Applied Toxicology* 5, 151-157.

Keeler, R.F. and Balls, L.D. (1978) Teratogenic effects in cattle of *Conium maculatum* and *Conium* alkaloids and analogs. *Clinical Toxicology* 20, 49-64.

Keeler, R.F. and Crowe, M.W. (1983) Congenital deformities in swine induced by wild tree tobacco, *Nicotiana glauca. Clinical Toxicology* 20, 47-58.

Keeler, R.F. and Crowe, M.W. (1984) Teratogenicity and toxicity of wild tree tobacco, *Nicotiana glauca* in sheep. *The Cornell Veterinarian* 74, 50-59.

Panter, K.E. and Keeler, R.F. (1992) Induction of cleft palate in goats by *Nicotiana glauca* during a narrow gestational period and the relation to reduction in fetal movement. *Journal of Natural Toxins* 1, 25-32.

Panter, K.E., Keeler, R.F. and Buck, W.B. (1985a) Induction of cleft palate in newborn pigs by maternal ingestion of poison hemlock (*Conium maculatum*). *American Journal of Veterinary Research* 46, 1368-1371.

Panter, K.E., Keeler, R.F. and Buck, W.B. (1985b) Congenital skeletal malformations induced by maternal ingestion of *Conium maculatum* (poison hemlock) in newborn pigs. *American Journal of Veterinary Research* 46, 2064-2066.

Panter, K.E., Bunch, T.D., Keeler, R.F., Sisson, D.V. and Callan, R.J. (1990) Multiple congenital contractures (MCC) and cleft palate induced in goats by ingestion of piperidine alkaloid-containing plants. Reduction in fetal movement as the probable cause. *Clinical Toxicology* 28, 69-83.

Panter, K.E., Gardner, D.R. and Molyneux, R.J. (1994) Comparison of toxic and teratogenic effects of *Lupinus formosus*, *L. arbustus* and *L. caudatus* in goats. *Journal of Natural Toxins* 3, 83-93.

Panter, K.E., Gardner, D.R., Gay, C.C., James, L.F., Mills, R., Gay, J.M. and Baldwin, T.J. (1997) Observations of *Lupinus sulphureus*-induced "Crooked Calf Disease." *Journal of Range Management* 50, 587-592.

Shupe, J.L., Binns, W., James, L.F. and Keeler, R.F. (1967) Lupine, a cause of crooked calf disease. *Journal of the American Veterinary Medical Association* 151, 198-203.

Chapter 69

Toxic Amines and Alkaloids from Texas Acacias

B.A. Clement[1] and T.D.A. Forbes[2]

[1]*Department of Veterinary Anatomy and Public Health, Texas A&M University, College Station, Texas 77843, USA*; [2]*Texas A&M University Agricultural Research and Extension Center, Uvalde, Texas 78801, USA*

Introduction

Acacias are native to southwest US and northern Mexico. *Acacia berlandieri* Benth. (guajillo) dominates the semiarid regions while *Acacia rigidula* Benth. (blackbrush) grows on rocky ridges and in arroyos. Sheep and goats grazing on guajillo during drought develop an ataxia referred to as "limberleg" or "guajillo wobbles" (Price and Hardy, 1953). Prior analyses of *A. berlandieri* identified *N*-methyl-β-phenethylamine (NMPEA), tyramine, *N*-methyl-tyramine, and hordenine (Camp and Lyman, 1956; Camp and Moore, 1960; Adams and Camp, 1966). Others showed that NMPEA caused adverse reproductive effects (Forbes *et al.*, 1994; Forbes *et al.*, 1995) and that guajillo has antinutrient effects (Camp, 1970).

Smith (1977) has shown that many plant species endemic to this region contain a variety of phenolic monoamines. Advances in GC/MS technology were used to investigate the amines in these plants and define thier mechanism of action.

Materials and Methods

About 500g of leaves, petioles and tender stems of *A. berlandieri* and *A. ridigula* were collected in the spring after new growth appeared and again in the late fall prior to frost and before color change was detected. The material was frozen at -20°C and stored until extracted. Voucher specimens were placed at the Texas A&M University Agricultural Research and Extension Center, Uvalde, Texas.

Frozen plant material was extracted continuously for 24hr with methanol (MeOH). The MeOH solution was removed and replaced with chloroform ($CHCl_3$)

and the extraction continued an additional 24hr. The MeOH extract was concentrated and the residue suspended in 100ml CHCl₃ and re-extracted 3x with 50ml portions of 10% HCl. The acid fractions were combined, the pH adjusted to 10.3 with NaOH, and the resulting solution was re-extracted 3x with 50ml of CHCl₃ and then 3x with 50ml of ethylacetate (EtOAc). These organic extracts were combined, dried over $MgSO_4$, filtered, concentrated under vacuum and stored under argon prior to analysis by GC/MS. The CHCl₃ solution from the Soxhlet extraction was treated the same way, except that it was not concentrated before acid extraction.

Mass spectra (EIMS at 70eV) were collected on Hewlett Packard Model 5988A and 5970C GC/quadrupole systems. Helium was the carrier gas and the transfer line was maintained at 280°C. Split injections were made at 200°C, with a purge flow to split vent of 60ml/min. The oven temperature was held at 200°C for 1min then ramped at 3°C/min to 270°C, where it was held for 20min. On-column injections were made at 60°C. The temperature was held for 1min, then ramped at 1.5°C/min to 270°C, where it was held for 35min. Cross-linked methyl silicone 12 (HP 5970 system) 36m x 0.2mm x 33μM capillary columns were used. Data were collected as total ion chromatograms in the operating range of 35-800amu. Preliminary identifications were made by comparison to the Wiley and NBS mass spectral libraries, and final identification was made by direct spectral comparison with the purchased or synthesized authentic samples.

Frozen samples of the foliage were lyophilized, pulverized in a mortar and pestle, weighed and subjected to Soxhlet extraction as above. The MeOH fraction was concentrated and combined with the CHCl₃ extract. The residue was diluted in MeOH and aliquots of each sample were quantified by GC/MS.

Quantitation was performed by determining the area under the *m/z* 44 peak *vs.* a linear standard curve for each analyte. Triplicate samples were prepared for early and late-season foliage, and each sample was analysed in triplicate by GC/MS.

Results and Discussion

Fifty amines and alkaloids, including four previously encountered compounds, were tentatively identified by GC/MS in the guajillo and blackbrush extracts (Table 69.1). On-column injection decreased thermal decomposition on the GC column. No derivatization was performed on the analytes; therefore GC/MS analysis would only be expected to detect the volatile amines and alkaloids and the volatile decomposition products of thermally labile compounds. No attempts were made to purify individual compounds for complete structure elucidation by other means or to evaluate the effect of the extraction process on the nature of the alkaloids subsequently identified.

Thirty-two of the alkaloids identified were related to β-phenethylamine, including the two 2-cyclohexylethylamines. In addition to the compounds reported in Table 69.1, several quaternary amines were postulated based on the presence of styrene and other decomposition products in guajillo extracts.

Table 69.1. Alkaloids and amines tentatively identified in Texas acacias.

Chemical found	Guajillo	Blackbrush
2-cyclohexylethylamine		+
N-2-cyclohexylethyl-N-methylamine	+	
phenethylamine	+	+
N-methylphenethylamine	+	+
N,N-dimethylphenethylamine	+	+
amphetamine	+	+
methamphetamine	+	+
N,N-dimethylamphetamine	+	+
p-hydroxyamphetamine	+	+
tyramine	+	+
N-methyltyramine	+	+
hordenine (anhaline)	+	+
dopamine	+	+
N-methyldopamine	+	+
N,N-dimethyldopamine		+
3-methoxytyramine	+	+
N-methyl-3-methoxytyramine		+
4-hydroxy-3-methoxyphenethylamine		+
N-methyl-4-hydroxy-3-methoxyphenethylamine		+
3,4-dimethoxyphenethylamine		+
N-methyl-3,4-dimethoxyphenethylamine		+
3,4,5-trihydroxyphenethylamine		+
N-methyl-3,4,5-trihydroxyphenethylamine		+
mescaline	+	+
N-methylmescaline	+	+
trichocereine	+	+
3,5-dimethoxytyramine	+	+
3,4-dimethoxy-5-hydroxy-β-phenethylamine	+	+
β-methoxy-3,4-dihydroxy-5-methoxy-β-phenethylamine	+	+
3,4-dimethoxy-5-hydroxy-α-methyl-β-phenethylamine	+	+
tryptamine	+	+
N-methyltryptamine		+
N,N-dimethyltryptamine		+
nicotine	+	+
nornicotine	+	+
anhalamine	+	+
anhalidine (N-methylanhalamine)	+	+
anhalonidine		+
peyophorine	+	+
mimosine, methyl ester	+	
pipecolamide		+
p-hydroxypipecolamide		+
1,4-benzezediamine		+
4-methyl-2-pyridinamine		+
nortriptyline	+	
musk ambrette	+	

Tryptamine, *N*-methyltryptamine and *N,N*-dimethyltryptamine were found in blackbrush. Tryptamine and *N,N*-dimethyltryptamine were detected in guajillo at relatively low levels. Early season levels of tryptamine were 90-124ppb, while late-season levels were 287-334ppb. *N,N*-dimethyltryptamine was only detected in late season guajillo (75-115ppb). An ataxia syndrome called "Phalaris staggers" occurs in Australia and is associated with consumption of various *Phalaris* species that contain *N,N*-dimethyltryptamine (Lee and Kuchel, 1953). The amides of the amino acids pipecolic acid and *p*-hydroxypipecolic acid were detected in blackbrush, while the methyl ester of mimosine was detected in guajillo. It is not known if the methyl ester of mimosine found is an artifact of the extraction procedure. Smith (1977) does not report the presence of amphetamines in any plant family, and they are unlikely to be natural products. The tricyclic antidepressant nortriptyline reported here has also never been identified from a plant source.

As was previously found with NMPEA, the foliage collected in the fall contained higher quantities of amines and alkaloids (Forbes *et al.*, 1991; Forbes *et al.*, 1995). There was also a distinct increase in the number and quantity of methylated analogs present (Table 69.2). Several other nitrogen-containing compounds have been detected but have not yet been identified. Levels of the more abundant amines are listed in Table 69.2.

Phenolic amines, as a group, impact the hypothalamic-pituitary-adrenal axis (Vera-Avila *et al.*, 1996), causing a release of ACTH and cortisol that results in sympathomimetic action. Animals consuming guajillo and blackbrush are probably consuming phenolic amines. None of the compounds identified has been associated with ataxia, but their presence may have caused a reduction in monamine oxidase activity (Mantle *et al.*, 1976).

Table 69.2. Amines from Texas acacias.

Chemical found	Guajillo		Blackbrush	
	Early season	Late season	Early season	Late season
phenethylamine	991.3	1,390.9	872.3	1,135.7
N-methylphenethylamine	1,702.7	3,742.2	2,314.6	5,264.8
N,N-dimethylphenethylamine	99.1	604.4	123.6	724.5
tyramine	367.2	1,263.4	459.1	1,699.2
N-methyltyramine	188.5	745.7	237.4	1,237.6
hordenine	9.2	333.1	6.4	533.8

Acknowledgments

We thank the Texas Advanced Technology Program Grant 010366-153 for partial support of this project.

References

Adams, H.R. and Camp, B.J. (1966) The isolation and identification of three alkaloids from *Acacia berlandieri. Toxicon* 4, 85-90.

Camp, B.J. (1970) Action of N-methyltyramine and N-methyl-*beta*-phenethylamine on certain biological systems. *American Journal of Veterinary Research* 31, 755-762.

Camp, B.J. and Lyman, C.M. (1956). The isolation of N-methyl-*beta*-phenethylamine from *Acacia berlandieri. Journal of the American Pharmaceutical Association* 45, 719-721.

Camp, B.J. and Moore, J.A. (1960) A quantitative method for the alkaloid of *Acacia berlandieri. Journal of the American Pharmaceutical Association* 49, 158-160.

Forbes, T.D.A., Pemberton, I.J., Smith, G.R., Hensarling, C.M. and Tolleson, D.R. (1991) Seasonal production of phenolic amines by *Acacia berlandieri. Texas Agricultural Experiment Station Consolidated Progress Report #4875.*

Forbes, T.D.A., Carpenter, B.B., Tolleson, D.R. and Randel, R.D. (1994) Effects of phenolic monoamines on gonadotropin-releasing hormone stimulated luteinizing hormone release and plasma adrenocorticotropic hormone, norepinephrine, and cortisol concentrations in wethers. *Journal of Animal Science* 72, 464-469.

Forbes, T.D.A., Pemberton, I.J., Smith, G.R. and Hensarling, C.M. (1995) Seasonal variation of two phenolic amines in *Acacia berlandieri. Journal of Arid Environment* 30, 403-415.

Lee, H.J. and Kuchel, R.E. (1953) The aetiology of Phalaris staggers in Sheep. I, preliminary observations on the preventative role of cobalt. *Australian Journal of Agricultural Research* 4, 88-99.

Mantle, T.J., Tiptons, K.F. and Garret, N.J. (1976) Inhibition of monoamine oxidase by amphetamine and related compounds. *Biochemical Pharmacology* 25, 2073-2077.

Price, D.A. and Hardy, W.T. (1953) Guajillo poisoning of sheep. *Journal of American Veterinary Medical Association* 122, 223-225.

Smith, T.A. (1977) Phenethylamine and related compounds in plants. *Phytochemistry* 16, 9-18.

Vera-Avila, H.R., Randel, R.D. and Forbes, T.D.A. (1996) Plant phenolic amines: potential effects on sympatho-adrenal medullary, hypothalamic-pituitary-adrenal and hypothalamic-pituitary-gonadal function in ruminants. *Domestic Animal Endocrinology* 13, 285-296.

Chapter 70

A Urushiol Component Isolated from the Australian Native Cashew (*Semecarpus australiensis*)

P.B. Oelrichs[1], J.K. MacLeod[2], A.A. Seawright[1] and J.C. Ng[1]
[1]*The National Research Centre for Environmental Toxicology, 39 Kessels Road, Coopers Plains, Queensland 4108, Australia;* [2]*Research School of Chemistry, Australian National University, Canberra, Australia*

Introduction

The native cashew *Semecarpus australiensis* (Anacardiaceae) is a large spreading tree natural to Queensland and the Northern Territory. It is found in the coastal areas of these states and in the Torres Strait Islands. The plant is dioecious and the small flowers are found clustered in bunches at the end of the branches. The leaves are alternate, 8-23cm long, dull green above, and almost white below (Everist, 1981). In summer, the female flowers are followed by large fruit, very like the Brazil cashew fruit (*Anacardium occidentale*) except smaller. What appears to be the fruit is a succulent stem, usually bright orange in color with the true fruit attached to it.

The fruit is a well known source of bush food for the aboriginal people that live in this area of northern Australia. It is also generally known that contact with any part of the tree can cause an allergic reaction in susceptible people. The almost colorless sap, which turns dark when exposed to the air, if touched, can cause inflammation and blisters very like a burn (Isaacs, 1987).

Some historical records relate the experiences of early explorers with edible native plants. Ludwig Leichardt engaged in this activity and sampled *S. australiensis*. He wrote that the fruit was "extremely refreshing," but the surrounding envelope contained an "acrid juice" that caused blisters, which were "followed by a deep and painful ulceration" (Low, 1985).

The aboriginal people baked the fruit before eating it, and the procedure was done in a fire well away from the camp, because the smoke induced dermatitis as well. The fruits were placed on coals and covered by more coals. After some time the fruits

356

were retrieved and the outside skins removed before eating. Another method used leaching in water for several days followed by roasting. In either case the end product was an edible cashew nut similar in flavor to the Brazil cashew nut *A. occidentale.*

The goal of this study was to determine the structures of the active principles in the fruit so as to determine the most effective way of rendering them suitable for human consumption. Using solvent extraction and silica-gel chromatography, one major urushiol component was isolated (Oelrichs *et al.*, 1997). Its structure was determined (Fig. 70.1) by comparison of its NMR and MS data with that previously reported (El Sohley *et al.*, 1982).

Isolation Procedure

Fruits of *S. australiensis* (19g) were extracted with methanol (MeOH) (200ml) by soaking for several days. The supernatant was removed and the same procedure repeated twice. The extract was concentrated to dryness, water added and the suspension extracted 3x with dichloromethane (DCM). After concentration to dryness, the residue was redissolved in DCM, added to a silica-gel (0.07-0.15mm) column (4x30cm) and eluted with DCM followed by 5% MeOH/DCM. The fractions collected from the column were monitored by TLC using 5% ethylacetate (EtOAc)/DCM. The fractions showing a positive ferric chloride test were combined, dried and the residue dissolved in n-hexane. The n-hexane solution was extracted 3x with acetonitrile and the extract dried with anhydrous sodium sulfate.

Results and Discussion

The method of isolation of the active principles was adapted from that employed by El Sohley *et al.* (1982), but different solvents gave a better chromatographic separation of individual components. The dried acetonitrile extract (311mg) was essentially one compound. Comparison of the ^{1}H and the ^{13}C-NMR spectra of this active fraction from silica-gel with data previously published (El Sohley *et al.*, 1982) identified it as a urushiol. This was a name given to 3-n-alkenyl catechols previously isolated from poison ivy (*Toxicodendron radicans*). Its direct probe EI mass spectrum

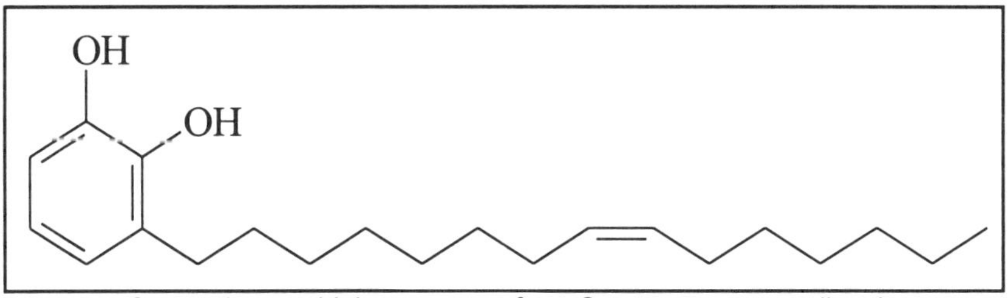

Fig. 70.1. One major urushiol component from *Semecarpus australiensis.*

revealed data identical to that reported for the compound in Fig. 70.1. Some related minor components were present in very small amounts.

The yield of urushiols (1.7%) from *S. australiensis* is considerably higher than that obtained from poison ivy (0.17%) (El Sohley *et al.*, 1982). The extraction of liquids from the Brazil cashew (*A. occidentale*) has been studied by Tyman *et al.* (1989). These authors used a number of extraction techniques, including static and continuous (Soxhlet), to remove the liquids. The percent recovery of cashew nut liquid was the same for static and Soxhlet extractions using both ether and petroleum ether. As the compounds involved in this exercise were very similar to the compound in Fig. 70.1 from *S. australiensis*, it is reasonable to assume static extractions with petroleum ether would be an effective way to remove the toxin. Further work will be carried out to determine if this method is successful in removing the toxin without removing or altering the flavor.

References

El Sohley, M.A., Adawadkar, P.D., Cheng-Yu, M.A. and Turner, C.E. (1982) Separation and characterization of poison ivy and poison oak urushiol components. *Journal of Natural Products* 45, 532-538.

Everist, S.L. (1981) *Poisonous Plants of Australia*. Angus and Robertson Publishers, Sydney, pp. 75-76.

Isaacs, J. (1987) *Aboriginal Food and Herbal Medicine*. Ure Smith Press, Weldon International, Sydney, p. 85.

Low, T. (1985) *Toxic plants and animals, a guide for Australia*. In: Covaevich, J., Davie, P. and Pearn. J. (eds), Queensland Museum Publishers, 65pp.

Oelrichs, P.B., MacLeod, J.K., Seawright, A.A. and Ng, J.C. (1997) The isolation and characterisation of urushiol components from the Australian native cashew (*Semicarpus australiensis*). *Natural Toxins* 5, 96-98.

Tyman, J.H.P., Johnson, R.A., Muir, M. and Rokhgar, R. (1989) The extraction of natural cashew nut-shell liquid from the cashew nut (*Anacardium occidentale*). *Journal of the American Oil Chemists Society* 66, 553.

Chapter 71

Plant Toxicology and Public Health: Critical Data Needs and Perspectives on Botanical (Herbal) Medicines

F.E. Leaders, Jr
Botanical Enterprises, Inc., Rockville, Maryland, USA

History of Botanical Medicines in the US

The stage may be set for considering the critical data needs concerning botanical (herbal) medicines by comparing the environment in the US for two 20yr intervals, 1930-1950 and 1980-2000 (Leaders, 1995a). In 1930, botanicals were still in widespread medical use. However, interest in single "active" ingredient products was growing. Analytical methods at the time were inadequate for botanicals and biological assays were sensitive, but too variable. The result was a limited ability to standardize botanical medicines. Single chemical entities standardized by relatively simple means were less variable and more cost effective than the extraction processes in use at that time. By 1950 many botanicals had fallen into disuse, a trend unique to the US. The shift away from botanical medicine was less pronounced in the rest of the world, including Europe.

The environment for medicinal botanicals in the US began to change in 1980. Recognition of a critical need for low-cost health care options fueled an increased public interest in natural products (botanicals) perceived to be more cost effective. Significant advances in chemical methodology dramatically enhanced the ability to characterize and standardize multi-component products. The botanical industry developed more effective extraction/control procedures. These changes provided an opportunity to reconsider the botanical products that had previously been bypassed by the US medical community. In mid-1994, the Office of Alternative Medicine (OAM) at the National Institutes of Health (NIH) took the first steps to establish a forum for scientific discussion on the uses of botanicals in medicine. The conference "Botanicals: A Role in US Health Care?" jointly sponsored by the OAM and the Food and Drug Administration (FDA) identified many issues.

Considerations for Botanical Medicines

Since botanicals are natural products, biological variation must be considered. The presence or absence of active principals and other constituents may vary significantly within the same plant, in plants grown in different geographic locations, at different seasons of the year, different climatic and soil conditions and with different cultivation practices (Leaders, 1996). This may result in a variation between batches during the extraction or manufacturing process even before the effects of the process itself are considered.

Single chemical entities are defined by chemical structure and identified by chemical analysis. For botanicals, however, the process defines the product. Compensating for inherent variability between batches is more complex, and in some cases may be achieved by extraction, processing and standardization. The primary extraction process determines which components are present and their ratios, and additional processing may change the concentrations and ratios of ingredients. Selection of standardization "markers" influences the composition and activity, and assumes a direct relationship to the desired therapeutic activity. Changes in content and ratios of multi-component botanicals may cause different pharmacological effects through inherent synergism and inhibition.

The chemical characterization of botanicals has improved dramatically over the past 50yrs. Modern methods may be used to obtain chemical "fingerprints" of the heterogeneous components that characterize a botanical produced by a specific process. Quantification/standardization of the botanicals between batches is done by quantifying one or more known active ingredients, or empirically selected markers if the active ingredients are unknown. The same methodologies used to identify and characterize plant toxins can be used to characterize and standardize medicinal botanicals.

Techniques for botanical characterization of medicinal plants have not changed over the last 50yrs. Organoleptic evaluation relies on sight, smell and touch, with macroscopic and microscopic examination for anatomical characteristics of the desired plant. This is straightforward when the whole plant is available, but becomes difficult when only plant parts are available and can be extremely difficult when only processed material is available. Therefore, the botanical identity assurance process must begin at the time the plants are harvested with an accurate record, to ensure that the proper plant or plant part is used. Recent rapid growth in this industry has resulted in an increasing number of schools reinstituting courses in pharmacognosy and plant identification. Two key support mechanisms for all botanical identification programs are reference collections and herbaria.

The botanical industry has found ways to compensate for inherent variability of botanicals. Dosage form selection and development, however, is still important. Many "traditional" medicines are administered in teas or soups that are not acceptable to the US population. Concentrating botanicals sufficiently to merge an effective dose into a manageable tablet or capsule can occasionally present a challenge. When that is overcome, the formulator must consider enhancing stability of the heterogeneous mixture to guarantee an adequate shelf life.

The most critical needs for scientific data to support the introduction of botanicals into mainstream medicine in the US are in the selection of appropriate therapeutic use(s), definition of dose, route and regimen of administration and determination of drug/botanical/food interactions. Sources of this information are historical "in-use" data from *in vitro*, animal and/or human data and carefully designed controlled, double-blind clinical trials.

Three primary sources of data used as predictors of safety of single chemical entities - *in vitro* data, animal data and human "in use" data - are applicable to botanicals, but the inherent differences between them require a shift in emphasis. Due to the potential presence of multiple active ingredients and many 'inactive' components, *in vivo* procedures are often more appropriate than *in vitro* methods. Even with their inherent variability, bioassays are the most direct means of evaluating biological activity. Measurement of "active ingredient" blood levels may be difficult or impossible, since some active botanical constituents are unknown. Standard animal toxicology testing may be used for safety assessment but, for botanicals with a long history of human use, they may be unnecessary. Since many botanicals are currently in human use, epidemiological techniques for evaluation of human "in use" data may be the most direct means of assessing long-term safety. Recommended dosing regimens for most botanicals are based on historical use in "traditional" settings. Methodology for scientific evaluation of dose response relationships and pharmacodynamic and pharmacokinetic properties to confirm and/or establish appropriate dosage regimens is being developed.

The Current Regulatory Environment for Botanicals

The regulatory category for a product in the US is determined by the intended use (product labeling). The two major product categories are foods and drugs. No legislative provision exits in the US for traditional medicine, as it does in many other countries. This absence reflects the difficulty of generating support for introducing medicines derived from traditional settings into a multicultural society.

The category "Dietary Supplements" came into being with passage of the Dietary Supplement Health and Education Act of 1994 (DSHEA), which influenced FDA policy for the testing of botanicals as drugs. In late 1994, the Office of Alternative Medicine assisted three applicants to file Investigational New Drug Applications (INDs) to clinically evaluate heterogeneous herbal formulations for the first time in many years. The FDA allowed clinical "pilot studies" to begin, waiving two key policies that had precluded clinical evaluation of botanicals in the US: non-clinical testing (animal safety data) and demonstration of the contribution of each ingredient to efficacy. If the ingredients could be purchased as dietary supplements, the public was not protected by the requirement for such information. The FDA made it clear that this change was only for INDs and not for New Drug Applications (NDAs).

Following the passage of DSHEA, a series of Botanical Workshops was developed under the sponsorship of the Drug Information Association (DIA), an organization of scientists interested in drug research, in cooperation with the FDA and other organizations such as the US Pharmacopoeia (USP), and NIHs OAM and Office of Dietary Supplements (ODS).

The first DIA-sponsored workshop, "Alternative Medicine Workshop on Botanicals" was in 1995. The theme "Botanicals: A Role in US Health Care?" was continued in the second workshop. A survey of the medicinal use of botanicals around the world confirmed the sense that the US lags behind the rest of the world in using and understanding medicinal botanicals. It confirmed that adequate technology and methods existed to support the use of "fingerprints" and "markers" to evaluate, manufacture and control botanical products (Leaders, 1995b).

The third DIA-Sponsored workshop on botanicals, "Botanical Testing: Developing the Scientific Evidence to Support the Medical Use of Heterogeneous Botanical Products," was held in 1997. Issues considered were the scientific testing needed to adequately support medical use of appropriately identified and characterized botanical products and whether epidemiological human "in-use" data was adequate for evaluation of safety. Epidemiological methodology appears adequate, but introducing such considerations into the toxicity/safety review process will require cooperation between epidemiologists, toxicologists and scientists.

The fourth DIA-sponsored workshop, "Botanical Testing: Developing Clinical Evidence of Safety, Activity and Efficacy for Heterogeneous Botanical Products" will be held in November, 1997. This session will consider issues for clinical evidence of safety, activity and efficacy for heterogeneous botanical products.

The neutral forum supplied by these workshops has fostered discussions of many of the issues raised by policy questions. A proposed regulatory working policy revolves around approval of NDAs for botanicals with a long use history based on: content and uniformity from "markers" and "fingerprints," safety from an uneventful "in use" history, efficacy from "in use" history plus clinical trials and Phase IV testing to answer possible specific concerns. There are some widely used botanicals to which these modest proposals might reasonably be applied.

References

Leaders. F. (1995a) *Botanicals and the FDA-A Report in Context.* First Annual International Congress on Alternative and Complementary Medicine, Arlington, Virginia.

Leaders, F. (1995b) *Getting Natural Products to Market by Working the U.S. Food and Drug Administration-A Report in Context.* Drug Discovery and Commercial Opportunities in Medicinal Plants and Natural Products, Washington, DC.

Leaders, F. (1996) *Working with Botanicals* vs. *Single Chemical Entities.* Botanical Quality: Workshop on Identification and Characterization, Second DIA-Sponsored Workshop on Botanicals, Washington, DC.

Chapter 72

Public Health and Risk Assessment

D.J. Wagstaff, P.M. Bolger and C.D. Carrington

US FDA, Center for Food Safety and Applied Nutrition, Washington, DC 20204, USA

Introduction

Public health, the health status of human populations, includes both preventative and remedial measures. Safety of plants and plant products used as foods or drugs is regulated in the US by the Food and Drug Administration (FDA), an agency of the Public Health Service, under the Food Drug and Cosmetic Act (FD&C Act). This law defines safety broadly and directs FDA to establish regulations to provide details. The US Department of Agriculture has authority for continuous inspection of meat by inspectors empowered to take restrictive action, while FDA inspection of food facilities is generally sporadic, by inspectors who must request restrictive action by a federal court on a case-by-case basis.

Specific plant products are regulated according to their category by intended use (product label) and promotional material instead of biologic or chemical properties. Product categories are drugs, foods, food and color additives and food supplements, and the same plant may be in different classes. For example, wormwood (*Artemisia absinthium* L.) could be a drug, a food additive or a dietary supplement depending on labeling and advertising statements.

Products within each category are divided into those that have been marketed and those that are new. Before a new drug can be marketed, a New Drug Application containing sufficient efficacy and safety data must be approved by FDA. The burden of proof of safety is upon the vendor.

The FD&C Act does not require safety testing for foods, or substances used in foods prior to 1958, that have been shown to be safe through "experience based on common use in food." Likewise, vendors of new food supplements such as a botanical are only required to notify the agency of intent to market and to state the basis for concluding that the product is safe.

In contrast to drugs and food additives, when safety concerns arise for particular foods or food supplements, the burden is on FDA to demonstrate hazard before regulatory action can be taken. The argument for this dichotomy is that new drugs and food additives have not been used by the public compared to foods and food supplements, which have a history of human use.

Risk Assessment Paradigm

Despite human experience with food and food supplement products, scientific data to resolve toxicologic concerns are sparse. The process of utilizing this data to infer the severity of public health problems is termed risk assessment. Risk assessments, coupled with policy decisions, are used to support enforcement actions (e.g., prosecutions, seizures, detentions), to provide public information, to identify research needs and to support harmonization of international regulatory activities.

Since it is recognized that absolute safety or zero risk is unattainable, except by total avoidance of exposure, standards of risk are given in the FD&C Act. Two standards apply to contaminants and potentially toxic components of foods, and a third standard applies to food supplements. The first standard, given in section 402(a)(1), includes the phrase "may render it injurious to health," referring to a substance that is added to food. The other food standard, in the same section, is "ordinarily render," which refers to natural substances that are not added. The standard for food supplements, given in section 402(f)(1)(A), is "significant or unreasonable risk of illness or injury." This latter portion of the FD&C Act was added by the Dietary Supplement Health and Education Act of 1994.

The law does not quantitatively define the different levels of acceptable risk. Practical considerations are the extent to which a risk is avoidable, limit of analyte detection, quantity of food loss and processing changes bearing on product safety.

There are several distinct steps in the risk assessment process: safety concerns, hazard identifications, safety, exposure, and dose response assessments, risk characterization and decision or action. If the evaluation at any step is sufficient to support an adequate regulatory decision, then subsequent steps are not taken. Rarely are all of the steps taken in a structured approach. Inferences learned in a subsequent step may be recycled to a prior step to improve the overall evaluation.

A safety/risk evaluation begins with exposure to the substance of concern being related to specific health effects. The limitations of animal toxicity testing are illustrated by the thalidomide disaster in which children of mothers who took the drug during pregnancy were born without limbs. Problems using past human or animal experience to identify hazards are not as well known. If people were exposed and if no adverse events occurred the conclusion might be drawn that the product is not a hazard. However, adverse effects could have occurred, and not been observed or not reported. A third, more likely, scenario is that a relationship of the exposure and a harmful outcome was not recognized.

Bracken is the world's most successful vascular plant. It is widely eaten and undoubtedly has always caused intoxication. The relationship of harm from this plant was unknown until 1893, when British veterinarians reported disease in cattle grazing this fern. Since then animal poisonings have been reported worldwide. Though bracken is eaten by some people, the carcinogenicity issue is unresolved.

When a hazard is identified, the next step is to determine a safe or tolerable level of exposure. Acceptable Daily Intake (ADI) was the first term coined by the FDA, followed by Tolerable Daily Intake (TDI) of the World Health Organization, the Reference Dose (RfD) of the Environmental Protection Agency and the Minimal Risk Level (MRL) of the Agency of Toxic Substances and Disease Registry. These are all different names for the same safety assessment approach.

Data from experimental animals are usually used to do safety assessments. The safe level is defined as the highest exposure level in animals that produced no observed adverse effects (NOAEL), which is then adjusted downward in response to various unknowns using safety factors (also known as uncertainty factors). A factor of ten is used to account for interspecies differences in susceptibility. Another factor of ten is used for intraspecies differences (susceptibility differences among people). Thus a total safety factor of 100 is often applied to the animal NOAEL to estimate the human TDI. Additional safety factors to account for other sources of uncertainty, such as differences in exposure length, have been used.

An example in the steps of risk assessment is a safety assessment for patulin, a substance in rotting apples. The NOAEL for patulin in a gastric intubation study of rats was 0.3mg/kg/wk, or 43µg/kg/d for the entire 109wk study (Becci *et al.*, 1981). The human TDI calculated using an uncertainty factor of 100 is 0.43µg/kg.

Exposure is estimated by multiplying the amount of each food eaten, or quantity of other product exposure, by the concentration of the toxic substance in that food or product. Food intake is sampled in national or local surveys. Quantity of each food eaten and statistical distribution by geographic region, age, race and sex are determined. Food supplement consumption data are sparse; future food intake surveys will have some broad questions partially to address this need.

Exposure to patulin is summarized in Table 72.1. The major human source of patulin is apple juice. Concentration of patulin in sampled apple juice averaged 32ppb. The greatest intake of apple juice is by infants and young children. When patulin intake is calculated on a body weight basis, infants and children less than 2yrs of age ingest on average 0.46µg/kg/d compared to 0.07 for people of all ages. Thus the TDI of 0.43 is slightly exceeded by children ingesting the average quantity of apple juice. Children in the upper tenth percentile of apple juice consumers exceed the TDI by about twice this extent. The TDI would be exceeded by apple juice containing more than 10ppb patulin. As seen in Table 72.1, the general population could consume apple juice products containing 203ppb patulin and not exceed the TDI. Officials with risk managment responsibilities will make the decision whether to regulate patulin in apple juice.

A safety assessment does not yield information about the expected effects of exceeding the ADI/TDI. That information comes from dose response studies, in which experimental animals are exposed at different concentrations. Table 72.2 summarizes chronic dose response data for rats given patulin by gastric intubation.

Table 72.1. Human patulin exposure from apple juice and maximum patulin (p) concentration to achieve a Tolerable Daily Intake (TDI) of 0.43µg/kg/d (Data are mean and 90th percentile).

Age group	Apple juice intake (g/p/d)*	Patulin exposure (µg/kg/d)	Max. patulin (ppb) to achieve TDI
>1	118 (227)	0.46 (1.06)	29 (15)
1-2	176 (496)	0.46 (1.01)	29 (10)
All ages	136 (258)	0.07 (0.14)	203 (107)

* Mean body weights used are 8kg for <1yr-olds, 12kg for 1-2yr-olds, and 64kg for the all age groups (Johnson, 1974).

Adverse effects increased at doses that are only five and 15 times greater than the NOAEL. The adverse effects, depressed growth and death, are severe. Thus consumption of apple juice containing patulin at concentrations near the NOAEL would be viewed more seriously than if the dose response curve were shallow and the adverse effects were minor, such as a small change in body weight.

Risk characterization can go beyond safety assessment and attempt to quantitate the risk. A quantitative evaluation requires responses for individual animals at each dose, which are then modeled. This quantitative approach estimates both the risk and the uncertainty related to it. Several alternative risk management options are generated to support decisions when the TDI is exceeded.

Table 72.2. Summary of dose response of rats given patulin by gastric intubation 3x/wk for 109wks (Becci *et al.*, 1981).

Effect	Dose (mg/kg)
Increased mortality	1.5
Decreased body weight gain	0.5
No observed adverse effects	0.1

References

Becci, P.J., Hess, F.G., Johnson, W.D., Gallo, M.A., Babish, J.G., Dailey, R.E., and Parent, R.A. (1981) Long-term carcinogenicity and toxicity studies of patulin in the rat. *Journal of Applied Toxicology* 1, 256-261.

Johnson, P.E. (1974) Nutrition standards: Manual Part I. United States: Children and adults. In: Altman, P.L. and Dittmer, D.S. (eds), *Biology Data Book, Vol 3*, 2nd ed. Federation of American Societies of Experimental Biology, Bethesda, MD, pp. 1447.

Chapter 73

Perspectives on Plant Toxicology and Public Health

J.M. Betz and S.W. Page
US FDA, Center for Food Safety and Applied Nutrition, Division of Natural Products, Washington, DC 20204, USA

Introduction

Disillusionment with Western medicine, coupled with the perception that natural substances are inherently safer and more healthful than synthetic ingredients, has led to a resurgence in the popularity of herbal remedies (Croom and Walker, 1995). Unfortunately, historical use does not guarantee safety and many of the herbs used medicinally are mentioned in monographs on poisonous plants.

In the US, the Dietary Supplement Health and Education Act (DSHEA) of 1994 (US Public Law 103-417) amended the Federal Food, Drug and Cosmetic Act by defining as a dietary supplement any product (other than tobacco) that contains a vitamin, mineral, herb or other botanical, or amino acid and is intended as a supplement to the diet. Dietary supplements marketed prior to 15 October, 1994, may remain on the market unless proven unsafe by the FDA. A dietary supplement is considered unsafe only if it "presents a significant or unreasonable risk of illness or injury under conditions of use recommended or suggested in labeling, or if no conditions of use are suggested or recommended in the labeling, under ordinary conditions of use."

Human Health Problems Caused by Botanicals

Despite knowledge that some plants may be harmful if ingested, reports of human intoxication continue to appear. Examples include poisonings by poison hemlock (*Conium maculatum*) (Frank *et al.*, 1995), tea tree (*Melaleuca alternifolia*) oil (Del Beccaro, 1995) and others. Recent retrospectives on human intoxications caused by particular plants include reviews on pennyroyal (*Mentha puleguim* or *Hedeoma pulegoides*) (Anderson *et al.*, 1996), oleander (*Nerium oleander, Thevetia peruviana*) (Langford and Boor, 1996), and eucalyptus oil (Tibballs, 1995).

Data on human poisonings by plants are collected in a fundamentally different manner from that of animals. Whereas both begin with reports of intoxications, toxicity of plants suspected of injuring animals may be confirmed by feeding studies, while toxicity to humans is extrapolated from animal data or by examination of the historical record for human poisonings. Reports of veterinary intoxications are valuable, but because of the inability to establish causality in humans experimentally, the literature remains inconsistent. Reviews of human cases such as those recently performed for mistletoe (*Phoradendron flavescens*) (Spiller *et al.*, 1996) and poinsettia (*Euphorbia pulcherrima*) (Krenzelok *et al.*, 1996) provide respite from the confusion.

Circumstances of Intoxication

In addition to medical, toxicological or chemical systems of classification, most adverse reactions to botanicals may be divided into four general categories based on the circumstances under which the intoxication occurred: misidentification, misuse, deliberate adulteration and inherent toxicity.

Many cases of human poisoning by plants have been caused by misidentification of plant species, especially from consumption of misidentified self-collected material. Examples include a Washington state couple who died after consuming a tea made from *Digitalis*, which they had collected in the belief that it was comfrey (Stillman *et al.*, 1977) and a man who died after he and his brother ingested "ginseng," which turned out to be water hemlock (*Cicuta maculata*) (Sweeney *et al.*, 1994).

Misuse (abuse, and incorrect or inappropriated use) of botanical products has also led to adverse reactions. This is especially important when discussing plants that have been used as traditional medicines and are now being used for conditions and in dosages and combinations that have no historical precedent.

Decoctions of the traditional Chinese medicine (TCM) *Má Huáng* (*Ephedra* spp.) have been used for thousands of years as a diaphoretic, stimulant and antiasthmatic (Bensky and Gamble, 1986). The plant contains (-)-ephedrine and several related alkaloids (Betz *et al.*, 1997). There have been a number of reports of adverse reactions (including deaths) to weight loss (Catlin *et al.*, 1993) and other products marketed as "safe, legal" herbal stimulants (Nightingale, 1996) that contain *Má Huáng* extracts and caffeine.

Adulteration of commercial products is another cause of intoxications. Three families suffered anticholinergic poisoning after consuming Paraguay tea (*Ilex paraguariensis*). Visual examination of the tea indicated that it contained non-*Ilex* plant material; and atropine, scopolamine, and hyoscyamine were subsequently identified by GC/MS (Hsu *et al.*, 1995).

In Belgium, more than 70 women were afflicted with progressive interstitial renal fibrosis after a slimming regimen which included the Chinese herbs *Stephania tetranda* and *Magnolia officinalis.* The nature of the nephropathy and the similarity of the Chinese names for *Stephania* (*Fangji*) and *Aristolochia* (*Fangchi*) led investigators to suspect substitution of *Aristolochia fangchi* for *Stephania.* A subsequent analysis identified the known nephrotoxin aristolochic acid (a constituent of *Aristolochia,* but not of *Stephania*) in 12 of 13 batches of the herbal material (Vanhaelen *et al.,* 1994).

Cases of deliberate substitution of one material for another date from the dawn of commerce. Neonatal androgenization associated with maternal ingestion of Siberian ginseng (*Eleutherococcus senticosus*) was subsequently discovered to have been caused by Chinese silk vine (*Periploca sepium*) (Awang, 1991) which had been substituted for the *Eleutherococcus* in the product.

Deliberate adulteration of botanicals is not limited to substitution of one plant for another. Various TCMs have been found to contain undeclared synthetic medicinal ingredients. Four patients developed agranulocytosis after ingesting an herbal arthritis product called *Chui Fong* which contained aminopyrine and phenylbutazone (Ries and Sahud, 1975). A TCM called "Black Pearl" was found to contain hydrochlorothiazide, diazepam, indomethacin, and mefenamic acid (By *et al.,* 1989). Corticosteroids (including hydrocortisone) have also been reported in herbal TCMs (Goldman and Myerson, 1991).

Substitution is not limited to herbal medicines or dietary supplements. A pink seeded cultivar of common vetch (*Vicia sativa*) called *blanche fleur* has been sold in the US as red lentils or *masoor dahl* (*Lens culinaris*). The human health aspects of this substitution are unclear, but *V. sativa* is known to produce the neurotoxic amino acids ß-cyanoalanine and γ-glutamyl-ß-cyanoalanine (Ressler, 1962), the favism factors vicine and convicine (Chevion and Navok, 1983) and the cyanogenic glycosides isolinamarin and vicianin (Poulton, 1983).

A number of plants should be avoided because the seriousness of potential adverse reactions outweighs any potential benefits. All members of the Boraginaceae and several members of the Fabaceae and Asteraceae have been found to contain pyrrolizidine alkaloids (PAs). Mass intoxications have been caused by bread cereals contaminated by the seeds of various species of *Heliotropium* and *Crotalaria* (Tandon *et al.,* 1978). A number of PA-containing plants (e.g. *Tussilago farfara, Cynoglossum officinale, Senecio* spp.) are still used medicinally in Europe (Roeder, 1995). Comfrey (*Symphytum* spp.) and its aqueous infusions contain a number of toxic PAs (Betz *et al.,* 1994). Hepatic veno-occlusive disease in humans has been associated with ingestion of *S. officinale* (Bach *et al.,* 1989), which is also a rodent carcinogen (Hirono *et al.,* 1978).

Chaparral or creosote bush (*Larrea tridentata*) is a dominant shrub in certain areas of the US and Mexico. Infusions of this plant were used by native Americans for a variety of diseases (Hutchens, 1992). Phenolic compounds account for 83-92% of the extractable dry weight of the plant, with nordihydroguaretic acid (NDGA) as the most abundant phenolic constituent. This is a potent antioxidant and was a generally recognized as safe (GRAS) food additive for this purpose until animal studies revealed evidence of toxicity (Grice *et al.,* 1968). When animal data and recent reports of hepatotoxicity were considered together, it was apparent that patients who ingested chaparral tablets or capsules suffered hepatic injuries (Sheikh *et al*, 1997). Consumption of whole plant instead of infusions greatly increases the exposure to its potenitally toxic phenolic constituents (Obermeyer *et al.*, 1995).

Conclusions

Beyond education, public health officials can do little about cases of poisoning by self-collected material except to offer expertise and occasionally laboratory assistance after the fact. Adequate quality assurance programs are needed for commercial botanical products, as are attempts to anticipate problem areas and alert the scientific community and consumers of problems which can be predicted by knowledge of botanicals in the marketplace. The size and complexity of the market for botanicals are difficulties which public health officials face when dealing with poisoning episodes. In North America alone, at least 700 plant species have been described as being poisonous in one way or another, and plant poisonings are often difficult to differentiate from environmental intoxications caused by pesticides and industrial chemicals and from adverse reactions to synthetic drugs (Der Marderosian and Liberti, 1988). Complicating the picture is the fact that symptoms of many intoxications mimic those of medical conditions not associated with a toxic exposure, and that establishing causality may therefore be difficult (Perrotta *et al.*, 1996).

The issue is further complicated by the use of common or local names for botanical products. A glance through the Index to Colloquial Names in the Eighth Edition of *Gray's Manual of Botany* (Fernald, 1950) has at least eight snakeroots in six families. Besides nomenclatural difficulties, problems exist because of the difficulty of determining exactly which plants are present in a finished product. Capsules which contain more than one plant and/or a dried crude extract make identification of the plant(s) by microscopy virtually impossible. Determination of the identity of a plant then becomes a matter of looking for specific toxic chemical constituents.

Reports of adverse reactions to botanical dietary supplements dominate this overview, but this is because changes in the legal status of these products have led to an explosion of the dietary supplement marketplace. The "cultural knowledge" which had historically protected the traditional consumer is often lacking in the US melting pot. Improved quality assurance and consumer education could significantly decrease the adverse health effects associated with these products.

References

Anderson, I.B., Mullen, W.H., Meeker, J.E., Khojasteh-Bakht, S.C., Oishi, S., Nelson, S.D. and Blanc, P.D. (1996) Pennyroyal toxicity: measurement of toxic metabolite levels in two cases and review of the literature. *Annals of Internal Medicine* 124, 726-734.

Awang, D.V.C. (1991) Maternal use of ginseng and neonatal androgenization. *Journal of the American Medical Association* 265, 1839.

Bach, N., Thung, S.N. and Schaffner, F. (1989) Comfrey herb tea-induced hepatic veno-occlusive disease. *American Journal of Medicine* 87, 97-99.

Bensky, D. and Gamble, A. (1986) *Chinese Herbal Medicine - Materia Medica.* Eastland Press, Seattle, WA. pp. 32-34.

Betz, J.M., Eppley, R.M., Taylor, W.C. and Andrzejewski, D. (1994) Determination of pyrrolizidine alkaloids in commercial comfrey products (*Symphytum* sp.). *Journal of Pharmaceutical Sciences* 83, 649-653.

Betz, J.M., Gay, M.L., Mossoba, M.M., Adams, S. and Portz, B.S. (1997) Chiral gas chromatographic determination of ephedrine-type alkaloids in dietary supplements containing *Má Huáng. Journal of the Association of Official Analytical Chemists International* 80, 303-315.

By, A., Ethier, J.C., Lauriault, G., LeBelle, M., Lodge, B.A., Savard, C., Sy, W.W. and Wilson, W.L. (1989) Traditional oriental medicines I. Black Pearl: identification and chromatographic determination of some undeclared medicinal ingredients. *Journal of Chromatography* 469, 406-411.

Catlin, D.H., Sekera, M. and Adelman, D.C. (1993) Erythroderma associated with ingestion of an herbal product. *Western Journal of Medicine* 159, 491-493.

Chevion, M. and Navok, T. (1983) A novel method for quantitation of favism-inducing agents in legumes. *Analytical Biochemistry* 128, 152-158.

Croom, E.M. Jr, and Walker, L. (1995) Botanicals in the pharmacy: New life for old remedies. *Drug Topics* November 6, 84-93.

Del Beccaro, M.A. (1995) Melaleuca oil poisoning in a 17-month-old. *Veterinary and Human Toxicology* 37, 557-558.

Der Marderosian, A.H. and Liberti, L. (1988) *Natural Product Medicine.* George F. Stickley, Philadelphia, PA, pp. 147-184.

Fernald, M.L. (1950) *Gray's Manual of Botany.* American Book, New York, NY, p. 1632.

Frank, B.S., Michelson, W.B., Panter, K.E. and Gardner, D.R. (1995) Ingestion of poison hemlock (*Conium maculatum*). *Western Journal of Medicine* 163, 573-574.

Goldman, J.A. and Myerson, G. (1991) Chinese herbal medicine: camouflaged prescription antiinflammatory drugs, corticosteroids, and lead. *Arthritis and Rheumatism* 34, 1207.

Grice, H.C., Becking, G. and Goodman, T. (1968) Toxic properties of nordihydroguaretic acid. *Food and Cosmetic Toxicology* 6, 155-161.

Hirono, L., Mori, H. and Haga, M. (1978) Carcinogenic activity of *Symphytum officinale. Journal of the National Cancer Institute* 61, 865-869.

Hsu, C.K., Leo, P., Shastry, D., Meggs, W., Weisman, R. and Hoffman, R.S. (1995) Anticholinergic poisoning associated with herbal tea. *Archives of Internal Medicine* 155, 2245-2248.

Hutchens, A.R. (1992) *A Handbook of Native American Herbs.* Shambala Publications, Boston, MA, pp. 76-78.

Krenzelok, E.P., Jacobsen, T.D. and Aronis, J.M. (1996) Poinsettia exposures have good outcomes...just as we thought. *American Journal of Emergency Medicine* 14, 671-674.

Langford, S.D. and Boor, P.J. (1996) Oleander toxicity: an examination of human and animal toxic exposures. *Toxicology* 109, 1-13.

Nightingale, S.L. (1996) Warning issued about street drugs containing botanical sources of ephedrine. *Journal of the American Medical Association* 275, 1534.

Obermeyer, W.R., Musser, S.M., Betz, J.M., Casey, R.E., Pohland, A.E. and Page, S.W. (1995) Chemical studies of phytoestrogens and related compounds in dietary supplements: flax and chaparral. *Proceedings of the Society for Experimental Biology and Medicine* 208, 6-12.

Perrotta, P.M., Coody, G. and Culmo, C. (1996) Adverse events associated with ephedrine-containing products - Texas, December 1993-September 1995. *Morbidity and Mortality Weekly Reports* 45, 689-693.

Poulton, J.E. (1983) Cyanogenic compounds in plants and their toxic effects. In: Keeler, R.F. and Tu, A.T. (eds), *Handbook of Natural Toxins, Vol 1-Plant and Fungal Toxins.* Marcel Dekker, New York, NY, pp. 117-157.

Ressler, C. (1962) Isolation and identification from common vetch of the neurotoxin ß-cyano-L-alanine, a possible factor in neurolathyrism. *Journal of Biological Chemistry* 237, 733-735.

Ries, C.A. and Sahud, M.A. (1975) Agranulocytosis caused by Chinese herbal medicines. Dangers of medications containing aminopyrine and phenylbutazone. *Journal of the American Medical Association* 231, 352-355.

Roeder, E. (1995) Medicinal plants in Europe containing pyrrolizidine alkaloids. *Pharmazie* 50, 83-98.

Sheikh, N.M., Philen, R.M. and Love, L.A. (1997) Chaparral-associated hepatotoxicity. *Archives of Internal Medicine* 157, 913-919.

Spiller, H.A., Willias, D.B., Gorman, S.E. and Sanftleban, J. (1996) Retrospective study of mistletoe ingestion. *Journal of Toxicology Clinical Toxicology* 34, 405-408.

Stillman, A.E., Huxtable, R.J., Fox, D.W., Hart, M.C., Bergeson, P.S., Counts, J.M., Cooper, L., Grunenfelder, G., Blackmon, J., Fretwell, M., Raey, J., Allard, J. and Bartleson, B. (1977) Poisoning associated with herbal teas - Arizona, Washington. *Morbidity and Mortality Weekly Reports* 26, 257-259.

Sweeney, K., Gensheimer, K.F., Knowlton-Field, J. and Smith, R.A. (1994) Water hemlock poisoning - Maine 1992. *Morbidity and Mortality Weekly Reports* 43, 229-231.

Tandon, B.N., Tandon, H.D. and Mattocks, A.R. (1978) Study of an epidemic of veno-occlusive disease in Afghanistan. *Indian Journal of Medical Research* 68, 84-90.

Tibballs, J. (1995) Clinical effects and management of eucalyptus oil ingestion in infants and young children. *Medical Journal of Australia* 163, 177-180.

United States Public Law 103-417. 103rd Congress, 25 October, 1994. *Dietary Supplement Health and Education Act of 1994.*

Vanhaelen, M., Vanhaelen-Fastre, R., But, P. and Vanherweghem, J.L. (1994) Identification of aristolochic acid in Chinese herbs. *Lancet* 343, 174.

Chapter 74

Evaluation of the Occurrence of Algae of the Genus *Prototheca* in Cheese and Milk from Brazilian Dairy Herds

E.O. da Costa, P.A. Melville, A.R. Ribeiro and E. Watanabe

Research Nucleus, on Mammary Gland and Milk Production/Faculdade de Medicina Veterinária e Zootécnia, Universidade de São Paulo, Brazil; Av Prof Dr Orlando Marques de Paiva, 87, CEP 05508-900, São Paulo-SP, Brazil

Introduction

The genus *Prototheca* comprises a group of colorless algae that is widespread throughout the environment, but is mostly found in humid areas rich in organic matter (particularly damp areas contaminated with manure) (Pore *et al.*, 1983). Members of the genus are known to cause systemic and cutaneous disease in humans and animals (Venezio *et al.*, 1982). In humans, these microorganisms have been associated primarily with gastroenteritis (Sudman, 1974; Iacoviello *et al.*, 1992) cutaneous lesions (Klintworth *et al.*, 1968; Iacoviello *et al.*, 1992), and bursitis (Iacoviello *et al.*, 1992). Protothecosis is manifested as clinical and subclinical mastitis in cattle. Recently, the incidence of bovine mastitis caused by this agent has increased (Kirk, 1991; Aalbaek *et al.*, 1994; Costa *et al.*, 1996a; Costa *et al.*, 1996b). *Prototheca*- induced mastitis is characterized by a sharp decrease in milk production, and affected cows are culled, since they usually do not respond to routine therapy (Anderson and Walker, 1988; Kirk, 1991; Costa *et al.*, 1996b).

This study evaluated the association of *Prototheca* spp. in milk samples from bulk tanks to mastitis of dairy cows. It also evaluated the occurrence of *Prototheca* in samples of cheese made with contaminated milk.

Materials and Methods

Milk from bulk tanks on six farms from large dairy production areas in the State of

São Paulo (Brazil) were examined. The herd sizes ranged from 64-1,286. Samples from each bulk tank were collected in sterile glass tubes, diluted in sterile physiological saline, and 1ml aliquots cultured on Sabouraud-dextrose agar plus chloramphenicol using the pour-plate technique (incubation at 25°C for 7d). After incubation, analysis for the presence of *Prototheca* spp. was carried out and the colony-forming units (cfu) were counted. The colonies of *Prototheca* spp. were identified by size, capsule formation, and assimilation of trehalose, dextrose and sucrose over 14d (Camargo and Fischman, 1979; Pore, 1985).

All lactating cows on the farms were examined for clinical or subclinical mastitis using the strip cup and the California mastitis test (CMT). When one or both of these tests were positive, milk samples from the mammary quarters were collected for microbiological examination and cultured in 5% sheep blood agar for 24-96hrs (38°C) and Sabouraud-dextrose agar plus chloramphenicol for 7d (25°C). *Prototheca* spp. were identified using the criteria above for milk from bulk tanks.

Samples of cheese made with milk from one of the dairy farms were also evaluated by plate count following the procedure described above. The cheese was made by filtering and coagulating the milk, removing the whey, salting and curing.

Results

A total of 3,043 milk samples from individual mammary quarters that showed clinical and subclinical mastitis were aseptically collected for microbiological examination. The presence of *Prototheca* spp. ranged from 0.5% to 14.9% (Table 74.1). Table 74.2 presents the results of *Prototheca* spp. isolation from bulk tank milk samples. The *Prototheca* spp. counts ranged from $1-3x10^4$cfu/ml of milk. The cheese sample contained *Prototheca* spp. at a level of $7.5x10$cfu/ml.

Table 74.1. Results of analysis of milk collected from mammary glands in cows on six dairy farms where *Prototheca* spp. was isolated from bulk tanks.

Dairy farm	Total cows examined n	Total milk samples collected from different quarters n	Quarters infected with *Prototheca* spp. n	%
1	82	186	15	8.0
2	177	202	5	2.5
3	1,286	2,159	13	0.6
4	107	202	1	0.5
5	64	187	3	1.6
6	91	107	16	14.9

Table 74.2. Results of the total count of colony-forming units per ml (cfu/ml) of *Prototheca* spp. in milk samples from bulk tanks on six dairy farms.

Dairy farm	Sabouraud-dextrose agar plus chloramphenicol (cfu/ml)
1	1.2×10^2
2	4.1×10
3	7.3×10
4	1.0×10
5	3.0×10^4
6	2.0×10^3

Discussion

Prototheca spp. has been considered an emergent mastitis pathogen, causing sporadic cases and large outbreaks (Aalbaek *et al.*, 1994; da Costa *et al.*, 1996a; da Costa *et al.*, 1996b). It can also be an important contaminant of milk. Milk and dairy products are considered important sources of nutrients for humans. However, when not properly treated for the elimination of pathogens, they represent a potential risk for human health (Sharp, 1987).

Contamination of milk by *Prototheca* spp. represents a risk to public health since this agent, in addition to causing non-systemic clinical manifestations, was isolated from the feces of patients with gastroenteritis (Iacoviello *et al.*, 1992).

The occurrence of these microorganisms in bulk tanks, as observed in the herds evaluated in the present study, is important because consumption of raw milk is common in several regions in Brazil. However, Melville *et al.* (1996) demonstrated that many *Prototheca* spp. strains isolated from cases of bovine mastitis are resistant to heat at the usual temperature/time combinations employed in pasteurization, leading to the conclusion that the control of *Prototheca* spp. and other microorganisms which cause mastitis in the dairy herds is one of the most important ways to protect consumer health.

References

Aalbaek, B., Stenderup, J., Jensen, H.E., Valbak, J., Nylin, B. and Huda, A. (1994) Mycotic and algal mastitis in Denmark. *Acta Pathologica Microbiologica Immunologica Scandinavica* 102, 451-456.

Anderson, K.L. and Walker, R.L. (1988) Sources of *Prototheca* spp. in a dairy herd environment. *Journal of the American Veterinary Medical Association* 193, 553-556.

Camargo, Z.P. and Fischman, O. (1979) Use of morpho-physiological characteristics for differentiation of the species of *Prototheca. Sabouraudia* 17, 275-278.

da Costa, E.O., Ribeiro, A.R., Watanabe, E.T., Pardo, R.B., Silva, J.A.B. and Sanches, R.B. (1996a) An increased incidence of mastitis caused by *Prototheca* species and *Nocardia* species on a farm in São Paulo, Brazil. *Veterinary Research Communications* 20, 237-241.

da Costa, E.O., Carciofi, A.C., Melville, P.A., Prada, M.S. and Schalch, U. (1996b) *Prototheca* sp. outbreak of bovine mastitis. *Journal of Veterinary Medicine* 43, 321-324.

Iacoviello, V.R., Degirolami, P.C., Lucarini, J., Sutker, K., Williams, M.E. and Wanke, C.A. (1992) Protothecosis complicating prolonged endotracheal intubation: case report and literature review. *Clinical Infectious Diseases* 15, 959-967.

Kirk, J.H. (1991) Diagnosis and treatment of difficult mastitis cases. *Agri-practice* 12, 15-20.

Klintworth, G.K., Fetter, B.F. and Nielsen, H.S. (1968) Protothecosis, an algal report of a case in man. *Journal of Medical Microbiology* 1, 211-216.

Melville, P.A., da Costa, E.O., Ribeiro, A.R., Watanabe, E.T., Silva, J.A.B. and Garino, F. Jr (1996) Evaluation of the susceptibility of *Prototheca zopfii* strains isolated from milk from cases of intramammarian infections in dairy cows and bulk tanks to different ratios of temperature/time employed in the pasteurizing (72-75°C/15 sec, 72-75°C/20 sec and 62-65°C/30 min) of milk. In: *Proceedings: XIX World Buiatrics Congress.* Edinburgh, pp. 199-200.

Pore, R.S. (1985) *Prototheca* taxonomy. *Mycopathologia* 90, 129-139.

Pore, R.S., Barnett, E.A., Barnes, W.C. Jr and Walker, J.D. (1983) *Prototheca* ecology. *Mycopathologia* 81, 49-62.

Sharp, J.C.M. (1987) Infections associated with milk and dairy products in Europe and North America 1980-1985. *Bulletin World Health Organization* 65, 397-406.

Sudman, M.S. (1974) Protothecosis. A critical review. *American Journal of Clinical Pathology* 61, 10-9.

Venezio, F.R., Lavoo, E., Williams, J.E., Zeiss, C.R., Caro, W.A., Mangkornkanok-Mark, M. and Phair, J.P. (1982) Progressive cutaneous protothecosis. *American Journal of Clinical Pathology* 77, 485-493.

Chapter 75

Shamans and Other Toxicologists: Current Approaches to the Discovery of New Pharmaceuticals in Traditional Herbal Medicines

M. Terry

Department of Animal Science, Texas A&M University, College Station, Texas 77843, USA

Interest in plants as sources of new pharmaceuticals has varied in this century as an inverse function of changing expectations about the capacity of organic synthesis and related technical developments to produce a steady flow of novel compounds suitable for pharmaceutical development. Several factors contribute to the current high level of interest in the screening of plants for biologically active compounds. There is the demanding problem of increasing bacterial resistance to existing antibiotics, which has led to increasing economic incentives for the pharmaceutical industry to search for new antibacterial agents. Large potential markets exist for safer, more effective chemotherapeutic agents for the treatment of neoplastic diseases, AIDS and other prevalent conditions that currently lack adequate means of treatment (Kemp, 1997).

On the supply side, traditional drug design, which is primarily computer-assisted synthetic modeling, is in its infancy and has thus far yielded only a few useful therapeutic agents (Cunningham *et al.*, 1996; Taylor, 1996). Empirical testing in biological systems, regardless of the source of the lead compound, is by far the best predictor of the safety and efficacy of a given compound in humans or target animals (Gustafson, 1997). In addition, useful plant species of tropical ecosystems and the indigenous knowledge of the therapeutic utility of those species are vanishing rapidly (Schultes, 1968; Mee, 1988). Such botanical and therapeutic knowledge is as fragile as human memory, and it has been observed that every time one of the old shamans dies, it is as if a whole library burned to the ground. Rainforest ecosystems are not much more durable, and there is a sense of urgency to salvage what remains of these irreplaceable resources.

A significant factor which bridges the supply and demand of pharmaceutical discovery, is the evolution of ethnobotany over the last 50yrs from an obscure academic curiosity into a mainstream operational approach to pharmaceutical discovery (Roberts, 1992). Ethnobotany may be defined as the study of plants utilized for economic purposes (food, fiber, medicine, poisons, etc.) by peoples of aboriginal cultures (Harshberger, 1896). In the basic ethnobotanical approach, the scientist becomes a student of the shaman in an indigenous culture and observes which plant is used to treat which clinical condition, how it is prepared, the dose employed and the frequency and route of administration.

The essence of pharmaceutical research and development is first to identify materials with desirable biological activity, and then to isolate, characterize and test the active compounds, with the ultimate objective of developing drugs that are shown to be both safe and effective in the treatment of some condition that would constitute a viable market. "Viable market" here means a market of sufficient size to provide a fair return on the financial investment and risk of developing a drug.

The discovery phase of drug development as applied to plants is collecting, identifying and screening as many plant species as possible from a wide range of different localities and seasons, since the production of active compounds may vary substantially within a single species as the environment fluctuates. This is a random process, although phylogenetically related plants have a higher-than-average probability of bearing structurally related compounds.

The screening process is ostensibly a numbers game. The probability of finding a plant whose extracts yield a new compound with potentially useful activity is low, on the order of about one in 100 (Feinsilver, 1996). The probability that an active compound isolated from such a plant will survive the rigors of market analysis, and of safety and efficacy testing, to emerge as a marketable product is extremely low, at one in 5,000 (Beary, 1996). The very small magnitude of the product obtained by multiplying these two low probabilities signifies that very large numbers of plants must be screened in order to find a compound that will actually complete the 10yr development process and become a commercial pharmaceutical product.

Ethnobotanists employed by pharmaceutical companies aspire to increase the first of these two low probabilities by reducing the randomness of the selection process of plants for screening and development. Other alternatives to truly random plant selection that compete with ethnobotany as a source of lead compounds include exploration of novel ecological niches, exploitation of chemical ecological relationships, and examination of genera phylogenetically related to species known to produce desirable constituents.

The central working assumption of the ethnobotanical approach is that the shaman utilizes plants that have been selected and manipulated in an evolving process of cumulative invention by a chain of practitioners in a shamanic tradition spanning several millennia, and that plants so selected are more likely than plants selected at random to contain compounds with properties that correspond to our Western concepts of biological activity. In theory, the selection of such plants for bioassay screening should raise the probability of positive screening results.

There is by no means a perfect correlation between indigenous medicinal use of a plant and a positive screening result based on our standard tests for common types of activity. Inattention to shamanic detail can result in collection of the plant at the wrong time of year, testing of the wrong part of the plant or testing the extract(s) of a single plant species when the plant is active only if used in combination with another plant. Instability of the active principle(s) may require reducing the time between plant collection and screening (preferably within the source country), and improving storage conditions. Failure to find recognizable activity may be due to use of an inappropriate screening test for the type of activity expected for the plant; for example, *in vitro* and whole-animal bioassays may be ineffective for detecting psychoactive compounds (Shulgin, 1996). The indigenous condition treated with the plant may have no counterpart in Western medicine. Some traditionally employed plants simply contain no active constituents.

Despite these caveats, even a modest positive correlation between indigenous medicinal use of plants and positive bioassay results with the plants so used, constitutes an economically valuable contribution to pharmaceutical discovery.

References

Beary, J.F., Duchaine, C.M. and Rhein, R.W. (1996) US drug and biologic approvals in 1994-1995. *Drug Development Research 1996* 37, 197-207

Cunningham, D., Zalcberg, J., Smith, I., Gore, M., Pazdur, R., Burris, H. 3rd, Meropol, N.J., Kennealey, G. and Seymour, L. (1996) 'Tomudex' (ZD1694): a novel thymidylate synthase inhibitor with clinical antitumour activity in a range of solid tumours. 'Tomudex' International Study Group. *Annals of Oncology* 7, 179-82.

Feinsilver, J.M. (1996) Prospección de la biodiversidad: Potenciales para los paises en desarrollo. *Revista de la Comision Economica para America Latina y el Caribe.* United Nations, December, pp. 111-128.

Gustafson, R.H. (1997) Personal communication.

Harshberger, J.W. (1896) Purposes of Ethnobotany. *Botanical Gazette* 21, 146-154.

Kemp, G. (1997) Personal communication.

Mee, M. (1988) *In Search of Flowers of the Amazon Forests.* Nonesuch Expeditions, Woodbridge, Suffolk, p. 295.

Roberts, L. (1992) Chemical prospecting: Hope for vanishing ecosystems? *Science* 256, 1142-1143.

Schultes, R.E. (1968) Personal communication.

Shulgin, A. (1996) Personal communication.

Taylor, G. (1996) Sialidases: structures, biological significance and therapeutic potential. *Current Opinion in Structural Biology* 6, 830-837.

Chapter 76

Morphologic Studies of Selenosis in Herbivores

M.F. Raisbeck and D. O'Toole

Department of Veterinary Sciences, University of Wyoming, Laramie, Wyoming 82070, USA

Introduction

Spontaneous selenosis in grazing animals has been historically divided into two distinct clinical entities: "alkali disease" and "blind staggers" (Olson, 1978; Beath, 1982). Surprisingly, although alkali disease (AD) was recognized as a distinct disease by livestock producers before the turn of the century (Franke and Potter, 1934), there are few morphologic studies of either AD or blind staggers (BS). There are no published histologic descriptions of the integumentary lesions of herbivores with AD, although Se-induced lesions have been characterized in the central nervous system of intoxicated swine and, to a lesser extent, in porcine integument (Wilson *et al.*, 1983). Neither are there any detailed pathophysiologic studies of selenosis in domestic herbivores due to selenomethionine (SEMET), the predominant form of Se in seleniferous grains and forage. The purpose of this report is to summarize 6yrs of morphologic investigation into experimental and spontaneous selenosis in horses and cattle.

Caveat Tinctor: the Search for a Reliable Means of Identifying Se in Tissues

Specimens taken from field cases or from chronic feeding trials in horses and cattle often reveal a variety of lesions, many of which are probably incidental to Se toxicity. Histochemical association between a given lesion and locally increased Se concentrations, while not absolute proof, would be strong evidence of a causal relationship. Autometallographic localization of Se in histological preparations was introduced by Danscher (1982) as a modification of the intravital Timm sulfide silver technique for detecting heavy metals. This method was used by the originator to demonstrate the cellular and subcellular distribution of Se in tissues of rats, and was later used to demonstrate Se in the tissues of acutely intoxicated sheep (Smyth *et al.*, 1990).

To determine whether the Danscher procedure was a reliable adjunct to chemical analysis prior to using it in pathophysiologic studies of chronic selenosis, adult male rates were dosed with four concentrations each of three different forms of Se: Na_2SeO_3, L-selenomethionine (L-SEMET) or selenocystine (SECYS) for 28d. Samples of liver, kidney, heart and brain were taken for autometallography and chemical analysis by fluorometry (Raisbeck *et al.*, 1996). Metallic silver precipitation was most frequent in kidneys of rats given SECYS, where it was concentrated in the apical cytoplasm of epithelial cells of the proximal convoluted tubules. There was no correlation between staining intensity and renal Se concentrations. Two of 24 Se-treated rats had silver precipitation in the hippocampus and dorsal cochlear nucleus of the brain, but the Se concentrations in these brains were among the lowest of the Se-treated rats. Hepatic Se concentrations were similar to the renal levels, but silver precipitation was absent in this organ (O'Toole *et al.*, 1995). A similar lack of correlation was found between tissue Se concentrations and autometallographic "staining" in cattle fed Na_2SeO_3, L-SEMET or SECYS grass hay and in mice fed Na_2SeO_3, L-SEMET or SECYS. Thus, this technique is not a reliable means of localizing Se in chronically exposed animals.

Elemental Se was identified by electron probe X-ray microanalysis as a component of silver-containing particles in tissues from human patients suffering from argyria (Matsumura *et al.*, 1992), but we are not aware of successful detection of more common physiologic forms, such as protein-bound Se, at a light or electron microscopic level. It was recently attempted to characterize the subcellular distribution of Se in Se-poisoned mallards by semiquantitative x-ray microanalysis. The instrument (JEOL 35CS, NORAN, Boston, MA) was insufficiently sensitive to detect biologically high Se concentrations of 30-50µg/gm (wwt) in liver and kidney (O'Toole and Raisbeck, unpublished data).

Integumentary Lesions in Horses and Cattle

The most distinctive gross lesions of chronic selenosis in horses and cattle involve the integument (Raisbeck *et al.*, 1993). A typical example involves a steer given 0.8mg Se/kg as L-SEMET daily by gavage for several months as part of a chronic feeding study (O'Toole and Raisbeck, 1995). After receiving SEMET for 96d, this steer became moderately lame and exhibed slight selling and erythema proximal to the coronary band of both front legs. The swelling and erythema subsided, but 10d after the initial signs, a hairline crack appeared parallel to and 0.5cm distal to the coronary band and the steer became markedly lame. After approximately 100d on L-SEMET,

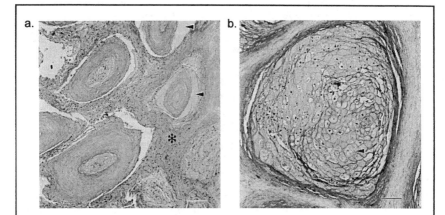

Fig. 76.1. Coronary papillae from a horse with spontaneous alkali disease. **a.** Transverse section to papillae at their distal tips, near a site of gross cleavage in hoof wall. Early changes include nucleated keratinocytes in inter-tubular matrix (parakeratosis; ✳), and multiple circumferential microfractures at the interface of incipient horn tubules with intertubular horn (➤). H and E stain. Bar: 100µm. **b.** Higher magnification of a papilla, sectioned near the tip. Marked ballooning degeneration of keratinocytes (➤), with karyrorrhexis. Bar: 50µm.

hair on the distal 6cm of the tail (the tail switch) became detached at the surface of the skin. Hoof separation and lameness increased in severity until, by 120d, the steer refused to walk to feed and water, and had to be hand fed. By the time the steer was euthanized at 134d, all hooves had cracks that extended through to the sensitive laminae (Fig. 77.4, Chapter 77). New, normal tubular horn was growing between dystrophic horn and the epithelium. In field cases, the dystrophic hoof wall is eventually displaced by new growth and is shed or, if it remains partially attached, results in an elongated, deformed hoof ("slipper toe").

Histologically, hoof damage begins as ballooning degeneration and necrosis of keratinocytes in the stratum spinosum of coronary papillary epidermis (Figs. 76.1 and 76.2) and near the tips of the primary laminae of the stratum internum covering the third phalanx. Similar degenerate changes occur in stratum spinosum of nail and beak of adult ducks with selenosis (O'Toole and Raisbeck, 1997a). Neutrophils and dyskeratotic cellular debris accumulate around epidermal papillae of the stratum corneum and, as cells are displaced from the papillae, in the lumen of horn tubules. The latter are dilated (300-500µm diameter *vs.* less than 100µm in normal hoof) and there is concomitant loss of intertubular keratin (Fig. 76.3). The normally highly organized spiral of keratinocytes around the tubule is disrupted. In more chronic cases of AD, injury within the epithelium of the hoof may be more extensive, involving more of the laminar epithelium and tubular horn papillae of the sole in addition to the

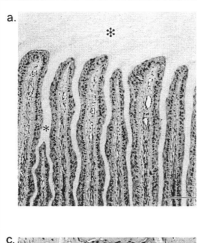

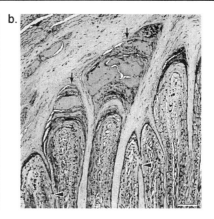

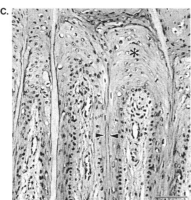

Fig. 76.2. Laminar portion of a hoof of a normal steer (a) and of a steer with experimentally-induced alkali disease (b and c) HE. **a.** Normal epidermal laminae. Normal horn matrix of stratum medium (✳) and stratum internum between laminae (✱). Bar: 100µm. **b.** Affected steer. Note irregularity in length of laminae. Parakeratotic caps overlie laminae (→). Keratin in stratum internum varies in thickness, with marked thinning in some sites (➤). Bar: 100µm. **c.** Higher magnification of the changes in the tip of a primary papilla. There is edema of the laminardermis, hyperplasia and disorganization of the papillary epithelium, including germinal epithelium. Stratum internum is irregularly thinned (between ➤). Bar: 50µm. **c.** Reprinted with permission of the American Association of Veterinary Diagnosticians (AAVLD).

coronary area. Moderate to severe hyperplasia, acanthosis and parakeratosis occur at the tips of the epithelial laminae (Fig. 76.2b, c).

Parakeratotic ridges of keratin derived from the tips of the laminae are interspersed with islands of relatively normal orthokeratotic keratin from the lateral walls of the laminae. In severe cases, there may be milder, more diffuse changes in the lateral laminar walls. "Keratin columns" of the stratum internum of normal orthokeratotic keratin between adjacent laminae are narrowed (Fig. 76.2b, c). In areas where changes are marked, the columnar germinal epithelium near laminar tips

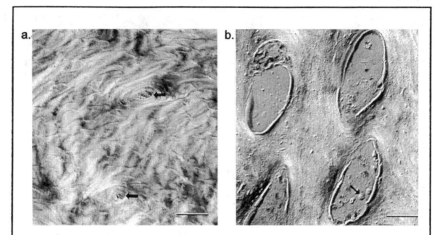

Fig. 76.3. Stratum medium from normal steer (a) and steer with selenosis (b). Differential interference contrast optics, H and E. Bar: 50µm. **a.** Note regularly spaced concentric arrays of keratinocytes in tubular horn (→), separated by intertubular keratinocytes. **b.** Distended tubule full of cellular debris, largely parakeratotic cells (→) are separated by apparently normal intertubular matrix. Reprinted with AAVLD permission.

becomes disorganized and attenuated, and dermis underlying the affected epithelium may be edematous (O'Toole and Raisbeck, 1995; 1997b).

Not surprisingly, chemical analysis of dystrophic hooves reveals that elevated Se concentrations are spatially correlated with the dystrophic hoof lesions (Fig. 77.3b, Chapter 77). In the past, several authors including ourselves speculated that substitution of Se for S in hard keratin results in weakened inter- and intraprotein disulfide bonds and thus softened keratin, which shears parallel to the grain of the hoof from the concussive forces of walking or running. However, the cytotoxic damages demonstrated in these animals suggests an alternate hypothesis. The marked tubular dilation demonstrated in Fig. 76.3b results from substitution of necrotic cellular debris, which has little or no structural strength, for a significant portion of the intertubular keratin normally present in a cross-section of stratum medium. Similarly, the tubule itself is weakened by the disorganization and thinning of keratinocytes in the tubular wall. These changes are concentrated in a small area parallel to the coronary band by virtue of the fact that tubular keratin formation is a continuous process, with cellular damage occuring simultaneously in all papillae when local Se concentrations reach a critical peak. Hooves are designed to divert dangerous proximally-directed cracks (along the plane of horn tubules) in a lateral direction (along the plane of intertubular material) (Bertram and Gosline, 1986).

Dilation of horn tubules combined with microfissures in the intertubular material of the coronary area (Fig. 76.1a) probably exaggerate this tendency, resulting in focused circumferential cracks in the proximal part of the hoof wall. This pathogenic mechanism may explain the failure of sheep to develop typical alkali disease hoof lesions. In spite of some excellent experimental studies of the systemic lesions of selenosis in sheep (Glenn *et al.*, 1964), ovine hooves have never been examined histologically for evidence of dystrophic hoof growth. The smaller body mass of sheep results in proportionately less stress on damaged hoof wall and decreases the likelihood of structural failure.

Selenium-induced alopecia most frequently affects the nape of the neck (mane) and the tail, but in severe cases may also involve other anatomic sites. At a histologic level, alopecia results from atrophy of primary hair follicles. Minimal or no changes are seen in secondary (non-medullated or undercoat) follicles. The ratio of atrophic to nonatrophic follicles in denuded areas may be as great as 1:3 as compared to less than 1:10 in normally haired areas. Most atrophic follicles are collapsed and lack a hair shaft (Fig. 76.4). The inner root sheath is atrophic or absent, the outer sheath contains poorly laminated or dyskeratotic (or apoptotic) keratinocytes (Fig. 76.4c). The hyaline membrane and connective tissue sheath surrounding the follicle are thickened. Follicles that are less severely affected may contain small dystrophic shafts. Accessory follicular structures such as sebaceous glands and arrector pili muscles are unaffected.

Other Lesions of Selenosis

To our knowledge a primary encephalopathy has never been reproduced as a direct effect of selenosis in cattle, sheep or horses, although "polioencephalomalacia" was reported recently in pigs fed a high-selenium diet containing *Astragalus bisulcatus* (Panter *et al.*, 1996). This term suggests that lesions in the brain were severe, since polioencephalomalacia implies a significant degree of necrosis and, in many instances, grossly evident changes (Gould, 1997). Selenium-intoxicated pigs can develop histological lesions in pontine, olivary, facial, reticular and motor trigeminal nuclei, in addition to well-characterized lesions in spinal intumescences (Wilson *et al.*, 1983; Wilson *et al.*, 1989). However, unlike the spinal lesions, which are responsible for clinical signs of ascending paralysis, lesions in the thalamas and brainstem do not generally result in grossly evident malacia. "Polioencephalomalacia" in intoxicated pigs does not appear to involve the cerebral cortex, unlike sulfur-induced polioencephalomalacia of ruminants.

Although the condition referred to as BS is often attributed to chronic selenosis in reviews and textbooks, only two original reports describe the condition (Draize and Beath, 1935; Rosenfeld and Beath, 1946). These reports attribute a virtual laundry list of lesions to Se intoxication. Despite a standard necropsy protocol that includes examining all major organ systems and histopathologic examination of at least 75 separate tissue specimens, none of the animals we have examined to date has shown

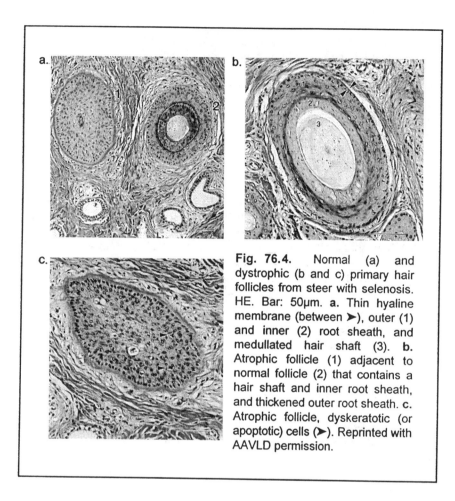

Fig. 76.4. Normal (a) and dystrophic (b and c) primary hair follicles from steer with selenosis. HE. Bar: 50µm. **a.** Thin hyaline membrane (between ➤), outer (1) and inner (2) root sheath, and medullated hair shaft (3). **b.** Atrophic follicle (1) adjacent to normal follicle (2) that contains a hair shaft and inner root sheath, and thickened outer root sheath. **c.** Atrophic follicle, dyskeratotic (or apoptotic) cells (➤). Reprinted with AAVLD permission.

any clinicopathologic or morphologic evidence of the lesions attributed to chronic selenosis by Beath's group (O'Toole *et al.*, 1996). The possible exception is myocardial necrosis. A single 16mo-old Angus-cross heifer fed 1mg Se/kg/d as Na_2SeO_3 for 81d developed disseminated myocardial necrosis and fibrosis, in addition to the integumentary lesions outlined above. This heifer developed a conditioned aversion to Se-treated diets early in the course of the experiment and, as a result, consumed the full Se dose somewhat irregularly. Myocardial necrosis is also prominent feature of acute Se intoxication in sheep (Maag and Glenn, 1967; Blodgett and Bevill, 1987), and there are numerous anecdotal reports of myocardial disease ("dishrag heart") in cattle with chronic selenosis. It thus seems reasonable that, under appropriate conditions of exposure, it is possible for an animal to sustain repeated subclinical toxic insults which culminate in a pattern of organ damage similar to that seen in acute Se poisoning. Nonetheless, the most distinctive and characteristic lesions of chronic selenosis in large ruminants involve the integument.

Acknowledgments

Portions of this work were supported by the Wyoming Abandoned Coal Mine Land Research Program, the University of Wyoming Agricultural Experiment Station, the UW Department of Veterinary Sciences and the Pharmaceutical Manufacturers Association.

References

Beath, O.A. (1982) *The Story of Selenium in Wyoming.* Bulletin No. 774, Agricultural Experiment Station, University of Wyoming, Laramie, WY.

Bertram, J.E. and Gosline, J.M. (1986) Fracture toughness design in horse hoof keratin. *Journal of Experimental Biology* 125, 29-47.

Blodgett, D.J. and Bevill, R.F. (1987) Acute selenium toxicosis in sheep. *Veterinary and Human Toxicology* 29, 233-236.

Danscher, G. (1982) Exogenous selenium in the brain. A histochemical technique for light and electron microscopical localization of catalytic selenium bonds. *Histochemistry* 76, 281-293.

Draize, J.H. and Beath, O.A. (1935) Observations on the pathology of blind staggers and alkali disease. *Journal of the American Veterinary Medical Association* 39, 753-763.

Franke, K.W. and Potter, V.R. (1934) A new toxicant occurring naturally in certain samples of plant foodstuffs. III. Hemoglobin in rats fed toxic wheat. *Journal of Nutrition* 8, 615-624.

Glenn, M.W., Jensen, R. and Griner, L.A. (1964) Sodium selenate toxicosis: The effects of extended oral administration of sodium selenate on mortality, clinical signs, fertility and early embryonic development in sheep. *American Journal of Veterinary Research* 25, 1479-1499.

Gould, D.H. (1998) Polioencephalomalacia. *Journal of Animal Science* (in press).

Maag, D.D. and Glenn, M.W. (1967) Toxicity of selenium: farm animals. In: Muth, O.H. (ed), *Selenium in Biomedicine.* AVI Publishing, Westport, CT, pp. 127-140.

Matsumura, T., Kumakiri, M., Ohkawara, A., Himeno, H., Numata, T. and Adachi, R. (1992) Detection of selenium in generalized and localized argyria: report of four cases with X-ray microanalysis. *Journal of Dermatology* 19, 87-93.

Olson, O.E. (1978) Selenium in plants as a cause of livestock poisoning. In: Keeler, R.F., VanKampen, K.R. and James, L.F. (eds), *Effects of Poisonous Plants on Livestock.* Academic Press, New York, NY, pp. 121-133.

O'Toole, D. and Raisbeck, M.F. (1995) Pathology of experimentally-induced chronic selenosis ("alklai disease") in yearling cattle. *Journal of Veterinary Diagnostic Investigation* 7, 364-373.

O'Toole, D. and Raisbeck, M.F. (1997a) Experimentally-induced selenosis of adult mallard ducks: clinical signs, lesions and toxicology. *Veterinary Pathology* 34, 330-340.

O'Toole, D. and Raisbeck, M.F. (1997b) Magic numbers, elusive lesions: comparative aspects of selenium toxicosis in herbivores and waterfowl. In: Frankenberger, W.T. and Engberg, R.A. (eds), *Environmental Chemistry of Selenium.* Marcel Dekker, New York, NY, pp. 355-395.

O'Toole, D., Castle, L.E. and Raisbeck, M.F. (1995) Comparison of histochemical autometallography (Danscher stain) to chemical analysis for detection of selenium in tissue. *Journal of Veterinary Diagnostic Investigation* 7, 281-284.

O'Toole, D., Raisbeck, M.F., Case, J.C. and Whitson, T.D. (1996) Selenium-induced "Blind Staggers" and related myths. A commentary on the extent of historical livestock losses attributed to selenosis on western rangelands. *Veterinary Pathology* 33, 104-116.

Panter, K.E., Hartley, W.J., James, L.F., Mayland, H.F., Stegelmeier, B.L. and Kechele, P.O. (1996) Comparative toxicity of selenium from seleno-DL-methionine, sodium selenate, and *Astragalus bisulcatus* in pigs. *Fundamental and Applied Toxicology* 32, 217-223.

Raisbeck, M.F., Dahl, E.R., Sanchez, D.A., Belden, E.L. and O'Toole, D. (1993) Naturally occurring selenosis in Wyoming. *Journal of Veterinary Diagnostic Investigation* 5, 84-87.

Raisbeck, M.F., O'Toole, D., Schamber, R.A., Belden, E.L. and Robinson, L.J. (1996) Toxicologic effects of a high-selenium hay diet in captive adult and yearling pronghorn antelope (*Antilocapra americana*). *Journal of Wildlife Diseases* 32, 9-16.

Rosenfeld, I. and Beath, O.A. (1946) *Pathology of Selenium Poisoning.* Bulletin No. 275. Agricultural Experiment Station, University of Wyoming, Laramie, WY, pp. 1-27.

Smyth, J.B.A., Wang, J.H., Barlow, R.M., Humphreys, D.J., Robins, M. and Stodulski, J.B. (1990) Experimental acute selenium intoxication in lambs. *Journal of Comparative Pathology* 102, 197-209.

Wilson, T.M., Scholz, R.W. and Drake, T.R. (1983) Selenium toxicity and porcine focal symmetrical encephalomalacia: description of a field outbreak and experimental reproduction. *Canadian Journal of Comparative Medicine* 47, 412-421.

Wilson, T.M., Cramer, P.G., Owen, R.L., Knepp, C.R., Palmer, I.S., deLahunta, A., Rosenberger, J.L. and Hammerstedt, R.H. (1989) Porcine focal symmetrical poliomyelomalacia: test for interaction between dietary selenium and niacin. *Canadian Journal of Veterinary Research* 53, 454-461.

Chapter 77

Chronic Selenosis in Ruminants

M.F. Raisbeck[1], D.O'Toole[1], E.L. Belden[1] and J.W. Waggoner[2]

Departments of [1]Veterinary Sciences and [2]Rangeland Ecology and Watershed Management, University of Wyoming, Laramie, Wyoming 82070, USA

Introduction

The first conclusive identification of spontaneous selenium (Se) toxicity was reported by the South Dakota and Wyoming Agricultural Experiment Stations (Beath *et al.*, 1934; Franke and Potter, 1934). These laboratories popularized a model of selenosis in which spontaneous selenosis in grazing animals takes one of three forms: acute selenosis, chronic selenosis "of the alkali disease type" and chronic selenosis "of the blind staggers type."

Acute selenosis is characterized by sudden death with few signs, or by gastroenteritis, myocardial necrosis, pulmonary edema, and hepatic and renal necrosis (Blodgett and Bevill, 1987; MacDonald *et al.*, 1981; Smyth *et al.*, 1990). Acute selenosis is of limited relevance under range conditions as plants which accumulate sufficient Se to be acutely toxic are extremely unpalatable. As noted by Beath (1920), "when cattle were placed in enclosures on *Astragalus* they would not touch the plant at all and after a few days could not be kept inside the fence." In our experience, near-acutely toxic seleniferous feedstuffs fed in confinement rapidly produce conditioned aversion in mammals and waterfowl.

"Alkali disease" (AD) in horses and cattle is characteristically manifested by alopecia and hoof dystrophy after prolonged consumption of non-accumulator plants (Olson, 1978). Most features of spontaneous AD have been reproduced by feeding either seleniferous vegetation or inorganic Se salts (Miller and Williams, 1940; Olson and Embry, 1973; James *et al.*, 1994; Raisbeck *et al.*, 1995). While not as prevalent as previously claimed, this condition occurs with some regularity and can be economically very important in localized areas (Raisbeck *et al.*, 1993).

"Blind staggers" (BS) purportedly results from ingestion of Se-accumulating weeds (Beath, 1982) and presents as blindness, circling, headpressing, dysphagia and paralysis. There are only two original reports of BS (Draize and Beath, 1935; Rosenfeld and Beath, 1946). There are no well-documented reports of BS resulting from controlled feeding experiments with either seleniferous plants or purified Se compounds. There is compelling evidence to question whether BS was actually related to Se (O'Toole *et al.*, 1996) yet the myth of massive mortalities seems permanently fixed in both the scientific and popular literature (Anonymous, 1991).

Much of the early dogma does not agree with the experience of ranchers and veterinarians in this area. Critical re-evaluuatuation of Se toxicity in grazing livestock and wildlife was begun in 1989.

Controlled Feeding Studies

Purified selenium compounds in cattle

Most experimental data regarding selenosis in grazing animals was derived from inorganic Se salts (Miller and Williams, 1940; Glenn *et al.*, 1964; Olson and Embry, 1973) whereas the predominant form in palatable vegetation is thought to be selenomethionine (Olson, 1978). Knowlege of the chronic toxicity of the latter in ruminants is scant, but there is evidence in other species (Heinz *et al.*, 1988) that it differs quantitatively and qualitatively from inorganic Se salts. The toxic effects of L-selenomethionine (SEMET) and Na_2SeO_3 were compared in cattle.

Details of trial diets and animals are in Chapter 53. Steers were observed daily for signs of illness; blood was collected at 21d intervals for CBC, clinical pathology and Se determination by a fluorometric method (Raisbeck *et al.*, 1996). At termination of the trial, steers were euthanized and tissues taken for Se analysis.

Both principles and controls gained weight and appeared clinically healthy throughout the 120d feeding study. Controls gained slightly but non-significantly more than steers receiving 25ppm Se as Na_2SeO_3 at 84, 105 and 120d. There were no abnormalities attributable to Se in blood urea nitrogen, serum creatinine, creatinine kinase, alkaline phosphatase, aspartate aminotransferase, bilirubin, glucose, albumin, total protein, blood hemoglobin, or CBC. In contradiction to early reports (Franke and Potter, 1934; Draize and Beath, 1935), there were no changes in the erythron that might indicate anemia.

Plasma Se concentrations increased more rapidly and were generally more labile than blood Se levels (Fig. 77.1). This suggests that whole blood is a better index of long-term Se status than plasma or serum in mammals. Blood Se concentrations plateaued between 60-80d. Blood glutathione peroxidase was non-significantly *decreased* in steers receiving 0.8mg/kg Se as Na_2SeO_3. Significantly higher blood Se concentrations resulted from SEMET than from corresponding dosages of $NaSeO_3$. Tissue Se concentrations, with the exception of liver, were also significantly greater in the SEMET group than the $SeO_3^=$ group (Fig 77.2).

One steer that received 0.8mg/kg Se as SEMET had grossly visible semicircular ridges and grooves parallel to the coronary band at necropsy. Both steers fed 0.08mg/kg, 1/2 steers fed 0.28 Se/kg as SEMET and 1/2 steers fed 0.8 mg Se/kg as

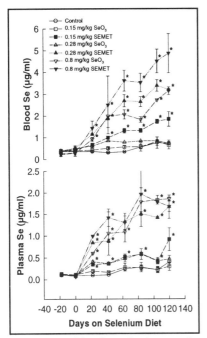

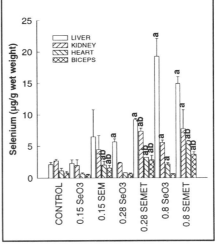

Fig. 77.1. Blood and plasma Se concentrations in steers fed three dosages of Na_2SeO_3 or SEMET. Asterisks denote significantly greater than controls (*P*<0.05).

Fig. 77.2. Tissue concentrations in steers fed Na_2SeO_3 or SEMET. **a** indicates significantly different than controls, **b** greater than the corresponding Na_2SeO_3 group (*P*<0.05).

Na_2SeO_3 had histologic lesions of dyskeratosis similar to those described in Chapter 76. Selenium deposition in hooves from these steers and from field cases with AD show a distinct morphologic relationship to necrotic lesions (Fig. 77.3) and varied by more than 10x. This is consistent with Se deposition in the hoof as keratin is formed.

A steer given 0.8mg Se/kg/d as SEMET *via* gelatin capsules for 4mos developed higher blood Se levels (≈7ppm) than steers fed a similar amount in their daily ration (≈4.5ppm), and his blood glutathione peroxidase activity declined precipitously from over 1,000 to 280mM/s/L just at the onset of clinical signs.

Antelope

It has been proposed that grazing species native to the Great Plains are genetically resistant to selenosis (Beath, 1982). Four pronghorn antelope fed a 15ppm Se diet prepared from seleniferous hay for more than 150d did not develop any clinical, morphologic or biochemical changes attributable to Se. The Se-treated group gained less than controls, but this was attributed to feed refusal rather than a direct

Se-toxic effect. They did, however, exhibit significantly decreased primary antibody response to hen egg albumin (Raisbeck *et al.*, 1996).

Blood and tissue Se concentrations were roughly equivalent to those of steers fed 0.28mg Se/kg/d as Na_2SeO_3 or 0.15mg Se/kg/d as SEMET, and were greater than previously suggested to be diagnostic of selenosis in this species (Williams, 1982). This observation reinforces the contention that tissue concentrations alone are not definitive evidence of selenosis (O'Toole and Raisbeck, 1997).

Cattle

To correlate biomarkers of experimental selenosis with naturally seleniferous hay, twelve 2yr-old steers were divided into three groups and housed in concrete pens. After a 60d acclimation period on normal hay, each group was fed a ration ground from seleniferous and non-seleniferous grass hay containing 3.33, 5.00 or 6.66ppm Se and balanced with corn and cottonseed meal to be isocaloric and isonitrogenous for 167d. Blood Se concentrations increased in a dose-related manner in all groups, and plateaued at 0.5, 0.8 and 1.4ppm, respectively. There were no significant differences in rate of gain between groups, nor did any steers exhibit clinical evidence of disease. Regression of blood concentration *vs.* dietary concentration yielded [blood]=0.2701*[hay]-0.449.

Field Study

To validate the experimental feeding trials under field conditions, four grazing sites were selected to represent the spectrum of Se concentrations available in eastern Wyoming and western Nebraska (Raisbeck *et al.*, 1997). Two (A and B) were mine reclamation projects suspected to be marginally high in Se. One (E) was a semi-improved pasture with a history of AD. The last (D) was known to contain merely adequate Se and served as a control. None contained any indicator plant species and all contained a few facultative accumulator specimens. Dietary botanical composition was determined by observing eating habits for 4hrs in the morning and evening at semi-monthly intervals. Representative 500g samples of grazed vegetation were collected onto dry ice at monthly intervals and frozen at -70°C until analyzed. Selenium concentrations were averaged by plant species, collection date and site (species mean). Selenium intake (dietary mean) was calculated from species means on the basis of the percentage of each plant species consumed.

Previous reports indicate that forage Se peaks in late spring. In this study Se concentrations peaked later in the growing season, typically in August. This discrepancy may be due to the fact that our samples were deliberately biased toward what cattle consumed rather than randomized to represent the biomass, or it may reflect site-specific differences in botany and geology between South Dakota and Wyoming. Possibly as a result of unusually lush growing conditions, the highest *dietary* mean Se concentrations were less than 1.0ppm at all sites except E, which reached nearly 8.0ppm. None of the cattle at any site developed overt signs of selenosis. Sites A, B and E each had individual forage samples that exceeded 2.0ppm, but only on site E did species mean concentrations exceed 2.0ppm. At the latter, several species mean concentrations approached 10ppm, and individual samples often exceeded 20ppm. Blood Se concentrations remained below 0.5ppm except at site E, where blood Se peaked at slightly more than 2.0ppm and showed much greater variability than in the other groups (Fig. 77.4). Regression of dietary forage concentrations against blood concentrations yielded a similar slope but lower constant ([blood]=0.2816*[grass]+0.0871) than steers fed seleniferous hay.

Summary

The most distinctive *chronic* toxic effect of Se in large ruminants is dyskeratosis, manifested as a particular pattern of hair loss and hoof dysplasia ("alkali disease") (Chapter 76). Decreased primary antibody response (Chapter 53) is not sufficiently specific to be diagnostically useful. This group has not recognized the syndrome termed BS or identified any lesions that might cause blindness and neurologic derangement in any animal fed any form of Se experimentally. In more than 40 field investigations since 1989, selenosis as an etiologic factor in any putative cases of BS has not been confimed. As suggested by James *et al.* (1994) and Raisbeck *et al.* (1993), the most common etiologies of BS in seleniferous areas of the Great Plains are sulfur-induced polioencephalomalacia and lead poisoning.

Many summaries of early selenosis research describe infertility as a subclinical

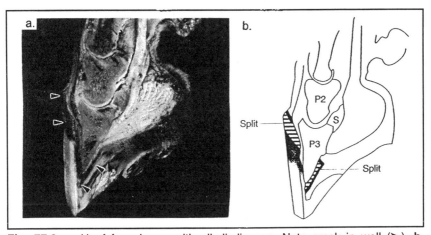

Fig. 77.3. a. Hoof from horse with alkali disease. Note crack in wall (➤). **b.** Schematic of same hoof. Note variation in Se concentration. Highest concentrations (darkest shading) correspond with defects (hatched) in hoof wall and in sole.

but major source of economic loss in "alkalied" livestock (Beath, 1982). Experimental studies in lab animals (Willhite, 1993), sheep (Panter *et al.*, 1995), cattle (Yeager, 1995) and non-human primates only produced infertility at maternally toxic doses. Veterinarians involved with early selenosis investigations hypothesized that the reported infertility was secondary to the crippling effects of selenosis and starvation (Tucker, 1996). A limited number of studies of ranches with histories of both reproductive failure and AD have revealed animals that were extremely Cu deficient as well as extremely high in Se. Intervention with Cu supplementation for 2yrs on one such premise increased pregnancy rates

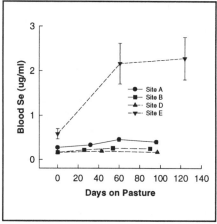

Fig. 77.4. Blood Se concentrations (mean±sd) in cattle grazing pastures with varying Se concentrations. Site E is significantly greater (*P*<0.05) at all times.

from less than 80% to nearly 100%. Given the geologic co-occurrence of Se and S in the Great Plains, it is possible that infertility attributed to Se by early studies was confused by S-induced Cu deficiency. Controlled experiments are needed to test this hypothesis.

Although ruminants are capable of synthesizing SEMET from toxic concentrations of SeO_3^- (Raisbeck and Belden, unpublished), dietary SEMET produced markedly higher tissue Se concentrations in ruminants than did Na_2SeO_3. This is presumably a result of nonspecific substitution for methionine in protein. These results suggest that Se from forages accumulates in a fashion more similar to SEMET than SeO_3^- at normal to low-toxic dietary concentrations. Selenomethionine is less cytotoxic than SeO_3^- (Spallholz, 1994) but must be metabolized to a reactive form to cause cell death. Although AD has been produced with inorganic Se salts, it was most likely the result of prolonged exposure to seleniferous forages or SEMET. Conversely, acutely lethal lesions such as myocardial necrosis are more likely due to exposure to inorganic salts or accumulator plants. Possibly, this is because the relatively low toxicity of the former permits Se to accumulate to locally toxic concentrations in keratinized tissues without first triggering feed aversion, severe illness and anorexia or death.

Acknowledgments

Portions of this work were supported by the Wyoming Abandoned Coal Mine Land Research Program, the University of Wyoming Agricultural Experiment Station, the

UW Department of Veterinary Sciences and the Pharmaceutical Manufacturers Association.

References

Anonymous (1991) Selenium poisoning. In: Fraser, C.M. (ed), *The Merck Veterinary Manual*, Merck and Company, Rahway, NJ, pp. 1727-1728.

Beath, O.A. (1920) *Chemical Examination of* Astragalus bisulcatus. University of Wyoming, Agricultural Experiment Station 31st Annual Report, p. 127.

Beath, O.A (1982) *The Story of Selenium in Wyoming*. University of Wyoming, Agricultural Experiment Station, bulletin No. 189, pp. 1-23.

Beath, O.A., Draize, J.H., Eppson, H.F., Gilbert, C.S. and McCreary, O.C. (1934) Certain poisonous plants of Wyoming activated by selenium and their association with respect to the soil. *Journal American Pharmaceutical Association* 23, 94-97.

Blodgett, D.J. and Bevill, R.F. (1987) Acute selenium toxicosis in sheep. *Veterinary and Human Toxicology* 29, 233-236.

Draize, J.H. and Beath, O.A. (1935) Observations on the pathology of blind staggers and alkali disease. *Journal of the American Veterinary Medical Association* 39, 753-763.

Franke, K.W. and Potter, V.R. (1934) A new toxicant occurring naturally in certain samples of plant foodstuffs. III. Hemoglobin in rats fed toxic wheat. *Journal of Nutrition* 8, 615-624.

Glenn, M.W., Jensen, R. and Griner, L.A. (1964) Sodium selenate toxicosis: The effects of extended oral administration of sodium selenate on mortality, clinical signs, fertility and early embryonic development in sheep. *American Journal of Veterinary Research* 23, 1479-1495.

Heinz, G.H., Hoffman, D.D. and Gold, L.G. (1988) Toxicity of organic and inorganic selenium to mallard ducklings. *Archives of Environmental Contamination and Toxicology* 17, 561-568.

James, L.F., Hartley, W.F., Panter, K.E., Stegelmeier, B.L., Gould, D. and Mayland, H.F. (1994) Selenium poisoning in cattle. In: Colegate, S.M. and Dorling, P.R. (eds), *Plant-Associated Toxins: Agricultural, Phytochemical and Ecological Aspects*. CAB International, Wallingford, Oxon, pp. 416-420.

MacDonald, D.W., Christian, R.G., Strausz, K.I. and Roff, J. (1981) Acute selenium toxicity in neonatal calves. *Canadian Veterinary Journal* 22, 279-281.

Miller, W.T. and Williams, K.T. (1940) Effects of feeding repeated small doses of selenium as sodium selenite to equines. *Journal of Agricultural Research* 61, 353-368.

Olson, O.E. (1978) Selenium in plants as a cause of livestock poisoning. In: Keeler, R.F., VanKampen, K.R. and James, L.F. (eds), *Effects of Poisonous Plants on Livestock*. Academic Press, New York, NY, pp. 121-133.

Olson, O.E. and Embry, L.B. (1973) Chronic selenite toxicity in cattle. *Proceedings of the South Dakota Academy of Sciences* 52, 50-58.

O'Toole, D. and Raisbeck, M.F. (1997) Magic numbers, elusive lesions: comparative aspects of selenium toxicosis in herbivores and waterfowl. In: Frankenberger, W.T. and Engberg, R.A. (eds), *Environmental Chemistry of Selenium*. Marcel Dekker, New York, NY, pp. 355-395.

O'Toole, D., Raisbeck, M.F., Case, J.C. and Whitson, T.D. (1996) Selenium-induced "Blind Staggers" and related myths. A commentary on the extent of historical livestock losses attributed to selenosis on western rangelands. *Veterinary Pathology* 33, 104-116.

Panter, K.E., James, L.F. and Mayland, H.F. (1995) Reproductive performance of ewes fed alfalfa pellets containing sodium selenate or *Astragalus bisulcatus* as a selenium source. *Veterinary and Human Toxicology* 37, 30-32.

Raisbeck, M.F., Dahl, E.R., Sanchez, D.A., Belden, E.L. and O'Toole, D. (1993) Naturally occurring selenosis in Wyoming. *Journal of Veterinary Diagnostic Investigation* 5, 84-87.

Raisbeck, M.F., O'Toole, D., Sanchez, D.A., Siemion, R.A. and Waggoner, J.W. (1995) Re-evaluation of selenium toxicity in grazing mammals. In: Schuman, G. and Vance, G. (eds), *Decades Later: A Time for Reassessment.* American Society for Surface Mining and Reclamation, Princeton, WV, pp. 372-383

Raisbeck, M.F., O'Toole, D., Schamber, R.A., Belden, E.L. and Robinson, L.J. (1996) Toxicologic effects of a high-selenium hay diet in captive adult and yearling pronghorn antelope (*Antilocapra americana*). *Journal of Wildlife Disease* 32, 9-16.

Raisbeck, M.F., Belden, E.L., O'Toole, D. and Waggoner, J.W. (1997) *Toxicologic Evaluation of Chronic Selenosis in Wyoming Herbivores.* University of Wyoming Abandoned Minelands Program, final report.

Rosenfeld, I. and Beath, O.A. (1946) *Pathology of Selenium Poisoning.* University of Wyoming Agricultural Experiment Station, bulletin No. 275.

Smyth, J.B.A., Wang, J.H., Barlow R.M., Humphreys, D.J., Robins, M. and Stodulski, J.B. (1990) Experimental acute selenium intoxication in lambs. *Journal of Comparative Pathology* 102, 197-209.

Spallholz, J.E. (1994) On the nature of selenium toxicity and carcinostatic activity. *Free Radical Biology and Medicine* 17, 45-64.

Tucker, J.O. (1996) personal communication.

Willhite, C.C. (1993) Selenium teratogenesis. *Annals of the New York Academy of Sciences* 678, 169-171.

Williams, E.S. (1982) Plant intoxication. In: Thorne, E.T., Kingston, N., Jolley W.R. and Bergstrom, R.C. (eds), *Diseases of Wildlife in Wyoming, Vol 2*, pp. 280-288.

Yeager, M.J. (1995) personal communication.

Chapter 78

Effects of Paxilline, Lolitrem B and Penitrem on Skeletal and Smooth Muscle Electromyographic Activity in Sheep

B.L. Smith[1] and L.M. McLeay[2]

[1]*Toxinology and Food Safety Research Group, New Zealand Pastoral Agriculture Research Institute Ltd, Ruakura Research Centre, Private Bag 3123, Hamilton 2001, New Zealand;* [2]*Department of Biological Sciences, University of Waikato, Private Bag 3105, Hamilton 2001, New Zealand*

Introduction

Several mycotoxins cause tremoring in animals (Gallagher *et al.*, 1981; Gallagher *et al.*, 1984; Cole and Dorner, 1985; Mortimer and Di Menna, 1985; Valds *et al.*, 1985). These tremorgenic mycotoxins are produced by a number of fungal genera including *Aspergillus*, *Penicillium*, *Claviceps* and *Neotyphodium*. The latter three produce respectively, penitrem, paxilline and lolitrem B, all associated with diseases of domestic animals. The modes of action and metabolic fates of some of these tremorgens have been studied (Mason, 1968; Mantle, 1983; Gallagher and Hawkes, 1986; Selala *et al.*, 1991; Smith and McLeay, 1997), but little is known about their basic modes of action. Some tremorgenic toxins that can be produced in culture have identical central indole diterpenoid moieties to lolitrem B (Gallagher *et al.*, 1984) and cause similar (although usually shorter-lasting) clinical signs in animals. Electromyography (EMG) was used to evaluate the tremorgenic activity of penitrem, paxilline and lolitrem B on different skeletal muscles and the influences of these tremorgens on the activity of smooth muscle of the gut.

Materials and Methods

A range of doses of tremorgens was given to yearling Romney sheep (20-45kg) maintained indoors in pens on a daily diet of dried chaffed meadow hay and concentrate pellets. Each sheep was prepared by suturing a set of three stainless steel electrode wires (Cooner Chatsworth, CA) into the smooth musculature of either the pyloric antrum and duodenum or the rumen and reticulum and exteriorized *via* the abdominal wall as described by McLeay *et al.* (1990). Three platinum cap electrodes (Devices UK) were sutured to the skin above the lateral neck muscles and another set was attached above either the shoulder, lumbar or hind limb musculature and contact with the skin maintained by electrode jelly. Electrodes were connected to AC preamplifiers (Grass P15), the EMGs integrated using a Devices 3520 integrator and recorded using an M19 chart recorder (Devices UK). In addition, the outputs of the integrators were also captured on disc and analyzed using a computer program, which provided total integrated volts under the trace over any time period. At least 24hrs prior to each experiment, each sheep had an in-dwelling intravenous catheter inserted into its jugular vein and anchored at its exit position.

A normal recording was collected for at least 30min, followed by a further 30min recording immediately after the administration of acetone and immediately prior to the administration of the tremorgen in the same volume of acetone *via* the indwelling jugular catheter. Effects of the substances on EMG activity of smooth and skeletal muscles were recorded for up to 7hrs after each administration.

Effects of Tremorgens on Electrical Activity of Skeletal Muscle

Recordings from neck, shoulder, lumbar and upper hind limb skeletal muscle were made. The shoulder was the most sensitive region as an indicator of mild, visually observed tremors. Acetone had no effect on skeletal EMG activity (Fig. 78.1).

Penitrem at 2.2-7.5µg/kg produced mild to moderate tremors in all five sheep, which began about 10-20min after injection, reached a maximum at 30min and persisted for up to 2.5hrs. Paxilline at 0.66-1.5mg/kg produced fine to moderate tremors in all five sheep. The onset was immediate, usually by the end of the 2min injection, and gradually diminished and disappeared over the next hour. Lolitrem B was given at 25, 45, 80 and 110µg/kg. Doses of 25µg/kg caused little or no visible tremors and were not detected by EMG recordings. Moderate tremors were produced by 45 and 80µg/kg, beginning after 15min and slowly increasing in severity, lasting several hours with the higher dose. The highest dose of 110µg/kg in one sheep caused tremors that built up over 30-45min and were severe for many hours. This animal continued to exhibit tremor on standing even after 24hrs.

Effects of Tremorgens on Electrical Activity of Smooth Muscle

The antral and duodenal EMG exhibited some variation during recording without administration of tremorgens, the most pronounced occurring with the onset of

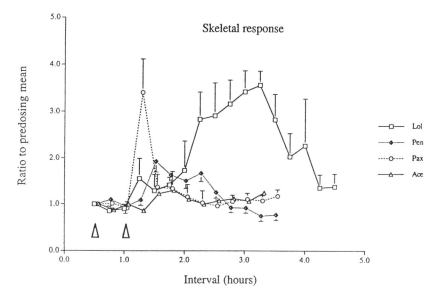

Fig. 78.1. The response in skeletal muscle EMGs to acetone and the tremorgens penitrem, paxilline and lolitrem B. The first arrow indicates administration of acetone and the second arrow administration of a second dose of acetone or tremorgen. The values (mean±SEM, n=5) are presented as changes every 15min expressed as a ratio of that activity averaged over the first 30min before acetone.

phase III of the migrating myoelectric complex (MMC), in which duodenal EMG activity increased, followed by cessation of activity for 20-30min. The antral EMG was also reduced during this period. In sheep this periodic activity normally occurs every 60-90min (Ruckebusch and Bueno, 1977). Two doses of 2ml acetone given 30min apart had no detectable effect on antral, duodenal, reticulum or rumen EMG activity compared to the usual variation. In one animal, acetone precipitated or coincided with the phase III MMC-like activity of the duodenum, but this was not repeated with a 2nd dose given 30min later in a control experiment.

Penitrem at 2.2-7.5µg/kg is clearly inhibitory on the antrum within 0-10min and for more than 3.5hrs duration. The duodenum was also inhibited over a similar time course, and occasionally stimulated to phase III MMC-like activity. Penitrem (approximately 5µg/kg) caused slowing of the regular contractions of the rumen and reticulum, starting at approximately 15min and lasting no longer than 1hr.

Paxilline in doses of 0.66-1.5mg/kg was inhibitory on the antrum in some animals with immediate onset and duration up to 1hr. A reduction in duodenal activity was observed in most animals. With 1mg/kg paxilline, cyclical contractions of the reticulum and rumen ceased immediately for about 2hrs.

Lolitrem B at doses of 110 and 80µg/kg inhibited activity of the antrum and duodenum. The onset of inhibition varied from 0-30min with a duration of up to 2hrs.

At lower doses of 25-45µg/kg, an inhibitory effect on the antrum was observed once and on the duodenum in two animals. One animal showed stimulation of the duodenum. For the rumen and reticulum, dose rates of 70µg/kg caused complete cessation of regular contractions starting approximately 30min after dosing and lasting up to 6hrs. Regular reticular contractions were replaced by marked local electromyographic activity.

Discussion

The EMG response of skeletal muscles of sheep showed that penitrem caused mild tremors at 2.2µg/kg, with moderate tremors occurring at 4.4µg/kg. Paxilline caused mild to moderate tremors at 0.7-1.5mg/kg. Lolitrem B caused moderate tremors between 45-80µg/kg and severe tremors at 110µg/kg. This confirms visually observed responses of sheep to these tremorgens and indicates a similar susceptibility and duration of response to that of mice and other animals (Sabotka *et al.*, 1978; Peterson *et al.*, 1982; Gallagher and Hawkes, 1986; Miles *et al.*, 1992). This method of recording responses may also provide an objective means of recording the effects of different blocking agents and neurotransmitter modulators or moderators on the tremorgenesis of the disease caused by these tremorgens. The *in vitro* effects of paxilline, together with penitrem B and verruculogen, have been previously described (Selala *et al.*, 1991). All three substances enhanced electrically stimulated contractions of guinea pig ileum, but did not influence contractions caused by exogenous acetylcholine, suggesting an enhanced release of acetylcholine.

This chapter also describes, for the first time *in vivo*, the inhibitory influence of these three tremorgens on the electrical activity of the smooth muscle of the pyloric antrum, duodenum, reticulum and rumen. This effect of tremorgens on smooth muscle of sheep may be significant, since any interference to gastrointestinal motility affects feed efficiency and animal thrift. Idiopathic scouring and ill thrift of ruminants are a common occurrence in livestock in New Zealand and elsewhere, often occurring at times of abundant feed, and mycotoxins have often been suggested as possible etiological agents. The three mycotoxic tremorgens investigated, in addition to their well-recognized effects on skeletal muscle, also have effects on the smooth muscle of the gut. The further investigation of these toxins is planned both in regard to smooth muscle functions of other organs and tissues and in relation to animal health.

References

Cole, R.J. and Dorner, J.W. (1985) Role of fungal tremorgens in animal disease. In: Steyn, P.S. and Vleggaar, R. (eds), *Mycotoxins and Phycotoxins*. Elsevier, Amsterdam, pp. 501-511.

Gallagher, R.T. and Hawkes, A.D. (1986) The potent tremorgenic neurotoxins lolitrem B and aflatrem - a comparison of the tremor response in mice. *Experientia* 42, 823-825.

Gallagher, R.T., Finer, J. and Clardy, J. (1980) Paspalinine, a tremorgenic metabolite from *Claviceps paspali* Stevens *et* Hall. *Tetrahedron Letters* 21, 235-238.

Gallagher, R.T., White, E.P. and Mortimer, P.H. (1981) Ryegrass staggers: isolation of potent neurotoxins lolitrem A and lolitrem B from staggers-producing pastures. *New Zealand Veterinary Journal* 29, 189-190.

Gallagher, R.T., Hawkes, A.D., Steyn, B.S. and Vleggaar, R. (1984) Tremorgenic neurotoxins from perennial ryegrass causing ryegrass staggers disorder of livestock: Structure elucidation of Lolitrem B. *Journal of the Chemical Society, Chemical Communications* 614-616.

Mantle, P.G. (1983) Amino acid neurotransmitter release from cerebrocortical synaptosomes of sheep with severe ryegrass staggers in New Zealand. *Research in Veterinary Science* 34, 373-375.

Mason, R.W. (1968) Axis cylinder degeneration associated with ryegrass staggers in sheep and cattle. *Australian Veterinary Journal* 44, 428.

McLeay, L.M., Comeskey, M.A. and Waters, M.J. (1990) Effects of epidermal growth factor on gastrointestinal electromyographic activity of conscious sheep. *Journal of Endocrinology* 124, 109-115.

Miles, C.O., Wilkins, A.L., Gallagher, R.T., Hawkes, A.D., Munday, S.C. and Towers, N.R. (1992) Synthesis and tremorgenicity of paxillines and lolitrems: Possible biosynthetic precusors of lolitrem B. *Journal of Agricultural and Food Chemistry* 40, 234-238.

Mortimer, P.H. and Di Menna, M.E. (1985) Research on the aetiology of perennial ryegrass staggers in New Zealand. In: Seawright, A.A., Hegarty, M.P., James, L.F. and Keeler, R.F. (eds), *Plant Toxicology, Proceedings of the Australia-USA Poisonous Plants Symposium* Queensland Poisonous Plants Committee, Brisbane, Qld, pp. 604-611.

Peterson, D.W., Penny, R.H.C., Day, J.B. and Mantle, P.G. (1982) A comparative study of sheep and pigs given the tremorgenic mycotoxins verruculogen and penitrem A. *Research in Veterinary Science* 33, 183-187.

Ruckebusch, Y. and Bueno, L. (1977) Origin of migrating myoelectric complex in sheep. *American Journal of Physiology* 233, E483-E487.

Sabotka,T.J., Brodie, R.E. and Spaid, S.L. (1978) Neurobehavioural studies of tremorgenic mycotoxins verrucullogen and penitrem A. *Pharmacology* 16, 287-294.

Selala, M.I., Laekeman, G.M., Loenders, B., Musuku, A., Herman, A.G. and Schepens, P. (1991) *In vitro* effects of tremorgenic mycotoxins. *Journal of Natural Products* 54, 207-212.

Smith, B.L. and McLeay, L.M. (1997) Effects of the mycotoxins penitrem, paxilline and lolitrem B on the electromyographic activity of skeletal and gastrointestinal smooth muscle of sheep. *Research in Veterinary Science* 62, 111-116.

Valds, J.J., Cameron, J.E. and Cole, R.J. (1985) Aflatrem: a tremorgenic mycotoxin with acute neurotoxic effects. *Environmental Health Perspectives* 62, 459-463.

Chapter 79

Tall Larkspur Poisoning in Cattle Grazing Mountain Rangeland

J.A. Pfister, D.R. Gardner, G.D. Manners, K.E. Panter, L.F. James, M.H. Ralphs, B.L. Stegelmeier, T.K. Schoch and J.D. Olsen
USDA-ARS, Poisonous Plant Research Laboratory, 1150 East 1400 North, Logan, Utah 84341, USA

Introduction

Tall larkspurs (*Delphinium barbeyi*, *D. occidentale*, *D. glaucescens*, *D. glaucum*) are the most serious toxic plant problem on mountain rangeland in the western US. Cattle losses average 2-5%, but may exceed 15% where tall larkspurs are abundant. Total costs to the livestock industry from all larkspurs exceed US$20m/yr. In addition to dead livestock, significant forage is wasted and management costs are increased as producers defer or avoid grazing larkspur-infested areas.

Tall Larkspur Toxins

Larkspurs contain numerous (>18) norditerpenoid alkaloids, two of which account for most of their toxicity: methyllycaconitine (MLA) and 14-deacetylnudicauline (DAN) (Manners *et al.*,1995). Deltaline is the dominant alkaloid in tall larkspurs, comprising ≥50% of the alkaloid composition, while MLA and DAN together comprise 20-50% of the alkaloid mix. The LD_{50} in mice for both MLA and DAN is near 4.5mg/kg, whereas the LD_{50} for deltaline is 200mg/kg. The effective intravenous (i.v.) dose for MLA is 2mg/kg in calves *vs.* 10mg/kg in sheep (Panter, unpublished data), while that for deltaline in calves and sheep is 10mg/kg. At these doses, clinical signs include labored breathing, rapid, irregular heartbeat and collapse.

Several methods have been developed to characterize toxic alkaloids in tall larkspur chemically. The current method of choice for rapid analysis is infrared spectroscopy (Gardner *et al.*, 1997).

The levels of MLA and DAN are highest in immature plant tissue (typically [dwt] for MLA: 5-12mg/g; DAN: 0.2-0.8mg/g), and generally decrease with maturation (MLA: 2-4mg/g; DAN: 0.1-0.4mg/g) (Pfister *et al.*, 1994a). Pods have relatively high toxicity (MLA+DAN=7-12mg/g), which declines rapidly as pods begin to shatter. Tall larkspur species vary in toxicity, ranking most to least toxic: *D. glaucum, D. barbeyi, D. glaucescens*, and *D. occidentale* (Ralphs *et al.*, 1997). Shade stress and inhibition of photosynthesis increase toxic alkaloid concentrations slightly due to reduction in other solutes, but absolute amounts remain constant (Ralphs, 1996).

The primary, life-threatening result of tall larkspur toxicosis is neuromuscular paralysis. Acetylcholine (Ach) receptors in the muscle and brain are blocked by MLA and related alkaloids (Dobelis *et al.*, 1993). Larkspur alkaloid binding to nicotinic Ach receptors is correlated to toxicity in various animal species, and may explain sheep tolerance to larkspur if larkspur toxins bind less avidly to sheep receptors (Chapter 57). Sheep are 4-5x more resistant to tall larkspur toxicosis than are cattle. Clinical signs of intoxication include muscular weakness and trembling, straddled stance, periodic collapse into sternal recumbency, respiratory difficulty and finally death while in lateral recumbency. Bloat is often a significant component of larkspur fatalities, as larkspur and other forbs are highly fermentable, while the eructation mechanism is inhibited. It is not known if larkspur provokes dry gas or frothy bloat. Key factors in larkspur intoxication are the concentration of toxic alkaloids and the amount and rate of ingestion. Toxicity, but not lethality, of MLA + DAN has been established using dried larkspur (Pfister *et al.*, 1997a). Assuming an MLA + DAN concentration of 5mg/g (dwt), a 450kg cow may show clinical signs after rapidly eating 1.8kg (dwt, 6.5kg wwt) of tall larkspur.

Cattle Grazing Patterns

Data from over 10yrs of grazing studies have shown two major findings: 1) cattle eat little or no tall larkspur before the plant has elongated flowering racemes (Pfister *et al.*, 1997a) and 2) weather patterns are very important determinants of larkspur consumption (Pfister *et al.*, 1988a; Ralphs *et al.*, 1994). Cattle often eat more larkspur during summer storms, and reduce larkspur consumption during drought. The reasons for these altered patterns of larkspur consumption are not explained by changes in plant alkaloid or sugar concentrations during these events.

Cattle generally begin grazing tall larkspur after flowering, and consumption increases as larkspur matures. Consumption usually peaks during the pod stage in late summer, when cattle may eat large quantities (25-30% of diet as herd average; >60% on some days by individual animals). On the Wasatch Plateau in central Utah, pods are especially preferred by grazing cattle (Pfister *et al.*, 1988b), whereas in western Colorado cattle select fewer seedheads (Pfister *et al.*, 1997a). Major differences in alkaloid concentrations have not been noted.

Because larkspur toxicity generally declines throughout the growing season and cattle tend to eat more larkspur after flowering, we have defined a 'toxic window' extending from the flower stage into the pod stage, or about 5wks depending on temperature and elevation. Many ranchers typically defer grazing on tall larkspur-infested ranges until the pod stage to avoid death losses, but this approach wastes much valuable forage. An additional 6-14wks of grazing may be obtained by grazing these ranges before larkspur flowering (Pfister *et al.*, 1997a). Once pods are mature and begin to shatter, larkspur ranges can usually be grazed with impunity because pod toxicity declines rapidly, and leaf toxicity is low.

Alkaloid Concentration and Palatability

Increasing consumption of larkspur by cattle as plants mature, coupled with decreasing alkaloid concentrations, suggests that alkaloid concentration may negatively influence larkspur palatability. In a series of trials using fresh and dried larkspur, there was no relationship between concentration of total or toxic alkaloids and larkspur palatability (Pfister *et al.*, 1996). Sheep consumption of larkspur, however, was negatively influenced by total and toxic alkaloid concentration. Whether or not larkspurs contain other secondary compounds that may influence palatability has not been determined. Predictions about the amount of larkspur likely to be eaten should be based primarily on plant phenology.

Early grazing studies suggested that larkspur consumption above about 25% of the diet for 1-2d leads to reduced consumption on subsequent days. Larkspur ingestion by cattle above a toxic threshold is regulated not by flavor, but by postingestive consequences (Pfister *et al.*, 1990). Cattle apparently limit ingestion of larkspur so that periods of high consumption are followed by periods of reduced consumption, allowing for detoxification (Pfister *et al.*, 1997b) and regulating larkspur consumption below a toxic threshold (14-18mg toxic alkaloid/kg), so that the cattle safely use an otherwise nutritious plant. Despite this, about 5% of range animals are killed by overingestion of larkspur each summer. Many cattle deaths occur after brief (30-90min) periods of overingestion (e.g., during or just after storms) (Ralphs *et al.*, 1994). Clinical signs of larkspur toxicity are most severe about 7-9hrs after a pulse dose of dried plant, and death may occur before negative postingestive feedback can occur. Larkspur-induced recumbency and bloat may kill cattle from posture-related pressure within 2 hrs (Pfister, unpublished.)

Preventive Measures

Sheep can be grazed before cattle in larkspur-infested pastures to reduce the risk to cattle (Ralphs *et al.*, 1991). Where larkspur grows as discreet patches, sheep can be herded into or bedded on the patches. In those areas where larkspur is uniformly spaced over a pasture, sheep must eat immature larkspur and leave sufficient feed for cattle. Early growth larkspur, however, may not be palatable to sheep.

When ranchers have found intoxicated animals, a variety of remedies have been applied, most without a solid scientific rationale. If less than a lethal dose were ingested, the animal would likely recover in spite of any treatment, unless bloat or vomition occurred during recumbency. Drugs increasing the Ach concentration at the neuromuscular junction have potential for reversing larkspur toxicity or reducing susceptibility. The cholinergic drug physostigmine (0.08mg/kg i.v.) has been used successfully under field and pen conditions to reverse larkspur intoxication (Pfister *et al.*, 1994b) for about 2hrs, and repeated injections of physostigmine are sometimes required. Despite some success under field conditions, the use of physostigmine-based treatments may actually aggravate losses in the absence of further treatment because suddenly mobile animals may later show increased muscular fatigue, dyspnea and death. Related drugs, such as neostigmine, may also have utility for treatment of larkspur toxicosis (Chapter 56).

Larkspur losses can be reduced greatly if dense larkspur populations are controlled by herbicides. Picloram, metsulfuron and glyphosate are effective in killing larkspur when applied at specific growth stages (Ralphs *et al.*, 1992). These herbicides do not reduce toxic alkaloid concentrations in treated larkspur plants, and metsulfuron may increase toxicity. Therefore, sprayed areas should not be grazed until larkspur has withered and decomposed (Ralphs *et al.*, 1998).

Some ranchers believe that cattle deaths may be reduced by feeding special mineral-salt supplements. Grazing studies showed that intraruminal dosing with mineral-salts (0, 0.5, 1.0g/kg) did not change the amount of tall larkspur eaten by cattle (Pfister and Manners, 1991; 1995), ruminal fluid passage rate or other fermentation characteristics, or consistently increase water intake. Mineral supplementation is sometimes used to counteract legume-induced pasture bloat, but whether larkspur-induced bloat would respond to such treatment is unknown. Cattle can also be trained to avoid eating tall larkspur by food aversion learning (Chapter 48).

Future research will focus on identifying and quantifying alkaloids in low and plains larkspurs, developing new analytical methods for larkspur alkaloids, including immunologic field tests to monitor toxin concentration, development of vaccines and toxin-binding antidotes. Further work is needed to understand the etiology of larkspur-related bloat, and examine possible biocontrol with insects.

References

Dobelis, P., Madl, J.E., Manners, G.D., Pfister, J.A. and Walrond, J.P. (1993) Antagonism of nicotinic receptors by *Delphinium* alkaloids. *Neuroscience Abstracts* 631, 12.

Gardner, D.R., Manners, G.D., Ralphs, M.H. and Pfister, J.A. (1997) Quantitative analysis of norditerpenoid alkaloids in larkspur (*Delphinium* spp.) by Fourier transform infrared spectroscopy. *Phytochemical Analysis* 8, 55-62.

Manners, G.D., Panter, K.E. and Pelletier, S.W. (1995) Structure-activity relationships of norditerpenoid alkaloids occurring in toxic larkspur (*Delphinium*) species. *Journal of Natural Products* 58, 863-869.

Pfister, J.A. and Manners, G.D. (1991) Mineral supplementation of cattle grazing larkspur-infested rangeland during drought. *Journal of Range Management* 44, 105-111.

Pfister, J.A. and Manners, G.D. (1995) Effect of carbachol administration in cattle grazing tall larkspur-infested rangeland. *Journal of Range Management* 48, 343-349.

Pfister, J.A., Manners, G.D., Ralphs, M.H., Hong, Z.X. and Lane, M.A. (1988a) Effects of phenology, site and rumen fill on tall larkspur consumption by cattle. *Journal of Range Management* 41, 509-514.

Pfister, J.A., Ralphs, M.H. and Manners, G.D. (1988b) Cattle grazing tall larkspur on Utah mountain rangelands. *Journal of Range Management* 41, 118-122.

Pfister, J.A., Provenza, F.D. and Manners, G.D. (1990) Ingestion of tall larkspur by cattle: Separating effects of flavor from postingestive consequences. *Journal of Chemical Ecology* 16, 1696-1705.

Pfister, J.A., Manners, G.D., Gardner, D.R. and Ralphs, M.H. (1994a) Toxic alkaloid levels in tall larkspur (*Delphinium barbeyi*) in western Colorado. *Journal of Range Management* 47, 355-358.

Pfister, J.A., Panter, K.E., Manners, G.D. and Cheney, C.D. (1994b) Reversal of tall larkspur (*Delphinium barbeyi*) toxicity with physostigmine. *Veterinary and Human Toxicology* 36, 511-514.

Pfister, J.A., Manners, G.D., Gardner, D.R., Price, K.W. and Ralphs, M.H. (1996) Influence of alkaloid concentration on acceptability of tall larkspur (*Delphinium* spp.) to cattle and sheep. *Journal of Chemical Ecology* 22, 1147-1168.

Pfister, J.A., Ralphs, M.H., Manners, G.D., Gardner, D.R., Price, K.W. and James, L.F. (1997a) Early season grazing by cattle of tall larkspur- (*Delphinium* spp.) infested rangeland. *Journal of Range Management* 50, 391-398.

Pfister, J.A., Provenza, F.D., Manners, G.D., Gardner, D.R. and Ralphs, M.H. (1997b) Tall larkspur ingestion: Can cattle regulate intake below toxic levels? *Journal of Chemical Ecology* 23, 759-777.

Ralphs, M.H. (1996) personal communication, USDA-ARS, Poisonous Plant Research Laboratory, Logan, UT.

Ralphs, M.H., Bowns, J.E. and Manners, G.D. (1991) Utilization of larkspur by sheep. *Journal of Range Management* 44, 619-622.

Ralphs, M.H., Evans, J.O. and Dewey, S.A. (1992) Timing of herbicide applications for control of larkspurs. *Weed Science* 40, 264-269.

Ralphs, M.H., Jensen, D.T., Pfister, J.A., Nielsen, D.B. and James, L.F. (1994) Storms influence cattle to graze larkspur: an observation. *Journal of Range Management* 47, 275-278.

Ralphs, M.H., Manners, G.D., Pfister, J.A., Gardner, D.R. and James, L.F. (1997) Toxic alkaloid concentration in tall larkspur species in the western United States. *Journal of Range Management* 50, 497-502.

Ralphs, M.H., Manners, G.D. and Gardner, D.R. (1998) Toxic alkaloid response to herbicides used to control tall larkspur. *Weed Science* (in review).

Chapter 80

Krimpsiekte, a Paretic Condition of Small Stock Poisoned by Bufadienolide-containing Plants of the Crassulaceae in South Africa

C.J. Botha[1], J.J. Van der Lugt[2], G.L. Erasmus[3], T.S. Kellerman[1, 3], R.A. Schultz[3] and R. Vleggaar[4]

[1]*Department of Pharmacology and Toxicology, Faculty of Veterinary Science, University of Pretoria, Onderstepoort, South Africa;* [2] *Department of Pathology, Faculty of Veterinary Science, University of Pretoria, Onderstepoort, South Africa;* [3]*Division of Toxicology, Onderstepoort Veterinary Institute, Onderstepoort, South Africa;* [4]*Department of Chemistry, Faculty of Science, University of Pretoria, Pretoria, South Africa*

Introduction

Poisoning of livestock by cardiac glycoside containing plants has the greatest economic impact of all plant poisonings in the Republic of South Africa (Kellerman *et al.*, 1996). Two major groups of cardiac glycosides, the cardenolides and bufadienolides, are recognized. Poisoning by bufadienolide-containing plants, which is more important than cardenolide poisonings, may be acute or chronic. Tulp poisoning (induced by various *Homeria* and *Moraea* species) and slangkop poisoning (caused by various *Urginea* species) induce only acute toxicity, as their bufadienolides are non-cumulative (Kellerman *et al.*, 1988; Kellerman *et al.*, 1996).

Three different genera of the Crassulaceae (*Cotyledon*, *Tylecodon* and *Kalanchoe*), generally referred to as plakkies, can cause either acute or chronic poisoning. Krimpsiekte, the chronic form, predominantly occurs in small stock. This toxicosis is believed to be caused by the ingestion of cumulative cardioactive bufadienolides with unique neurotoxic properties (Kellerman *et al.*, 1988; Naudé *et al.*, 1992). In general, the cardiac, respiratory and intestinal symptoms typical in acute poisoning diminish, and the neuromuscular signs increase. Affected stock tire easily, lag behind the flock, assume a characteristic posture, with feet together and back arched ("krimpsiekte"), display torticollis, become recumbent and suffer protracted paralysis. Paralyzed sheep are fully conscious in lateral recumbency, sometimes for weeks, until they die or are destroyed (Kellerman *et al.*, 1988).

Vahrmeijer (1981) stated that krimpsiekte or 'nenta' has been a serious problem since 1775, but only in 1891 did Soga reproduce the condition, feeding *Tylecodon ventricosus* to goats, the first plant poisonings to be reproduced experimentally in South Africa. Krimpsiekte is the only plant poisoning known to cause secondary poisoning (Kellerman *et al.*, 1988). This is difficult to explain in the light of our current knowledge of the toxicity of bufadienolides, as huge quantities of residue in the meat must be ingested at a single sitting to result in relay poisoning.

In recent dosing trials, *T. ventricosus* administered to a sheep on three consecutive days (at an incremental dose rate of 2.5, 5 and 10g/kg) induced signs of asphyxiation (apparently due to paralysis of intercostal muscles) and tachycardia (otherwise normal ECG). The animal died after 4d, following a respiratory crisis. Another sheep displayed dyspnea after receiving 10g/kg of the plant material. No ECG abnormalities, even at this relatively high dose, were noticed. Removal of the rumen contents and dosing with activated charcoal, normally very effective in the treatment of cardiac glycoside poisoning, gave disappointing results. Typical krimpsiekte finally developed from days 4-23. Since there is no indication of a cumulative effect in this instance, it was suggested that an unidentified neurotoxin may be the causative agent. All the aforementioned uncertainties and questions prompted the re-evaluation of the role of cumulative bufadienolides in the etiology of krimpsiekte.

Isolation of the Toxic Principle(s)

Tylecodon wallichii is one of the most important causes of krimpsiekte. Semi-dried plant material was milled and extracted twice with ethyl acetate (EtOAc). The plant residue was discarded, and the extract evaporated and defatted with hexane. Using flash chromatography (silica gel, particle size 0.040-0.063mm, Merck) with different mobile phases, the eluents were collected. Fractions were analyzed by TLC on silica gel $60F_{254}$ plates (Merck), and those containing compounds with similar R_f values were combined. These different fractions were dosed orally to male albino guinea-pigs (233-488g) to ascertain toxicity. Only fractions that induced a neuromuscular syndrome (weakness, tremors, neck paresis) in guinea-pigs were isolated. Final purification of the toxins was achieved by reversed phase column chromatography (LiChroprep RP-8 (40-63μm), Merck) using 80% aqueous methanol (MeOH), and crystallization from chloroform $(CHCl_3)$/MeOH. A UV absorption spectrum was recorded, and the toxin was further analyzed by NMR spectroscopy. The ^1H and ^{13}C NMR spectra were measured for solutions in $CDCl_3$ on a Bruker AC-300 (7.0 T) spectrometer operating at 300MHz for ^1H and 75MHz for ^{13}C. One toxic principle was isolated from *T. wallichii*. Its maximum UV absorption occurred at 298nm, indicating a gamma-lactone ring. All NMR signals matched those of an authentic sample of cotyledoside (Van Wyk, 1975; Steyn *et al.*, 1984).

Since a novel neurotoxic substance was not isolated from *T. wallichii*, it was considered essential to determine whether the active principles of *T. ventricosus* were also cardiac glycosides. Despite the fact that *T. ventricosus* was the first plant incriminated by Soga (1891), the toxic principles were never isolated. Fresh plant material was obtained and, following the extraction and chromatographic techniques described above, one toxic compound was purified from *T. ventricosus*. Another fraction was also toxic, but could not be purified. For the former compound, UV absorption maximum occurred at 284nm, indicating a gamma-lactone ring. All NMR signals matched those of an authentic sample of tyledoside D, which was previously isolated from *T. grandiflorus* (Anderson *et al.*, 1983b).

Cotyledoside-induced Toxicosis

Clinical signs of krimpsiekte may be confused with general weakness due to heart failure, and detailed description of the pathology of this condition was undertaken in sheep experimentally. Two yearling South African Mutton Merino sheep were housed individually and fed chopped hay and a pelleted maize concentrate. During adaptation, clinical examinations and ECG recordings were made to establish baseline values. Cotyledoside (0.01-0.015mg/kg) was administered i.v. on consecutive days, except during weekends. ECG recordings were done immediately prior to and 5, 10 and 15min after the infusion of cotyledoside. Once clinical signs were elicited, cotyledoside administration ceased and ECG recordings were carried out with greater frequency.

Both sheep developed typical krimpsiekte on Day 9 of the experiment, which lasted until they were euthanized. In both animals, rumen movements were weak from the first signs of krimpsiekte and lasted for 3d, although transient rumen stasis developed in one sheep on Day 10. The appetite was depressed when clinical signs were most severe, but the feces were always well formed and defecation occurred regularly. Throughout the experiment the habitus of the sheep remained essentially normal. No significant ECG abnormalities were detected during the trial. Tachycardia, slight increases in the amplitudes of the QRS- and T-waves and a depression of the ST-segment were infrequently recorded.

At necropsy no significant gross lesions were observed. In both sheep, histology of the brain revealed mild edema with vacuolation of the neuropil and astrocyte swelling in the cerebral and cerebellar grey and white matter. Vacuolation was particularly pronounced in the white matter of the ventrolateral nucleus and optic radiation in one sheep and in the dorsolateral and ventrorostral nuclei in the other. In these thalamic areas the vacuolation was bilateral. Vacuoles were spherical to ovoid or elliptical and up to 300µm in diameter. In one sheep the vacuoles were generally larger, but less numerous than in the other. They were empty, sometimes multilocular and coalescing and partitioned by thin myelin septa. Scattered myocardial fibers showed increased eosinophilia associated with multifocal loss of cross-striations, and the interstitium contained a few lymphocytic infiltrates. No significant lesions were detected in other organs.

Discussion

The neurotoxic principles of *T. wallichii* (cotyledoside) and *T. ventricosus* (tyledoside D) were specifically isolated by purifying only chemical fractions that induced nervous signs in guinea-pigs. No corroborating evidence could be found for the contention that a neurotoxin other than a cardiac glycoside might be involved in the etiology of krimpsiekte. It was difficult to explain how a cardiac glycoside such as cotyledoside could cause secondary intoxication and how high doses of *T. ventricosus* could exert a purely neurological effect in animals.

No clear-cut relationship between the structure and cumulative effects of these bufadienolides has yet been discerned. However, a strongly bonded levorotatory sugar in the 3-position of the A-ring is common to all cumulative bufadienolides so far identified. An unusual epoxy group over the 7,8-position in the B-ring of the aglycone is common to some of the cumulative bufadienolides. Neurospecificity of some of the cumulative bufadienolides is not related to differences in Na^+K^+-pump inhibition (Van der Walt *et al.*, 1997)

The previous findings that krimpsiekte can be induced by consecutive i.v. injection of small doses of cotyledoside to sheep was confirmed. The most prominent clinical findings recorded in this study were an inability to stand, tachypnea and tachycardia, mild to severe tremors, especially of the hindquarters, and difficulty in lying down (Naudé and Schultz, 1982). Krimpsiekte has also been induced with other cumulative bufadienolides, tyledosides D and A, isolated from *T. grandiflorus* (Anderson *et al.*, 1983b), and two isolated from *Kalanchoe lanceolata* (Anderson *et al.*, 1983a).

No significant ECG abnormalities were noticed in the present investigation. Anderson *et al.* (1983b) reported no unusual ECG changes following i.v. administration of tyledosides D and A. Anderson *et al.* (1983a) described ECG changes only with the K28A bufadienolide isolated from *K. lanceolata* (subsequently characterized as lanceotoxin A by Anderson *et al.*, 1984), but not with K28B (lanceotoxin B). These findings indicate that the electrical activity of the heart is usually not affected in krimpsiekte (Kellerman *et al.*, 1988). This apparent lack of cardiac involvement may be reflected in the relatively minor and non-specific myocardial changes seen in the two sheep used in this trial. Mild, multifocal infiltrations of predominantly lymphocytes in the myocardium, not associated with necrosis, as well as mild, multifocal areas of hyaline degeneration and necrosis scattered throughout the myocardium, have previously been reported in krimpsiekte (Anderson *et al.*, 1983a; Anderson *et al.*, 1983b).

Nervous lesions in the two sheep comprised mild edema of the cerebrum and cerebellum with particularly pronounced vacuolation, interpreted as edema, in some thalamic nuclei. A moderate brain edema in the periventricular white matter, brainstem and cerebellum was also noted in a sheep that received small, repeated doses of K28B (lanceotoxin B) i.v. (Anderson *et al.*, 1983a). The thalamus participates in regulation of motor activity, arising from the cerebral cortex, by relaying cerebellar feedback to the cerebral cortex in order to coordinate the intended motor action. Lesions in the ventrolateral thalamic nucleus may lead to dysmetria and asynergy (King, 1987). Both sheep with krimpsiekte made repeated attempts to lie down and their lack of coordination could be ascribed to asynergy. The basal nuclei, important motor centers of the extrapyramidal system, collaborate directly through the thalamus with the motor areas of the cerebral cortex, and lesions in the basal nuclei could lead to locomotory and postural abnormalities (King, 1987). Grinding of the teeth as well as the repeated, strenuous attempts to lie down observed in both sheep could be interpreted as an indication of pain. de Lahunta (1983) reported that "animals with thalamic lesions occasionally act as if they were experiencing pain."

The sheep, however, also showed signs of neuromuscular dysfunction (tremors, fatigue, frequently lying down, stiff gait) and a general proprioceptive deficit (crossing of the hind limbs). A more diffuse neurologic disorder, possibly a lower motor neuron/muscle dysfunction, may thus form the basis of krimpsiekte. Further studies are needed to locate the basic lesion(s) and to determine the precise pathogenesis of krimpsiekte.

Acknowledgments

We would like to extend our gratitude to Drs Alex de Kock and Giel van Aardt who arranged the collection and sending of the plant material. We would also like to thank Dr Alexander de Lahunta for his valuable comments and Mrs Leonie Labuschagne for her assistance.

References

Anderson, L.A.P., Schultz, R.A., Joubert, J.P.J., Prozesky, L., Kellerman, T.S., Erasmus, G.L. and Procos, J. (1983a) Krimpsiekte and acute cardiac glycoside poisoning in sheep caused by bufadienolides from the plant *Kalanchoe lanceolata* Forsk. *Onderstepoort Journal of Veterinary Research* 50, 295-300.

Anderson, L.A.P., Joubert, J.P.A., Prozesky, L., Kellerman, T.S., Schultz, R.A., Procos, J. and Olivier, P.M. (1983b) The experimental production of krimpsiekte in sheep with *Tylecodon grandiflorus* (Burm. F.) Toelken and some of its bufadienolides. *Onderstepoort Journal of Veterinary Research* 50, 301-307.

Anderson, L.A.P., Steyn, P.S. and Van Heerden, F.R. (1984) The characterization of two novel bufadienolides, lanceotoxins A and B from *Kalanchoe lanceolata* [Forssk.] Pers. *Journal of the Chemical Society, Perkin Transactions I* 1573-1575.

de Lahunta, A. (1983) Diencephalon. In: *Veterinary Neuroanatomy and Clinical Neurology.* WB Saunders, Philadelphia, PA, pp. 433-350.

Kellerman, T.S., Coetzer, J.A.W. and Naudé, T.W. (1988) Heart. In: *Plant Poisonings and Mycotoxicoses of Livestock in Southern Africa.* Oxford University Press, Cape Town, pp. 83-130.

Kellerman, T.S., Naudé, T.W. and Fourie, N. (1996) The distribution, diagnosis and estimated economic impact of plant poisonings and mycotoxicoses in South Africa. *Onderstepoort Journal of Veterinary Research* 63, 65-90.

King, A.S. (1987) Extrapyrimidal feedback: extrapyrimidal disease and upper motor neuron disorders. In: *Physiological and Clinical Anatomy of the Domestic Mammals, Vol 1: Central Nervous System.* Oxford University Press, Oxford, pp. 158-165.

Naudé, T.W. and Schultz, R.A. (1982) Studies on South African cardiac glycosides II. Observations on the clinical and haemodynamic effects of cotyledoside. *Onderstepoort Journal of Veterinary Research* 49, 247-254.

Naudé, T.W., Anderson, L.A.P., Schultz, R.A. and Kellerman, T.S. (1992) "Krimpsiekte": A chronic paralytic syndrome of small stock caused by cumulative bufadienolide cardiac glycosides. In: James, L.F., Keeler, R.F., Bailey, E.M. Jr, Cheek, P.R. and Hegarty, M.P. *Poisonous Plants: Proceedings of the Third International Symposium.* Iowa State University Press, Ames, pp. 392-396.

Soga, J.F. (1891) Disease 'Nenta' in goats. *Agricultural Journal of the Cape of Good Hope* 3, 40-142.

Steyn, P.S., Van Heerden, F.R. and Van Wyk, A.J. (1984) The structure of cotyledoside, a novel toxic bufadienolide glycoside from *Tylecodon wallichii* (Harv.) Toelken. *Journal of the Chemical Society, Perkin Transactions I* 965-967.

Vahrmeijer, J. (1981) Types of stock poisoning and the plants responsible. In: *Poisonous Plants of Southern Africa that cause Livestock Losses.* Tafelberg Publishers Ltd, Cape Town, pp. 15-24.

Van der Walt, J.J., Van Rooyen, J.M., Kellerman, T.S., Carmeliet, E.E. and Verdonck F. (1998) Neurospecificity of some phytobufadienolides is not related to differences in Na^+/K^+-pump inhibition. *European Journal of Pharmacology* 329, 201-211.

Van Wyk, A.J. (1975) The chemistry of *Cotyledon wallichii* Harv. Part III. The partial constitution of cotyledoside, a novel bufadienolide. *Journal of the South African Chemical Institute* 28, 281-283.

Chapter 81

A Toxicity Assessment of *Phalaris coerulescens*: Isolation of a New Oxindole

N. Anderton[1], P.A. Cockrum[1], S.M. Colegate[1], J.A. Edgar[1], K. Flower[2] and R.I. Willing[3]

[1]*Plant Toxins Unit, CSIRO Division of Animal Health, Australian Animal Health Laboratory, Private Bag 24, Geelong, Victoria 3220, Australia;* [2]*Kybybolite Research Centre, PO Box 2, Kybybolite, South Australia 5262, Australia;* [3]*NMR Facility, CSIRO Division of Chemicals and Polymers, Private Bag 10, Rosebank MDC, Clayton, Victoria 3169, Australia*

Introduction

Phalaris grasses are a useful pasture component that can contain gramine-, tyramine-, tryptamine- and/or β-carboline alkaloids. They have been implicated in toxic syndromes following ingestion of the grass by livestock (Anderton *et al.*, 1994). *Phalaris coerulescens*, a lesser-known species of *Phalaris*, is a winter-growing perennial with rapid germination, vigorous seedling growth and high production in its seedling year. Trials in Australia in the late 1950s and the 1960s produced conflicting reports on persistence and productivity (Rhodes, 1967). The introduced strains were not commercialized but have since become volunteer species in higher rainfall areas, particularly in the state of Victoria.

A recent collection of grasses on the Iberian Peninsula (SE Portugal and SW Spain) has resulted in the reintroduction of *P. coerulescens* to Australia in an attempt to provide a more persistent, productive grass for low water-holding, acid soils than perennial veldt grass (*Ehrharta calycina*). Pastures on these soils in low to moderate rainfall areas (475-550mm/yr) are generally poor and readily become infested with the problem annual, silver grass (*Vulpia* spp.).

In 1994, trial *P. coerulescens* accessions were established in southeastern South Australia. Under trial conditions, they have persisted and performed better than the control *P. aquatica* cv. Sirosa and are the most productive of the grass accessions evaluated. As a consequence of the potential toxicity of *Phalaris* spp. and a suspected association of *P. coerulescens* with horse fatalities on a farming property in Victoria in the late 1980s, it was considered crucial to investigate the accessions for alkaloid content before further agronomic development is pursued.

Collection and Extraction of Plant Samples

The trial site was sand (1.5m) over clay and had received 45mm of rain in the 5d prior to sampling. The weather conditions at the time of sampling were cold, overcast, raining and windy. Since the *Phalaris* plants in the paddock had been grazed closely by sheep, leaf samples (1-5cm long, about 20g wet weight) for analysis were cut from the ungrazed side of the plants and immersed immediately in 100ml of 0.1N HCl.

Three days after collection, the plant/aqueous acid samples (which were stored at ambient temperature) were filtered to yield orange/pink solutions. Strong cation exchange solid-phase extraction (SPE) cartridges were conditioned with 2M HCl (3ml) followed by distilled water (10ml). The filtered extract was applied to the SPE column, and the column washed with water and eluted with 6M NH_4OH. The eluates were immediately concentrated to dryness using a rotary evaporator. Residues were dissolved in methanol (MeOH) (1ml) and stored at -5°C.

Chromatographic Analysis

Samples were analyzed by thin layer chromatography (TLC), high performance liquid chromatography (HPLC) and by gas chromatography/mass spectrometry (GC/MS). The standards used for comparison were tyramine, N-methyltyramine, N,N-dimethyltyramine (hordenine), indolomethyl-N,N-dimethylamine (gramine), N,N-dimethyltryptamine (DMT), 5-hydroxy-N,N-dimethyltryptamine (bufotenine), 5-methoxy-N,N-dimethyltryptamine (5MeODMT) and 2-N-methyl-1,2,3,4-tetra-hydro-β-carboline (THBC, 1, Fig. 81.1).

An aliquot of each extract (about 5µl) was applied to fluorescent silica gel TLC plates, which were developed using chloroform ($CHCl_3$)/ammoniated MeOH (80/20). The plates were visualized under UV light (254nm, dark spots), or by lightly spraying with Dragendorff's reagent or acidified anisaldehyde reagent. The TLC results demonstrated the absence of the tyramine and tryptamine alkaloids normally associated with toxic varieties of *Phalaris*. There were two major components, alkaloid A, which turned a deep purple color with anisaldehyde spray, and alkaloid B, which was visualized with Dragendorff's reagent. There were a number of minor components that turned pink/purple with anisaldehyde spray but did not match any of the available standards. Conceivably, these compounds are similar in structure to the indole alkaloids and may contribute to some toxicity.

Reversed phase HPLC was performed on the diluted extract. Separation was achieved over 95min with a water/MeOH/0.1% trifluoroacetic acid gradient. The

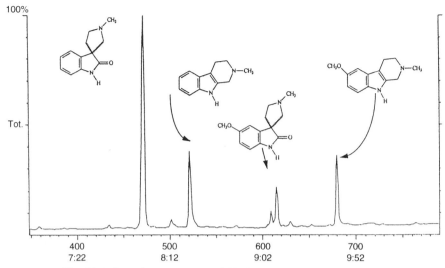

Fig. 81.1. Gas chromatogram of *Phalaris coerulescens* extract.

complete UV spectrum of each eluted peak was obtained. The samples were qualitatively similar to each other at 214nm, with a major component (t_R=22min, λ_{max} 250nm) corresponding to alkaloid B, and six to ten minor components. Tyramine, N-methyltyramine, hordenine, bufotenine, gramine and DMT were not detected. Small quantities of 5MeODMT (t_R=48min) and, in larger amounts, THBC (1) (t_R=56min, λ_{max} 260-350nm) were found. Alkaloid A eluted at 52min (λ_{max} 218, 271nm).

The reconstituted extract was diluted 1:20 with MeOH and 5µl was injected onto a 15m x 0.22mm SE30 GC column monitored using a Finnigan GCQ mass spectrometer in both electron impact (EI) and chemical ionization (CI) modes. Standards were treated as above. The *P. coerulescens* extracts were qualitatively similar with two major (7.59min and 8.22min) and two main minor peaks (9.09min and 9.41min) with several very minor peaks observed (Fig. 81.1). The major peak at 8.22min co-eluted with and had the same mass spectrum as THBC (1). The other peaks did not co-elute with any of the standards. Mass spectral analysis indicated that the peak at 9.41min could be 5OMe-THBC (2) (Gander *et al.*, 1976). Some of these samples showed trace amounts of DMT. It was subsequently shown that the major GC/MS component at 7.59min was equivalent to the HPLC major component with t_R=22min and to alkaloid B in the TLC experiment.

Structural Identification of Alkaloid B

Chemical ionization MS indicated that alkaloid B had a molecular weight of 202, while MS/MS yielded valuable information on the fragmentation of the molecule. High resolution MS of the molecular ion suggested a formula of $C_{12}H_{14}N_2O$, and high resolution MS of several daughter ions confirmed the association of the oxygen atom with the indole portion of the molecule. Analysis of the MS data revealed a remarkable similarity between alkaloid B and that expected of a tetrahydro-β-carboline (e.g. the 5-hydroxy derivative of THBC). However, when alkaloid B was methylated (methyl iodide/acetone/potassium carbonate), GC/MS results for the derivative were different than expected for the β-carboline (2).

Extensive nuclear magnetic resonance (NMR) spectroscopic investigation of the isolated molecule (^{1}H, ^{13}C, ^{1}H-^{1}H correlation spectroscopy (COSY), ^{1}H-^{13}C distortionless enhancement polarization transfer (DEPT), ^{1}H-^{13}C heteronuclear multiple quantum coherence (HMQC) and ^{1}H-^{13}C heteronuclear multiple bond correlation (HMBC) experiments) (Byrne, 1993) indicated that alkaloid B was the oxindole (3). The oxindole structure was differentiated from the isomeric pseudo-oxindole by the chemical shift of the spiro carbon and an observed 3-bond correlation between the spiro carbon and H-4 on the aromatic ring.

A related oxindole (4), along with the β-carboline (2) and 5MeODMT, had been previously isolated from *Horsfieldia superba* and named horsfiline (Jossang *et al.*, 1991). Mass spectral data indicate that horsfiline may be the minor alkaloid observed at 9.09min in the GC/MS work on the *P. coerulescens* accessions.

Comparison with Equine Death Associated *Phalaris* Extracts

In 1988-89, several incidences of equine nervous disorder and death occurred on a single property in central Victoria. In the last case, a horse was put into the same paddock in which four horses had died the previous year. After about 22d, the horse died 10min after collapsing following spontaneous wild galloping. There was no obvious gross or microscopic pathology, and tests for strychnine, lead, arsenic, barbiturate, cyanide and organophosphates were negative. The pasture was largely *P. coerulescens* (Oram, 1989). Similar, but infrequent, equine deaths have been observed over the past 15yrs in New South Wales, and have been attributed to *Phalaris* spp. which cause the cardiac form of the *Phalaris*-related sudden death syndrome (Bourke, 1996).

Extracts of the *Phalaris* spp. obtained from the property in central Victoria were examined at the time by TLC and HPLC, and the amounts of tryptamines or tyramines observed were not sufficient to account for the sudden deaths. The extracts had been stored at -10°C until the current examination by GC/MS. Two major components were observed with retention times and mass spectra identical to those of the two minor peaks observed in the *P. coerulescens* accessions. These were assigned as the oxindole (4) and its corresponding tetrahydro-β-carboline (2). Small amounts of the oxindole (3) and THBC (1) were also detected.

Conclusions and Further Work

The *P. coerulescens* samples do not contain the tryptamine and tyramine alkaloids that have been previously associated with the syndromes of *P. aquatica* cultivar intoxication. However, the samples appear to contain appreciable quantities of oxindoles and related tetrahydro-β-carboline alkaloids. The ovine oral doses of β-carbolines required to initiate neurological signs (280mg/kg) are greater than those for tryptamines (40mg/kg) (Bourke, 1994) but less than the oral dose of N-methyltyramine required to produce pronounced cardiac effects (300-400mg/kg) (Anderton *et al.*, 1994). Bourke has reported the association of a locomotor disturbance in sheep with the prevalence of *Tribulus terrestris* during drought in New South Wales (Bourke, 1984). It was subsequently demonstrated that large subcutaneous doses (54mg/kg) of harmane and norharmane, the major β-carbolines isolated from *T. terrestris*, also induced locomotor disturbance in sheep. Based upon the estimated total alkaloid content of *T. terrestris* (44mg/kg dwt) it was suggested that many months of exposure would be required in order for a toxic dose of β-carbolines to accumulate. This is consistent with field observations of *T. terrestris* intoxication (Bourke *et al.*, 1992). Bourke (1994) has also suggested that the delayed onset form of the neurological syndrome of *Phalaris* intoxication (up to 5mos after being removed from the *Phalaris* pasture) may be due to β-carbolines. These observations of long-term, low-level exposure of sheep to the alkaloids in the *P. coerulescens* pasture may have human health implications if the alkaloids accumulate and can be translocated to humans ingesting animal products.

The tetrahydro-β-carboline and oxindole content of the *P. coerulescens* samples analyzed were approximately 260mg/kg (dwt). It has been estimated that the daily intake of *Phalaris* by sheep is around 1,047g (dwt) (Hogan *et al.*, 1969). A 40kg sheep feeding on *P. coerulescens* could thus be subjected to about 65mg/kg/d of tetrahydro-β-carbolines and oxindoles.

Because of the presence of tetrahydro-β-carbolines, the unknown toxicity of oxindoles and the presence of unidentified tryptamine-like alkaloids, it seems that the *P. coerulescens* accessions analyzed may have the potential to be toxic to grazing animals. Before further agronomic work is pursued, the initial quantitation of β-carboline content and comparison with other *Phalaris* spp. should be confirmed, and the potential for toxicity of the oxindoles, especially the potential for producing cardiac effects, should be investigated. In addition, the major unknown tryptamine-like alkaloid should be isolated, identified and quantitated and the potential contribution to overall toxicity from the minor, unidentified components should be assessed. Finally, cross-seasonal screening for alkaloid content should be used in order to select appropriate accessions for agronomic development.

Acknowledgments

The high resolution mass measurements were made by I. Vit at the CSIRO Division of Chemicals and Polymers. SM Colegate gratefully acknowledges useful discussions with Dr C.A. Bourke about the neurotoxicology of indole-related alkaloids.

References

Anderton, N., Cockrum, P.A., Walker, D.W. and Edgar, J.A. (1994) Identification of a toxin suspected of causing sudden death in livestock grazing *Phalaris* pastures. In: Colegate, S.M. and Dorling, P.R. (eds), *Plant-Associated Toxins: Agricultural, Phytochemical and Ecological Aspects*. CAB International, Wallingford, Oxon, pp. 269-274.

Bourke, C.A. (1984) Staggers in sheep associated with the ingestion of *Tribulus terrestris*. *Australian Veterinary Journal* 61, 360-363.

Bourke, C.A. (1994) The clinico-toxicological differentiation of *Phalaris* spp. toxicity syndromes in ruminants. In: Colegate, S.M. and Dorling, P.R. (eds), *Plant-Associated Toxins: Agricultural, Phytochemical and Ecological Aspects*. CAB International, Wallingford, Oxon, pp. 523-528.

Bourke, C.A. (1996) personal communication.

Bourke, C.A., Stevens, G.R. and Carrigan, M.J. (1992) Locomotor effects in sheep of alkaloids identified in Australian *Tribulus terrestris*. *Australian Veterinary Journal* 69, 163-165.

Byrne, L. (1993) Nuclear magnetic resonance spectroscopy strategies for structural determination. In: Colegate, S.M. and Molyneux, R.J. (eds), *Bioactive Natural Products: Detection, Isolation and Structural Determination*. CRC Press, Boca Raton, FL, pp. 75-104.

Gander, J.E., Marum, P., Marten, G.C. and Hovin, A.W. (1976) The occurrence of 2-methyl-1,2,3,4-tetrahydro-β-carboline and variation in alkaloids in *Phalaris arundinacea*. *Phytochemistry* 15, 737-738.

Hogan, J.P., Weston, R.H. and Lindsay, J.R. (1969) The digestion of *Phalaris tuberosa* at different stages of maturity. *Australian Journal of Agricultural Research* 20, 925-940.

Jossang, A., Jossang, P., Hadi, H.A., Sévenet, T. and Bodo, B. (1991) Horsfiline, an oxindole alkaloid from *Horsfieldia superba*. *Journal of Organic Chemistry* 56, 6527-6530.

Oram, R. (1989) personal communication.

Rhodes, I. (1967) Growth and development of *Phalaris coerulescens* Desf. *CSIRO Field Station Records Division Plant Industry* 6, 69-76.

Chapter 82

Two New Alkaloids of *Conium maculatum*, and Evidence for a Tautomeric Form for "γ"-Coniceine

D.G. Lang and R.A. Smith

Livestock Disease Diagnostic Center, Veterinary Science Department, University of Kentucky, 1429 Newtown Pike, Lexington, Kentucky 40511, USA

Introduction

Poison hemlock, *Conium maculatum*, known for millennia to be toxic, has many publications on its toxic and teratogenic effects in recent years. The chemistry and biosynthesis of the conium alkaloids were recently reviewed (Panter and Keeler, 1989). The plant grows in patches along streams, fencelines and disturbed areas. It is considered highly dangerous to livestock, and piperidine alkaloid patterns have been reported to fluctuate hourly in the living plant (Kingsbury, 1964). Sometimes the Livestock Disease Diagnostic Center (LDDC) receives submissions of rumen contents of dead cattle, and a specimen of the plant suspected of having caused the death to determine whether alkaloids in the plant are also present in the rumen contents. In suspected poison hemlock intoxication, alkaloids in the rumen have generally been those of *Conium maculatum*, but they differ from those in submitted plant material. This is probably because the plants gathered in the field have often been sampled several hours or even days after ingestion by the cattle. This investigation was prompted by the desire to explore short-term variation in the identity and quantity of the alkaloids present in these plants. Specimens of wild *C. maculatum* growing on the University of Kentucky Coldstream Research Park in Lexington, Kentucky, were collected and examined for alkaloids. Results of rapidly changing identity of the alkaloids were in agreement with prior reports. However, during the investigation, two new compounds, 2-n-pentyl-3,4,5,6-tetra-hydropyridine and 5-hydroxy-2-n-pentyl-piperidine were tentatively identified. Data accumulated during the extraction and derivatization imply that the "γ" form of coniceine is a tautomer rather than a stable molecule.

Materials and Methods

The extraction of piperidine alkaloids involved grinding plant material in distilled water and adding enough 1N NaOH to make the material basic. The sample was then extracted in a separatory funnel with dichloromethane (DCM). The organic and aqueous layers were separated and the DCM extract was dried by filtering through $NaSO_4$. The DCM extract was washed with 1N H_2SO_4 and the layers were separated. Sufficient 1N NaOH was added to the aqueous extract to make it basic, and it was then extracted with fresh DCM. The DCM extract was then separated and treated with acetic anhydride. Conium alkaloids have low molecular weights and are thus too volatile to be well retained by most GC columns. The amine functionalities also cause these compounds to tail. Acetylation increases molecular weight and decreases polarity, which allows them to be more readily retained and separated by GC or GC/MS.

The DCM extract was split into two portions; the first was not acetylated, and the second acetylated by adding a few drops of acetic anhydride prior to the evaporation step. The DCM in each case was evaporated under a stream of nitrogen and 1ml of methanol was added to the concentrated extracts. Aliquots (1µl) of the samples were injected on a Varian 3400 GC coupled to a Finnigan Mat Incos 50 mass spectrometer. The GC column used was a 12m x 0.2mm x 0.33µm film HP Ultra 2 (cross-linked 5% phenylmethylsilicone).

Results

Mass spectral data for the acetylated conium alkaloids (including "γ"-coniceine and the alkaloids tentatively identified as 2-n-pentyl-3,4,5,6-tetrahydropyridine and 5-hydroxy-2-n-pentylpiperidine) are given in Table 82.1. When the acetylated samples were injected, the masses of the parent ions of the compounds that have been acetylated gain in molecular weight by 42 (C_2H_2O). There is also a *m/z* 43 ion added to the mass of each compound (C_2H_3O, the acetyl group). Like coniine, the major ion for acetyl coniine was *m/z* 84, indicating that acetylation does not affect the major ion of this compound. The acetyl-"γ"-coniceine retained *m/z* 97 as its major ion and displayed the *m/z* 43 peak and the increase of 42 in its molecular weight. Even though "γ"-coniceine acetylates, it takes several minutes to react fully. This is in contrast to the acetylation of coniine, which is instantaneous. This delay may be due to the slowness of the tautomeric shifts of the "γ"-coniceine molecule. Once the tautomeric form was acetylated, "γ"-coniceine could not be reformed. It could also be due to a reaction rate for tertiary amines, which is slower than that for secondary amines such as coniine.

The major alkaloid found in the extracts in April, 1996, was Ψ-conhydrine, which is considered a minor alkaloid in poison hemlock. It has been proposed that the level of Ψ-conhydrine formed is directly associated with the level of coniine in the

Table 82.1. Mass spectral data for the alkaloids of *Conium maculatum* and their acetyl derivatives.

Alkaloid or derivative	Mass(% abundance)
"γ"-coniceine	97(100), 110(25), 41(20), 70(15), 125(12)
2-n-pentyl-3,4,5,6-tetrahydropyridine	97(100), 110(20), 82(15), 124(10), 139(5), 153(1)
5-hydroxy-2-n-pentylpiperidine	100(100), 114(30), 142(20), 157(20), 171(2)
acetyl "γ"-coniceine	97(100), 43(60), 110(30), 82(25), 167(25), 124(22), 152(20)
acetyl coniine	84(100), 43(30), 126(25), 56(15), 169(10)
acetyl conhydrine	143(100), 40(60), 185(40), 43(40), 115(25)
acetyl conhydrinone	84(100), 126(35), 43(10), 56(10), 183(10)
acetyl Ψ-conhydrine	100(100), 43(99), 72(45), 56(30), 114(15), 142(15), 185(10)
acetyl-2-n-pentyl-3,4,5,6-tetrahydropyridine	97(100), 43(35), 139(35), 110(30), 139(30), 195(10)
acetyl 5-hydroxy-2-n-pentylpiperidine	100(100), 43(90), 72(35), 114(30), 156(30), 157(25), 170(15), 213(2)

plant. Leete and Adityachaudhury (1967) grew poison hemlock in a greenhouse and found that Ψ-conhydrine was the major alkaloid while, in plants grown outside, coniine and "γ"-coniceine were the major alkaloids. This suggested that alkaloid production changes due to variances in temperature and moisture.

Discussion

Leete (1971) reported the presence of 2-methyl piperidine in the plant."γ"-coniceine is believed to be synthesized from four acetate molecules (Leete, 1963). Synthesis of 2-methyl piperidine from three acetates could be accomplished by the same mechanism. Extending this line of reasoning, the new alkaloids could presumably be synthesized from five acetate moieties. The molecular weight of coniine is 127, and it has no double bond in the ring structure. The mass spectrum of coniine contains a major ion with *m/z* 84, indicating that the primary product produced by electron impact has lost the entire n-propyl side-chain. The mass spectrum of "γ"-coniceine has a molecular weight of 125 with a major peak at *m/z* 97. This peak suggests that the major cleavage point for "γ"-coniceine is at the terminal two carbon atoms of the propyl group, yielding a $(C_5H_8N)CH_3^+$ ion. The presence of a *m/z* 97 ion as the major fragmentation product of the alkaloid suggests that it, too, produces a 2-methyl-3,4,5,6-tetrahydropyridine $(C_5H_8N)CH_3^+$ fragment. The interpretations of the other major ions are as follows: $(C_5H_8N)CH_2CH_2^+$ (*m/z* 110), also the mass spectrum of "γ"-coniceine; $(C_5H_8N)CH_2CH_2CH_2^+$ (*m/z* 124); another $(C_5H_8N)CH_2CH_2CH_2CH_3^+$ (*m/z* 139); also $(C_5H_8N)CH_2CH_2CH_2CH_2CH_3^+$ as the molecular ion (*m/z* 153).

Leete and Adityachaudhury (1967) have discussed tautomeric shifts of the double bond to explain the synthesis of higher molecular weight molecules from coniceine. They proposed that a tautomeric form, 2-n-propyl-2,3,4,5-tetrahydro-pyridine exists. Like "γ"-coniceine, this tautomer also contains a teriary amine. Acetylation of a tertiary amine is unlikely, but "γ"-coniceine readily acetylates. The ability to acetylate "γ"-coniceine suggests that there may be a tautomeric shift to a form with a double bond between carbons two and three, allowing the acetyl group to react with a nitrogen that is now a secondary amine. Further analysis to confirm the structures of the new alkaloids and "γ"-coniceine will be performed using X-ray crystallography or some alternative technique.

Acknowledgments

The investigations reported in this paper (96-14-112) are published by permission of the Dean and Director of the Kentucky Agricultural Experimental Station and the College of Agriculture, University of Kentucky, Lexington, KY 40546-0076, USA.

References

Kingsbury, J.M. (1964) *Poisonous Plants of the United States and Canada.* Prentice-Hall, Englewood Cliffs, NJ, pp. 379-383.

Leete, E. (1963) The biosynthesis of coniine from four acetate units. *Journal of the American Chemical Society* 86, 3523.

Leete, E. (1971) Biosynthesis of hemlock and related piperidine alkaloids. *Accounts of Chemical Research* 4, 100.

Leete, E. and Adityachaudhury, N. (1967) Biosynthesis of the hemlock alkaloids. II. The conversion of γ-coniceine to coniine and pseudoconhydrine. *Phytochemistry* 6, 219-223.

Panter, K.E. and Keeler, R.F. (1989) Piperidine alkaloids of poison hemlock (*Conium maculatum*). In: Cheeke, P.R. (ed), *Toxicants of Plant Origin, Vol I: Alkaloids.* CRC Press, Boca Raton, FL, pp. 109-132.

Chapter 83

Probable Interaction between *Solanum eleagnifolium* and Ivermectin in Horses

T. Garland[1], E.M. Bailey Jr[1], J.C. Reagor[2] and E. Binford[1]
[1]Department of Veterinary Physiology and Pharmacology, College of Veterinary Medicine, Texas A&M University, College Station, Texas 77843, USA; [2]Texas Veterinary Medical Diagnostic Laboratory, 1 Sippel Road, College Station, Texas 77843, USA

Introduction

The antiparasitic drug ivermectin was introduced in 1981. Its efficacy against nematode and arthropod parasites was unprecedented in potency and spectrum (Campbell, 1989). Ivermectin received acceptance in the health care of companion animals and in livestock production on a worldwide basis. Also, its efficacy in human onchocerciasis has further extended its impact on health care.

Ivermectin is available for parenteral use in other species, but is normally administered to horses as either an oral paste or a liquid for administration through a nasogastric tube. The liquid formulation (1%) attains its peak concentration within 4-5hrs while the paste product takes 15hrs to reach its maximum plasma concentration. The bioavailablity is 20% higher following the nasogastric dosing.

Normal canine doses of ivermectin have ranged from 68-272µg/kg, with most collie and collie-type dogs requiring 136-272µg/kg for the prevention of heart-worm disease. Collie dogs and collie-type dogs were reported to have an increased sensitivity to ivermectin at doses of 100-2500µg/kg. Clinical signs included ataxia, depression, tremors, recumbency and mydriasis (Campbell, 1989). Excessive ivermectin has been detected in neurological tissues of fatally poisoned dogs.

In horses dosed consecutively for 2d with the paste formulation, at 10x (2.0mg/kg) the normal treatment level, 5/11 had transient impairment of vision, depression and ataxia. Two of these horses were dehydrated for 5d after initial treatment (Campbell, 1989).

In 1989 in southern Texas, horses developed a clinical disease syndrome within 20hrs to several days following treatment with reported safe dosage levels of ivermectin. The first case consisted of a group of 14 horses, all of which were wormed with ivermectin paste. Eight horses confined to stalls were fed grass hay that contained *Solanum eleagnifolium* (silverleaf nightshade). The other six animals were on pasture where silverleaf nightshade grew but showed no evidence of being consumed. Those animals consuming the hay became ill, while those on pasture were unaffected. The clinical signs included ataxia, drooling, drooping lips and ears, head pressing, severe depression, muscle fasciculations and absence of menace response. Of this group of horses, one died within 24hrs and ivermectin was found at 115ppb in the brain. A second fatal case involved a horse that had consumed *S. eleagnifolium* in the hay and was treated with ivermectin. The brain level of ivermectin was 672ppb.

A third case involved a group of show horses given ivermectin paste every 2wks. These horses were released into a pasture that contained *S. eleagnifolium*. Six horses grazed the plant, and all of them showed the same clinical signs as previously described. The owner noticed the ataxia and drooping lips and ears and removed the animals from the pasture. All six animals had elevated liver enzymes and displayed an associated weight loss that manifested itself over several weeks. All the signs persisted after their removal from the pasture. The horses were removed from the show circuit, given clean hay, and their regular antiparasitic program was suspended. All six horses eventually recovered over a period of about 5mos (Fiske, 1990).

Solanum eleagnifolium is a perennial that is 0.5-1m tall with white-haired leaves and stem. The leaves are simple, thick, lanceolate to linear and entire to sinuate. Leaf ribs and stems usually have short stiff, spines. The flowers are violet or blue, while the berries are yellow or occasionally orange (Kingsbury, 1964). It is a prolific plant in Texas and other areas of the southwestern United States. The toxic agent suspected of causing gastrointestinal toxicosis is the glycoalkaloid solanine. The aglycone solanidine (a steroidal alkaloid) is the primary form acting upon the nervous system and is the suspected toxic agent. Ripe fruit of silverleaf nightshade is more toxic than the green berries. The ripe berries have produced moderate to severe poisoning in cattle when consumed at 0.1-0.3% of the animal's weight (Kingsbury, 1964). Leaves may be on the same order of toxicity as the green berries, or slightly less. Clinical signs range from irritation of the gastrointestinal tract to neurological signs that include trembling, progressive muscular weakness, labored breathing, CNS depression and incoordination, often accompanied by paralysis of the rear legs. Coma may occur without other neurological signs.

Since these cases, there have been several others with similar histories where consumption of *S. eleagnifolium* and treatment with ivermectin resulted in clinical signs of ivermectin poisoning in horses (Fiske, 1990; Reagor, 1990; Scroggs, 1990). These clinical cases would indicate an apparent interaction between ivermectin and *S. eleagnifolium*. Therefore, an investigation was initiated to examine the feasibility of utilizing rabbits as a model for the toxicity studies involving these two agents.

Model Justification

The rabbit is an appropriate laboratory model for the horse for several reasons. Both the horse and the rabbit are entirely herbivorous, and microbial fermentation occurs in the posterior part of the digestive tract. The ratio of body length to the length of the intestine is similar in both species. In the horse the ratio is 1:12 while the rabbit has a ratio of 1:10 (Hill, 1970). As in the rabbit, the stomach of a horse is never empty. Fermentation digestion follows enzymic digestion in these herbivores, so that only the fermentation products and not the bacterial bodies are available for digestion and absorption by the host. The proportion of volatile fatty acids as a percentage of total acid in the large intestine is similar in the horse and the rabbit. Percentages of acetic acid, propionic acid and butyric acid in the cecum of the horse are 73.1, 19.6 and 7.2 as compared with percentages in the cecum of the rabbit of 78.3, 9.3,and 12.5 (Hill, 1970). There is a difference between the percentages of propionic acid in the two species, but no other laboratory animal comes as close to the horse as does the rabbit. Regarding these and other physiological parameters the horse and rabbit are analogous (Meier, 1963).

Experimental Trial One

New Zealand White rabbits used for this study ranged between 2.88-3.99kg. The rabbits were purchased from a commercial supplier and acclimatized for 14d. Each rabbit was weighed and randomly assigned to one of five groups of five animals each. The groups consisted of a control group and four ivermectin and *S. eleagnifolium* treatment groups, with the ivermectin dose varying geometrically.

Solanum eleagnifolium was collected and air dried while it was in bloom. Immediately following dosing, all of the groups, including the control group, were given *ad libitum* access to fresh, air-dried *S. eleagnifolium* in addition to their normal ration of rabbit feed. Although horses will eat the yellow berries on the plants (Reagor 1990; 1995; 1997), the rabbits did not eat the berries or the stems of the plant, but eagerly consumed the leaves and the blooms.

The control group was dosed once *via* oral gavage with the inactive ingredients of liquid ivermection, a mixture of propylene glycol, polysorbate 80, butylated hydroxytoluene, ethylene diamine tetraacetic acid in phosphate buffer, equivalent to a horse dose on a weight basis. The treatment groups were dosed only once *via* oral gavage with 1.5, 3, 6 and 12mg/kg of ivermectin, respectively. The rabbits were weighed and observed daily for clinical signs.

Clinical signs developed in the three higher dosage groups within 24hrs of dosing the ivermectin and consumption of the plant. The most prominent clinical signs were depression and anorexia. Affected rabbits initially decreased feed and nightshade intake by half, and were completely anorexic within 48hrs. The group dosed with 12mg/kg was the most severely affected. Anorexia and depression occurred in a dose-related manner. Although weight loss was seen in all dosed groups, the loss was less than 10%. On the fourth day after dosing, the most severely depressed and lethargic rabbits ostensibly were less depressed, but refused to eat any of their normal ration or any of the silverleaf nightshade.

No signs of anorexia or lethargy were observed in the control group in the 7d trial. The two lower dosed groups, 1.5 and 3mg/kg, consumed more than half their normal ration and half the original amount of the silverleaf nightshade by the end of the 7d. Some rabbits were still slightly depressed, but were much improved compared with the first 24-48hrs. Rabbits in the 6 and 12mg/kg groups were still moderately depressed and lethargic, consuming less than half the normal ration and less than a quarter of the original amount of silverleaf nightshade.

All the rabbits were euthanized, the brains immediately removed and frozen at -70°C. The brains were then analyzed for ivermectin by the Illinois Department of Agriculture. There was essentially no difference in brain ivermectin levels (0-3ppb) between any of the groups.

Experimental Trial Two

Analysis of the rabbit brains indicated that the observation time had been too long, especially considering that actual cases with horses had occurred within 24hrs of being dosed and allowed access to silverleaf nightshade. The trial was repeated in an abbreviated fashion using nine rabbits.

The same procedure was followed as for the first experiment. The rabbits were divided randomly into three groups of three animals each, a control and two dose groups at 3 and 6mg/kg, respectively. These rabbits were dosed once *via* oral gavage and allowed access to air-dried *S. eleagnifolium*. A major difference between this trial and trial one was that the control rabbits were given 1mg/kg of ivermectin and no *S. eleagnifolium*. Rabbits were euthanized 40hrs after dosing, the brains immediately removed and and frozen at -70°C.

Clinical signs were obvious within 24hrs of dosing with the ivermectin and consumption of the plant. Although the control group was given ivermectin, there were no observed clinical signs. Rabbits treated with ivermectin and *S. eleagnifolium* displayed the same clinical signs of severe depression, anorexia and lethargy as those rabbits in the first experiment. No measurable weight loss occurred because the time from dosing to euthanasia was short.

The levels of ivermectin in the brain were much higher in trial two (Table 83.1) than in trial one. Likely this was due to euthanizing the rabbits when they appeared most depressed and completely anorectic. These analyses indicated that some ivermectin may normally cross the blood brain-barrier (Table 83.1), as the ivermectin control rabbits had very low levels of ivermectin in the brain. However, there was a significant ($P<0.05$) difference in brain concentration of ivermectin in the two treated groups *vs.* the control group.

Table 83.1. Brain ivermectin levels 40hrs after dosing the rabbits with ivermectin and exposure to *Solanum eleagnifolium* and ivermectin. Each block in the table represents an individual rabbit. Detection limit was 1ppb.

Ivermectin control 1mg/kg	3mg/kg ivermectin and *S. eleagnifolium*	6mg/kg ivermectin and *S. eleagnifolium*
2ppb	5ppb	31.7ppb
1ppb	6ppb	20.8ppb
2ppb	12.5ppb	61.7ppb

Conclusions

Intracellular toxicants may be transported into the extracellular spaces in the brain. Brain capillary endothelial cells contain an ATP-dependent membrane transporter known as the multidrug-resistance (MDR) protein or P-glycoprotein, which contributes to the blood/brain barrier and extrudes chemicals, such as the neurotoxic pesticide ivermectin. When the MDR apparatus is disrupted, there are higher brain levels of and sensitivity to ivermectin (Gregus and Klaassen, 1996). It is possible that *S. eleagnifolium* may promote absorption of ivermectin into the brain. Clinical signs in the rabbits were analogous to those seen in clinically affected horses and are similar to those seen in collie dogs that have been intoxicated by ivermectin. Results from the rabbits experiments suggest that they are an adequate laboratory model for the horse. The levels of ivermectin found in the brains of the rabbits in this study are comparable to the levels found in the clinical cases involving brains of horses. The evidence of these two trials and the clinical cases in horses further underscores the need to repeat this work in horses.

References

Campbell, W.C. (1989) *Ivermectin and Abamectin.* Springer-Verlag, New York, NY.

Fiske, J. (1990) Personal communication, College Station, TX.

Gregus, Z. and Klaassen, C.D. (1996) Mechanisms of Toxicity. In: Klaassen, C.D., Amdur, M.O. and Doull, J. (eds), *Casarett and Doull's Toxicology: The Basic Science of Poisons*, 5th ed. McGraw-Hill, New York, NY, pp. 37-39.

Hill, K.J. (1970) Developmental and comparative aspects of digestion. In: Swenson, M.J. (ed), *Dukes' Physiology of Domestic Animals*, 8th ed. Comstock Publishing Associates of Cornell University Press, Ithaca, NY, pp. 409-423.

Kingsbury, J.M. (1964) *Poisonous Plants of the United States and Canada.* Prentice-Hall, Englewood Cliffs, NJ, pp. 290-291.

Meier, H. (1963) Fluids and electolytes. In: Cornelius, C.E. and Kaneko, J.J. (eds), *Clinical Biochemistry of Domestic Animals.* Academic Press, New York, NY, pp. 500-553.

Reagor, J.C. (1990; 1995; 1997) Personal communications, College Station, TX.

Scroggs, M.G. (1990) Personal communication, Amaruillo, TX.

Chapter 84

Ipomoea carnea: The Cause of a Lysosomal Storage Disease in Goats in Mozambique

K.K.I.M. de Balogh[1], A.P. Dimande[1], J.J. van der Lugt[2], R.J. Molyneux[3], T.W. Naudé[4] and W.G. Welman[5]

[1]*Department of Clinical Studies, Veterinary Faculty, Eduardo Mondlane University, C.P.257, Maputo, Mozambique;* [2]*Onderstepoort Veterinary Institute, Onderstepoort, South Africa. Present address: Department of Pathology, Faculty of Veterinary Science, University of Pretoria, Private Bag X04, Onderstepoort, 0110 South Africa;* [3]*USDA-ARS, Western Regional Research Center, Albany, California 94710, USA;* [4]*Section of Toxicology, Onderstepoort Veterinary Institute, Private Bag X5, Onderstepoort 0110, South Africa;* [5]*National Botanical Institute, Private Bag X101, Pretoria 0001, South Africa*

Introduction

Most lysosomal storage diseases are genetic disorders (Summers *et al.*, 1995), but a few can be induced by the ingestion of toxic plants such as the locoweeds (*Astragalus* and *Oxytropis* spp.) from North America, South America and China (van Kampen and James, 1970) and the poison peas (*Swainsona* spp.) found in Australia (Dorling *et al.*, 1978). These plants of the Fabaceae (Leguminosae) contain the indolizidine alkaloid swainsonine, an inhibitor of lysosomal α-mannosidase and mannosidase II (Dorling *et al.*, 1980; Molyneux and James, 1982). Histologically, cytoplasmic vacuoles in cells of the nervous system and other tissues are seen. Neurons are most consistently affected, as in most lysosomal storage diseases (Summers *et al.*, 1995).

Animals consuming these plants exhibit a variety of clinical signs reflecting derangement of the nervous system including depression, rough hair coat, staggering gait, muscle tremors, ataxia and nervousness, especially when stressed (James *et al.*, 1981; Huxtable and Dorling, 1982). Generally these plants are only grazed during periods of food shortage, but some individual animals apparently develop a taste for the plants and selectively eat them despite other feed being available (Huxtable and Dorling, 1982).

An outbreak of a novel plant-induced lysosomal storage disease in goats caused by *Ipomoea carnea* is described here. Experimental reproduction of the condition and the isolation and identification of swainsonine and two calystegine glycosidase inhibitors from the plant are also reported.

Description of the Outbreak

During the dry season (June to September) of 1993, a nervous condition was noted in goats of the local Landin breed in a village 35km south west of Maputo, Mozambique. At the time of the outbreak, all villagers were involved in small scale subsistence farming. The village houses were built of reed, mud and a few stones and were separated by hedges of growing plants. Few people in the village kept cattle, but about 50 of 620 families kept goats. Goat rearers kept their animals kraaled in rudimentary pens at night. During the day, the animals were taken out by young boys to the restricted grazing areas.

The most prominent clinical signs in the affected goats were ataxia, head tremors and nystagmus. Their appetite was maintained for a long time, but the progressive course of the nervous symptoms finally rendered the animals unable to walk and graze normally. Goats of all ages and both sexes were affected. Most animals died within a few weeks of showing the first clinical signs, although spontaneous recovery was occasionally reported. A survey conducted in July, 1994, showed that approximately 10% of the animals had been lost due to this condition during the previous year. Cases had been observed as early as 1987, but a drastic increase in the number of affected animals occurred in 1992.

Affected goats were seen consuming a plant grown as a hedge around the homesteads and kraals. In the dry season, this plant was one of the few that maintained its leaves. After ruling out other probable local plants or mycotoxicoses, the plant identified as *I. carnea* was examined for its toxicity in goats.

Material and Methods

Samples of the incriminated hedge plant were sent for identification to the Herbarium of the Instituto Nacional de Investigação Agronómica (INIA) in Maputo, Mozambique. Identification was confirmed by staff of the Botanical Research Institute, Pretoria, South Africa.

Ipomoea carnea plant material was collected from the village in December, 1995, for isolation and identification of alkaloids as described by Molyneux *et al.* (1988). The identities of the alkaloids were established by GC/MS analysis.

Necropsies were done on three natural cases (Goats 1-3) originating from the village. All three goats were yearlings. One was male, and the others were female. From each goat necropsied, the brain, spinal cord, a portion of the sciatic nerve and samples of liver, kidney, spleen, lung and myocardium were fixed in 10% formalin for light microscopy. Specimens were routinely prepared and stained with hematoxylin and eosin (HE).

Results

Ipomoea carnea Jacq. subsp. *fistulosa* (Mart. ex Choisy) D Austin is a member of the Convolvulaceae family and is also known by the synonyms *I. fistulosa* Mart. ex Choisy and *I. crassicaulis* (Benth.) Robinson. Its common name is Shrubby Morning Glory. The plant is densely leaved, erect to scrambling and almost unbranched up to 3m high. The leaves are alternate, entire, ovate to lanceolate, truncate to shallowly cordate at the base, acuminate apically and 10-25cm long. The petioles are 2-10cm long. At the branch tips occur clusters of flowers, each with five persistent sepals and a corolla that is funnel shaped, five-lobed, deep pink to rose-purple and 5-9cm long. A complete botanical description of the plant is in Austin (1977). A voucher specimen No. 51806 was deposited at the Herbarium of the INIA under the name *Ipomoea carnea* Jacq. subsp. *fistulosa* (Mart. ex Choisy) D Austin.

Ipomoea carnea is of tropical American origin, but one subspecies is now pantropical. The subsp. *carnea* seems to be confined to its natural distribution from Peru to Mexico. The subsp. *fistulosa* has its natural distribution from Argentina to Florida and Texas and has been introduced to the tropics of the Eastern Hemisphere and Hawaii, where it has often escaped from cultivation (Austin and Huaman, 1996). In southern Africa it is found in Zimbabwe, Mozambique and South Africa and according to Gonçalves (1987), also in Zambia and Malawi. *Ipomoea carnea* flowers throughout the year. The subspecies *carnea* prefers dry habitats while the subspecies *fistulosa* prefers wet habitats. In southern Africa, *I. carnea* subsp. *fistulosa* is cultivated as an ornamental and in hedges and wind-breaks and often occurs as culture relics and escapes from cultivation. It will then occur in disturbed areas and along roadsides, but also to a limited extent in grassland, along river banks and in other moist areas.

Extraction of the milled leaves of *I. carnea* with methanol (MeOH), followed by ion-exchange chromatography on Dowex 50W-X8, produced a fraction containing alkaloids and basic amino acids as a pale yellow, partially crystalline solid. Analysis of the extract identified the alkaloid components as swainsonine, calystegine B_2 and calystegine C_1.

The major component, with retention time (R_t) 15.94min on GC/MS, had an R_t identical to that of the trimethyl silyl (TMSi) derivative of authentic calystegine B_2. The secondary component peak at R_t 17.17min corresponded in both its R_t and mass spectrum to penta-TMSi calystegine C_1. The other major component was identical in its R_t (13.90min) to tri-TMSi swainsonine.

The three field cases (Goats 1-3) had head tremors and ataxia. Additionally, Goat 2 exhibited hyperesthesia and a high stepping gait and Goat 3 showed nystagmus.

Edema of the lumbar spinal cord and meningeal congestion were seen in Goat 1. Goat 2 had marked asymmetry and atrophy of the cerebellum. No significant gross lesions were observed in Goat 3.

In Goat 1, neurons and glial cells in the brain and spinal cord were distended with numerous small, spherical, intracytoplasmic storage vacuoles. Cytoplasmic vacuolation was present in the cerebral cortex, thalamus, brain stem, Purkinje cells in the cerebellum and dorsal and ventral horn neurons in the spinal cord.

Eosinophilic, round, homogenous, occasionally finely granular axonal spheroids were conspicuous in the brain and spinal cord in areas of cytoplasmic vacuolation, often in close association with larger neurons. Other lesions in the brain of Goat 1 comprised gliosis and capillary accentuation in the caudal colliculi and pronounced vacuolation of the white matter of the rostral caudate nucleus. Cytoplasmic vacuolation was also detected in renal tubular epithelial cells and in phagocytic cells in the spleen and lymph nodes.

Histological lesions in Goats 2 and 3 were similar in nature, but they were more widespread in Goat 3. There was degeneration and loss of Purkinje neurons, affecting most folia in Goat 3. Some persisting Purkinje neurons were shrunken and possessed condensed cytoplasm and hyperchromatic nuclei. Cytoplasmic vacuolation of neurons and glial cells was rarely observed in sections stained with HE, and few axonal spheroids were present.

Accumulations of small amounts of yellowish-green to yellowish-brown, fine granular pigment in the cytoplasm of glial cells (and apparently free-lying in the neuropils) associated with mild gliosis, were noted in some brain stem nuclei in Goats 2 and 3. Phagocytic cells in the spleen of Goat 2 were vacuolated.

Discussion

The earliest reports on poisoning with *I. carnea* came from Sudan and India (Idris *et al.*, 1973; Tirkey *et al.*, 1987). Although the toxicity of the plant and its effects on the nervous system have been reported by these and other authors, a detailed histopathological examination of the nervous system was either inconclusive or not done (Idris *et al.*, 1973; Tirkey *et al.*, 1987) and the toxic principle(s) and mechanisms of intoxication were not determined.

The histological findings of cytoplasmic vacuolation in cells in the nervous system, particularly neurons, and in other tissues in the acute cases of *I. carnea* poisoning are consistent with a lysosomal storage disorder and closely resemble those of the induced storage diseases in livestock following ingestion of the locoweeds (*Astragalus* and *Oxytropis* spp.) (van Kampen and James, 1970) and poison peas (*Swainsona* spp.) (Dorling *et al.*, 1978; Huxtable and Dorling, 1982). In these and the disorder reported here, storage vacuoles in histologic sections are empty, reflecting the solubility of their contents in water or organic solvents used in tissue processing. Granular or membranous remnants may be detected in the vacuoles with the electron microscope (Walkley and Siegel, 1989).

Clinical signs of acute and chronic *I. carnea* poisonings reflect impairment of tissues by vacuolar lesions, particularly nervous tissue. The signs vary from general depression, ruminal and digestive disorders, to staggering gaits, reluctance to move, abduction and weakness of the hind limbs, generalized weakness, tremors, dullness, incoordination, posterior paresis and paralysis, diarrhea, lacrimation, nasal discharge and pallor of visible mucous membranes (Tirkey *et al.*, 1987). Hyperesthesia, head tremors and nystagmus were observed in the field and experimental cases.

Clinical recovery and disease reversal can occur after animals are withdrawn from the toxic plants if the exposure was brief or if the animals are in the early stages of intoxication (Huxtable and Dorling, 1982; Huxtable *et al.*, 1982). In goats 6 and 12mos after intoxication, neuronal cytoplasmic vacuolation was rare, while only few spheroids and no significant loss of neurons in the cerebrum or spinal cord occurred. However, residual nervous lesions in these goats were conspicuous in the cerebellum, and included loss of Purkinje neurons with degeneration and atrophy of remaining Purkinje cells and gliosis of the Purkinje cell layer. These findings agree with previous observations suggesting that Purkinje cells appear particularly vulnerable to injury in α-mannosidosis (Walkley and Siegel, 1989). Pigmentation in certain brain areas was also noted in chronic cases of *Swainsona* toxicity (Hartley, 1971; Walkley and Siegel, 1989).

The identification of swainsonine and calystegines B_2 and C_1 as the biologically active substances in *I. carnea* provided further evidence of a lysosomal storage disorder in the goats. Swainsonine is a potent, reversible inhibitor of α-mannosidase present in the locoweeds (*Astragalus* and *Oxytropis* spp.) and *Swainsona* species (Colegate *et al.*, 1979; Molyneux and James, 1982).

Calystegines have so far been identified in members of the plant families Convolvulaceae, Solanaceae and Moraceae. Their co-occurrence with swainsonine has previously been reported only in two *Ipomoea* species of very limited distribution in Australia, namely *I.* sp. Q6 [aff. *calobra*] (Weir vine) and *I. polpha* (Molyneux *et al.*, 1995). Weir vine has been associated with induction of a lysosomal storage disease in livestock.

Although the analysis did not permit an accurate quantitative measurement of the alkaloid content in the *I. carnea* sample, the GC/MS detector response was comparable to that observed for the Australian samples, indicating a combined level of calystegines and swainsonine of about 0.1%. Ostensibly, locoweeds containing at least 0.001% swainsonine can produce neurological damage if consumed regularly over a sufficient period of time (Molyneux *et al.*, 1994), and the content in *I. carnea* is therefore far more than the level necessary to induce poisoning. The exceptional potency of swainsonine as an α-mannosidase inhibitor does not require high levels for toxicity to occur, but the length of the grazing period is highly significant, since continuous suppression of enzyme activity will lead to the cellular vacuolation characteristic of the poisoning (Molyneux *et al.*, 1994).

Except for a single outbreak of a fatal posterior paralysis affecting exotic goats in Tanzania (Masselle *et al.*, 1989), no reference is made to the toxicity of *I. carnea* in the literature from eastern and southern Africa (Kellerman *et al.*, 1988). *Ipomoea carnea* is known to have been planted in Zambia, Zimbabwe, Malawi and Mozambique (Gonçalves, 1987), and it appears widely distributed in southern Africa, especially in the southern part of Mozambique.

The circumstances that prevailed in this village at the time of the outbreak, especially the ongoing civil-war and the drought, had restricted the grazing area. Goats were forced to consume plants in the village that were otherwise not browsed, including *I. carnea*. Even after the drought was over, goats showing central nervous symptoms continued consuming *I. carnea*, suggesting that once the animals were forced to eat the plant, some developed a liking for it, consuming it even after other food sources became available.

After the toxicity of the plant was confirmed, the villagers and extension workers were informed about the potential danger of *I. carnea*. This led to the removal of many hedges of *I. carnea* from the village. A drastic decrease in the number of new cases ensued.

Acknowledgments

The Netherlands Project at the Eduardo Mondlane University in Maputo is thanked for its financial support and the pathologists Dra Rosa Costa and Prof Patrocinio da Silva for the postmortems performed. Also, we thank Mr Arlindo Pene and Albino Panguene for their support during the feeding trials and the processing of samples.

References

Austin, D.F. (1977) *Ipomoea carnea* Jacq. v. *Ipomoea fistulosa* Mart. ex Choisy. *Taxon* 26, 235-238.

Austin, D.F. and Huaman, Z. (1996) A synopsis of *Ipomoea* (Convolvulaceae) in the Americas. *Taxon* 45, 3-38.

Colegate, S.M., Dorling, P.R. and Huxtable, C.R. (1979) A spectroscopic investigation of swainsonine: an α-mannosidase inhibitor isolated from *Swainsona canescens*. *Australian Journal of Chemistry* 32, 2257-2264.

Dorling, P.R., Huxtable, C.R. and Vogel, P. (1978) Lysosomal storage in *Swainsona* spp. toxicosis: an induced mannosidosis. *Neuropathology and Applied Neurobiology* 4, 285-295.

Dorling, P.R., Huxtable, C.R. and Colegate, S.M. (1980) Inhibition of lysosomal α-mannosidase by swainsonine, an indolizidine alkaloid isolated from *Swainsonine canescens*. *Biochemical Journal* 191, 649-651.

Gonçalves, M.L. (1987) Convolvulaceae-Cuscutaceae. In: Launert, E. (ed), *Flora Zambesiaca, Vol 8, 1*. Flora Zambesiaca Managing Committee, London, pp. 117-118.

Hartley, W.J. (1971) Some observations on the pathology of *Swainsona* spp. poisoning in farm livestock in Eastern Australia. *Acta Neuropathologica* (Berl.) 18, 342-355.

Huxtable, C.R. and Dorling, P.R. (1982) Poisoning of livestock by *Swainsona* spp.: current status. *Australian Veterinary Journal* 59, 50-53.

Huxtable, C.R., Dorling, P.R. and Walkley, S.U. (1982) Onset and regression of neuroaxonal lesions in sheep with mannosidosis induced experimentally with swainsonine. *Acta Neuropathologica* (Berl.) 58, 27-33.

Idris, O.F., Tartour, G., Adam, S.E.I. and Obeid, H.M. (1973) Toxicity to goats of *Ipomoea carnea*. *Tropical Animal Health and Production* 5, 119-123.

James, L.F., Hartley, W.J. and van Kampen, K.R. (1981) Syndromes of *Astragalus* poisoning in livestock. *Journal of the American Veterinary Medical Association* 178, 146-150.

Kellerman, T.S., Coetzer, J.A.W. and Naudé, T.W. (1988) *Plant Poisonings and Mycotoxicosis of Livestock in Southern Africa*. Oxford University Press, Cape Town, pp. 47-82.

Maselle, R.M., Mbassa, G.K., Jiwa, S.F.H. and Ndemanisho, E. (1989) An epidemic of fatal posterior paralysis and sudden death in goats: a case report. *Beitrage zur Tropischen Landwirtschaft und Veterinarmedizin* 27.H.2, 211-215.

Molyneux, R.J. and James, L.F. (1982) Loco intoxication: indolizidine alkaloids of spotted locoweed (*Astragalus lentiginosus*). *Science* 216, 190-191.

Molyneux, R.J., Benson, M., Wong, R.J., Tropea, J.E. and Elbein, A.D. (1988) Australine, a novel pyrrolizidine alkaloid glucosidase inhibitor from *Castanospermum australe*. *Journal of Natural Products* 51, 1198-1206.

Molyneux, R.J., James, L.F., Ralphs, M.H., Pfister, J.A., Panter, K.E. and Nash, R.J. (1994) Polyhydroxy alkaloid glycosidase inhibitors from poisonous plants of global distribution: analysis and identification. In: Colgate, S.M. and Dorling P.R. (eds), *Plant-Associated Toxins: Agricultural, Phytochemical and Ecological Aspects*. CAB International, Wallingford, Oxon, pp. 107-112.

Molyneux, R.J., McKenzie, R.A., O'Sullivan, B.M. and Elbein, A.D. (1995) Identification of glycosidase inhibitors Swainsonine and Calystegine B2 in Weir vine (*Ipomoea* Sp.Q6 [aff.Calobra]) and correlation with toxicity. *Journal of Natural Products* 58, 878-886.

Summers, B.A., Cummings, J.F. and de Lahunta, A. (1995) *Veterinary Neuropathology*. Mosby, St Louis, MO, pp. 208-350.

Tirkey, K., Yadava, K.P. and Mandal, T.K. (1987) Effect of aqueous extract of *Ipomoea carnea* on the haematological and biochemical parameters in goats. *Indian Journal of Animal Science* 57, 1019-1023.

van Kampen, K.R. and James, L.F. (1970) Pathology of locoweed (*Astragalus lentiginosus*) poisoning in sheep. *Pathologia Veterinaria* 7, 503-508.

Walkley, S.U. and Siegel, D.A. (1989) Comparative studies of the CNS in swainsonine-induced and inherited feline α-mannosidosis. In: James, L.F., Elbein, A.D., Molyneux, R.J. and Warren, C.D. (eds), *Swainsonine and Related Glycosidase Inhibitors*. Iowa State University Press, Ames, IA, pp. 57-75.

Chapter 85

Investigation of the Neurotoxic Compounds in *Asclepias subverticillata*, Western-whorled Milkweed

G.H. Robinson[1], G.E. Burrows[1], E.M. Holt[1], R.J. Tyrl[1] and A.D. Jones[2]

[1]Oklahoma State University, Stillwater, Oklahoma 74078, USA; [2]University of California, Davis, California 95617, USA

Introduction

Since the early 1900s, species of *Asclepias* (Asclepiadaceae) have caused devastating losses to livestock in western portions of the US. These highly toxic plants are not normally eaten, but may be consumed by very hungry animals. Drying increases palatability but does not diminish toxicity. All plant parts are considered toxic (Marsh *et al.*, 1920; May, 1920; Couch, 1937; Clark, 1979; Edwards *et al.*, 1984).

Initial studies of *Asclepias* led toxicologists to divide them into narrow-leaved and broad-leaved groups. Based on clinical observations, it was believed that the broad-leaved group was cardiotoxic and the narrow-leaved group neurotoxic (Fleming *et al.*, 1920; Kingsbury, 1964; Edwards, *et al.*, 1984). Later it was found that several narrow-leaved species produced cardiotoxic effects and that only verticillate-leaved species were neurotoxic (Ogden, 1989; Ogden *et al.*, 1992a; Ogden *et al.*, 1992b). The 16 verticillate-leaved species were classified by Woodson (1954) in the subgenus *Asclepias* and series *Incarnatae*. Five of these are distinctly verticillate-leaved: *A. subverticillata, A. fascicularis, A. verticillata, A. pumila*, and *A. mexicana*; and one appears nearly so, *A. incarnata*.

The cardiotoxic effects of *Asclepias* are produced by cardenolides (Roeske *et al.*, 1976; Benson *et al.*, 1978; Seiber *et al.*, 1983; Seiber *et al.*, 1985). The basic cardenolide aglycone is a C_{23} steroidal genin with a five-membered, singly unsaturated lactone ring at C-17, a hydroxyl group at C-14 and methyl groups at C-10 and C-13. Glycosidic linkage occurs at C-3 to one or more sugar moieties. Cardenolides inhibit Na^+K^+-ATPase *via* the structural features residing in the unsaturated lactone ring at C-17 and the hydroxyl group at C-14 (Joubert, 1989).

Although the neurotoxins of *Asclepias* had not been identified, investigations had provided clues to their character (Marsh *et al.*, 1920; Ogden *et al.*, 1992a; Ogden *et al.*, 1992b) and described symptoms similar to those produced by *Cynanchum africanum*. This South African species of the family synthesizes neurotoxic pregnane glycosides that cause a disorder called cynanchosis in domestic ruminants (Kellerman *et al.*, 1988). Pregnanes are the putative biological precursors to cardenolide genins. Their basic structure is a C_{21} steroid with methyl groups at C-10 and C-13, and an ethyl side-chain at C-17 (Deepak *et al.*, 1989).

This study, the first step in a series, was to test the hypothesis that a unique neurotoxicant(s) occurs in *Asclepias* and is restricted in distribution to the verticillate-leaved species. The first plant investigated was *A. subverticillata* (A. Gray) Vail, commonly known as western-whorled or horsetail milkweed, which has caused heavy livestock losses and has been recognized as one of the most poisonous plants in the US (Marsh *et al.*, 1920; May, 1920; Marsh and Clawson, 1921; Couch, 1937). It is an erect, herbaceous perennial 15-120cm tall. Its stems bear small, sterile branches. The leaves are petiolate, 1-4mm wide, and whorled with 3-5 at each node. Usually solitary at the upper nodes, the umbellate inflorescences are several to many flowered. The flowers are small with a white or sometimes lightly tinged greenish purple corolla (Woodson, 1954). Flowering from June to August, *A. subverticillata* is found in Arizona, New Mexico, Colorado, Oklahoma, Texas, and Utah, occurring on sandy, rocky plains and flats. Its stout, woody rootstalk allows it to spread rapidly in damp pastures and ditches, where it often forms dense stands (Fleming, *et al.*, 1920; Woodson, 1954; Kingsbury, 1964). The study's objectives were to extract and identify its toxic fractions and to determine the structure of the neurotoxicant(s).

Materials and Methods

Plants were collected, air-dried, ground, and frozen or refrigerated until use (Robinson, 1997). With a previously developed bioassay model (Ogden, 1989), plant material, crude extracts and their fractions were placed in gel capsules and force-fed to chickens. Male and female white leghorn chickens, weighing 0.4-2.6kg, were housed indoors, individually or in pairs, in wire cages at a constant temperature, and allowed free access to commercial feed and water. The chickens were observed at periodic intervals during the tests. Fractions that did not produce neurologic symptoms within 24hrs were considered non-toxic (Robinson, 1997).

Extraction, isolation, and structural determination methods were as previously reported (Robinson, 1997). Plant material was extracted with dichloromethane (DCM), and filtrates were concentrated and extracted with hot petroleum ether. The insoluble residue was subjected to a series of silica gel chromatography, always proceeding in the series with the most toxic fraction. Selected toxic fractions were analyzed using reverse-phase HPLC coupled with diode array UV detector, low resolution electrospray MS (ESMS), MS/MS, GC/electrospray ionization MS

(GC/EIMS), ^{1}H NMR, and ^{1}H correlated spectroscopy (COSY) NMR.

Results

The signs of intoxication in chickens fed dried plant material resembled those in chickens fed the plant extracts, and were typically, apparent 6-12hrs after administration. Later fractions produced intoxication at 2-4hrs. Neurologic symptoms included ataxia, excitement, head tremors, intermittent seizures, torticollis, depression and sometimes death, and were consistent with those previously described (Ogden, 1989; Ogden *et al.*, 1992b).

Ultraviolet spectra indicated a cinnamoyl group ($\lambda_{max,CH3CN}$=277nm) and a lactone ring ($\lambda_{max,CH3CN}$=220nm). The low resolution ESMS positive ion spectrum of analytes with the greatest UV absorbance intensity at 277nm and 220nm showed a dominant peak at 1,067. Daughters of 1,067 closely corresponded to the loss of rhamnose (m/z 921), thevetose (m/z 761), cymarose (m/z 617), and a cinnamoyl group (m/z 471). Spectra from GC/EIMS supported the presence of a cinnamoyl group and a genin moiety. The ^{1}H NMR spectrum confirmed the cinnamoyl group at δ6.08 (d, *J*=15.8 Hz) and 7.39 (d, *J*=15.8). Other signals originating from the genin moiety appeared at δ1.14 (C-19 Me), 1.52 (C-18 Me) and 5.38 (C-6 olefinic proton).

Complexity of the various spectra showed that the analytes were a mixture of several compounds. The data were consistent with the presence of a group of glycosides, which all seemed to be cinnamate ester derivatives (Robinson, 1997).

Discussion

This study suggests the presence of a cinnamate-containing cardenolide in the neurotoxic DCM extract of *A. subverticillata*. The proposed structure, named verticenolide (Fig. 85.1), has a molecular weight of 1,066 and a molecular formula of $C_{57}H_{78}O_{19}$. The position of the tigloyl group at C-8 was determined by difference in molecular weight and labile character in acid. Although cinnamoyl groups occur in some pregnane glycosides of Asclepiadaceae (Hayashi *et al.*, 1981; Yoshimura *et al.*, 1983; Tsukamoto *et al.*, 1985) and in a cardenolide isolated from *Asclepias asperula* (Martin *et al.*, 1991), this is the first report of a cinnamate-containing cardenolide with the cinnamoyl group attached directly to the genin moiety.

Evidence of a cinnamate-containing cardenolide in *A. subverticillata* supports the hypothesis that a unique neurotoxicant(s) occurs in a verticillate-leaved species. This finding is consistent with other studies (Ogden, 1989; Ogden *et al.*, 1992a; Ogden *et al.*, 1992b) describing myocardial infiltrates in sheep and chickens fed *A. subverticillata* and *A. verticillata*. This work provides a basis for future studies to determine the nature and mode of action of the neurotoxins present in the other verticillate-leaved species.

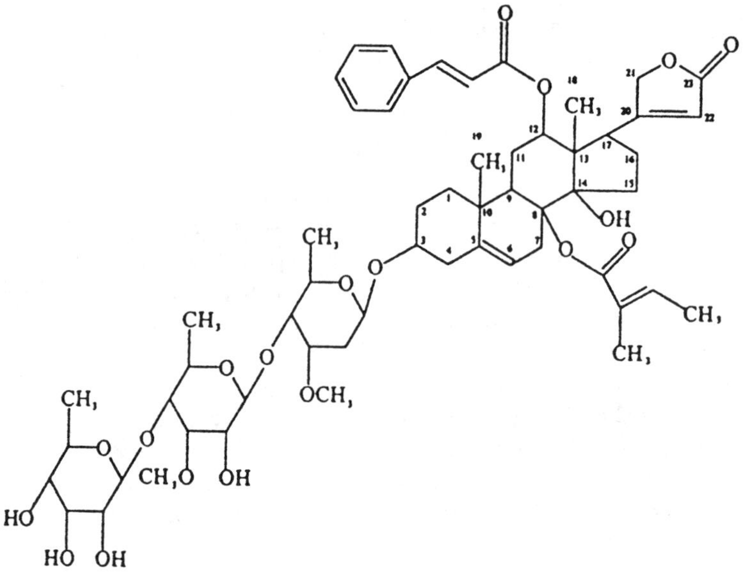

Fig. 85.1. Proposed cardenolide (verticenolide) in the neurotoxic extract of *Asclepias subverticillata*.

References

Benson, J.M., Seiber, J.N., Keeler, R.F. and Johnson, A.E. (1978) Studies on the toxic principle of *Asclepias eriocarpa* and *A. labriformis*. In: Keeler, R.F., Van Kampen, K.R. and James, L.F. (eds), *Effects of Poisonous Plants on Livestock*. Academic Press, New York, NY, pp. 273-284.

Clark, J.G. (1979) Whorled milkweed poisoning. *Veterinary and Human Toxicology* 21, 431.

Couch, J.F. (1937) Chemistry of stock-poisoning plants. *Journal of Chemical Education* 14, 16-30.

Deepak, D., Khare, A. and Khare, M.P. (1989) Review article number 49, plant pregnanes. *Phytochemistry* 28, 3255-3263.

Edwards, W.C., Burrows, G.E. and Tyrl, R.J. (1984) Toxic plants of Oklahoma - milkweeds. *Oklahoma Veterinary Medical Association* 36, 74-79.

Fleming, C.E., Peterson, N.F., Miller, M.R., Vawter, L.R. and Wright, L.H. (1920) *The Narrow-Leaved Milkweed and the Broad-Leaved or Showy Milkweed*. University of Nevada Agricultural Experiment Station Bulletin 99.

Hayashi, K., Wada, K., Mitsuhashi, H., Bando, H., Takase, M., Terada, S., Koide, Y., Aiba, T., Narita, T. and Mizuno, D. (1981) Further investigation of antitumor condurango-glycosides with C-18 oxygenated aglycone. *Chemical and Pharmaceutical Bulletin* 29, 2725-2730.

Joubert, J.P.J. (1989) Cardiac glycosides. In: Cheeke, P.R. (ed), *Toxicants of Plant Origin, Vol II*. CRC Press, Boca Raton, FL, pp. 61-96.

Kellerman, T.S., Coetzer, J.A.W. and Naudé, T.W. (1988) *Plant Poisonings and Mycotoxicoses of Livestock in Southern Africa*. Oxford University Press, Cape Town, pp. 51-53.

Kingsbury, J.M. (1964) *Poisonous Plants of the United States and Canada*. Prentice-Hall, Englewood Cliffs, NJ, pp. 267-270.

Marsh, C.D. and Clawson, A.B. (1921) *The Mexican Whorled Milkweed (Asclepias mexicana) as a Poisonous Plant*. USDA Bulletin 969.

Marsh, C.D., Clawson, A.B., Couch, J.F. and Eggleston, W.W. (1920) *The Whorled Milkweed (Asclepias galioides) as a Poisonous Plant*. USDA Bulletin 800.

Martin, R.A., Lynck, S.P., Schmitz, F.J., Pordesimo, E.O., Toth, S. and Horton, R.Y. (1991) Cardenolides from *Asclepias asperula* subsp. *capricornu* and *A. viridis*. *Phytochemistry* 30, 3935-3939.

May, W.L. (1920) *Whorled Milkweed: the Worst Stock-Poisoning Plant in Colorado*. USDA Bulletin 255.

Ogden, L. (1989) Toxic Effects of Milkweeds (*Asclepias*) in Sheep and Chickens. MS Thesis, Oklahoma State University, Stillwater, OK.

Ogden, L., Burrows, G.E., Tyrl, R.J. and Ely, R.W. (1992a) Experimental intoxication in sheep by *Asclepias*. In: James, L.F., Keeler, R.F., Bailey, E.M. Jr, Cheeke, P.R. and Hegarty M.P. (eds), *Poisonous Plants: Proceedings of the Third International Symposium*. Iowa State University Press, Ames, IA, pp. 495-499.

Ogden, L., Burrows, G.E., Tyrl, R.J. and Gorham, S.L. (1992b) Comparison of *Asclepias* species based on their toxic effects on chickens. In: James, L.F., Keeler, R.F., Bailey, E.M. Jr, Cheeke, P.R. and Hegarty M.P. (eds), *Poisonous Plants: Proceedings of the Third International Symposium*. Iowa State University Press, Ames, IA, pp. 500-505.

Robinson, G.L.H. (1997) Investigations of Toxic Plants: *Albizia julibrissin* and *Asclepias subverticillata*. MS Thesis, Oklahoma State University, Stillwater, OK.

Roeske, C.N., Seiber, J.N., Brower, L.P. and Moffitt, C.M. (1976) Milkweed cardenolides and their comparative processing by monarch butterflies. In: Wallace, J.W. and Mansell, R.L. (eds), *Biochemical Interaction Between Plants and Insects*. Plenum Press, New York, NY, pp. 93-167.

Seiber, J.N., Lee, S.M. and Benson, J.M. (1983) Cardiac glycosides (cardenolides) in species of *Asclepias* (Asclepiadaceae). In: Keeler, R.F. and Tu, A.T. (eds), *Handbook of Natural Toxins, Vol I, Plant and Fungal Toxins*. Marcel Dekker, NewYork, NY, pp. 43-83.

Seiber, J.N., Lee, S.M., McChesney, M.M., Watson, T.R., Nelson, C.J. and Brower, L.P. (1985) New cardiac glycosides (cardenolides) from *Asclepias species*. In: Seawright, A.A., Hegarty, M.P., James, L.F. and Keeler, R.F. (eds), *Plant Toxicology: Proceedings of the Australia-USA Poisonous Plants Symposium, Brisbane, Australia, May 14-18, 1984*. Queensland Poisonous Plants Committee, Yeerongpilly, Qld, pp. 427-437.

Tsukamoto, S., Hayashi, K. and Mitsuhashi, H. (1985) Studies on the constituents of Asclepiadaceae plants. LX. *Chemical and Pharmaceutical Bulletin* 33, 2294-2304.

Woodson, R.E. (1954) The North American species of *Asclepias*. *Annals of the Missouri Botanical Gardens* 41.

Yoshimura, S., Narita, H., Hayashi, K. and Mitsuhashi, H. (1983) Studies on the constituents of Asclepiadaceae plants. LVI. *Chemical and Pharmaceutical Bulletin* 31, 3971-3983.

Chapter 86

Isolation of Karwinol A from Coyotillo (*Karwinskia humboldtiana*) Fruits

H.L. Kim[1] and R.D. Stipanovic[2]

[1]*Department of Veterinary Physiology and Pharmacology, Texas A&M University, College Station, TX 77843, USA;* [2]*USDA-ARS, Cotton Pathology Research Unit, Southern Crops Research Laboratory, College Station, TX 77845, USA*

Introduction

Karwinskia humboldtiana Zucc. (Rhamnaceae), commonly known as coyotillo or tullidora, is a shrub growing along the Rio Grande into the semi-desert areas of Mexico. Accidental poisoning of children and grazing animals by ingestion of coyotillo fruits has been known for over 200yrs (Marsh and Clawson, 1928). Fifty-six confirmed cases of coyotillo intoxication in humans were reported in Mexico between January 1991 and December 1993 (Bermudez-de Rocha *et al.*, 1995). Four polyphenolic compounds were previously isolated and their gross structures determined. They are known as compounds T-496 ($C_{30}H_{24}O_7$), T-514 ($C_{30}H_{26}O_8$), T-516 ($C_{30}H_{28}O_8$), and T-544 ($C_{32}H_{32}O_8$) according to their molecular weights (Dreyer *et al.*, 1975). Oral administration of T-544 caused quadriplegia in a monkey (Kim and Camp, 1972), and T-544 and T-514 caused LDH leakage from primary rat hepatocyte cultures (Garza-Ocanas *et al.*, 1992).

A sample of T-544 separated on silica gel and thought to be pure exhibited four peaks when subjected to analytical HPLC. The UV spectra of the four peaks were nearly superimposable. One of the four was isolated by preparative HPLC and named karwinol A. Reported herein is the isolation and determination of the relative stereochemistry of karwinol A.

Materials and Methods

Carbon-13 magnetic resonance spectra, heteronuclear multiple bond correlation

(HMBC) spectra and nuclear Överhauser effect spectroscopy (NOESY) spectra were recorded in CDCl$_3$ on a Bruker AMR 300 instrument. Proton magnetic resonance spectra were recorded in CDCl$_3$ on a Bruker AMR 500 instrument. Conditions for NOESY experiment were: relaxation delay 6sec, mixing time 700msec. Conditions for HMBC experiment were: relaxation delay 2sec, long-range coupling evaluation delay 65msec. Fast atom bombardment mass spectra (FAB MS) were obtained with a VG analytical 70S (Manchester, UK) double focusing instrument equipped with a VG 11/250J data system.

Dried and ground coyotillo fruits were extracted with CHCl$_3$. T-544 was isolated by silica gel chromatography (Kim and Camp, 1972) and further purified by preparative HPLC with an Inertsil 10μm ODS-2, 20x250mm column (GL Science, Tokyo, Japan). The mobile phase was 10mM phosphate buffer (pH 3.0) -ACN azeotrope (15:85). The flow rate was 4ml/min. Fractions were collected and karwinol A eluted between 29.5-33min. Crude T-544 solution (0.3ml, 150mg/ml acetone) was injected onto the preparative column and elution of karwinol A and other isomers was monitored by analytical HPLC. The analytical system was of a equipped with an Inertsil ODS-2, 4.6x150mm column and a variable wave-length UV monitor or diode array detector. Separation was achieved with a mobile phase of 10mM phosphate buffer (pH 3.0):acetonitrile (ACN) (25:75) and a flow rate of 0.4ml/min. The karwinol A fractions were pooled, concentrated to remove ACN, then extracted twice with dichloromethane (DCM). The combined DCM soluble portion was concentrated to a light brown gum under reduced pressure and karwinol A crystallized upon trituration with methanol, m. 203°C (dec).

Results and Discussion

The chloroform extract of dried and ground mature coyotillo berries consisted primarily of T-544, which eluted between 12.15-15.07min as four peaks on the HPLC chromatogram (Fig. 86.1a). Their UV spectra were nearly superimposable (Fig. 86.1b). Karwinol A, the earliest eluting isomer of the four T-544 peaks (Fig. 86.1c), was separated under the preparative conditions stated above. The UV spectrum of karwinol A is shown in Fig. 86.1d. Separation of the other isomers was not easily achieved, and the proton NMR spectra of these isomers were very similar to that of karwinol A. The FAB MS spectrum of karwinol A showed a pseudomolecular ion at m/z545 [M+H]$^+$ as expected for C$_{32}$H$_{33}$O$_8$. Assignment of proton NMR signals of karwinol A (Table 86.1) was based on chemical shifts and coupling constants, and agrees with those reported by Dreyer *et al.* (1975). The assignment of ^{13}C-NMR signals (Table 86.1) is based on chemical shift and on other experiments including distortionless enhancement polarization transfer (DEPT), HMBC (Table 86.2) (Bax and Summers, 1986) and heteronuclear multiple quantum coherence (HMQC) (Bax and Suhramanian, 1986). The relative stereochemistry at C-1' and C-3' was established by NOE difference spectra. The following NOESY interactions were found (Fig. 86.2): between H-1' and both H-3', CH$_3$-1'; between H-3' and CH$_2$-4', H-1'

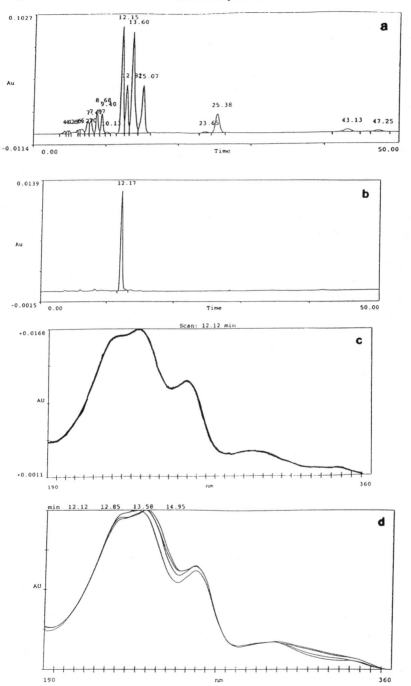

Fig. 86.1. Chromatograms of **a**) coyotillo fruit extract and **b**) karwinol A; and UV spectra of **c**) karwinol A and its isomers **d**) found in coyotillo fruit.

Table 86.1. Carbon and proton chemical shift assignments for karwinol A.

Carbon		Proton			
#	δ	δ	#H	Multiplic	Coupling Constants (H$_z$)
1	203.1(s)				
2	51.2(t)	2.87	2	d	JC2-C2=17.5
3	71.0(s)				
3-Me	29.1(q)	1.48	3	s	
4	43.2(t)	3.13	2	d of d	JC4-C4=16.0, JC4-C10=1.1
4a	134.7(s)				
5	118.5(d)	7.27	1	d	JC5-C6=8.2
6	136.5(d)	7.39	1	d	JC5-C6=8.2
7	121.5(s)				
8(OH)	154.8(s)	9.82	1	s	
8a	112.9(s)				
9(OH)	165.7(s)	16.00	1	s	
9a	109.5(s)				
10	118.6(d)	7.09	1	bd	JC4-C10=1.1
10a	139.2(s)				
1'	71.2(d)	5.22	1	q	JC1'Me-C1'=6.0
1'-Me	21.9(q)	1.66	3	d	JC1'Me-C1'=6.0
3'	69.4(d)	3.69	1	d of d of q	-C3Me=6.2, J4'ax-3'=10.3, J4'eq-3'=2.5
3'-Me	21.7(q)	1.18	3	d	JC3'Me-C3'=6.2
4'	36.7(t)	2.29	2	d of d of d	-C4'=16.0, J4'ax=3'=10.3, J4'eq-3'=2.5
4'a	135.1(s)				
5'	123.0(s)				
5'a	134.3(s)				
6'	97.7(d)	6.23	1	d	JC6'-C8'=2.1
7'	157.1(s)				
7'-OMe	55.1(q)	3.55	3	s	
8'	96.9(d)	6.40	1	d	JC6'-C8'=2.1
9'	157.4(s)				
9'-OMe	56.2(q)	4.01	3	s	
9'a	109.4(s)				
10' (OH)	150.2(s)	9.62	1	s	
10'a	119.7(s)				

and CH$_3$-3'; and between the CH$_2$-4' and both H-3' and H-6. No NOESY interaction was found between H-1' and CH$_3$-3', and between H-3' and CH$_3$-1' (Fig. 86.2). Thus the protons at C1' and C3' are in axial orientation.

The relative stereochemistry of CH$_3$-3 is difficult to establish. The two protons on C-2 and those on C-4 are equidistant to the protons on CH$_3$-3 regardless of whether the methyl group is axial or equatorial. A NOESY experiment which showed equal interaction between CH$_3$-3 and the protons on C-2 and C-4 confirmed this assumption. However, the 4-bond coupling between this methyl group and the protons on C-2 and C-4 provides some guidance. At 500MHz the CH$_3$-3 appears as a broad singlet with a width at half height of 1.3Hz (widths at half height of 0.9Hz and 0.8Hz were observed for the O-Me groups). A CH$_3$-3 is predicted to exhibit coupling constants of 0.8Hz and 0.3Hz with the axial and equatorial protons, respectively, on C-2 and C-4 (Barfield, 1964). An equatorial CH$_3$-3 is predicted to have coupling constants of 0.3Hz and 0.1Hz with axial and equatorial protons, respectively, on C-2 and C-4. Since the coupling is less than 0.6 Hz, the CH$_3$-3 is tentatively assigned to the equatorial position. The proposed structure for karwinol A is shown in Fig. 86.2.

Karwinol A appears to be stable in coyotillo fruits stored in a freezer for over ten years at -20° C or in an extract as yellow powder mixed with isomers and other phenolic compounds stored in a refrigerator. The initial chloroform extract of dried and ground berries consisted primarily of karwinol A and its isomers. The extracts exhibited four peaks in chromatograms on both analytical and preparative columns, and the chromatographic appearance of purified T-544 stored in a refrigerator for years was essentially identical to those of freshly extracted samples. The detection

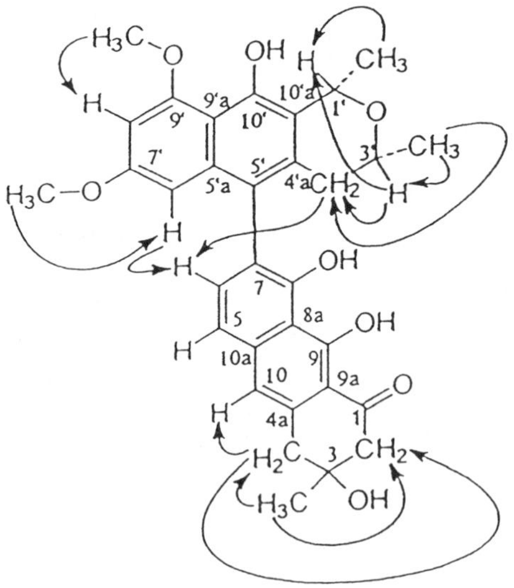

Fig. 86.2. Proposed structure of karwinol A and NOESY interaction.

Table 86.2. HMBC correlation of karwinol A.

Proton	²J	³J	⁴J
2	71.0 (C-3), 203.1 (C-1)		
3-Me	71.0 (C-3)	43.2 (C-4), 51.2 (C-2)	
4	134.7 (C-4a)	9.1 (C3-Me), 51.2 (C-2), 109.5 (C-9a), 118.6 (c-10)	
5	139.2 (C-10a)	112.9 (C-8a), 118.6 (C-10), 121.5 (C-7)	
6		123.0 (C-5'), 139.2 (C-10a), 154.8 (-8)	
8-OH	254.8 (C-8)	112.9 (C-8a), 121.5 (C-7)	
9-OH	165.7 (C-9)	109.5 (C-9a), 112.9 (C-8a)	203.1 (C -1)
10	139.2 (-10a)	43.2 (C-4), 109.5 (C-9a), 112.9 (C-8a), 118.5 (C-5)	203.1 (C -1)
1'	21.9 (C-1'-Me), 119.7 (C -10'a)	135.1 (C-4'a)	
1'-Me	71.2 (C-1')	119.7 (C-10'a)	
3'			
3'-Me	69.4 (C-3')	36.7 (C-4')	
4'	69.4 (C-3'), 135.1 (C-4'a)		
6'	157.1 (C-7')	96.9 (C-8'), 109.4 (C-9'a), 123.0 (C-5')	
7'-OMe		157.1 (C-7')	
8'	157.1 (C-7'), 157.4 (C -9')	97.7 (C-6')	
9'-OMe		157.4 (C-9')	
10'-OH	150.2 (c-10')	109.4 (C-9'a), 119.7 (C-10'a)	

limit of karwinol A was less than 5ng and the method may be applicable for the analyses of biological samples such as animal and plant tissues. Reversed-phase columns resolved karwinol A isomers, but the same sample showed a single peak when subjected to chromatography on a polymer or a cyano column. Further studies are in progress to isolate and characterize karwinol A and and its isomers, and to determine the biological activities of these compounds.

Acknowledgments

The work was supported, in part, by the Texas Agricultural Experiment Station, Texas A&M University System.

References

Barfield, M. (1964) Angular dependence of long-range proton coupling constants across four bonds. *Journal of Chemical Physics* 41, 3825-3832.
Bax, A. and Suhramanian, S. (1986) Sensitivity-enhanced two-dimensional heteronuclear shift correlation NMR spectroscopy. *Journal of Magnetic Resonance* 67, 565-569.
Bax, A. and Summers, M.F. (1986) H and ^{13}C assignments from sensitivity-enhanced detection of heteronuclear multiple-bond connectivity by 2D multiple quantum NMR. *Journal of the American Chemical Society* 108, 2093-2094.
Bermudez-de Rocha, M.V., Lozano-Melendez, F.E., Tamez-Rodriguez, V.A., Diaz-Cuello, G. and Pineyro-Lopez, A. (1995) Frecuencia de intoxicacion con *Karwinskia humboldtiana* en Mexico. *Salud Publica de Mexico* 37, 57-62.
Dreyer, D.L., Arai, I., Bachman, C.D., Anderson, W.R. Jr, Smith, R.G. and Daves, G.D. Jr (1975) Toxins causing noninflammatory paralytic neuronopathy. Isolation and structure elucidation. *Journal American Chemical Society* 97, 4985-4990.
Garza-Ocanas, L., Hsieh, G.C., Acosta, D., Torres-Alanis, O. and Pineyro-Lopez, A. (1992) Toxicity assessment of toxins T-514 and T-544 of buckthorn (*Karwinskia humboldtiana*) in primary skin and liver cell cultures. *Toxicology* 73, 259-267.
Kim, H.L. and Camp, B.J. (1972) Isolation of a neurotoxic substance from *Karwinskia humboldtiana* Zucc. (Rhamnaceae). *Toxicon* 10, 83-84.
Marsh, C.D. and Clawson, A.B. (1928) *Coyotillo (*Karwinskia humboldtiana*) as a poisonous plant*. USDA Technical Bulletin No. 29.

Chapter 87

The Toxicity of the Australian Cycad *Bowenia serrulata* to Cattle

A.A. Seawright[1], P.B. Oelrichs[1], J.C. Ng[1], Y. Sani[1] and C.C. Nolan[2]

[1]*National Research Centre for Environmental Toxicology, 39 Kessels Rd, Coopers Plains, Queensland 4108, Australia;* [2]*Medical Research Council Toxicology Unit, University of Leicester, Leicester 9HN LEI, UK*

Introduction

Cycad neurotoxicity occurs in Australia, the West Indies and the southern islands of Japan. Poisonings are caused by *Zamia* spp. in the West Indies (Mason and Whiting, 1968), *Cycas revoluta* in Japan (Kobayashi *et al.*, 1983) and *Macrozamia* and *Cycas* spp. in southern and northern Australia, respectively (Hall, 1957).

Cycad neurotoxicity (Zamia staggers) affecting cattle has been recognized from the earliest days of European settlement of Australia, and feeding trials confirm the cause of the condition (Seawright *et al.*, 1993). The plants usually have to be fed daily for several months before signs of the characteristic posterior ataxia occur. However, Hall and McGavin (1968) reported that the condition was induced in a Shorthorn steer calf within 2-3wks of feeding *Bowenia serrulata* leaf added to a chaffed oat/lucern hay diet over 20d. Ataxia was first observed at 18d, and became progressively worse. Lesions observed at necropsy on day 40 were confined to the central nervous system, and included bilaterally symmetrical degeneration in all funiculi, but mainly in the fasciculus gracilis of the dorsal columns and the dorsal spinocerebellar tract of the lateral columns. This pattern of neuropathological changes was confirmed in cattle fed leaves of *C. revoluta* in Japan (Yasuda *et al.*, 1985). Since *B. serrulata* induced neurotoxicity in a calf in only 18d compared with at least 50d for most other cycad feeding trials, this plant was considered to be the most appropriate for investigation of the toxic principle. Preliminary trials in which calves were dosed or fed with leaves of this plant were carried out in order to confirm reported neurotoxicity.

Materials and Methods

Friesian bull calves (~120kg) were fed chaffed oat and lucerne hay (Hall and McGavin, 1968). Rumen fistulas were inserted in some animals by a standard procedure so that plant material could be delivered directly into the rumen. All calves were kept in individual pens and provided with water *ad libitum*.

Bowenia serrulata fronds were collected in the Byfield Forest in Central Queensland, packed in airtight plastic parcels and sent overnight to Brisbane. Fronds were stored in 50g lots in airtight plastic bags at -30°C.

Plasma samples were collected by venous puncture before and during dosing or feeding trials for determination of AST. After signs of toxicity appeared, calves were euthanized and necropsied. Tissues collected included liver, brain and spinal cord, which were fixed and prepared for routine histopathological examination.

Results

Initially, one calf with a rumen fistula was dosed with the frozen fresh *Bowenia* leaf at 200g/d, similar to the dosing schedule used by Hall and McGavin (1968). This animal died due to acute severe hepatotoxicity within 10d, after six doses of plant were given. No lesions characteristic of cycad neurotoxicity were found in the brain and cord, but there was extensive hepatocellular necrosis with hemorrhage.

A second animal was dosed with the cycad at 50, 100, 150 or 200g/d depending on the plasma AST (Fig. 87.1). A total of 3,400g of leaf were dosed over 36d. There was no sign of posterior ataxia, but the condition of the calf deteriorated despite the return of plasma AST to normal levels, and the animal was euthanized in a moribund condition after 45d. Necropsy revealed ascites, edema of the mesentery and a hard, greyish liver. Histologically, there was marked cirrhosis characterized by loss of parenchyma, hepatocellular disorganization, extensive deposition of connective tissue and moderate biliary hyperplasia (Fig. 87.2). The central nervous system revealed extensive spongiform vacuolation of white matter characteristically surrounding the grey matter of the spinal cord (Fig. 87.3). There were no changes in spinal cord typical of cycad neurotoxicity.

In further attempts to induce *B. serrulata* toxicity, the plant was included in the diet as described by Hall and McGavin (1968). Initially, food was withdrawn overnight so that the animals would be hungry when offered hay containing the *Bowenia* leaf. After the first day, calves reluctantly consumed the hay and some animals consistently refused it. All calves that continued to consume the supplemented hay equivalent to a daily ingestion of 100g of *Bowenia* leaf for at least 7d subsequently developed hepatotoxicity, followed by death 3-6wks later. None of these calves had neurological changes that resembled those reported for cycad neurotoxicity. Clearly, the samples of frozen fresh *B. serrulata* leaf used in this study were acutely hepatotoxic and seemed not to contain any neurotoxicant.

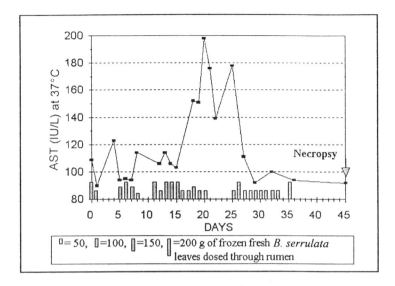

Fig. 87.1. Periodic measurement of plasma AST in a calf after multiple intra-ruminal dosing with freshly frozen *Bowenia serrulata* leaves.

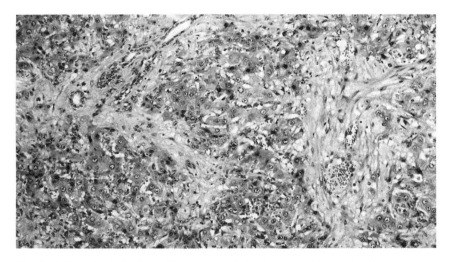

Fig. 87.2. Hepatic cirrhosis and parenchymal atrophy caused by dosing freshly frozen *Bowenia serrulata* leaf to a calf.

Discussion

Two classes of toxins occur in cycads: methylazoxymethanol (MAM) glycosides and the non-protein amino acid L-beta-methylaminoalanine (BMAA). *Bowenia serrulata*

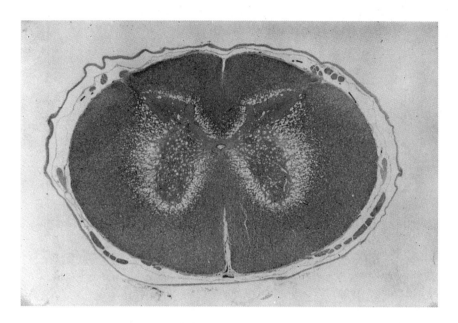

Fig. 87.3. Pattern of spongiform change surrounding the gray matter of the spinal cord typically seen in chronic hepatoencephalopathy.

has been shown to contain MAM glycosides (Moretti *et al.*, 1983), but this plant contains only trace amounts of BMAA (Duncan *et al.*, 1992). The pathological changes observed in all calves necropsied in the present study are typical of those in mammals intoxicated with MAM glycosides (Hooper, 1978; Hirono, 1987), with the pattern of spongiform changes in the white matter that are typically associated with hepatoencephalopathy in ruminants (Hooper, 1975). Glycosides of MAM are known to cause neuronal necrosis in the fetal and neonatal period in some species, but do not affect differentiated neurones *in vivo*, nor are they known to give rise to other degenerative changes in the central nervous system (Cattabeni and Di Luca, 1997).

Prior to the present study, it had been presumed that Zamia staggers in cattle grazing in Byfield was caused by *B. serrulata*. It was believed that experimental neurotoxicity could be induced rapidly if the plant was dosed or fed to calves in as fresh a state as possible. Since the *B. serrulata* used in this study produced hepatotoxicity rather than neurotoxicity, it was thought that this sample might not be representative of the *Bowenia* plants of the Byfield Forest.

The original *B. serrulata* feeding experiments (Hall and McGavin, 1968) produced typical cycad neurotoxicity but not hepatotoxicity, and it therefore seemed reasonable to assume that some *Bowenia* plants were low in MAM glycosides but contained an unknown neurotoxin. Collections of *B. serrulata* were made in 1996 at the same time of year and at the same Byfield Forest site (and at several other sites in the forest) as the 1960s collection. Within a few hours after the 1996 collection, the leaves were frozen in liquid nitrogen and stored at -80°C for MAM glycoside analysis. All 25 of the new *Bowenia* leaf samples (including those from the original site) were similar and contained 1-2% wet weight of MAM glycosides (Seawright *et al.*, 1997). Thus, the leaves used by Hall and McGavin should have caused fatal hepatotoxicity in calves. That this did not happen indicates that the leaves fed in the earlier study had a different toxic potential than those now growing at the Byfield Forest site.

In the 1968 study, leaf samples were stored at 2-5°C for 7-16d before feeding. When *Bowenia* is stored in this way, MAM glycosides are reduced to 7.5% of the original level after 7d and are undetectable by 10d. No MAM glycosides were detected in *Bowenia* samples that had been collected in the early 1990s and allowed to desiccate in open bags in the freezer at -30°C. Leaves stored in this manner were not hepatotoxic or neurotoxic when fed to a calf at the rate of 10kg over 3wks. Samples of the 1990 collection stored in airtight plastic bags contained appreciable amounts of MAM glycosides and so retained their hepatotoxicity.

Observations to date indicate that all fresh *Bowenia* leaf contains hepatotoxic amounts of MAM glycosides and that such leaves are not palatable to cattle. The plant is therefore not likely to be grazed in the field, and is unlikely to be a cause of natural cycad toxicity. In the Byfield area, cycad neurotoxicity is probably due to consumption of another cycad, *Macrozamia miquelii*, the leaves of which are very low in MAM glycosides (Seawright *et al.*, 1997).

Studies of leaves of *C. revoluta* growing in Queensland revealed that MAM glycoside levels vary considerably between individual plants (Seawright *et al.*, 1997). Cattle in Japan have grazed *C. revoluta* plants in the field and developed neurotoxicity without hepatotoxicity. Variability in MAM glycoside levels has also been shown in the leaves of Western Hemisphere cycads such as *Zamia* spp. (Seawright *et al.*, 1997). All Australian *Macrozamia* spp. tested to date have had low MAM glycoside levels, and all have been shown to cause cycad neurotoxicity.

The cause of the neurotoxicity of *B. serrulata* leaf in Hall and McGavin's (1968) trial remains unknown. One possibility is that the MAM glycosides originally present were converted into a neurointoxicant during cool storage for a week or more. If so, the unknown compound also seems to lose activity after desiccation of the plant in prolonged frozen storage. Daily oral dosing of goats with the MAM glucoside cycasin over 100d produced spinal cord degeneration and substantial hepatotoxicity (Shimizu *et al.*, 1986). No neurotoxicant derived from MAM has been discovered in cycad leaves. Shimizu *et al.* (1986) suggests that such a neurotoxicant may be produced in the animal itself. Rumen microbial metabolism may give rise to a potential neurotoxicant from MAM as long as the levels are not inhibitory to the rumen flora required for metabolic conversion of the compound. This could apply to neurotoxic, non-hepatotoxic foliage of *Macrozamia* and *Cycas* spp., as well as to the stored leaves of *B. serrulata* fed by Hall and McGavin (1968). This aspect of the toxicity is currently under investigation.

References

Cattabeni, F. and Di Luca, M. (1997) Developmental models of brain dysfunctions induced by targeted cellular ablations with methylazoxymethanol. *Physiological Reviews* 77, 199-215.

Duncan, M.W., Markey, S.P., Levy, M., Kopin, I.J., Marini, A. and Norstog, K. (1992) Cycad-induced hind limb paralysis in cattle is unrelated to the neurotoxin 2-amino-3-(methylamino)-propanoic acid (BMAA). *Phytochemistry* 31, 3429-3432.

Hall, W.T.K. (1957) Toxicity of the leaves of *Macrozamia* spp. for cattle. *Queensland Journal of Agricultural Science* 14, 41-52.

Hall, W.T.K. and McGavin, M.D. (1968) Clinical and neuropathological changes in cattle eating the leaves of *Macrozamia lucida* or *Bowenia serrulata* (family Zamiaceae). *Pathologia Veterinaria* 5, 26-34.

Hirono, I. (1987) Cycasin. In: Hirono, I. (ed), *Naturally Occurring Carcinogens of Plant Origin. Toxicology, Pathology and Biochemistry.* Elsevier, Tokyo, pp. 3-24.

Hooper, P.T. (1975) Spongy degeneration in the central nervous system of domestic animals. Part 1 Morphology. *Acta Neuropathologia* (Berlin) 31, 325-334.

Hooper, P.T. (1978) Cycad poisoning in Australia - etiology and pathology. In: Keeler, R.F., Van Kampen, K.R. and James, L.F. (eds), *Effects of Poisonous Plants on Livestock.* Academic Press, New York, NY, pp. 337-347.

Kobayashi, A., Tadera, K., Yagi, F., Kono, I., Sakamoto, T. and Yasuda, N. (1983) Cattle poisoning due to ingestion of cycad leaves. *Toxicon* Supplement 3, 229-232.

Mason, M.M. and Whiting, M.G. (1968) Caudal motor weakness and ataxia in cattle in the Caribbean area following ingestion of cycads. *Cornell Veterinarian* 58, 541-544.

Moretti, A., Sabato, S. and Gigliano, G.S. (1983) Taxonomic significance of methyl-azoxymethanol glycosides in the cycads. *Phytochemistry* 22, 115-117.

Seawright, A.A., Brown, A.W., Nolan, C.C. and Cavanagh, J.B. (1993) Cycad toxicity in domestic animals - what agent is responsible? In: Stevenson, D.W. and Norstog, K.J. (eds), *Proceedings of Cycad 90, Second International Conference on Cycad Biology.* Palm and Cycad Societies of Australia, Brisbane, Qld, pp. 61-70.

Seawright, A.A., Ng, J.C., Oelrichs, P.B., Sani, Y., Nolan, C.C., Lister, T., Holton, J., Ray, D.E. and Osborne, R. (1998) Recent toxicity studies in animals with chemicals derived from cycads. In: Chen, C.R. (ed), *Proceedings of Cycad 96, Fourth International Conference on Cycad Biology.* Panzhihua, China, May, 1996, Academia Sinica, Beijing (in press).

Shimizu, T., Yasuda, N., Kono, I., Yagi, F., Tadera, K. and Kobayashi, A. (1986) Hepatic and spinal lesions in goats chronically intoxicated with cycasin. *Japan Journal of Veterinary Science* 48, 1291-1295.

Yasuda, N., Kono, I. and Shimizu, T. (1985) Pathological studies on cycad poisoning of cattle experimentally caused by feeding with leaves of the cycad, *Cycas revoluta* Thunb. *Bulletin Faculty of Agriculture, Kagoshima University* 35, 171-178.

Chapter 88

Evaluation of the Toxic Effects of the Legumes of Mimosa (*Albizia julibrissin*) and Identification of the Toxicant

G.H. Robinson, G.E. Burrows, E.M. Holt, R.J. Tyrl and R.P. Schwab
Oklahoma State University, Stillwater, Oklahoma 74078, USA

Introduction

The legumes of *Albizia tanganyicensis* Baker f. and *A. versicolor* Welw. ex Oliver (Mimosaceae) have poisoned livestock for 30yrs in Malawi, South Africa, Zambia and Zimbabwe (Needham and Lawrence, 1966; Basson *et al.*, 1970; Kellerman *et al.*, 1988; Soldan *et al.*, 1996). Steyn *et al.* (1987) attributed the neurologic symptoms and mortalities caused by *Albizia* to two neurotoxic alkaloids (Fig. 88.1) that they hypothesized act as pyridoxine antagonists. It was also hypothesized that their effects could be counteracted by dosing with pyridoxine, based on the similarity of 4-methoxypyridoxine to that of pyridoxine (vit. B$_6$) and on prior work on pyridoxine antagonists (Kamrin and Kamrin, 1961; Mizuno *et al.*, 1980). Pyridoxine and its HCl were effective therapeutic agents for guinea-pigs and sheep intoxicated by the legumes of *A. versicolor* (Gummon and Erasmus, 1990; Gummon *et al.*, 1992).

Introduced into North America as an ornamental, *A. julibrissin* Durazz is commonly known as mimosa or silk-tree. It is native to tropical Asia, and is cultivated in the southeastern US where it has escaped and naturalized. The tree is 3-6m tall and blooms from May to August, with globose clusters of pink flowers 2.5-5cm in diameter occurring in corymbose racemes. The mature legumes are oblong, flat, 12-20cm long and 1.5-2.5cm wide (Fernald, 1950; Isely, 1973).

Few reports exist of intoxications in North America, dosing legumes of *A. julibrissin* to sheep produced neurointoxication and death (Dollahite, 1978). This investigation was to confirm its neurotoxicity, to determine if its legumes have alkaloids identical or similar to those present in the South African species and to evaluate the effect of pyridoxine HCl as a preventive/antagonist of intoxication.

Fig. 88.1. Structures of neurotoxins in *Albizia tanganyicensis* and of pyridoxine: a) 5-acetoxymethyl-3-hydroxy-4-methoxymethyl-2-methylpyridine; b) 3-hydroxy-5-hydroxymethyl-4-methoxymethyl-2-methylpyridine (= 4-methoxypyridoxine); c) 5-hydroxy-6-methyl-3,4-pyridinedimethanol (= pyridoxine).

Materials and Methods

Mature green legumes were collected in August and September, 1994, and frozen until used. In addition, green legumes and more mature, brown legumes, stored at room temperature for approximately 1yr, were fed to three animals. Six mature cross-bred, white-faced female sheep (42-74kg) were prepared with ruminal fistulae and maintained on chopped corn, commercial pelleted feed and alfalfa cubes. The sheep were housed indoors at a constant temperature and allowed free access to water.

The legumes were thawed, dried, ground and stored at room temperature or refrozen until administered. Doses of ground legumes (5-25g/kg; 0.5-2.5%) were dosed directly into the rumen and followed by 3L of tap water. Blood samples for serum chemistry were taken before, and 24 and 48hrs after dosing, and sheep were visually monitored. Because there were so few animals, some were used twice, but 2wks were allowed between exposures. Pyridoxine HCl was given intramuscularly (i.m.) and/or subcutaneously (s.c.) simultaneously with the ground legumes at 20mg/kg or intravenously (i.v.) after the onset of seizures at 10-15mg/kg.

Extraction and isolation procedures were similar to those of Steyn *et al.* (1987). Legumes (19kg) were coarsely ground in a blender with ethyl acetate (EtOAc) and extracted at room temperature for 20hrs. The extract was concentrated to a thick syrup under vacuum and extracted with petroleum ether at 40-50°C. The insoluble residue was extracted with chloroform (CHCl$_3$), and this extract subjected to silica gel chromatography. The column was eluted with CHCl$_3$, CHCl$_3$-methanol (MeOH) (90:10, v/v), CHCl$_3$-MeOH (50:50) and MeOH. The CHCl$_3$-MeOH (90:10) eluate was rechromatographed with increasingly polar solvents ($CHCl_3$ to MeOH). Fractions (10ml) were characterized by GC/EIMS. Pure alkaloids of *A. tanganyicensis* were obtained from Robert Vleggaar (South Africa).

Results

Results of administration of the legumes are shown in Table 88.1. The lethal dose was $\geq$15g/kg and the toxic dose was 10-15g/kg. Symptoms became apparent 12-14hrs after administration of a toxic dose. There was an exaggerated response to tactile, auditory, and visual stimuli, followed by muscular twitching that lasted briefly or for several minutes. Body temperature increased slightly in some animals. Severe signs included convulsive seizures with deep, labored respiration, excessive salivation, tremors or shaking, backing-up or turning, torticollis, opisthotonus, collapse, outstretched forelimbs, and paddling hindlimbs. Seizures lasted about 2min, followed by quiescence in lateral and then sternal positions. In mild cases, seizures were infrequent, at intervals of 1hr or more. In severe cases, the seizures occurred every few minutes. Two sheep given 15g/kg or 20g/kg of green 1yr-old legumes exhibited similar signs and died, as did an animal given 25g/kg of the 1yr-old brown legumes.

A sheep given 15g/kg of legumes in combination with 20mg/kg of pyridoxine HCl (half i.m., half s.c.) failed to develop any signs of intoxication. A second animal given 20g/kg of legumes with 20mg/kg of s.c. pyridoxine HCl also failed to develop

Table 88.1. Results of administration of various dosages of dried, ground legumes of *Albizia julibrissin* to sheep.

Dosage (g/kg)	Number of Animals	Results Observed
5	2	no signs of intoxication
10	1	no signs of intoxication
10	1	mild seizures
15	2	death
15	2	seizures, recovery when given i.v. pyridoxine HCl
15	1	no signs when given legumes with i.m. and i.v. pyridoxine HCl
20	2	death
20	1	no signs when given legumes with s.c. pyridoxine HCl
25	2	death

signs. Two weeks later, these sheep were given the same amount of legumes alone and died. Two other sheep developed severe seizures when given 15g/kg of legumes, but prompt relief was provided by 10-15mg/kg of i.v. pyridoxine HCl. However, the animals remained depressed for 1-2d.

Serum chemistries of sheep that did not die within 24hrs showed no significant or consistent alterations from baseline values. Parameters evaluated were glucose, urea nitrogen, creatinine, total protein, albumin, Na^+, K^+, Cl^-, Ca^{2+}, P, cholesterol, total and direct bilirubin, alkaline phosphatase, LDH, CPK, AST, and GGT.

A compound that possessed a fragmentation pattern and retention time matching that of the alkaloid 5-acetoxymethyl-3-hydroxy-4-methoxymethyl-2-methylpyridine found in *A. tanganyicensis* was revealed by GC/MS (Fig. 88.2).

Discussion

Legumes of *A. julibrissin* are clearly toxic. Although in some instances a substantial amount of material was ingested, availability of large amounts of the legumes on the pendant branches of these trees is hazardous. Neurological signs (tremors and seizures) appear abruptly several hours after ingestion.

It has been shown (Gummon and Erasmus, 1990; Gummon *et al.*, 1992) that the toxins in *A. versicolor* legumes are pyridoxine antagonists, and their neurotoxic effects are prevented or counteracted by pyridoxine or pyridoxine HCl. In this study, 10-15mg/kg of pyridoxine HCl provided prompt relief of seizure activity but the animals remained depressed for 1-2d. When legumes were accompanied by parenteral pyridoxine HCl, no signs of intoxication were observed. Thus, the protective effects of 10-15mg/kg of pyridoxine HCl were confirmed.

A compound present in the legumes of *A. julibrissin* displays the same fragmentation pattern and retention time of an alkaloid present in *A. tanganyicensis*; it is therefore tentatively identified as the same alkaloid.

The toxicity of *A. julibrissin* (15g/kg) seems to be less than that of the South African species (5g/kg) (Gummon *et al.*, 1992). However, the mode of action seems similar. The specific mechanism of activity is not fully understood, but pyridoxine antagonists such as 4-deoxypyridoxol and 4-methoxymethylpyridoxol cause seizures in laboratory animals (Holtz and Palm, 1964; Horton and Meldrum, 1973; Dakshinamurti, 1977). Pyridoxine serves as a cofactor with glutamic acid decarboxylase in the formation of GABA and with GABA transaminase in the breakdown of GABA to succinic acid (Kosower and Rock, 1968). These roles suggest that seizures may be due to impairment of GABA synthesis (Sawaya *et al.*, 1978). In mice, 4-deoxypyridoxine caused a decrease in glutamic acid decarboxylase and GABA in the brain. There was also a decrease in L-3,4-dihydroxyphenylalanine (DOPA) decarboxylase associated with a decrease in DOPA formation. Furthermore, 4-deoxypyridoxine caused a decrease in GABA transport, an effect that was counteracted by pyridoxal phosphate (Snodgrass and Iversen, 1973). The antagonists also impaired activity of GABA transaminase and the degradation of GABA.

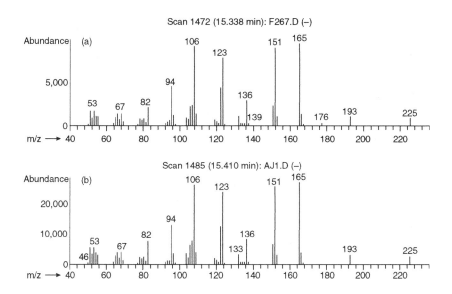

Fig. 88.2. Compounds extracted from *Albizia* spp. a) Mass spectrum of compound at R$_t$=15.338min from North American *A. julibrissin*. b) Mass spectrum of compound at R$_t$=15.410min from South African *A. tanganyicensis*.

However, it is of interest that while the effects of the toxicants in the South African species of *Albizia* are ameliorated by pyridoxine, they are not by pyridoxal (Gummon and Erasmus, 1990).

References

Basson, P.A., Adelaar, T.F., Naudé, T.W. and Minne, J.A. (1970) *Albizia* poisoning: report of the first outbreak and some experimental work in South Africa. *Journal of the South African Veterinary Medical Association* 41, 117-130.

Dakshinamurti, K. (1977) Neurobiology of pyridoxine. *Advances in Nutritional Research* 4, 143-179.

Dollahite, J.W. (1978) *Poisonous Plants:* Albizia julibrissin *(Silktree).* Texas Agricultural Experimental Station Technical Report 78-1, 2.

Fernald, M.L. (1950) *Gray's Manual of Botany*, 8th ed. American Book, New York, NY.

Gummon, B. and Erasmus, G.L. (1990) Pyridoxine (a vitamin B$_6$) and its derivative pyridoxal as treatment for *Albizia versicolor* poisoning in guinea-pigs. *Onderstepoort Journal of Veterinary Research* 57, 109-114.

Gummon, B., Bastianello, S.S., Labuschagne, L. and Erasmus, G.L. (1992) Experimental *Albizia versicolor* poisoning in sheep and its successful treatment with pyridoxine HCl. *Onderstepoort Journal of Veterinary Research* 59, 111-118.

Holtz, P. and Palm, D. (1964) Pharmacological aspects of vitamin B₆. *Pharmacology Review* 16, 113-178.

Horton, R.W. and Meldrum, B.S. (1973) Seizures induced by allylglycine, 3-mercapto-propionic acid and 4-deoxypyridoxine in mice and photosensitive baboons, and different modes of inhibition of cerebral glutamic acid decarboxylase. *British Journal of Pharmacology* 49, 52-63.

Isely, D. (1973) *Leguminosae of the United States: I. Subfamily Mimosoideae. Memoirs of the New York Botanical Garden, Vol. 25, #1*. The New York Botanical Garden, Bronx, NY.

Kamrin, R.P. and Kamrin, A.A. (1961) The effects of pyridoxine antagonists and other convulsive agents on amino acid concentrations of the mouse brain. *Journal of Neurochemistry* 6, 219-225.

Kellerman, T.S., Coetzer, J.A.W. and Naudé, T.W. (1988) *Plant Poisonings and Mycotoxicosis of livestock in Southern Africa*. Oxford University Press, Cape Town.

Kosower, N.S. and Rock, R.A. (1968) Seizures in experimental porphyria. *Nature* 217, 565-567.

Mizuno, N., Kawakami, K. and Morita, E. (1980) Competitive inhibition between 4'-substituted pyridoxine analogues and pyridoxal for pyridoxal kinase from mouse brain. *Journal of Nutritional Science and Vitaminology* 26, 535-543.

Needham, A.J.E. and Lawrence, J.A. (1966) The toxicity of *Albizia versicolor*. *Rhodesian Agricultural Journal* 63, 137-140.

Sawaya, C., Horton, R. and Medlrum, B. (1978) Transmitter synthesis and convulsant drugs; effects of pyridoxal phosphate antagonists and allylglycine. *Biochemical Pharmacology* 27, 475-481.

Soldan, A.W., van Inzen, C. and Edelsten, R.M. (1996) *Albizia versicolor* poisoning in sheep and goats in Malawi. *Journal of the South African Veterinary Medical Association* 67, 217-221.

Snodgrass, S.R. and Iversen, L.L. (1973) Effects of amino-oxyacetic acid on [³H]GABA uptake by rat brain slices. *Journal of Neurochemistry* 20, 431-439.

Steyn, P.S., Vleggaar, R. and Anderson, L.A.P. (1987) Structure elucidation of two neurotoxins from *Albizia tanganyicensis*. *South African Journal of Chemistry* 40, 191-192.

Chapter 89

The Development of Lupinosis in Weaner Sheep Grazed on Sandplain Lupins

K.P. Croker, J.G. Allen, S.P. Gittins and G.H. Doncon

Animal Research and Development Services, Agriculture Western Australia, South Perth, Western Australia 6151, Australia

Introduction

Sandplain lupins (*Lupinus cosentinii*) grow wild on 0.5×10^6 ha of infertile sands in the West Midland area of Western Australia. These plants are often colonized by the fungus *Diaporthe toxica* (anamorph *Phomopsis* spp.) (Williamson *et al.*, 1994), which under certain conditions produces phomopsin toxins (van Warmelo *et al.*, 1970; Culvenor *et al.*, 1977), which can lead to liver damage resulting in poor growth and death of sheep. Usually only older sheep are grazed on the sandplain lupins. Farmers who use them successfully for summer grazing maintain that they do not have large losses of sheep from either alkaloid poisoning or the lupinosis that can develop after toxic pods or lupin stalks are eaten (Gardiner, 1967). The experiments described here were conducted to measure the performances of Merino weaner sheep grazed on stands of sandplain lupins at different stocking rates and to assess the development of lupinosis in the sheep.

Materials and Methods

The experiments were sited in paddocks containing self-regenerating stands of sandplain lupins in a typically Mediterranean climate. The Lancelin paddock (Site 1) was a pale yellow and the Badgingarra paddock (Site 2) a white, non-wetting siliceous sand, with clay contents <5%. Site 1 plots were fertilized with 100kg superphosphate/ha in June, 1995 and 1996, while Site 2 plots received 125kg/ha in July, 1996. The average annual rainfalls (1989-1996) were 632mm and 560mm for Sites 1 and 2, respectively, with 85% of the rain falling from April to October.

Five randomly sited replicates/site of three stocking rates (5, 10 and 20sh/ha) were grazed by five sheep (75 sheep/experiment), giving plot areas of 0.25, 0.5 and 1ha. The Merino weaner sheep (June and July born, about 6mos) were allocated to groups based on stratified liveweights and subjected to normal husbandry practices. New sheep were used for each grazing period.

The plots were only grazed in summer/autumn, soon after the lupins and other plants had senesced. At Site 1, plots were grazed during four consecutive summers (Summer 1, 1993-94; Summer 2, 1994-95; Summer 3, 1995-96; Summer 4, 1996-97); Site 2 was grazed for the latter three summers. The weaners were removed when the average liveweight of the sheep in a plot fell to 90% or less of the heaviest average weight measured, or when lupinosis developed (at least one jaundiced sheep/plot) or the green feed germinated.

In the first three seasons, 10cm-diameter cores 10cm deep were collected on each plot in the spring and following the break of the next growing season. This material was combined for each plot and the lupin seeds present were counted and weighed to estimate seed reserves (kg/ha). Plant material, excluding the lupin seeds in the first three seasons, was collected from ten equally distributed $0.25m^2$ quadrats on the plots and combined for each plot. After drying at 60°C for 48hrs, the material was weighed and dry matter (DM) calculated (kg DM/ha).

The sheep were weighed every 14d on the plots until near the end of the grazing time, and were then weighed weekly. Blood samples were collected from all sheep when grazing ended. Plasma glutamate dehydrogenase (GLDH) and γ-glutamyl transferase (GGT) activities were measured using standard kit procedures. The results were examined by analysis of variance with the plasma GLDH and GGT values subjected to a log transformation.

Results

At the start of grazing there was 3.3t/ha or more of DM on the plots, with no statistically significant differences between stocking rate plots in any year (Table 89.1). There tended to be more DM at Site 2. A high proportion of the material was grasses and broad-leaved weeds at both sites. At the end of grazing, the amounts of DM available had decreased to between 1.4 and 3.5t/ha.

The amounts of seed recovered before the start of grazing in Summers 1-3 varied between 59-1617kg/ha (Table 89.1), with much more seed at Site 2 where the stands of lupins were more dense. The seed reserves were not measured in Summer 4 because of the heavy vegetation cover. In most treatments there were decreased amounts of seed recovered at the end of grazing. It is assumed that they were eaten by the sheep, but only at Site 1/Summer 3 was a high proportion of the seed consumed (Table 89.1). There was a high intake of seed on some plots in Site 2/Summer 2, but a large quantity remained after grazing.

Apart from Site 2/Summer 2, the rates at which the weaners' liveweights changed appeared to be related to stocking rate (Table 89.2), with sheep at 5/ha doing the

Table 89.1. The amounts of dry matter (DM, t/ha) and seed (kg/ha) available before and after grazing of sandplain lupins at different stocking rates (SR) (n=5).

	Summer 1 (1993-94)				Summer 2 (1994-95)			
	DM		Seed		DM		Seed	
SR (sh/ha)	Start	End	Start	End	Start	End	Start	End
Site 1 - Lancelin								
5	3.7	2.9	105	70	4.5	2.3	62	56
10	3.8	2.6	94	73	4.5	2.4	74	86
20	3.3	1.9	97	98	3.6	1.6	59	35
Site 2 - Badgingarra								
5	-	-	-	-	5.7	3.0	1,371	1,516
10	-	-	-	-	6.1	2.6	1,530	1,261
15	-	-	-	-	5.3	1.4	1,617	1,033
	Summer 3 (1995-96)				Summer 4 (1996-97)			
Site 1 - Lancelin								
5	4.8	3.5	215	45	7.8	6.6		
10	4.8	3.4	256	50	8.2	5.9		
20	4.7	2.6	200	56	6.7	5.2		
Site 2 - Badgingarra								
5	8.3	6.2	1,600	2,729	4.3	4.8		
10	8.2	4.2	2,902	1,870	4.3	4.1		
20	7.0	1.8	2,301	1,409	4.2	3.6		

best over longer periods than those at the higher stocking rates, although this was only statistically significant at Site 1/Summer 1. Sheep at Site 2 gained weight for longer periods than at Site 1 due to more available DM.

The changes at the development of lupinosis were not consistent (Table 89.3). At Site 1, plasma GLDH and GGT activities indicated that the sheep grazed at the two lower stocking rates in Summer 2, and virtually all the sheep in Summer 3, developed lupinosis by the end of grazing. At Site 1/Summer 4, a high proportion of the sheep had developed lupinosis by the end of grazing even though there had been no significant amounts of rain during the summer. At Site 2/Summer 2, the GLDH and GGT activities indicated lupinosis was present, and that liver damage was increased with increasing stocking rate. In Summer 3, there was no indication that phomopsins had affected the livers before grazing was stopped.

Discussion

The results from this study show that on very poor sandy soils, where it is difficult to grow productive improved pastures, weaner sheep can be grazed on regenerated stands of sandplain lupins and other volunteer species at low stocking rates for 3-4mos without being provided with supplementary feed. Where plenty of DM is

Table 89.2. The lengths of grazing and changes in liveweights of weaner sheep grazed on sandplain lupins at different stocking rates (SR) (n=5).

	Summer 1 (1993-94)			Summer 2 (1994-95)		
SR (sh/ha)	Grazing (days)	Days to heaviest wt	Wt change[1] (g/hd/d)	Grazing (days)	Days to heaviest wt	Wt change[1] (g/hd/d)
Site 1 - Lancelin						
5	95	81	36	112	74	10
10	98	18	- 1	131	30	- 18
20	78	14	- 23	93	21	- 38
Site 2 - Badgingarra						
5	-	-	-	126	112	87
10	-	-	-	126	112	96
20	-	-	-	112	112	94
	Summer 3 (1995-96)			Summer 4 (1996-97)		
Site 1 - Lancelin						
5	40	17	- 16	79	79	24
10	40	13	- 39	65	52	- 18
20	40	8	- 66	52	0	- 29
Site 2 - Badgingarra						
5	135	127	122	107	93	47
10	135	127	102	107	79	21
20	135	127	88	93	63	- 6

[1]Rate of change in liveweight while on the plots.

Table 89.3. The activities of plasma glutamate dehydrogenese (GLDH, U/L) and gamma-glutamyl transferase (GGT, U/L) in weaner sheep grazed on sandplain lupins at different stocking rates (SR) (n=5).

	Summer 1 (1993-94)		Summer 2 (1994-95)		Summer 3 (1995-96)		Summer 4 (1996-97)	
SR	GLDH[1]	GGT[2]	GLDH[1]	GGT[2]	GLDH[1]	GGT[2]	GLDH[1]	GGT[2]
Site 1 - Lancelin								
5	12.3	69.3	39.5	121	37.4	159	44.1	96
10	9.8	66.0	94.1	108	26.2	134	60.1	85
20	11.2	58.2	12.8	82	27.2	137	35.0	106
Site 2 - Badgingarra								
5	-	-	35.1	73	32.2	64	17.1	47
10	-	-	40.7	81	30.3	68	19.8	54
15	-	-	60.4	113	21.1	63	18.1	56

[1]GLDH - normal range 0-30 (U/L); [2]GGT - normal range 0-70 (U/L).

available, weaner sheep can increase weights at up to 120g/hd/d. Therefore, these pastures can be a valuable resource during the summer/autumn period when annual based pastures usually decrease in quantity and quality (Biddiscombe *et al.*, 1980).

Where lupinosis does not develop and there is enough DM available, young Merino sheep can reach liveweights during summer/autumn that make them suitable for either the prime lamb market or for shipping live to overseas markets.

This study indicates that the understorey in these pastures makes an important contribution to the performance of the sheep. With careful management, enough DM can be left at the end of grazing to minimize soil erosion.

Clearly, the development of lupinosis can severely reduce the length of grazing of sandplain lupins in some years. However, it is difficult to predict when lupinosis will develop, as in the sheep at Site 1/Summer 4 even though there was no rain during the summer. Sea mists may have triggered the production of phomopsins. In addition, there was more lupin material available in Summer 4, giving a higher potential for intake of lupin and consumption of phomopsins high enough to produce lupinosis. An effective vaccine to protect against the phomopsins would increase the efficient use of sandplain lupins (Allen *et al.*, 1994).

Acknowledgments

The cooperation of Jim Mazza, Bob Wilson, and Harry and Glen Wilkinson in providing the sites and assisting with the study is acknowledged. This investigation was financed by the Wool Program of Agriculture Western Australia.

References

Allen, J.G., Than, K.A., Edgar, J.A., Doncon, G.H., Dragicevic, G. and Kosmac, V.H. (1994) Field evaluations of vaccines against lupinosis. In: Colegate, S.M. and Dorling, P.R. (eds), *Plant-Associated Toxins: Agricultural, Phytochemical and Ecological Aspects.* CAB International, Wallingford, Oxon, pp. 427-432.

Biddiscombe, E.F., Arnold, G.W., Galbraith, K.A. and Briegel, D.J. (1980) Dynamics of plant and animal production of a subterranean clover pasture grazed by sheep. 1. Field measurements for model calibration. *Agricultural Systems* 6 , 3-22.

Culvenor, C.C.J., Beck, A.B., Clarke, M., Cockrum, P.A., Edgar, J.A., Frahn, J.L., Jago, M.V., Lanigan, G.W., Payne, A.L., Peterson, J.E., Petterson, D.S., Smith, L.W. and White, R.R. (1977) Isolation of toxic metabolites of *Phomopsis leptostromiformis* responsible for lupinosis. *Australian Journal of Biological Sciences* 30, 269-277.

Gardiner, M.R. (1967) Lupinosis. *Advances in Veterinary Science* 11, 85-138.

van Warmelo, K.T., Marasas, W.F.O., Adelaar, T.F., Kellerman, T.S., van Rensburg, I.B.J. and Minne, J.A. (1970) Experimental evidence that lupinosis of sheep is a mycotoxicosis caused by the fungus *Phomopsis leptostromiformis* (Kuhn) Bubak. *Journal of the South African Veterinary Medical Association* 41, 235-247.

Williamson, P.M., Highet, A.S., Gams, W., Sivasithamparam, K. and Cowling, W.A. (1994) *Diaporthe toxica* sp. nov., the cause of lupinosis in sheep. *Mycological Research* 98, 1364-1368.

Chapter 90

Mycotoxin Contamination of Australian Pastures and Feedstuffs

W.L. Bryden

Department of Animal Science, University of Sydney, Camden, New South Wales 2570, Australia

Introduction

Mycotoxins are fungal metabolites that may cause acute and fatal toxicoses when ingested. More often, the concentration of mycotoxins in animal feeds is low and their presence goes undetected, resulting in production losses. In Australia, as elsewhere, this insidious aspect of mycotoxicoses is likely to have a great economic impact on animal production. Ruminant production in Australia has largely been based on extensive grazing of native or improved pastures, but has recently moved toward feeding concentrate to dairy cows and feedlot beef. This change in production systems will expose animals to a different array of mycotoxins (Table 90.1), similar to those encountered by intensively farmed pigs and poultry.

Mycotoxins Associated with Pastures

Black soil blindness was first reported from northwestern Australia in 1994 (Jubb *et al.*, 1996), and was linked to the grazing of Mitchell grass (*Astrebla* spp.) on which fugal stromata of *Corallocyostroma* spp. were present. Detailed accounts of its causation and effects can be found in Chapters 94, 95 and 97.

Diplodia maydis has worldwide distribution, but except for one bovine case in Queensland (Qld) (Darvall, 1964) the disease has only been reported in South Africa. Animals develop neurological signs, and the unknown toxin is teratogenic in sheep exposed during the second and third trimesters (Kellerman *et al.*, 1991).

Classical ergotism (fescue foot) is characterized by lameness followed by gangrene of the extremities usually occurring when cool ambient temperatures exacerbate vasoconstriction caused by the alkaloids (Bryden, 1994b). *Claviceps purpurea* is widespread in Australia, but only a few case of classical ergotism, caused by ergovaline, have been reported in cattle from Western Australia (WA) (Fraser and Dorling, 1983), South Australia (SA), Victoria (Vic) and New South Wales (NSW) (Culvenor, 1974).

Facial eczema (pithomycotoxicosis) occurs occasionally in sheep in Vic, WA and NSW (Greenwood and Williamson, 1985). Spores of *Pithomyces chartarum* contain the mycotoxin sporidesmin, which causes a hepatogenous photosensitization in animals ingesting them. Sporidesmin toxicosis has been reported in alpacas introduced into the southern tablelands of NSW (Coulton *et al.*, 1997).

Lupinosis has been recorded in Australia for the last 30yrs (Culvenor, 1974), and is caused by the ingestion of toxins produced by the fungus *Phomopsis leptostromiformis* colonizing dead lupin plants or stubble. It is primarily a hepatogenous photosensitization disease of sheep, especially in WA, but natural outbreaks have been reported in cattle, goats, donkeys and horses (Allen, 1987).

Paspalum or dallisgrass (*Paspalum dilatatum*) staggers and perennial ryegrass (*Lolium perenne* L.) staggers (PRGS) result from ingestion of tremorgenic mycotoxins of *Claviceps paspali* and *Acremonium lolii* (Bryden, 1994a, b).

Mycotoxins Associated with Cereal Grains

Aflatoxicoses have occurred in poultry, sheep, pigs, cattle and dogs in Australia (Bryden, 1982). *Aspergillus flavus* and *A. parasiticus* are widespread in nature and local isolates are highly toxigenic, but due to climatic conditions and agronomic practices of this country this mycotoxin is not a major problem. Exceptions are locally produced peanuts in drought years (Graham, 1982) and also maize on some occasions (Blaney and Williams, 1991; Ravindran *et al.*, 1996).

Alternariol and its monomethyl ether, secondary metabolites of *Alternaria*, have been associated with poor production in broilers (Bryden *et al.*, 1984), but have such low toxicity in poultry (Bryden *et al.*, 1985) that their presence would not explain the production drops observed. The much more toxic compound, tenuazonic acid, was also present in the moldy sorghum from the 1984 episode (Andrews and Lukas, 1984). Many Australian isolates of *Alternaria* from wheat, barley and sorghum are toxic in a chick bioassay (Bryden *et al.*, 1987a).

Vulvo-vaginitis occurs in pigs consuming maize-based diets infected with *Fusarium graminearum* Group 2. The toxin zearalenone was detected at 1-8ppm in an outbreak of hyperestrogenism in pigs in Qld (Blaney *et al.*, 1984). There were very early reports of the syndrome in Vic (Pullar and Lerew, 1937) and a recent NSW report of the syndrome in pigs consuming moldy maize. Several other mycotoxins are also produced by *F. graminearum* Group 2, including deoxynivalenol (DON; vomitoxin), a trichothecene which is toxic to pigs but much better tolerated by cattle and chickens. It has been found in locally grown wheat and triticale associated with feed refusal and vomiting in pigs (Bryden *et al.*, 1987b; Tobin, 1988).

Equine leukoencephalomalacia (ELEM) has been recently diagnosed in Australia

(Christley *et al.*, 1993; Shanks *et al.*, 1995), but there have been no reports of porcine pulmonary edema, the clinical expression of fumonisin intoxication in pigs. The occurrence of *F. moniliforme* and fumonisins in Australian maize is detailed in Chapter 92.

Hyperthermia (41-42°C) of cattle, increased respiration rate and excessive salivation exacerbated by daily temperatures in excess of 35°C was shown to be due to *C. purpurea*-contaminated diets (Burgess *et al.*, 1986; Ross *et al.*, 1989). There was no evidence of gangrene in any cattle, and the clinical expression of ergotism was mediated by ambient temperature (Bryden, 1990; 1994b). Death following hyperthermia of feedlot cattle has been reported from WA after consumption of rations contaminated with *C. purpurea* (Peet *et al.*, 1991).

Penicillium spp. are also widespread and produce a variety of toxins including penicillic acid, cyclopiazonic acid, penitrem A, ochratoxin A, patulin and roquefortine. Ochratoxin A causes a porcine nephropathy that has not been reported in Australia, although ochratoxin has been found in Australian feed samples (Connole *et al.*, 1981). Penitrem A, a tremorgenic mycotoxin, was found in a moldy hamburger bun (*P. crustosum*) consumed by a dog that developed severe muscle tremors and had great difficulty standing (Hocking *et al.*, 1988).

Table 90.1. Mycotoxins and mycotoxic disorders reported in Australia.

Toxin	Fungal genus	Disorder	Grass/grain
Grazing animals			
Unknown	Corallocytostroma	Black soil blindness	Mitchell grass
Unknown	Diplodia	Diplodiosis	Maize stubble
Ergot alkaloids	Claviceps	Ergotism	Ryegrass
Sporidesmin	Pithomyces	Facial eczema	Pasture litter
Ergot alkaloids	Acremonium	Fescue foot	Tall fescue
Phomopsins	Phomopsis	Lupinosis	Lupin stubble
Paspalanine	Claviceps	Paspaslum staggers	Paspalum
Lolitrem B	Acremonium	Ryegrass staggers	Ryegrass
Grain-fed animals			
Aflatoxins	Aspergillus	Aflatoxicosis	Peanuts, maize
Alternariols	Alternaria	Poor performance	Sorghum
Deoxynivalenol	Fusarium	Feed refusal	Wheat
Ergot alkaloids	Claviceps	Bovine hyperthermia	Ryegrass
Fumonisins	Fusarium	Leukoencephalomalacia	Maize
Zearalenone	Fusarium	Vulvo-vaginitis	Maize, sorghum

Conclusions

Outlined above are 14 distinct mycotoxicoses that were acute, sporadic disease episodes that would have been obvious to farmers and disrupted farm income. However, in many situations mycotoxins cause insidious losses: production drops, illthrift, increased returns to first service and reduced pathogen resistance. It is likely that mycotoxins will be overlooked in the diagnoses of such cases. If various aspects of loss from mycotoxicoses are considered, one is faced with a multitude of possible economic impacts (CAST, 1989). Mycotoxicoses in Australia are ill-defined and improperly assessed because of the lack of systematic surveys of the occurrence of mycotoxin contamination of pastures and feedstuffs, and inadequate knowledge of the effects of mycotoxins on animal health.

References

Allen, J.G. (1987) Lupinosis. *The Postgraduate Committee in Veterinary Science, University of Sydney, Proceedings No. 103*. Veterinary Clinical Toxicology, pp. 113-131.

Andrew, S. and Lukas, S. (1984) personal communication.

Blaney, B.J., Bloomfield, R.C. and Moore, C.J. (1984) Zearalenone intoxication of pigs. *Australian Veterinary Journal* 61, 24-27.

Blaney, B.J. and Williams, K.C. (1991) Effective use in livestock feeds of mouldy and weather-damaged grain containing mycotoxins: Case histories and economic assessments pertaining to pig and poultry industries of Queensland. *Australian Journal of Agricultural Research* 42, 993-1012.

Bryden, W.L. (1982) Aflatoxin and animal production: an Australian perspective. *Food Technology in Australia* 34, 216-223.

Bryden, W.L. (1990) Ambient temperature and the clinical expression of ergotism and ergot-like syndromes. In: Quisenburg, S.S. and Joost, R.E. (eds), *Acremonium/Grass Interactions*. Louisiana State University Press, Baton Rouge, LA, pp. 209-211.

Bryden, W.L. (1994a) Neuromycotoxicoses in Australia. In: Colegate, S.M. and Dorling, P.R. (eds), *Plant-Associated Toxins: Agricultural, Phytochemical and Ecological Aspects*. CAB International, Wallingford, Oxon, pp. 363-368.

Bryden, W.L. (1994b) The many guises of ergotism. In: Colegate, S.M. and Dorling, P.R. (eds) *Plant-Associated Toxins: Agricultural, Phytochemical and Ecological Aspects*. CAB International, Wallingford, Oxon, pp. 381-386.

Bryden, W.L., Suter, D.A.I. and Jackson, C.A.W. (1984) Response of chickens to sorghum contaminated with *Alternaria*. *Proceedings of the Nutrition Society of Australia* 9, 109.

Bryden, W.L., Barrow, K.D. and Suter, D.A.I. (1985) Toxicity of *Alternaria* to chickens. *Sixth International IUPAC Symposium on Mycotoxins and Phycotoxins*. Pretoria, South Africa, p. 60.

Bryden, W.L., Bakau, B.J.K. and Burgess, L.W. (1987a) Toxicity of Australian *Alternaria* isolates in a chick bioassay. *Proceedings of the Nutrition Society of Australia* 12, 171.

Bryden, W.L., Love, R.J. and Burgess, L.W. (1987b) Feeding grain contaminated with *Fusarium graminearum* and *Fusarium moniliforme* to pigs and chickens. *Australian Veterinary Journal* 64, 225-226.

Burgess, L.W., Bryden, W.L., Jessup, T.M., Scrivener, C.J. and Barrow, K.D. (1986) Role for ergot alkaloids in bovine hyperthermia. *Proceedings of the Nutrition Society of Australia* 11, 120.

CAST (1989) *Mycotoxins: Economic and Health Risks*. Council for Agricultural Science and Technology. Task Force Report No. 116, Ames, IA.

Christley, R.M., Begg, A.P., Hutchins, D.R., Hodgson, D.R. and Bryden, W.L. (1993) Leukoencephalomalacia in horses. *Australian Veterinary Journal* 70, 225-226.

Connole, M.D., Blaney, B.J. and McEwan, T. (1981) Mycotoxins in animal feeds and toxic fungi in Queensland 1971-80. *Australian Veterinary Journal* 57, 314-318.

Coulton, M.A., Dart, A.J., McClintock, S.A. and Hodgson, D.R. (1997) Sporidesmin toxicoses in an alpaca. *Australian Veterinary Journal* 75, 136-137.

Culvenor, C.C.J. (1974) The hazard from toxigenic fungi in Australia. *Australian Veterinary Journal* 50, 69-78.

Darvall, P.M. (1964) Mouldy corn cobs, a danger to cows. *Queensland Agricultural Journal* 90, 692-693.

Fraser, D.M. and Dorling, P.R. (1983) Suspected ergotism in two heifers. *Australian Veterinary Journal* 60, 303-305.

Graham, J. (1982) The occurrence of aflatoxin in peanuts in relation to soil type and pod splitting. *Food Technology in Australia* 34, 208-212.

Greenwood, P.E. and Williamson, G.N. (1985) An outbreak of facial eczema in sheep. *Australian Veterinary Journal* 62, 65-66.

Hocking, A.D., Holds, K. and Tobin, N.F. (1988) Intoxication of tremorgenic mycotoxin (penitrem A) in a dog. *Australian Veterinary Journal* 65, 82-85.

Jubb, T.F., Main, D.C., Mitchell, A.A., Shivas, R.G. and de Witte, D.E. (1996) Black soil blindness: a new mycotoxicosis of cattle grazing *Corallocytostroma*-infected Mitchell grass (*Astrebla* spp.). *Australian Veterinary Journal* 73, 49-51.

Kellerman, T.S., Prozesky, L., Schultz, R.A., Rabie, C.J., Van Ark, H., Maartens, B.P. and Lubben, A. (1991) Perinatal mortality in lambs of ewes exposed to cultures of *Diplodia maydis (Stenocarpella maydis)* during gestation. *Onderstepoort Journal of Veterinary Research* 58, 297-308.

Peet, R.L., McCarthy, M.R. and Barbetti, M.J. (1991) Hyperthermia and death in feedlot cattle associated with the ingestion of *Claviceps purpurea*. *Australian Veterinary Journal* 68, 121.

Pullar, E.M. and Lerew, W.M. (1937) Vulvovaginitis of swine. *Australian Veterinary Journal* 13, 28-31.

Ravindran, G., Gill, R.J. and Bryden, W.L. (1996) Aflatoxin, fumonisin and zearalenone contamination of Australian maize. *XX World's Poultry Congress, Vol IV*. New Delhi, India, p. 273.

Ross, A.D., Bryden, W.L., Bakau, W.J.K. and Burgess, L.W. (1989) Induction of heat stress in beef cattle by feeding the ergots of *Claviceps purpurea*. *Australian Veterinary Journal* 66, 247-249.

Shanks, G., Tabak, P., Begg, A.P. and Bryden, W.L. (1995) An outbreak of acute leuko-encephalomalacia associated with fumonisin intoxication in three horses. *Australian Equine Veterinarian* 13, 17-18.

Tobin, N.F. (1988) Presence of deoxynivalenol in Australian wheat and triticale - New South Wales Northern Rivers Region, 1983. *Australian Journal of Experimental Agriculture* 28, 107-110.

Chapter 91

Control of the Mycotoxic Hepatogenous Photosensitization, Facial Eczema, in New Zealand

B.L. Smith, N.R. Towers, R. Munday, C.A. Morris, and R.G. Collin
Toxicology and Food Safety Research Group, New Zealand Pastoral Agriculture Research Institute, Ruakura Research Center, Private Bag 3123, Hamilton 2001, New Zealand

Introduction

Control methods for facial eczema include the recognition and avoidance of dangerous pastures, fungicidal spraying of pastures, facial eczema prophylaxis and breeding resistant animals. The distinctive spores of *Pithomyces chartarum* are easily recognized, and spore counting of pastures is used to identify dangerous conditions. Stock and pasture management combined with the provision of alternative feeds controls the disease. The benzimidazole fungicides have been effectively used in the past to reduce fungal growth but are now used only for valuable livestock because of expense. Prophylaxis using zinc salts has proved to be effective and safe, and recently an intraruminal zinc bolus has been developed for sheep; a bolus for cattle is being developed. Breeding sheep for facial eczema resistance has been successfully applied in New Zealand. Encouraging results have been obtained recently in research fostering 'biocontrol,' inoculating pastures with non-toxic strains of *P. chartarum* to compete with toxigenic strains.

Facial Eczema

Facial eczema, a hepatogenous photosensitization, is one of the most important ruminant diseases in New Zealand. It also occurs in most other warm temperate areas of the Southern Hemisphere and in some warm temperate areas of Europe. The saprophytic fungus *P. chartarum* proliferates in the warm, humid conditions of autumn and forms spores containing the hepatotoxin, sporidesmin.

Pithomyces chartarum grows in the litter at the base of ryegrass pastures. During the late summer and autumn when humidity levels are high (>90%) and grass minimum temperatures are greater than 13°C (especially when they range up to 20°C) the fungus proliferates, and high spore counts (50,000-300,000 spores/g of fresh grass) are obtained. In New Zealand the spore of *P. chartarum* is easily recognized and most spores are toxigenic. Spore counts are easily performed and there is a good correlation between spore counts and toxicity. Depending on the length of time that the count remains high and on the grazing pressure (Smith *et al.*, 1987), these spores can result in liver injury. Counts of 50,000 over several weeks or 200,000-300,000 over just a few days are considered dangerous. Liver injury occurs, and about 7-10d later phylloerythrin (from the breakdown of chlorophyll in the rumen) concentrations in the skin are sufficiently high to cause photosensitivity after relatively brief exposure to strong sunlight. The sporidesmin molecule, which is concentrated in the biliary system, contains a disulfide bridge, and this has been shown (Munday, 1982) to undergo cyclic oxidation and reduction resulting in the formation of free radicals, which may cause the biliary injury.

All farmed ruminants in New Zealand are susceptible to facial eczema. Alpacas and fallow deer are the most susceptible and goats, especially the fiber breeds, are most resistant. Horses appear to be unaffected, and sheep, red deer and cattle are of intermediate susceptibility. Facial eczema only causes occasional deaths, but there are substantial losses due to reductions in animal weights or weight gains, carcass and offal production, milk production, pelt and wool yields, lamb birth and weaning weights and reproductive performance. The severity of facial eczema outbreaks varies from year to year, depending on the particular weather patterns. Consequently, estimates of its cost to the New Zealand economy vary, with figures for the mean annual cost ranging from NZ$30m (Smith and Towers, 1985) for all species (1981 costs) to NZ$69m (Anonymous, 1990) for sheep production alone (1989 costs). A figure of NZ$50m represents approximately 0.15% of New Zealand's Gross Domestic Product (1982 prices).

High litter levels in pasture tend to increase spore numbers. Weather patterns that favor several cycles of warmth and humidity, usually from tropical low pressure systems (Smith and Towers, 1985) moving south onto the North Island of New Zealand, produce very high numbers of animals showing photosensitization. Light rains and dews also contribute to the humidity at the base of the plant. Lower latitudes and altitudes, northerly aspects and shelter from the wind and trampling by stock are known to increase the risk of facial eczema (di Menna and Bailey, 1973).

In New Zealand the diagnosis of facial eczema is based on the seasonality of the outbreak, the previous presence of high spore numbers, typical climatic conditions and characteristic liver lesions or clinical pathology profile suggesting obstructive pericholangitis. Exacerbating factors can include the presence in the pasture of *Panicum dichotomiflorum*. In countries outside of Australia an initial diagnosis requires, in addition to the foregoing, a proper identification of toxigenic strains of *P. chartarum* because many strains may be non-toxigenic.

Control of Facial Eczema

Strategies for control are based on reducing sporidesmin intake, protecting animals from the effects of sporidesmin and increasing genetic resistance of the animals.

Identifying dangerous pastures and times is based on recognizing dangerous climatic conditions, identifying areas, farms or paddocks known from past experience to be more at risk, and conducting spore counting. Many farmers in areas prone to facial eczema collect their own grass samples and either do their own spore counts or send the samples to a local laboratory. Others may rely on spore counts reported in local newspapers or on the radio. Intake of spores can then be avoided or reduced by reducing grazing pressure through an 'open gate' set stock regime, feeding conserved feed or crops or by selling off stock or drying off milking cows. All methods aim to reduce the grazing of the pasture base.

Benzimidazole fungicides sprayed at the rate of 140-280g/ha will reduce spore rises by 55-65% and provide safe pastures for 4-6wks. However, farmers must respray if >25mm rain falls within 24hrs of spraying. This method, while effective (Parle and di Menna, 1972a, b) is relatively expensive and tends to be used for more valuable farm animals.

Zinc protects the liver from the effects of sporidesmin (Smith *et al.*, 1977; Towers, 1977; Towers and Smith, 1978), reducing the severity of the liver injury by 50-90% depending on the strategies used. High dose rates of zinc are needed to produce prophylaxis (not treatment) (Smith, 1977; Smith *et al.*, 1979), and at recommended rates the margin of safety is adequate; the use of zinc has been adopted for facial eczema control in sheep and cattle (Smith *et al.*, 1983a; Towers and Smith, 1983). Parenteral administration of zinc causes local reactions at the injection site, but oral drenching of zinc oxide slurries and use of a slow release zinc bolus afford no notable adverse effects (Munday *et al.*, 1997). Zinc sulfate is introduced into the drinking water of dairy cattle for prevention of facial eczema (Smith *et al.*, 1983a, b). Recipes and dose rates have been published for farmers and professionals, and commercial preparations of zinc have been registered for facial eczema control with the national Animal Remedies Board.

Resistance of sheep to facial eczema is inherited, and susceptible and resistant flocks of sheep have been established (Morris *et al.*, 1995). Performance tests and selective matings increase the resistance of specific commercial flocks, and a business for performance-testing rams, "Ramguard," has been established in New Zealand. "Ramguard" testing dose rates used to produce minor liver injury from a single dose of sporidesmin have risen over 12yrs from 0.07 to 0.48mg/kg.

Field trials introducing strains of *P. chartarum* that do not produce sporidesmin (Collin and Towers, 1995; Collin *et al.*, 1996) have been carried out during the last three facial eczema seasons. The 1994-95 trial used one field with maximum counts of 80,000 spores of *P. chartarum*/g of grass. When treated with the biocontrol fungus, pasture sporidesmin toxicity was reduced by up to 80% compared to untreated areas. The 1995-96 trials used three fields with three different rates of application of the biocontrol fungus. Maximum spore counts varied from 44,000-240,000 spores/g of grass, and pasture toxicity was reduced by 81%, 47% and 19% at the high, medium and low application rates, respectively. In another trial, a field with high levels of sporidesmin was divided, and one half of the field was treated with the biocontrol fungus. At the end of grazing, lambs on the untreated area showed substantial liver damage (average GGT value of 280U/L), while those on the treated area were healthy and free of liver injury. Future trials will extend these results to determine optimal treatment rates and formulations of the biocontrol strain of *P. chartarum*.

References

Anonymous (1990) *Estimated Cost of Facial Eczema to Sheep Production*. Paper No. T124 New Zealand Meat and Wool Boards' Economic Service, Wellington, NZ.

Collin, R.G. and Towers, N.R. (1995) Competition of a sporidesmin-producing *Pithomyces* strain with a non-toxigenic *Pithomyces* strain. *New Zealand Veterinary Journal* 43, 149-152.

Collin, R.G., Smith, B.L. and Towers, N.R. (1996) Lack of toxicity of a nonsporidesmin-producing strain of *Pithomyces chartarum* in cell culture and when dosed to lambs. *New Zealand Veterinary Journal* 144, 131-134.

di Menna, M.E. and Bailey, J.R. (1973) *Pithomyces chartarum* spore counts in pasture. *New Zealand Journal of Agricultural Research* 16, 343-351.

Morris, C.A., Towers, N.R., Wheeler, M. and Wesselink, C. (1995) Selection for or against facial eczema susceptibility in Romney sheep, as monitored by serum concentrations of a liver enzyme. *New Zealand Journal of Agricultural Research* 38, 211-219.

Munday, R. (1982) Studies on the mechanism of toxicity of the mycotoxin, sporidesmin. I. Generation of superoxide radical by sporidesmin. *Chemico-Biological Interactions* 41, 361-374.

Munday, R., Thompson, A.M., Fowke, E.A., Wesselink, C., Smith, B.L., Towers, N.R., O'Donnell, K., McDonald, R.M., Stirneman, M. and Ford, A.J. (1997) A zinc-containing intraruminal device for facial eczema control in lambs. *New Zealand Veterinary Journal* 45, 93-98.

Parle, J.N. and di Menna, M.E. (1972a) Fungicides and the control of *Pithomyces chartarum*. I: Laboratory trials. *New Zealand Journal of Agricultural Research* 15, 48-53.

Parle, J.N. and di Menna, M.E. (1972b) Fungicides and the control of *Pithomyces chartarum*. II: Field trials. *New Zealand Journal of Agricultural Research* 15, 54-63.

Smith, B.L. (1977) Toxicity of zinc in ruminants in relation to facial eczema. *New Zealand Veterinary Journal* 25, 310-312.

Smith, B.L. and Towers, N.R. (1985) Pithomycotoxicosis (facial eczema) in New Zealand and the use of zinc salts for its prevention. In: Seawright, A.A., Hegarty, M.P., James, L.F. and Keeler, R.F. (eds), *Plant Toxicology: Proceedings of the Australia-USA Poisonous Plants Symposium, Brisbane, Australia, May 14-18, 1984.* Queensland Poisonous Plant Committee, Yeerongpilly, Qld, pp. 70-79.

Smith, B.L., Embling, P.P., Towers, N.R., Wright, D.E. and Payne, E. (1977) The protective effect of zinc sulphate in experimental sporidesmin poisoning of sheep. *New Zealand Veterinary Journal* 25, 124-127.

Smith, B.L., Reynolds, G.W. and Embling, P.P. (1979) The effect of method of oral administration on acute zinc toxicity in sheep. *New Zealand Journal of Agricultural Research* 7, 107-110.

Smith, B.L., Towers, N.R., Jordan, R.B. and Mills, R.A. (1983a) Facial eczema control in dairy cattle by zinc in drinking water. *AgLink*: FPP 789. New Zealand Ministry of Agriculture and Fisheries, Wellington, NZ.

Smith, B.L., Embling, P.P. and Pearce, M.G. (1983b) Zinc sulphate in the drinking water of lactating dairy cows for facial eczema control. *Proceedings of the New Zealand Society of Animal Production* 43, 217-219.

Smith, B.L., Embling, P.P. and Gravett, I.M. (1987) *Pithomyces chartarum* spore counts in rumen contents and faeces of sheep exposed to autumn pasture at three different grazing pressures. *Journal of Applied Toxicology* 7, 179-184.

Towers, N.R. (1977) Effect of zinc on the toxicity of the mycotoxin sporidesmin to the rat. *Life Science* 20, 413-418.

Towers, N.R. and Smith, B.L. (1978) The protective effect of zinc sulphate in experimental sporidesmin intoxication of lactating dairy cows. *New Zealand Veterinary Journal* 26, 199-202.

Towers, N.R. and Smith, B.L. (1983) Facial eczema: zinc dosing for prevention. Revised recommendations 1984. *AgLink*: FPP 496. New Zealand Ministry of Agriculture and Fisheries, Wellington, NZ.

Chapter 92

Occurrence of *Fusarium moniliforme* and Fumonisins in Australian Maize in Relation to Animal Disease

W.L. Bryden[1], G.J. Shanks[2], G. Ravindran[1], B.A. Summerell[3] and L.W. Burgess[4]

[1]*Department of Animal Science, University of Sydney, Camden, New South Wales 2570, Australia;* [2]*Veterinary Clinic, 57 Maize Street, Tenambit, NSW, Australia;* [3]*Royal Botanic Gardens, Mrs Macquarie's Road, Sydney, NSW, Australia;* [4]*Department of Crop Sciences, University of Sydney, NSW, Australia*

Introduction

The fungal genus *Fusarium* is one of the most economically important plant pathogens. It includes many species that are toxigenic (Marasas *et al.*, 1984; Joffe, 1986), causing several diseases such as root and crown rots, vascular wilts, stem and stalk rots, and diseases of the inflorescences such as cob rot of maize and head blight of wheat and other winter cereals (Nelson *et al.*, 1981). Some inflorescence pathogens are aggressive colonizers of grain pre- or post-maturity, and produce mycotoxins in the grain. *Fusarium* spp. are isolated regularly from Australian grain exposed to wet conditions between maturity and harvest (Burgess *et al.*, 1981). *Fusarium moniliforme*, the most studied member of section Liseola, is one of the most prevalent fungi associated with human dietary staples such as corn and sorghum throughout the world (Marasas *et al.*, 1984; Nelson *et al.*, 1981). It can be internally seed-borne in symptomless, apparently healthy corn kernels.

In 1988, fumonisins, a new group of mycotoxins from *F. moniliforme*, were structurally characterized (Gelderblom *et al.*, 1988). Fumonisin B_1 is the major fumonisin present in culture and naturally contaminated samples. It causes equine leukoencephalomalacia (ELEM) (Marasas *et al.*, 1988; Kellerman *et al.*, 1990) and porcine pulmonary edema (Ross *et al.*, 1990), is hepatocarcinogenic in rats (Gelderblom *et al.*, 1991) and is associated with human esophageal cancer (Norred and Voss, 1994). Fumonisin carcinogenicity and toxicity are due to inhibition of sphinganine (sphingosine) N-acyl transferase (ceramide synthase); this disruption of sphingolipid metabolism is an early event in the onset and progression of diseases associated with fumonisins (Merrill *et al.*, 1996; Riley *et al.*, 1994). The toxigenicity of Australian isolates of section Liseola is explored here.

Mycogeography

Fusarium moniliforme and *F. proliferatum* are widespread in eastern Australia, being more common in the warmer subtropical areas, and are most commonly found on maize. *Fusarium subglutinans* grows in the cooler subtropical and temperate regions of eastern Australia and is also associated with maize diseases. *Fusarium anthophilum* is rarely isolated in Australia. *Fusarium nygamai* is more abundant in hot dry areas, especially in western New South Wales (NSW) and Queensland (Qld), and in the arid parts of the Northern Territory (NT), where it is associated with grassland vegetation. *Fusarium babinda* has been recovered from eastern coastal Australia, from Tasmania, and Norfolk and Lord Howe Islands.

Maize is a relatively minor crop in eastern Australia compared to wheat and grain sorghum. It is mainly grown under dryland conditions on the north coast of NSW, the southeast of Qld and the Atherton Plateau in north Qld in areas with a relatively high summer rainfall. It is also grown under irrigation in drier areas of inland NSW. A wide range of *Fusarium* spp. have been associated with cob rot and grain damage in dryland maize in the wetter areas. Various reports indicate that *F. graminearum* Group 2, *F. subglutinans* and *F. moniliforme* are the most common species associated with maize grain from eastern Australia, occurring in most seasons, but with severity fluctuating in relation to rainfall during the pre-harvest period (Burgess, 1985). Where there is obvious mold damage to maize kernels, *F. moniliforme* is the species most frequently isolated (Williams *et al.*, 1992).

Grain sorghum is grown mainly as a dryland summer crop in the summer dominant rainfall region of the wheat belt and the central highlands (Emerald area) of Qld. The crop is normally harvested in late summer or autumn, but is delayed in some seasons by prolonged wet conditions, which favor the occurrence of head blight and pre-harvest grain damage. There have been no systematic surveys of the fungi associated with these problems, but minor studies indicate that a wide range of fungi are associated with damaged grain. *Fusarium moniliforme* and *Alternaria* spp. are particularly common on rachis tissue and damaged grain (Ali *et al.*, 1991; Burgess *et al.*, 1981). *Fusarium moniliforme* is commonly associated with basal stalk rot, the most important disease of grain sorghum in NSW and Qld, and thus can provide abundant inoculum to infect aerial plant parts.

Of 675 visually normal maize kernels, 60% were infected with *F. moniliforme*, 15% with *Acremonium*, and less than 2% with *F. subglutinans, F. proliferatum* and *F. nygamai* (Benyon *et al.*, unpublished). This result is similar to previous studies in Australia (Williams *et al.*, 1992) and USA (Arino and Bullerman, 1994).

The mating population (MP) of the cultures of *F. moniliforme* (Leslie *et al.*,

1992) isolated from maize kernels examined above was predominantly MP-A, with occasional occurrence of MP-F (Summerell *et al.*, unpublished). High levels of fumonisins are produced by MP-A, which is common on corn, while MP-F does not produce fumonisins and is more common on sorghum (Leslie *et al.*, 1992). Where these crops are grown together, the interactions will be more complex.

Toxigenicity

The toxigenicity of Australian isolates in section Liseola was examined in 50 isolates of *F. moniliforme* and related species from the culture collection of the Fusarium Research Laboratory, University of Sydney, after growth on Weetbix® media for 21d at 26°C. A chloroform ($CHCl_3$)/methanol (MeOH) extract of each culture was subjected to a chick bioassay (Wing *et al.*, 1993), and fumonisin production was determined by ELISA. Isolates of *F. moniliforme, F. napiforme, F. nygamai* and *F. proliferatum* produced fumonisins. Some isolates of these species and those of *F. anthophilum, F. babinda, F. beomiforme* and *F. subglutinans* were toxic in the chick bioassay, but no fumonisins were detected (Bryden *et al.*, 1994); the other toxic secondary metabolites were not identified.

A survey of 100 samples of apparently normal Australian maize obtained from human and animal food manufacturers and grain merchants found only seven samples negative for fumonisin B_1 determined by ELISA (Veratox®; Neogen) with a sensitivity of 200µg/kg (Table 92.1). Corn destined for human consumption was obtained from breakfast cereal manufacturers, and the level of contamination is consistent with previous studies from other parts of the world (Bullerman and Tsai, 1994). Almost half (46%) of the animal feed samples contained fumonisin levels of 5mg/kg and above, with 23% containing 10mg/kg or more. The two highest concentrations of fumonisins in this survey, 40.6mg/kg and 35mg/kg, were detected in corn used in equine and pig feeds, respectively. Both showed visible signs of fungal infection. The equine feed was associated with two deaths from ELEM. In the pigs, there were reduced growth rates and feed conversion efficiencies, but no clinical signs of pulmonary edema. In addition, 18 samples of corn gluten all contained fumonisins (maximum, 28.8mg/kg).

During 1994-95, two cases of ELEM involving the deaths of three and two

Table 92.1. Fumonisin contamination (mean±SE) of corn destined for human or animal consumption.

Consumer	No. of samples	No. positives	% contaminated	Mean±SE mg/kg	Range mg/kg
Human	40	35	88	1.7±0.6	0-13.5
Animal	60	58	97	7.8±1.01	0-40.6

horses respectively were reported in NSW, with maize fumonisins of 164mg/kg and 40.6mg/kg (Shanks *et al.*, 1995). Three separate outbreaks of ELEM from the last decade were confirmed by retrospective study of the clinical records of the Rural Veterinary Center, University of Sydney (Christley *et al.*, 1993). There have been no Australian reports associating maize consumption and porcine pulmonary edema. Maize infected with *F. moniliforme* and/or *Diplodia maydis* was without ill effect when fed to ruminants and chickens (Bryden *et al.*, 1987).

Conclusions

These study results demonstrate that occurrence of *F. moniliforme* mating type A with maize and the associated fumonisin contamination is similar to that in the northern hemisphere and southern Africa. The lower incidence of animal disease related to fumonisins in Australia reflects the lower inclusion rates of maize in animal diets. However, the levels of fumonisin contamination suggest that this toxin should be monitored routinely in Australia.

Acknowledgments

This paper is based on research supported in part by the Grains Research and Development Corporation.

References

Ali, H., Summerell, B.A. and Burgess, L.W. (1991) An evaluation of three media for the isolation of *Fusarium, Alternaria* and other fungi from sorghum grain. *Australasian Plant Pathology* 20, 134-138.

Arino, A.A. and Bullerman, L.B. (1994) Fungal colonization of corn grown in Nebraska in relation to year, genotype and growing season. *Journal of Food Protection* 57, 1084-1087.

Bryden, W.L., Love, R.J. and Burgess, L.W. (1987) Feeding grain contaminated with *Fusarium graminearum* and *Fusarium moniliforme* to pigs and chickens. *Australian Veterinary Journal* 64, 225-226.

Bryden, W.L., Van Wel, P.W., Salahifar, H. and Burgess, L.W. (1994) Fumonisin production by *Fusarium moniliforme* and related species. *Proceedings of the Nutrition Society of Australia* 18, 67.

Bullerman, L.B. and Tsai, W.J. (1994) Incidence and levels of *Fusarium moniliforme, Fusarium proliferatum* and fumonisins in corn and corn-based foods and feeds. *Journal of Food Protection* 57, 541-546.

Burgess, L.W. (1985) Mycotoxigenic species of *Fusarium* associated with grain diseases in eastern Australia. In: Lacey, J. (ed), *Trichothecenes and Other Mycotoxins*. John Wiley and Sons, London, pp. 15-19.

Burgess, L.W., Dodman, R.L., Pont, W. and Mayers, P. (1981) *Fusarium* diseases of wheat, maize and grain sorghum in eastern Australia. In: Nelson, P.E., Tousson, T.A. and Cook, R.J. (eds), Fusarium: *Diseases, Biology and Taxonomy*. Pennsylvania State University Press, University Park, PA, pp. 64-76.

Christley, R.M., Begg, A.P., Hutchins, D.R., Hodgson, D.R. and Bryden, W.L. (1993) Leukoencephalomalacia in horses. *Australian Veterinary Journal* 70, 225-226.

Gelderblom, W.C.A., Jaskiewicz, K., Marasas, W.F.O., Thiel, P.G., Horak, R.M., Vleggaar R. and Kriek, N.P.J. (1988) Fumonisins - novel mycotoxins with cancer-promoting activity produced by *Fusarium moniliforme*. *Applied Environmental Microbiology* 54, 1806-1811.

Gelderblom, W.C.A., Kriek, N.P.J., Marasas, W.F.O. and Thiel, P.G. (1991) Toxicity and carcinogenicity of the *Fusarium moniliforme* metabolite, fumonisin B_1, in rats. *Carcinogenesis* 12, 1247-1251.

Joffe, A.Z. (1986) *Fusarium Species: Their Biology and Toxicology*. John Wiley and Sons, New York, NY.

Kellerman, T.W., Marasas, W.F.O., Thiel, P.G., Gelderblom, W.C.A., Cawood, M. and Coetzer, A.W. (1990) Leukoencephalomalacia in two horses induced by oral dosing of fumonisin B_1. *Onderstepoort Journal of Veterinary Research* 57, 269-275.

Leslie, J.F., Plattner, R.D., Desjardins, A.E. and Klittich, C.J.R. (1992) Fumonisin B_1 production by strains from different mating populations of *Gibberella fujikuroi* (*Fusarium* section Liseola). *Phytopathology* 82, 341-345.

Marasas, W.F.O., Nelson, P.E. and Tousson, T.A. (1984) *Toxigenic* Fusarium *species*. The Pennsylvania State University Press, University Park, PA.

Marasas, W.F.O., Kellerman, T.S., Gelderblom, W.C.A., Coetzer, J.A.W., Thiel, P.G. and Van der Lugt, J.J. (1988) Leukoencephalomalacia in a horse induced by fumonisin B_1 isolated from *Fusarium moniliforme*. *Onderstepoort Journal of Veterinary Research* 55, 197-203.

Merrill, A.H. Jr, Liotta, D.C. and Riley, R.T. (1996) Fumonisins: fungal toxins that shed light on sphingolipid function. *Trends in Cell Biology* 6, 218-223.

Nelson, P.E., Tousson, T.A. and Cook, R.J. (1981) Fusarium: *Diseases, Biology and Taxonomy*. Pennsylvania State University Press, University Park, PA.

Norred, W.P. and Voss, K.A. (1994) The toxicity and role of fumonisins in animal diseases and human esophageal cancer. *Journal of Food Protection* 57, 522-527.

Riley, R.T., Voss, K.A., Yoo, H.S., Gelderblom, W.C.A. and Merrill, A.H. Jr (1994) Mechanism of fumonisin toxicity and carcinogenesis. *Journal of Food Protection* 57, 528-535.

Ross, P.F., Nelson, P.E., Richard, J.L., Osweiler, G.D., Rice, L.G., Plattner, R.D. and Wilson, T.M. (1990) Production of fumonisins by *Fusarium moniliforme* and *Fusarium proliferatum* isolates associated with equine leukoencephalomalacia and a pulmonary edema syndrome in swine. *Applied Environmental Microbiology* 56, 3225-3226.

Shanks, G., Tabak, P., Begg, A.P. and Bryden, W.L. (1995) An outbreak of acute leukoencephalomalacia associated with fumonisin intoxication in three horses. *Australian Equine Veterinarian* 13, 17-18.

Williams, K.C., Blaney, B.J., Dodman, R.L. and Palmer, C.L. (1992) Assessment for animal feed of maize kernels naturally infected predominantly with *Fusarium moniliforme* and *Diplodia maydis*. I. Fungal isolations and changes in chemical composition. *Australian Journal of Agricultural Research* 43, 773-782.

Wing, N., Lauren, D.L., Bryden, W.L. and Burgess, L.W. (1993) Toxicity and trichothecene production by *Fusarium acuminatum* subsp. *acuminatum* and *Fusarium acuminatum* subsp. *armeniacum*. *Natural Toxins* 1, 229-234.

Chapter 93

Equine Leukoencephalomalacia in Brazil

F. Riet-Correa[1], M.A. Meireles[1], C.S.L. Barros[2] and A. Gava[3]

[1]*Laboratório Regional de Diagnóstico, Faculdade de Veterinária, Pelotas University, 96010, Pelotas RS, Brazil;* [2]*Departamento de Patologia, UFSM, 97119, Santa Maria RS, Brazil;* [3]*Centro de Ciências Agro-Veterinárias, Lages SC, 88500, Brazil*

Introduction

An acute disease in Equidae related to the ingestion of moldy corn was observed for the first time in Brazil during May and June, 1949, in the state of São Paulo (Rego Chaves, 1950). Clinically the disease was characterized by severe nervous signs, and most affected animals died after being sick for 6-72hrs. Five horses and three donkeys died, and one donkey survived. Lesions of the nervous system were not mentioned in this report, but mortalities ceased when the corn was withdrawn from the ration. Thirty years later, in 1979, two outbreaks were reported in the state of Rio Grande do Sul. Typical lesions of equine leukoencephalomalacia (ELEM) were found at necropsies, and *Fusarium moniliforme* was isolated from the corn fed to the horses (Riet-Correa *et al.*, 1982). Between 1979 and 1996, the disease had been reported from the states of Rio Grande do Sul (Riet-Correa *et al.*, 1982; Barros *et al.*, 1984), Paraná (Hirooka *et al.*, 1988; Hirooka *et al.*, 1990), Santa Catarina (Meireles, 1993), São Paulo (Xavier *et al.*, 1991; Meireles, 1993) and Minas Gerais (Brito *et al.*, 1982).

The epidemiology, clinical signs and pathology of 93 outbreaks of ELEM observed in Brazil between 1979 and 1996 are reviewed in this chapter. Forty-five outbreaks were diagnosed by the Regional Diagnostic Laboratories from Pelotas and Santa Maria Universities in Rio Grande do Sul, and 16 outbreaks were observed by the Diagnostic Laboratory of the University of Santa Catarina in Lages. Other reported cases are also included (Brito *et al.*, 1982; Hirooka *et al.*, 1988; Hirooka *et al.*, 1990; Xavier *et al.*, 1991; Meireles, 1993). Results obtained by microbiological and fumonisin analysis are also reviewed.

Epidemiology

All outbreaks of ELEM occurred from March to December, but were more frequent between June and September (late autumn to early spring) (Fig. 93.1).

The mean morbidity rate in 28 outbreaks was 17.9%, ranging from 4% to 100%. Case fatality rate was 100%. At least 230 horses (male and female) died in the 93 outbreaks. The yearly distribution of 90 outbreaks is presented in Fig. 93.2.

Clinical Signs

Clinical signs began abruptly and included anorexia, somnolence and depression or hyperexcitability, impaired prehension and mastication, ataxia, tremors, head pressing, circling, dullness, unilateral or bilateral blindness and recumbency. The clinical signs varied from 2-72hrs following onset, but most affected horses died in 6-24hrs after the onset of signs (Riet-Correa *et al.*, 1982; Barros *et al.*, 1984; Hirooka *et al.*, 1988; Hirooka *et al.*, 1990). In one outbreak the horses died after 1-7d (Xavier *et al.*, 1991). Some animals did not develop clinical signs for periods up to 12d following withdrawal of corn from the diet (Riet-Correa, 1993).

Pathology

At necropsy one cerebral hemisphere was usually enlarged with flattened gyri. On the cut surface there were yellowish and hemorrhagic areas of malacia in the cut

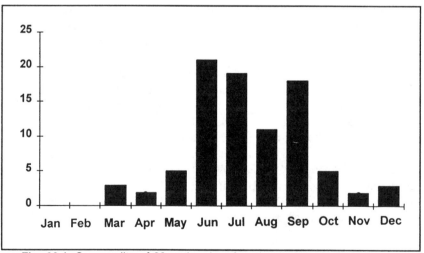

Fig. 93.1. Seasonality of 89 outbreaks of equine leukoencephalomalacia (ELEM) in Brazil.

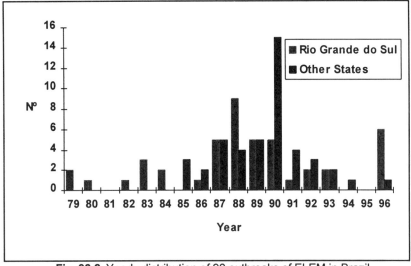

Fig. 93.2. Yearly distribution of 90 outbreaks of ELEM in Brazil.

centrum semi-ovale and corona radiata of the cerebral hemispheres. Fluid-containing cavities were frequently observed within these areas. The internal capsule of the thalamus was also frequently affected. Yellowish or hemorrhagic areas were occasionally observed in the colliculi, cerebellar peduncles, pons and medulla oblongata. Lesions were usually unilateral, but occasionally they were bilateral, being more marked on one side. After fixation the yellow areas were less evident and the hemorrhagic lesions appeared brown to gray (Brito *et al.*, 1982; Hirooka *et al.*, 1988; Meireles, 1993).

Histologic examination revealed malacic areas surrounded by edematous and hemorrhagic neuropil. Swollen astrocytes with eosinophilic cytoplasm and dark marginated nuclei (sometimes called clasmadendrocytes) were within the malacic areas. Hypertrophied and degenerative changes in the vascular endothelium, perivascular edema, hemorrhages and eosinophilic globules were also seen. Some vessels had perivascular cuffs consisting of eosinophils, neutrophils or mononuclear cells (Riet-Correa *et al.*, 1982; Barros *et al.*, 1984; Xavier *et al.*, 1991).

Microbiological Studies and Fumonisin Analysis

Fusarium moniliforme was consistently isolated from corn collected during outbreaks (Riet-Correa *et al.*, 1982; Barros *et al.*, 1984; Hirooka *et al.*, 1988; Hirooka *et al.*, 1990). Microbiological examination of 39 feed samples associated with 29 outbreaks of ELEM indicated *Fusarium* spp. as the most frequently isolated mold, occurring in 97.4% of samples, followed by *Penicillium* spp. in 61.5% and *Aspergillus* spp. in

35%. *Fusarium moniliforme* was the predominant species with an occurrence of 82%. Two additional *Fusarium* spp. not associated with ELEM were isolated in low frequency: *F. proliferatum* in 12% of samples and *F. subglutinans* in 2.6%. Fumonisin B_1 ranging from 10-500μg/g was detected in 24 (82.1%) of 32 corn samples examined (Meireles *et al.*, 1994).

Sydenham *et al.* (1992) examined the fumonisin levels in 13 feed samples collected in farms from the state of Parana where ELEM was occurring. Fumonisin was detected in all samples and varied between 0.2-38.5μg/g for fumonisin B_1 and between 0.1-12μg/g for fumonisin B_2 with mean concentration of 11.9μg/g and 4μg/g, respectively, and *F. moniliforme* was isolated from the 13 samples. All isolates were acutely toxic for ducks and produced fumonisins on corn cultures at levels of 270-2260μg/g for fumonisin B_1 and 40-825μg/g for fumonisin B_2.

References

Barros, C.S.L., Barros, S.S., Santos, M.N. and Souza, M.A. (1984) Leucoencefalomalacia em equinos no Rio Grande do Sul. *Pesquisa Veterinária Brasiliera* 4, 101-107.

Brito, L.A.B., Nogueira R.H.G., Pereira, J.J., Chiquiloff, M.A.G. and Biondini, J. (1982) Leucoencefalomalacia em equino associada à ingestão de milho mofado. *Arquivos Esc. Veterinária* UFMG 34, 49-53.

Hirooka, E.Y., Viotti, N.M.A., Soares L.M.V. and Alfieri, A.L. (1988) Intoxicação em equinos por micotoxinas produzidas por *Fusarium moniliforme* no norte do Paraná. *Semina* 9, 128-135.

Hirooka, E.Y., Viotti, N.M.A., Marochi, M.A., Ishii, K. and Ueno, Y. (1990) Leuco-encefalomalacia em equinos no norte do Paraná. *Revista Microbiological* 21, 223-227.

Meireles, M.A. (1993) Leucoencefalomalacia equina no Brasil. PhD Dissertation, University of São Paulo, São Paulo.

Meireles, M.A., Corrêa, B., Fischman, O., Gambale, W., Paula, C.R., Chacon-Reche, N.O. and Pozzi, C.R. (1994) Mycoflora of the toxic feeds associated with equine leuco-encephalomalacia (ELEM) outbreaks in Brazil. *Mycopathologia* 127, 183-188.

Rego Chaves, L. (1950) Doença de sintomatologia nervosa causada pela intoxicação pelo milho. *Revista do Medicina Veterinária* 10, 199-217.

Riet-Correa, F. (1993) Intoxicação por milho contaminado por *Fusarium moniliforme* (leucoencefalomalacia). In: Riet-Correa, F., Méndez, M.C. and Schild, A.L. (eds), *Intoxicações por Plantas e Micotoxicoses em Animais Domésticos*. Editorial Hemisferio Sur, Montevideo, pp. 146-153.

Riet-Correa, F., Meireles, M.A., Soares, J.M., Machado, J.J. and Zambrano, A.F. (1982) Leucoencefalomalacia associada à ingestão de milho mofado. *Pesquisa Veterinária Brasiliera* 2, 27-30.

Sydenham, E.W., Marasas, W.F.O., Shephard, G.S., Thiel, P.G. and Hirooka, E.Y. (1992) Fumonisin concentration in Brazilian feeds associated with field outbreaks of animal mycotoxicoses. *Agricultural and Food Chemistry* 40, 994-997.

Xavier, J.G., Brunner, C.H.M., Sakamoto, M., Correa, B., Fernandes, W.R. and Dias, J.L.C. (1991) Equine leukoencephalomalacia: report of five cases. *Brazilian Journal of Veterinary Research and Animal Science* 28, 185-189.

Chapter 94

Isolation of an Extract of *Corallocytostroma ornicopreoides* sp. nov., which Causes the Rumenitis Associated with Black Soil Blindness

J.G. Allen[1], S.M. Colegate[2], P.R. Dorling[2], T.F. Jubb[3], D.C. Main[1], A.A. Mitchell[1] and R.G. Shivas[1]

Plant Toxins Unit, CSIRO Division of Animal Health, Australian Animal Health Laboratory, Private Bag 24, Geelong, Victoria 3220, Australia; [1]Agriculture Western Australia, South Perth, Western Australia 6151, Australia; [2]School of Veterinary Studies, Murdoch University, Murdoch, Western Australia, Australia; [3]Agriculture Western Australia, Kununurra, Western Australia 6743, Australia

Introduction

Black Soil Blindness (BSB) is a mycotoxicosis of cattle recently recognized in northwest Australia (Chapter 95), but it is suspected to have occurred occasionally throughout tropical northern Australia. It results from ingestion of the conidiomata of the fungus *Corallocytostroma ornicopreoides* sp. nov. These structures are up to 3cm in diameter, and are referred to as fungal corals because of their physical resemblance to pieces of coral (Jubb *et al.*, 1996; Shivas *et al.*, 1996).

The fungus grows predominantly on Mitchell grass (*Astrebla* spp.), a native species that is extremely valuable to Australia's pastoral industry. Fungal growth occurs relatively infrequently when two wet seasons are separated by a moderately dry season. The fungus has also been observed to grow on blue grass (*Dicanthium* spp.) and Flinders grass (*Iseilema* spp.).

A sample of the fungal corals was analyzed and found to contain water (6.4%), sugars (7.2%), protein (9.4%) and phosphorus (0.15%). Selective appetite for the corals has been observed with some animals, which may find the sweet taste and smell of the fungus desirable. The high levels of protein and phosphorus, which are 2x and 3x higher, respectively, than normally found in Mitchell grass (Holm and

Elliot, 1980) might indicate a potential attraction for animals later in the season when the pasture becomes deficient in protein and phosphorus.

Thus far, overt intoxication of cattle in the field has only been observed following ingestion of the fungal growth on Mitchell grass. The corals remain on the grass until the first rains wash them away.

Bioassay Selection and Procedure

Attempts were made to identify bioactivity within a laboratory-manageable species and to link the activity to the field syndrome (Dorling *et al.*, 1993), but toxicity tests undertaken in brine shrimp or mice were unproductive. Sheep were found to provide an adequate model, and intoxication could be produced experimentally by drenching them with an aqueous slurry of 400g of the fungal corals.

Kidney damage occurs readily in cattle exposed to the fungal corals, but damage to the forestomachs is the most consistent gross and microscopic pathological change in sheep. Specifically, severe rumenitis is accompanied by sloughing of the epithelium in many areas (Fig. 94.1). This effect is repairable and reversible if the exposure to the fungal conidiomata is stopped. Despite being less susceptible to the BSB-inducing toxins than cattle, the toxic effect in sheep formed the basis of the bioassay for toxins.

The bioassay procedure included monitoring feed intake and extensive pathological examination of euthanized sheep. Because it has been suggested that dehydration might be a factor in the intoxication, access to water was denied for 48hrs prior to dosing with test samples. Sheep (30-42kg) were fed 300g of sheep cubes and 200g of chaff per day, and were housed in individual pens for 5d prior to administration of test samples. Aqueous or vegetable oil solutions of test samples were dosed to sheep *via* intraruminal tube, and access to water was reinstated 24hrs post-dosing. Feed intake was recorded at zero (always 500g), one and 2d. Sheep were euthanized after 2d, the rumens were removed and the contents washed out gently to allow close examination of the lining of the rumen, reticulum and omasum. Samples for histological examination of affected or suspected affected areas were taken from the dorsal and ventral sacs of the rumen, from the reticulum and from the omasum. Liver and kidney samples were routinely taken.

Extraction of Fungal Corals

Approximately 400g of finely milled fungal corals were extracted through a series of non-polar to polar organic solvents. A renal toxin was found in the choroform ($CHCl_3$) soluble material, and a rumen epithelial toxin in the more polar fractions. Since the forestomach lesions were the most prominent effects in sheep, efforts were focused on characterization of these polar toxins. Specifically, the rumen epithelial

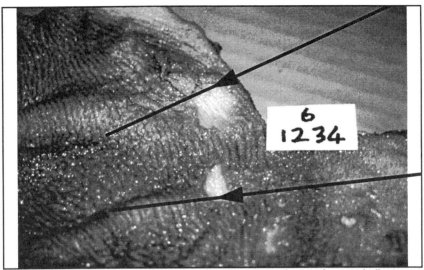

Fig. 94.1. The exposed rumen of a bioassay sheep. Arrows indicate sloughing of the epithelium.

toxicity was monitored using the sheep bioassay of extracts and subsequent fractions.

The milled fungal corals (400g) were extracted 4x with water (1L) by gentle agitation over 2hrs. The filtrate of the aqueous extract was concentrated under reduced pressure. Addition of methanol (MeOH) to the concentrated combined extracts yielded an orange solution from an off-white solid. The solid was recrystallized from aqueous MeOH to afford white needles subsequently identified by GC/MS and NMR spectroscopy as mannitol, which represented about 30% of the weight of the fungal corals.

The remaining MeOH-soluble material was adsorbed onto silica gel and applied to a silica gel column in $CHCl_3$. Elution of the column with $CHCl_3$ (80-0%)/MeOH (20-100%) solvents yielded several minor compounds in addition to the major component, a pale yellow oil, the bulk of which was identified by GC/MS and NMR spectroscopy as arabitol (approximately 25% of the bulk of the fungal corals). Final elution of the column with aqueous MeOH yielded a fraction, of <5% of the weight of the fungal corals, which elicited the rumenitis associated with ingestion of intact corals by sheep.

Preliminary Analysis of the Toxic Fraction

Silica gel TLC of the toxic fraction did not afford any meaningful separation, as most of the material (visualized by sulfuric acid charring) failed to migrate from the origin. Examination of the acetylated toxic fraction by GC/MS indicated the major presence of mannitol and arabitol in addition to some minor unidentified peaks.

Reverse phase (C18) chromatography of small aliquots of the toxic mixture and GC/MS examination of the acetylated fractions indicated that some separation of components was possible. Mannitol and arabitol, which were not retained by the column, were separated by GC from an unknown compound that occurred at a similar retention time to arabitol. Neither mannitol nor arabitol was responsible for inducing the forestomach lesions observed in sheep.

Further Work

Future work will be aimed at identifying the bioactive component(s) of the polar mixture. Bioassay-directed fractionation will be used for the isolation and characterization of the component that causes the forestomach lesions. Various fractions or solvent partitions of the toxic fraction will be tested in the bioassay as a preliminary step to isolating, purifying, identifying and confirming the toxicity of the toxic component. The kidney toxins discovered in the $CHCl_3$ extracts of the milled corals will also be investigated. In addition, assay techniques will be developed to investigate the pharmacodynamics of the toxin(s) as a prelude to assessing food safety issues associated with subclinically intoxicated animals.

References

Dorling, P.R., Colegate, S.M. and Huxtable, C.R. (1993) Plants affecting livestock: an approach to toxin isolation. In: Colegate, S.M. and Molyneux, R.J. (eds), *Bioactive Natural Products: Detection, Isolation and Structural Determination*. CRC Press, Boca Raton, FL, pp. 481-506.

Holm, A.McR. and Elliot, G.J. (1980) *Australian Rangeland* 2, 175.

Jubb, T.F., Main, D.C., Mitchell, A.A., Shivas, R.G. and De Witte, K.W. (1996) Black soil blindness: A new mycotoxicosis of cattle grazing *Corallocytostroma*-infected Mitchell grass (*Astrebla* spp.). *Australian Veterinary Journal* 73, 49-51.

Shivas, R.G., Mitchell, A.A. and Jubb, T.F. (1997) *Corallocytostroma ornicopreoides* sp. nov., an unusual toxic fungus on *Astrebla* and *Dicanthium* in north-western Australia. *Mycological Research* 101, 849-852.

Chapter 95

Corallocytostroma ornicopreoides: A New Fungus Causing a Mycotoxicosis of Cattle Grazing Mitchell Grass Pastures in Australia

T.F. Jubb[1], D.C. Main[2], A.A. Mitchell[2], R.G. Shivas[2] and K.W. de Witte[3]

[1]*Agriculture Western Australia, Kununurra, Western Australia 6743, Australia;*
[2]*Agriculture Western Australia, South Perth, Western Australia 6151, Australia;*
[3]*Department of Primary Industry and Fisheries, Katherine, Northern Territory 0851, Australia*

Introduction

A new mycotoxicosis of cattle was diagnosed in northwestern Australia in the dry season of 1994 (Jubb *et al.*, 1996). It was caused by an unusual fungus that produces stromata up to 2.5cm in diameter on the stems of Mitchell grass (*Astrebla* spp.). Over 500 cattle died on two properties in the southern part of the East Kimberley District, about 20 cattle died on two properties in the adjacent southwest Victoria River District, and many more became illthrifty. The grazing areas of these districts are composed of simple mosaics of black soils dominated by Mitchell grass and red soils dominated by desert shrubs and spinifex (*Triodia* spp.). Blindness in affected cattle and the common history of having grazed black soil areas, provided the name "black soil blindness" (BSB). The appearance of the disease caused grave concern because large areas of cattle production in northern Australia utilized Mitchell grass pastures.

The herds in the area are extensively managed. Stocking rates of five to ten breeding cows/km^2 and the paddock sizes of 200-400km^2 are standard. Mean annual rainfall is about 450mm occurring during the summer months. The average pasture growing season is about 12wks. After this, a decline in herbage mass caused by grazing and dying-off of annual plants occurs during the dry season.

Outbreak Patterns

Cattle deaths were first noticed at the beginning of the cattle mustering season. Large numbers of mustered cattle were reported becoming blind then dying, with all except suckling calves affected. Initially only mustered cattle were thought to be affected. Between 1-5% of some groups died after mustering, and other groups were unaffected. As the dry season progressed, increasing numbers of sick and dead cattle, many of which had not been mustered, were found in paddocks and confirmed to have BSB. Ill animals were notable by their separation from other cattle; they had congregated around water points, or become trapped in water yards, unable to find the open gates. The appearance of BSB in unmustered cattle paralleled increasing exposure and consumption of stromata as the dry season progressed. Both paddock and mustering-related deaths continued until wet-season rain storms caused the fungus to disappear in early November.

Clinical Features

Acute onset of blindness and rapid progression to death were the major clinical features of BSB. In a few cases subsequently found to have autopsy lesions of BSB there was depression, weakness and illthrift without blindness. Occasional cases occurred 1-3d after mustering, but most cattle died after 4-7d. Yarded cattle were usually first noticed because they would walk into fences. Their negotiation of obstacles was consistent with partial blindness, demonstrated by swerving away at the last moment. Menace reflexes were absent, and some became obviously sunken-eyed. These cattle would quietly assume sternal, then lateral recumbency and die without signs of struggle 1-24hrs after appearing blind. Extra stress such as forced movement in stockyards seemed to hasten death. A few cases were seen to progress within an hour from being almost normal with partial blindness, to being completely blind, very disorientated, sunken-eyed and weak, and finally laterally recumbent and dead.

Initially it was thought that all cattle that showed signs of blindness died, but some may have recovered, at least temporarily. Some cattle noticed blind but left undisturbed were not found sick or dead in the next few days, but were lost to individual follow-up when released into large paddocks. Sick and dead cattle were found for many weeks after some musters.

Necropsy Findings

At necropsy, some affected cattle had extensive perirenal and mesenteric edema with pale swollen kidneys and swollen, friable, orange-yellow livers with varying areas of red mottling on the surface. Others exhibited no obvious gross lesions. Carcasses were usually very dehydrated with severely sunken eyes. Slight yellowing of connective tissues was present in some cases. Partly digested, whole and fragmented fungal stromata could be found in the rumen contents, appearing as hard, rubbery, white, rounded, pebble-like structures.

Microscopic examination of kidney sections from affected cattle demonstrated the presence of acute to subacute renal tubular necrosis, which varied in degree from moderate to severe. There was Kupffer cell and bile duct hyperplasia, as well as diffuse, mild to moderate centrilobular vacuolar change and occasional individual hepatocellular necrosis. Oil red O (ORO) stain on frozen sections revealed that the vacuoles in hepatocytes were not due to fat accumulation, but some small ORO positive globules were seen within the cytoplasm of many hepatocytes. In some cases, there was a mild to moderate interstitial pneumonia. There was widespread focal, and often severe, epithelial necrosis in sections of reticulum. No lesions were seen in routine light microscopic examinations of hematoxylin and eosin sections of brain, eyes and optic nerves including the optic tracts from chiasma to lateral geniculate bodies. No evidence of vascular leakage was found in periodic acid Schiff-stained sections of cerebellum or cerebrum.

The Fungus and Host Plants

The stromata were hard, dry, white, coral-like structures ranging in size from 0.5-2.5cm diameter and were found on all three species of Mitchell grass (*Astrebla elymoides*, *A. pectinata* and *A. squarrosa*). The stromata were firmly attached to growing points and flower heads throughout the tussocks, but mainly on the eastern side. The fungus was tentatively identified as a new species of *Corallocytostroma*. Areas with high densities of stromata occurred in paddocks where sick and dying cattle had grazed, while only low densities occurred where cattle were unaffected.

The fungus has been named *Corallocytostroma ornicopreoides* (Shivas *et al.*, 1997). It is the second known member of the *Corallocytostroma* genus within the Clavicipitaceae. The species name, *ornicopreoides*, is the Latin derivation for bird dung, which the fungus resembles. The only other species of *Corallocytostroma* that has been described affected rice crops in China (Yu and Zhang, 1980).

In some areas, nearly all Mitchell grass tussocks were infected and each bore up to ten stromata. In other areas, only 0.1-1% of tussocks were infected, mostly with single stroma. Other grasses found with infected Mitchell grass included *Iseilema* spp., *Aristida* spp., *Chrysopogon fallax* and *Dichanthium* spp.

Stromata morphologically similar to those on Mitchell grass were also found on *Dichanthium* spp. (bundle bundle, bluegrass) across the Kimberley region, but were never associated with animal disease. Monitoring sites were established in 1994 throughout the infected Mitchell grass area on uninfected and infected sites, and also at sites of infected bluegrass across the region.

The sexual stages of the fungi were eventually found on Mitchell grass but not on *Dichanthium* spp. Stromata on both plant species disappeared with the onset of the wet season.

Feeding Trial

Stromata were fed to three steers (2yrs, 280kg). One of these cattle was fed daily a maintenance ration of sorghum hay and 500g of stromata, for which it had a ravenous appetite. After 4d, it began to exhibit loss of appetite, lethargy and dragging of its hind feet when walking, which coincided with increases in serum glutamate dehydrogenase (GLDH), aspartate aminotransferase (AST), and gamma glutamyl transferase (GGT) activities and in bilirubin, creatinine and urea concentrations. It was euthanized and necropsied. The other two steers would not eat the stromata and were drenched daily for 6d with 500g ground stromata mixed with water. No adverse effects were noticed and drenching was discontinued. Over the next few days, pathological increases in serum GLDH, AST and GGT activities, and bilirubin, and creatinine and urea concentrations occurred in one steer, and in GLDH, creatinine and urea in the other. The fifth day after dosing stopped, the steers showed inappetance and lethargy and were necropsied. Although blindness was not observed, the histological findings in all three steers matched those of field cases of BSB.

Subclinical Disease

Subclinical disease was extensive. Cattle grazing heavily infected pastures were rougher-coated and in poor condition, despite plentiful feed, compared with cattle grazing lightly infected pastures. A higher than usual number of aborted calves were observed in the yards after mustering. In one paddock where 1,533 cows had been grazing for 8wks, 14 died. After mustering, a further 32 died in the yards. Of 220 apparently healthy cull cows from this mob sent to slaughter, three died in transit and three became ill and depressed, necessitating an emergency kill. Fragments of stromata were recovered from the rumens of the three pre-slaughter fatalities, and histological lesions consistent with BSB were found in all three sick cows. Of 156 sera collected at slaughter, 36 showed pathological elevations of at least one of serum creatinine, GLDH and GGT.

Risk Factors and Control Options

Factors affecting exposure and expression of the disease in cattle were not readily apparent. Based on the first of the trial steers, the stromata may have been selectively grazed by some animals. A sample of stromata contained 6.4% water, 7.2% sugars, 9.4% protein and 0.15% phosphorus. The sugar content of the fungus, and the protein and phosphorus concentrations 2x and 3x times higher, respectively, than the concentrations found in mostly dry Mitchell grass (Holm and Eliot, 1980) may have made them attractive. As pastures become deficient in protein and phosphorus, supplementation may play a critical role in controlling the disease.

There are probably no practical economical control methods for the disease because of the large areas that were involved and the variable distribution of infected pasture. Because mustering stress was important in the development of clinical disease, a test muster on a sample of cattle from a herd grazing an infected area could be done to estimate the risk of further mustering losses. Burning heavily infected pastures has been suggested, but conditions suitable for effective burning put other pastures at risk. Exclusion, from heavily infected areas by temporary electric fencing may be an option where grazing control by turning off water supplies is unavailable. Disappearance of the stromata with the first heavy rains of the wet season indicated natural removal by rainfall.

Analysis of satellite imagery and rainfall records of the affected East Kimberley properties showed that for the 2yrs leading up to the outbreak in 1994, many of the areas of heavily infected pasture had above average rainfall in the wet season and extended growing seasons.

After 1994

In the dry season of 1995, which followed an average rainfall wet season, a small amount of new growth of the fungus occurred, but only at previously infected Mitchell grass sites. A few weathered remnants of the fungus and a rarer, weather-resistant sclerotial form, speculated to be the dormant reservoir form of the fungus, were also found at sites that were previously heavily infected. By 1996, there was no evidence of the fungus to be found anywhere.

Early in 1997, after one of the highest rainfall wet seasons on record, the fungus re-emerged. It infected *Atrebla* spp. and *Dichanthium* spp., at monitoring sites in previously infected areas but, as in 1994, both species were not found to be infected simultaneously at the same site. The additional presence of the fungus on Flinders grass (*Iseilema vaginiflorum*) intermingled with *Dichanthium* was a surprise. Flinders grass is an annual species that also grows on black soils. At the time of writing (March 1997), the extent of infected pastures and their danger to stock had yet to be defined.

Anecdotal reports of previous occurrence of the disease in the Northern Territory and Queensland were investigated. However, these areas had suffered low rainfall wet seasons for a number of years and were heavily grazed. Finding *Corallocytostroma* in the Northern Territory and Queensland would have supported the view that the fungus is endemic, and that the disease BSB is a rare event. The negative finding makes it uncertain as to whether the fungus had been present for many years but not found, or whether the fungus and disease were new. If the disease occurred again but in another area, it would be uncertain whether the fungus had spread there from the known infected areas in the Kimberley and southern Victoria River District, or had developed from widespread, low level, endemic infection detected because of increased surveillance.

The fungus that appeared on *Dichanthium* spp. was morphologically similar to that on Mitchell grass, but where the two grasses coexisted no fungus grew on the latter. Clinical BSB only occurred in a localized area, and only with infected Mitchell grass. There is concern that the fungus may be an endemic pathogen of *Dichanthium* that has adapted itself to Mitchell grass, and now to Flinders grass. The failure of the fungus to substantially reappear in 1995, and not to appear in 1996 after average wet seasons, had allayed fears that the fungus was going to be a serious and ongoing problem and suggested that its development was subject to rare combinations of environmental factors. The reappearance of the fungus in 1997, and its presence on a new host (Flinders grass) have caused some concern.

A great deal remains unknown about BSB. Existing monitoring sites on the previously affected stations will be maintained for at least the next few years. If the disease occurs again, it will be important to learn as much as possible about the circumstances of recurrence. Defining what constitutes a toxic pasture over the range of fungal densities that occur in pastures should be a research priority to reduce the impact of the disease by controlled grazing. The dangers of toxic residues in meat must also be defined. Research to define the nature of the fungus and its toxins further may lead to discoveries that could reduce the impact of the disease on the Australian meat and livestock industry.

References

Holm, A.McR. and Eliot, G.J. (1980) Seasonal changes in the nutritional value of some native pasture species in north-western Australia. *Australian Rangeland Journal* 2, 175-182.

Jubb, T.F., Main, D.C., Mitchell, A.A., Shivas, R.G. and De Witte, K.W. (1996) Black soil blindness: A new mycotoxicosis of cattle grazing *Corallocytostroma*-infected Mitchell grass (*Astrebla* spp). *Australian Veterinary Journal* 73, 49-51.

Shivas, R.G., Mitchell, A.A. and Jubb, T.F. (1997) *Corallocytostroma ornicopreoides* sp. nov., an unusual toxic fungus on *Astrebla* and *Dichanthium* in north-western Australia. *Mycological Research* 101, 849-852.

Yu, Y. and Zhang, Z. (1980) *Corallocytostroma* Yu et Zhang, gen. nov., a stromatic Coelomycete. *Acta Microbiologica Sinica* 3, 230-235.

Chapter 96

Ergotism and Feed Aversion in Poultry

W.J.K. Bakau and W.L. Bryden

Department of Animal Science, University of Sydney, Camden, New South Wales 2570, Australia

Introduction

Claviceps purpurea-induced ergotism has been known for many centuries (Bryden, 1994; Rehacek and Sajdl, 1990). Many economically important plant species are infected by *C. purpurea*, particularly rye and ryegrass, where it forms an ergot or sclerotium in the ovary of the host plant (Lorenz, 1979). The ergot is a capsule of dehydrated conidia containing as many as 100 compounds. The main compounds are amines, amino acids, glucans, pigments, enzymes and fatty acids. The principal toxic components of the ergots are alkaloids, that can be divided structurally into three groups; simple lysergic acid amines, peptide derivatives of lysergic acid and the clavines. The most biologically active are those with a peptide moiety, especially ergotamine and ergotoxine (Gilman *et al.*, 1985). These, along with amine types (ergine and ergometrine), are the main alkaloids of *C. purpurea* and may account for up to 1% w/w of the ergot (Mantle, 1977).

The actions of ergot alkaloids are varied and complex. In general, the effects result from actions as partial agonists or antagonists at adrenergic, dopaminergic and serotonergic receptors. The spectrum of alkaloid effects and the differences in activity among individual compounds are a function of chemical configuration resulting in differences in potencies at various receptors and differences in the manner in which each interacts with different receptors (Gilman *et al.*, 1985). Ergot alkaloids have many effects in animals and birds including neurohormonal, peripheral and central nervous effects (Rehacek and Sajdl, 1990). The peptide derivatives of lysergic acid are stimulatory of smooth muscles (both vascular and non-vascular). This direct action is often expressed in the form of vasoconstriction, local ischemia and gangrene (Bryden, 1994). The ergotoxins cause interference with implantation (Kraicer and Shelesnyak, 1965) and prolactin secretion (Shaar and Clemens, 1972; Clemens *et al.*, 1974). The latter two effects have attracted much interest because these alkaloids appear to interfere with processes controlled by the hypothalamic-pituitary axis (Thorner *et al.*, 1980). Other alkaloids such as ergometrine induce oxytocin-like effects, but exert no adrenergic blockage effects (Gilman *et al.*, 1985).

Most studies of the toxicity of the ergots of *C. purpurea* have evaluated toxicity in ruminants, especially the development of gangrenous ergotism (Bryden, 1994). Young birds appear to be most susceptible, and tolerate dietary levels up to 0.3-0.8% (Bragg *et al.*, 1970; Rotter *et al.*, 1985) before growth, feed intake and feed conversion efficiency are affected. Ingestion of higher dietary levels can result in poor feathering, nervousness, incoordination and gangrene of the foot and beak (Bragg *et al.*, 1970). Laying hens can tolerate dietary levels as high as 0.6-1.2% of dietary ergot (Johnson and Sell, 1976) although egg production, egg shape and egg shell thickness may be adversely affected at these higher levels. The level of dietary ergot that may be tolerated depends not only on the total ergot alkaloid content but also on the constituent alkaloids (Rotter *et al.*, 1985).

Toxicity Studies

Clinical expression of ergotism is affected by ambient temperature (Bryden, 1990). Chickens fed ergots of *C. purpurea* developed gangrene of the feet except when subjected to high ambient temperatures (Bakau and Bryden, 1987). In the studies, this observation extended to the alkaloid, ergotamine.

In the first trial, 1d-old male broiler chicks and layer strain male chicks were fed diets containing ergot at 0, 5, 10, 15 or 20g/kg. In the second trial, 3wk-old broiler chicks were fed diets containing ergotamine at 0, 150 or 300mg/kg. Half of the birds in the second study were maintained at a constant 22°C while the remainder were maintained at 22°C for 12hrs each day and the temperature was then increased to 35°C for the following 12hrs.

Dietary levels of ergot above 5g/kg depressed growth performance while higher levels (10g/kg) induced necrosis and gangrene of the digits in both chick strains. Birds fed ergotamine and exposed to a high temperature were most affected with respect to growth rate, feed intake and feed conversion efficiency. All the birds ingesting ergotamine exhibited poor feathering and those maintained at a constant 22°C had severe gangrenous lesions. In contrast, foot lesions did not develop in those exposed to the high daily temperature. Plasma prolactin levels decreased ($P<0.05$) as the levels of dietary ergotamine increased.

Feed Discrimination Studies

The ability of birds to select a diet free of the sclerotia of *C. purpurea* or ergotamine was assessed. In the studies, birds were offered feed in divided feeders in which one half contained normal feed and the other, feed with the toxin.

In the first experiment, layer strain chicks were given feeders with one half containing basal grower diet and the other half filled with the basal diet with ergot at 0, 10 or 20g/kg. The second experiment was essentially the same, except that half the feeder was filled with basal diet containing ergotamine at 0, 150 or 300mg/kg. In both studies the birds were 3wks old and allowed free access to the feeders for 21d. Chicks fed diets containing graded levels of ergot or ergotamines developed gangrenous lesions of the digits and had depressed feed intake and reduced weight gain. In contrast, chicks offered a choice between contaminated and uncontaminated feed grew at the normal rate and had no clinical signs of ergotism. These birds consumed more of the uncontaminated diet.

In the third experiment, commercial laying hens, aged 55wks and housed in single cages, were offered a layer diet containing ergot at 0, 5, 15 or 40g/kg. Other birds were offered two feed containers: one with uncontaminated feed and the other with feed containing ergot. The hens did not develop gangrenous lesions, but those ingesting a high level of ergot (40g/kg) ate less feed, laid fewer eggs and ceased laying after 7d. Those offered a choice of diets continued laying at a normal rate and maintained feed intake by increasing their intake of the toxin-free diet.

Conclusions

The results demonstrate that both ergot and ergotamine are toxic to chickens, and that the temperature to which birds are exposed significantly influences the outcome of ergotism. Most likely high temperatures counteract the vasoconstrictive effects of ergot alkaloids on peripheral blood vessels, preventing ischemia, necrosis and gangrene (Bryden, 1990).

Feed refusal or aversion is the first response animals exhibit when suffering from a toxic insult following feed ingestion. Domestic fowl are able to select a diet that meets nutrient requirements for commercial rates of growth and egg production if offered a choice (Cumming *et al.*, 1987; Emmans, 1977). The results of this study show that young or older birds given a choice select against a diet containing ergot or ergotamine, limiting their intake of alkaloid to a tolerable level (Nolan *et al.*, 1996). A number of the ergot alkaloids are agonists/antagonists of neurotransmitters, including prolactin, which, among other activities, regulate feed intake to alter feeding behavior (Buntin, 1992).

References

Bakau, W.J.K. and Bryden, W.L. (1987) Toxicity of ergots of *Claviceps purpurea* in chickens subjected to heat stress. *Proceedings of the Nutrition Society of Australia* 12, 170.

Bragg, D.B., Salem, H.A. and Devlin, T.J. (1970) Effect of dietary *Triticale* ergot on the performance and survival of broiler chicks. *Canadian Journal of Animal Science* 50, 259-264.

Bryden, W.L. (1990) Ambient temperature and the clinical expression of ergotism and ergot-like syndromes. In: Quisenberg, S.S. and Joost, R.E. (eds), *Acremonium/Grass Interactions*. Louisiana State University Press, Baton Rouge, LA, pp. 209-211.

Bryden, W.L. (1994) The many guises of ergotism. In: Colegate, S.M. and Dorling, P.R. (eds), *Plant-Associated Toxins: Agricultural, Phytochemical and Ecological Aspects*. CAB International, Wallingford, Oxon, pp. 381-386.

Buntin, G.D. (1992) Neural substrates for prolactin-induced changes in behavior and neuroendocrine function. *Poultry Science Reviews* 4, 275-287.

Clemens, J.A., Shaar, C.J., Smalstig, E.B., Bach, N.J. and Kornfeld, E.C. (1974) Inhibition of prolactin secretion by ergolines. *Endocrinology* 9A, 1171-1176.

Cumming, R.B., Mastika, I.M. and Wodzicka-Tomaszewska, M. (1987) Practical aspects of choice feeding in poultry and its future role. In: Farell, D.J. (ed), *Recent Advances in Animal Nutrition in Australia*. University of New England, Publishing Unit, Armidale, pp. 283-289.

Emmans, G.C. (1977) The nutrient intake of laying hens given a choice of diets, in relation to their production requirements. *British Poultry Science* 18, 227-236.

Gilman, A.G., Goodman, L.S., Rall, T.W. and Murad, F. (1985) *The Physiological Basis of Therapeutics*, 7th ed. MacMillan, New York, NY, pp. 931-940.

Johnson, R.L. and Sell, J.L. (1976) Effects of ergot on the productive performance of laying hens. *Poultry Science* 55, 2049-2050.

Kraicer, P.F. and Shelesnyak, M.G. (1965) Studies on the mechanism of nidation. XIII. The relationship between chemical structure and biodynamic activity of certain ergot alkaloids. *Journal of Reproduction and Fertility* 10, 221-226.

Lorenz, K. (1979) Ergot on cereal grains. *Critical Reviews in Food Science and Nutrition* 11, 311-335.

Mantle, P.G. (1977) The genus *Claviceps*. In: Wyllie, T.D. and Morehouse, L.G. (eds), *Mycotoxic Fungi, Mycotoxins, Mycotoxicoses - An Encyclopedic Handbook, Vol 1*. Marcel Dekker, New York, NY, pp. 83-89.

Nolan, J.V., Hinch, G.N. and Lynch, J.J. (1996) Voluntary food selection. *Proceedings of the Nutrition Society of Australia* 20, 31-37.

Rehacek, Z. and Sajdl, P. (1990) *Ergot Alkaloids: Chemistry, Biological Effects, Biotechnology*. Elsevier, Amsterdam.

Rotter, R.G., Marquardt, R.R. and Young, J.C. (1985) Effect of ergot from different sources and of fractionated ergot on the performance of growing chicks. *Canadian Journal of Animal Science* 65, 953-961.

Shaar, C.J. and Clemens, J.A. (1972) Inhibition of lactin and prolactin secretion in rats by ergot alkaloids. *Endocrinology* 90, 285-288.

Thorner, M.O., Fluckiger, E. and Calne, D.B. (1980) *Bromocriptine: A Clinical and Pharmacological Review*. Raven Press, New York, NY.

Chapter 97

Experimental Black Soil Blindness in Sheep

J.G. Allen[1], D.C. Main[1], T.F. Jubb[2] and R.G. Shivas[1]

[1]Agriculture Western Australia, 3 Baron-Hay Court, South Perth, Western Australia 6151, Australia; [2]Agriculture Western Australia, Durack Drive, Kununurra, Western Australia 6743, Australia

Introduction

Black soil blindness (BSB) is a newly discovered mycotoxicosis of cattle. The disease is often fatal, and death is associated mainly with a severe nephrosis, although the liver, lungs and reticulum may also be affected. It is caused by the consumption of conidiomata, or "corals," of the fungus *Corallocytostroma ornicopreoides* that infects Mitchell grass (*Astrebla* spp.) in northern Australia. Reported blindness in affected cattle and the common history of having grazed black soil areas resulted in the name of the disease (Jubb *et al.*, 1995; 1996; Shivas *et al.*, 1997).

A series of experiments was conducted to determine if sheep might also be susceptible to this disease. Merino wethers (28.2-43.2kg) were kept in individual pens in an animal house and fed a ration of commercial sheep pellets and chaff. During all experiments blood samples were collected from the sheep each day and subjected to an extensive panel of assays. Conidiomata (referred to as corals because of their physical appearance) collected in October, 1994, from Mitchell grass at several sites in the infected area were kept dry and in the dark until required in the experiments.

Preliminary Toxicological Studies

Eighteen sheep were offered only corals to eat, and they consumed 0-435g within 48hrs, then refused to eat any more. They also refused to eat any normal feed that had corals mixed with it. Only two sheep showed any effects.

The sheep that consumed 435g (9.7g/kg) was subdued from days two to four after first receiving the corals, had smelly diarrhea or passed pale, soft, smelly large stools from days three to five, and did not regain its full appetite for normal feed until day 11. Clinical chemistry results for this sheep remained within normal ranges.

Another sheep that consumed 150g (4.6g/kg) was subdued by the end of the first day and depressed for the next 2d, regaining a more normal disposition on the fourth day. During the second and third days, it was reluctant to stand and urinated frequently. It had diarrhea on the fourth day. Plasma urea concentrations were increased on days two, three and four (11.73, 19.02 and 16.93mM/L) and, although the animal was recovering, it was necropsied on day four. The only gross change was a mild thickening and yellowish discoloration of the mucosa in areas of the rumen and reticulum, which were almost unnoticeable until the organs were washed clean of contents. Microscopic examination of tissues revealed mucosal necrosis of the rumen and reticulum. There was diffuse mild hepatocellular vacuolar change and very mild bile duct hyperplasia. In the kidney, there was a mild tubular nephropathy characterized by acute epithelial cell necrosis in occasional tubules.

This feeding experiment indicated that few sheep are attracted to eat the corals, and those that do stop eating them before a lethal dose is consumed. It would be necessary to drench the corals to sheep to achieve a sufficient intake to produce the disease as seen in the field. Corals were hammer-milled to a powder and mixed with 800-1,000ml of water to be administered as a slurry *via* an intraruminal tube.

It was determined that a severe reticulitis, rumenitis and omasitis could consistently be produced in sheep dosed 400g of corals if the sheep were deprived of water for 48hrs before dosing. This damage was most apparent 48hrs after dosing, and had usually completely repaired by 5d later. A feature of the pathological changes in the forestomachs was quick and effective regeneration.

Despite consistently producing pathological changes in the forestomachs, changes in the kidney and liver could not be produced, even with total intakes of up to 1,032g (28g/kg) over 11d. Black soil blindness was experimentally produced in steers with a total dose rate of only 7.1-12.5g corals/kg (Jubb *et al.*, 1995; 1996). The disease as seen in the field was produced in only one sheep by drenching corals, and it required giving 2,564g (70g/kg) over 11d.

This particular sheep became subdued 8d after drenching started, and depressed on day 12. In the hours before it was terminated, it bumped into gates and rails when being moved, but could focus on people 5-20m away who were moving quietly. Its pupillary light reflex appeared unimpaired. Plasma aspartate aminotransferase activity and plasma creatinine and urea concentrations increased substantially over days 11 and 12 (205IU/L, 350µM/L and 14.49 mM/L, respectively). At necropsy it had a very severe necrotic and diphtheritic reticulitis, rumenitis and omasitis, affecting virtually 100% of the epithelial lining in each of these organs. Furthermore, the urine was pH 6 and contained protein (500mg/dl) and glucose (300-1,000mg/dl). Microscopic examinations confirmed the gross changes seen in the forestomachs, and revealed an acute to sub-acute tubular nephrosis characterized by epithelial cell necrosis, amorphous or granular eosinophilic casts and evidence of some regeneration of epithelial cells in cortical tubules. There was also centrilobular and midzonal vacuolar change in the liver, necrosis of occasional hepatocytes and mild Kupffer cell hyperplasia. There were no significant microscopic lesions in the eyes, brain, adrenal glands, spleen, pancreas, abomasum or gall bladder.

Observations Made on the Corals

During the handling of the corals, it became obvious that there were four fungal structures being fed. The first was the true coral; the second was a structure that looked like a coral, but it contained a very hard core of material similar to that in a sclerotium; the third was the true sclerotium; and the fourth looked like a coral but had the remnants of a sclerotium attached, giving the appearance of conidiophores and conidia germinating or growing out of a sclerotium (Fig. 97.1). There were different proportions of these four structures in different batches of the collected material. Corals were always the major component, but there were always large numbers of the corals with hard cores. Sclerotia and germinating sclerotia were present (0-3%) in the material being fed. This finding raised the question as to whether these different structures had different toxicities.

Drenching with Conidia, Sclerotia and Cultures of the Fungus

The treatments used in this experiment were prepared as follows:
1) Corals were dosed as found in the field.
2) Dry corals were placed in a food processor and rolled using dough blades. The conidia and conidiophores crumbled away and formed a powder. True corals completely disintegrated, corals with hard cores were reduced to the core, and the conidia and conidiophores on germinated sclerotia were removed to leave the remnant sclerotia. The material was separated using a 1.5mm screen. Only material that passed through the screen was used as the "conidia" treatment.
3) Sclerotia were mostly removed from the corals by manual sorting before the corals were processed. These were added to the sclerotial remnants and other hard material recovered from the processor that would not pass through a 5mm screen. The resultant mix was hammer-milled to a powder and used as the "sclerotia" treatment.
4) An isolate of the fungus (WAC8705) was grown on potato dextrose agar for

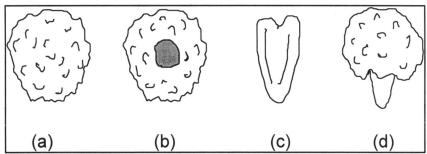

Fig. 97.1. Diagrammatic representation of the four types of fungal structure administered to sheep: **a)** true coral; **b)** coral with a hard core; **c)** sclerotium; **d)** germinated sclerotium.

10wks in the dark. Several 0.5cm cubes of this culture were dropped into 1L flasks containing 200g of moist, autoclaved wheat or millet seed, the flasks sealed, shaken, and then incubated at 25°C for 53 or 87d in the dark. After these periods, the grain/fungal culture was removed from the flasks, placed in sealed plastic bags and frozen at -40°C. Prior to administration the required amount of grain/fungal culture was weighed out, placed in a Waring commercial blender and homogenized in added water. Sixty per cent of the material used was incubated for 53d, and the rest for 87d.

The treatments were all dosed as aqueous slurries *via* an intraruminal tube at the amounts and intervals listed in Table 97.1.

Treatment Results

One sheep dosed with corals stopped eating after 2d, then only ate 140g of feed during the last 3d of the experiment. The other three had transient reductions in appetite but had regained full appetite by the end of the experiment. At necropsy the sheep that did not regain its appetite had gross and microscopic evidence of reticulitis, rumenitis and omasitis. There were no pathological changes in the other three sheep.

Of the animals dosed with conidia, one sheep fed 49g/kg stopped eating after 3d and only ate a 75g in the remainder of the experiment, while the other had only a brief reduction in appetite. The sheep dosed 20 and 10g/kg maintained their full appetite and were alert throughout. None of the animals had any significant changes in clinical chemistry. At necropsy the sheep that received 49g/kg and did not regain its appetite had grossly apparent reticulitis, rumenitis and omasitis. The second sheep dosed 49g/kg had microscopic evidence of multifocal necrosis of the mucosa in all forestomachs. There were no pathological changes in the other two sheep.

All three sheep dosed with sclerotia stopped eating by day two and remained inappetant until death. All became depressed in the last 12hrs of their lives, and during this period the two dosed with 8-9g/kg walked with a staggering gait and

Table 97.1. Treatment of sheep with various fungal forms.

Number sheep	Fungal form	Treatment (g/kg) day												
		0	1	2	3	4	5	6	7	8	9	10	11	12
4	corals	7	7					T						
2	conidia	7	7	7	7				7	7	7	T		
1	conidia	2	2	2	2				2	2	2	2	2	T
1	conidia	2	2	2	2	2		T						
1	sclerotia	7	7			D								
1	sclerotia	2	2	2	1.5	1.5	D							
1	sclerotia	2	2	2	2	D								
4	culture	7	7	7	7				7	7	7	T		

T, terminated; D, died

bumped into rails and unexpected small objects as they were moved. They were able to avoid large objects placed on the ground in their path and to focus on people moving quietly around them, and they both had a pupillary light reflex. The sheep dosed with14g/kg had considerable elevations in plasma γ-glutamyl transferase activities, and bilirubin, creatinine and urea concentrations over days three and four (up to 161IU/L, 37.6μM/L, 237μM/L and 17.66mM/L, respectively), but the clinical chemistry changes were not so marked in the other two sheep. The plasma bilirubin concentration became elevated (13.6μM/L) in the sheep dosed 8g/kg. There were definite upward trends in the plasma creatinine and urea concentrations in both sheep, but only the urea levels became abnormally high (11.61 and 11.02mM/L). The urine of the sheep dosed 8g/kg had a pH of 6 and contained protein (100mg/dl), while in the sheep dosed 9g/kg, the pH was 6.5, and it contained protein (100-500mg/l) and glucose (300-1,000mg/dl). All three sheep had severe reticulitis, rumenitis and omasitis affecting the entire surface of these organs, and the sheep dosed 14 and 9g/kg had substantial acute to subacute nephropathies. The sheep dosed 8g/kg had acute necrosis of scattered individual hepatocytes. There were no pathological changes in the brains or eyes.

The four sheep dosed with fungal culture material maintained a full appetite and remained clinically normal throughout. There were no clinical chemistry or pathological changes.

Discussion

This experiment presented clear evidence that most, if not all, of the toxins responsible for BSB reside in the sclerotia or sclerotia-like material in the center of many corals. Dose rates of just 8-14g sclerotia/kg were sufficient to cause death or severe clinical disease within 3.5-5.5d. Although sheep dosed with sclerotia had very severe reticulitis, rumenitis and omasitis, death was likely due to the nephrosis, and possibly hepatopathy, that developed rapidly towards the end, as indicated by clinical chemistry and urinalysis. Microscopic evidence of these was not always present because failure of the metabolic processes within these organs progressed faster than associated morphological changes.

The fact that two sheep dosed with conidia developed pathological changes in their forestomachs suggests that some of the toxins do occur in the conidia. However, it is possible that some small chips of sclerotial material passed through the screens with the conidia. There was no evidence that the fungus grown on wheat was toxic. This supports the finding that most toxicity is in the sclerotia, since the hyphae within the culture were similar to those hyphae within the corals that are associated with the conidiophores and conidia. Hyphae within the sclerotia are densely packed and presumably have different metabolic processes.

Conclusions

Sheep are less susceptible than cattle to the lethal toxin(s) involved in BSB. In sheep, lethal intakes were 70g corals/kg or 8-14g sclerotia/kg, compared to 7.1-12.5g corals/kg for cattle. Black soil blindness is unlikely to be a problem in sheep in the field, since few sheep are attracted to the corals, and those that do eat them stop before they have consumed a lethal dose. Sheep appear to be particularly susceptible to toxins within the corals that damage the mucosa of the forestomachs. This early damage to the forestomachs may be what causes sheep to stop eating the corals before they have consumed a lethal intake. Liver and/or kidney damage is required for a lethal toxicosis in BSB. In sheep, terminal metabolic dysfunction of these organs may occur so rapidly that death occurs before microscopic changes have become apparent.

Sheep in the terminal stages of BSB stumbled and walked into some objects, giving the appearance of being blind. However, close observation suggested they were not blind; their pupillary light reflex was still present and no microscopic lesions consistent with blindness could be found in the brain, eyes or optic nerve in those sheep examined. This apparent blindness may have been associated with their discomfort and overall metabolic disruption leading to a state of lack of interest in their surroundings. It is worth noting the description of the clinical signs of rumenitis associated with grain poisoning recorded by Blood and Henderson (1968). "Severely affected animals have a staggery, drunken gait and appear blind. They bump into objects and have no eye preservation reflex. The pupillary light reflex appears to be unimpaired." However, the blindness in cattle in field cases of BSB seems real, having been witnessed by at least seven veterinarians. The blindness seen in the cattle occurred initially without other signs of illness or discomfort being observed.

Most toxicity resides in the sclerotia or the sclerotia-like material in the center of many corals. The conidia and conidiophores have little if any toxicity and the fungus cultured on cereal grain is not toxic. It seems unlikely that livestock would deliberately eat sclerotia because they are so hard. Therefore, field intoxication is most likely to result from consumption of corals that contain central cores of hard sclerotia-like material.

Acknowledgments

This work was funded by the Australian Meat Research Corporation and the Industry Resource Protection Program of Agriculture Western Australia.

References

Blood, D.C. and Henderson, J.A. (1968) *Veterinary Medicine*, 3rd ed. Baillière, Tindall and Cassell, London, p. 87.

Jubb, T.F., Main, D.C., Mitchell, A.A., Shivas, R.G. and De Witte, K.W. (1995) Black soil blindness: a new fungal toxicosis of cattle. *Proceedings of the Epidemiology Chapter of the Australian College of Veterinary Scientists*. pp. 104-108.

Jubb, T.F., Main, D.C., Mitchell, A.A., Shivas, R.G. and De Witte, K.W. (1996) Black soil blindness: A new mycotoxicosis of cattle grazing *Corallocytostroma*-infected Mitchell grass (*Astrebla* spp). *Australian Veterinary Journal* 73, 110-112.

Shivas, R.G., Mitchell, A.A. and Jubb, T.F. (1997) *Corallocytostroma ornicopreoides* spp. nov., an unusual toxic fungus on *Astrebla* and *Dichanthium* in north-western Australia. *Mycological Research* 101, 849-852.

Chapter 98

Transient Hepatotoxicity in Sheep Grazing *Kochia scoparia*

J.G. Kirkpatrick[1], R.G. Helman[4], G.E. Burrows[2], D. vonTungeln[5], T. Lenenbauer[3] and R.J. Tyrl[6]

[1]*Departments of Medicine and Surgery, Anatomy,* [2]*Pathology and Pharmacology,* [3]*Infectious Diseases and Physiology, College of Veterinary Medicine,* [4]*Animal Disease Diagnostic Laboratory,* [6]*Department of Botany, Oklahoma State University, Stillwater, Oklahoma 74078, USA;* [5]*USDA, El Reno, Oklahoma, USA*

Introduction

Kochia scoparia (L.) Roth (Chenopodiaceae), kochia, mexican fireweed, burning bush, is a drought-resistant, rapidly growing plant with a wide geographic distribution. Although an undesirable weed in most areas, it has been investigated as a possible forage crop because of its drought resistance and high volume of forage production. Nutritional values of kochia at pre-full bloom and alfalfa at 20% bloom are similar (Knippfel *et al.*, 1989). Plant composition is dependent on soil type, weather conditions and genetic variability (Mir *et al.*, 1991).

Although of nutritional value, kochia may cause adverse effects when fed as the total forage. Lambs fed a ration of 35% kochia hay and 65% alfalfa hay for 4wks showed no signs of intoxication. Increasing kochia to 50% did not adversely affect consumption or digestibility, but weight gains were negligible (Rankins and Smith, 1991). Kochia has been associated with icterus and photosensitization in cattle, sheep and horses (Sprowls, 1981).

The etiology of the adverse effects caused by kochia is uncertain. Oxalate concentrations may be high enough to cause adverse effects, but in those cases where nephrosis occurred, few oxalate crystals were present (Sprowls, 1981; Dickie *et al.*, 1989). Thilsted and Hibbs (1989), in a 3yr study of steers grazing kochia, found that the clinical signs were not consistent with those caused by oxalate, and that plant oxalate concentrations were inversely proportional to animal morbidity and mortality.

It is unclear whether alkaloids found in kochia are in sufficient concentration to represent a risk. β-carbolines are present in such low amounts as to make them unlikely toxicants (Drost-Karbowska and Kowalewski, 1978). Alkaloids of unknown structure and toxicity have been reported for kochia, their concentrations changing with maturity of the plant (Thilsted and Hibbs, 1989). The severity of kochia toxicity in rats correlated positively with the forage content of substances reacting to Dragendorff's reagent (Rankins, 1987).

Saponins, steroidal glycosides with gastrointestinal and/or hepatotoxic effects, have been identified in the fruits of *K. scoparia* (Souto and Milano, 1966), but there is no report of the kochia saponins producing toxicity in animals.

A preliminary study was carried out in 1995, to investigate changes in serum γ-glutamyl transferase (GGT) concentrations and in hepatic ultrastructural morphology in sheep grazing nearly pure *K. scoparia* for 83d. Serum GGT levels in the experimental group were significantly increased at 25-55d ($P<0.03$), and GGT concentrations were significantly increased overall, compared to controls ($P<0.03$) (Fig. 98.1). No microscopic changes in liver morphology were observed. The present study was undertaken to further describe and evaluate hepatic changes and weight gains in sheep grazing kochia as the total diet.

Materials and Methods

Twenty yearling western whiteface ewes were randomly selected from the flock at the Grazinglands Research Center in El Reno, OK. Baseline liver biopsy samples were obtained from all sheep on June 19 (Day 1) and post-exposure liver biopsy samples were obtained on July 31 and August 30, 1996. Liver specimens were fixed, processed and stained with hematoxylin and eosin for routine histopathology and with Masson's trichrome method for collagen.

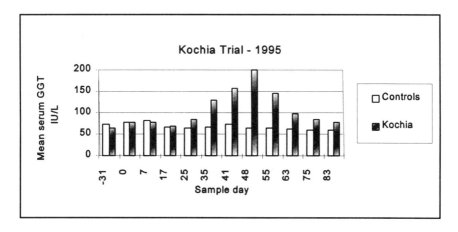

Fig. 98.1. Serum GGT concentrations in kochia-fed and control sheep at various sampling days. Sample Day 0 is July 21, 1995 and is first day of exposure to kochia.

The sheep were randomly divided into two groups of eight kochia-fed and four control sheep. Baseline blood samples for 17 component serum chemistry profiles were collected from Group 1 and repeated at weekly intervals. Group 1 sheep were placed on a >95% pure stand of *K. scoparia* on June 19 (Day 0), while the four controls were placed on an adjacent weedy bermudagrass pasture. Baseline blood samples were collected from Group 2 on July 31 (Day 14), after which they were placed on kochia and bled weekly with Group 1. The delay between Groups 1 and 2 sheep grazing kochia was used to evaluate effects of plant maturity on serum chemistry parameters, liver ultrastructural morphology and body weight changes.

Kochia scoparia samples were collected at weekly intervals, dried and stored in brown paper bags. Feed analysis, oxalate, sulfate and nitrate concentrations were obtained on samples collected on June 5, July 9 and August 20, 1996.

Results

The only parameter in which there was a consistent change was serum GGT concentration. In Group 1 sheep, duration of exposure to kochia had a significant effect upon serum GGT concentration for individual ewes. In contrast, there was no significant change in GGT levels for individual ewes in Group 2 during their exposure to kochia (Fig. 98.2).

The ultrastructural morphology of livers from control and kochia-fed sheep were similar for both groups. Mean body weight gains in Groups 1 and 2 sheep were significantly greater than the controls (Table 98.1). Results of nutritional, oxalate, sulfate and nitrate analyses of the three dried plant samplings are presented in Table 98.2.

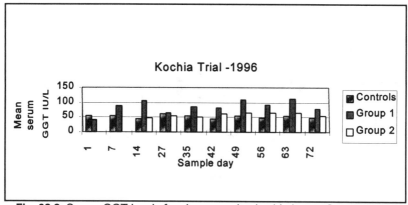

Fig. 98.2. Serum GGT levels for sheep grazing kochia in test Groups 1 and 2 and control sheep. Group 2 exposure to kochia began 14d after Group 1 (6/19 and 7/3/96, n=8; Control, n=4).

Table 98.1. Body weight (bw) gains in sheep grazing *Kochia scoparia* and controls, 1996.

Group	Mean bw gain (kg)	95% confidence interval for mean bw gain
Control (n=4)	2.7†	(-5.4, 17.2)
#1 (n=8)	7.8*	(12.9, 21.6)
#2 (n=8)	9.3*	(14.9, 26.9)

* *P*<0.005, † *P*>0.05; Both groups; combined weight gains were significantly greater than controls (*P*<0.003). Weight gains did not significantly differ between Groups 1 and 2 (*P*>0.50).

Table 98.2. Feed oxalate, sulfate and nitrate in *K. scoparia* samples, 1996.

Component	June 5	July 9	August 20
Dry matter % [1]	94.0	93.2	93.7
Crude protein % [1]	22.3	17.1	18.9
ADF % [1]	32.2	31.3	32.7
NDF% [1]	45.1	49.4	46.2
TDN% [1]	66.6	67.4	66.1
Relative feed value*[1]	132.0	122.0	128.0
Total oxalate %[2]	3.8	3.7	3.4
Soluble oxalate %[2]	2.5	1.82	2.1
Nitrate %[2]	3.3	1.0	0.46
Sulfate %[2]	0.42	0.52	0.44

[1] Servi-Tech Laboratories, Dodge City, KS; [2] Oklahoma Animal Disease Diagnostic Laboratory, Stillwater, OK; * Comparisons used to evaluate alfalfa hay; Premium ->175; Prime - 152-175; #1Dairy hay - 125-151; #2Good feed lot hay - 103-124; #3Average feed lot hay - 87-102; #4Poor hay - 75-86; Very poor hay - <75.

Discussion

Preliminary studies in 1995 used the same kochia and control pastures, same breed, sex and age sheep as were used in the 1996 trial. Serum GGT, an important marker enzyme of hepatobiliary disease in the ruminant, was measured weekly. The time interval from exposure to significant increase in serum GGT concentrations and plant maturity and palatability were considered to be possibly important factors. The 1996 study addressed plant maturity by exposing Group 1 sheep 14d earlier in the grazing season, and there was a significant increase in mean GGT concentration by 7d. Group 2 sheep did not show a significant increase in serum GGT until day 42. No significant differences in ultrastructural hepatic morphology were noted. Considerable cell damage occurs prior to histopathologic change, and measuring enzyme leakage provides a more sensitive and earlier measure of cell pathology. Sorbitol dehydrogenase is a better identifier of ongoing hepatocellular necrosis, but its instability over time precluded its use in this study (Smith, 1996).

The body weight changes in the 1996 trials were encouraging if considering kochia as a source of forage. Body weight gains were greater than those reported in other studies with rations containing up to 50% kochia, but pure kochia forage did not produce livestock weight gains expected from measured nutrient levels. Although oxalates, alkaloids, saponins, nitrates and sulfates all occur in the plant, sometimes at toxic levels, there was no indication by physical examination, periodic observations or in the serum chemistry profile of adverse effects which could be related to these compounds. Because of its possibilities as a livestock forage, kochia warrants further study to identify its toxic principle(s) and to determine ways to manage those principles.

References

Dickie, C.W., Gerlach M.L. and Hamar, D.L. (1989) *Kochia scoparia* oxalate content. *Veterinary and Human Toxicology* 31, 240-242.

Drost-Karbowska, K. and Kowalewski, Z. (1978) Isolation of harmane and harmine from *Kochia scoparia. Lloydia* 41, 289-290.

Knippfel, J.E., Kerman, J.A., Coxworth, E.C. and Cohen, R.D.H. (1989) The effect of stage and maturity on the nutritive vale of Kochia. *Canadian Journal of Animal Science* 69, 1111-1114.

Mir, Z., Bittman, S. and Townley-Smith, L. (1991) Nutritive value of Kochia (*Kochia scoparia*) hay or silage grown in a black soil zone in northeastern Saskatchewan for sheep. *Canadian Journal of Animal Science* 71, 107-114

Rankins, D.L. Jr (1987) Evaluations of treatments to improve tolerance of toxicants in herbage of *Kochia scoparia* (L) Schrad by rats and sheep. MS Thesis. New Mexico State University, Las Cruces, NM.

Rankins, D.L. Jr and Smith, G.S. (1991) Nutritional and toxicological evaluations of Kochia hay (*Kochia scoparia*) fed to lambs. *Journal of Animal Science* 69, 2925-2931.

Smith, B.P. (1996) *Large Animal Internal Medicine*. Mosby, St Louis, MO, pp. 460-464.

Souto, J.V. and Milano, A. (1966) Triterpenic saponin in ripe fruits of *Kochia scoparia. Review of Investigations in Agriculture* 3, 367-383.

Sprowls, R.W. (1981) Problems observed in horses, cattle and sheep grazing kochia. *Proceedings of the 24th Annual Meeting of the American Association of Veterinary Laboratory Diagnosticians*, pp. 397-403.

Thilsted, J. and Hibbs, C. (1989) Kochia (*Kochia scoparia*) toxicosis in cattle: Results of four experimental grazing trials. *Veterinary and Human Toxicology* 31, 34-41.

Chapter 99

Calliandra calothyrsus Leaf Extracts' Effects on Microbial Growth and Enzyme Activities

M.B. Salawu[1,2], T. Acamovic[1] and C.S. Stewart[2]

[1]SAC, 581 King Street, Aberdeen AB24 5UD, UK; [2]Rowett Research Institute, Bucksburn, Aberdeen AB21 9SB, UK

Introduction

Proanthocyanidins (PACNs) in *Calliandra calothyrsus* have been implicated as the cause of its low degradability by rumen microbes (Salawu *et al.*, 1997). Inhibition of microbial activities by PACNs may occur by both direct and indirect effects. It seems that rumen microbes do not produce enzymes capable of degrading PACNs, but rumen microbes may show some adaptation to condensed tannins depending on their molecular weight and degree of polymerization (Field and Lettinga, 1992). This chapter describes *in vitro* studies of the inhibitory effects of *Calliandra* leaves and leaf extracts on selected microbes.

Materials and Methods

Calliandra leaves harvested from the central highlands of Kenya were dried at 50°C for 48hrs and ground to pass a 1mm screen. Rumen bacteria and the rumen fungus *Neocallimastix frontalis* strain RE1 were from the culture collection of the Rowett Research Institute. *Butyrivibrio fibrisolvens* strain 16/4 was isolated from a fermenter inoculated with human feces (Rumney *et al.*, 1995).

For determination of PACN, ground samples (200mg) were extracted with 70% aqueous methanol (MeOH) by continuous vortexing for 20min. The supernatants were filter-sterilized through 0.2μm single use, low protein binding, non-pyrogenic sterile acrodiscs (Gelman Science, UK) into sterilized 16x125mm Hungate culture tubes under O_2-free CO_2 and capped. Separate samples were extracted with water as described by Smart *et al.* (1961). Free and bound PACNs were measured using the butanol-HCl method (Jackson *et al.*, 1996).

The *in vitro* degradability of *Calliandra* leaves was measured with mixed rumen microbes in a consecutive batch culture system (Theodorou *et al.*, 1987). For measurement of bacterial growth, filter-sterilized (22μm pore-size membrane filters, NML, Gottingen) PACN extracts (50, 100 or 200μl) were added to 9ml of anaerobic nutrient medium with cellobiose (0.3% w/v) as the energy source. The basal medium used was that of Hungate and Stack (1982), except for *Prevotella bryantii* (strain $B_1 4$) and *B. fibrisolvens* (strain D6/1), which were grown on medium M10 (Caldwell and Bryant, 1966) at 39°C. Cultures were inoculated with 0.3ml of a pre-grown (16-18hrs) culture of the relevant bacterium. Growth was monitored at 600nm.

The carboxymethyl cellulase (CMCase) and xylanase activities of the cell free extracts (CFE) were determined using the methods of Williams and Orpin (1987), with 200μl CFE (Marvin-Sikkema *et al.*, 1993), 800μl buffer containing substrate and 200μl *Calliandra* extract or solvent control incubated at 39°C for 1hr.

Data were subjected to analysis of variance using the one-way design model. Means were separated for comparison using the least significant difference method (Steele and Torrie, 1980).

Results and Discussion

The PACN contents of *Calliandra* leaves are in Table 99.1. Extractable PACNs were lowest ($P<0.05$) in the 1.5yr-old leaves, but the amounts of PACN bound to different fractions varied.

In vitro dry matter (DM) losses of *Calliandra* in CBC (Table 99.2) were low for all of the leaves, the 4mos-old leaves having lowest DM loss over the seven consecutive cultures. Adaptation by rumen microbes to the leaf constituents was seen and DM disappearance tended to plateau after the third consecutive culture.

Water extracts of the 1.5yr-old leaves did not inhibit the CMCase activity, but reduced ($P<0.05$) the xylanase activity more than the water extracts of the coppiced

Table 99.1. Proanthocyanidin (PACN) content of *Calliandra*.

Leaves	PACN (g/kg DM equivalent)					
	E	CP-B	NDF-B	ADF-B	total	% E
Coppice	80.9[a]	20.7[a]	9.0[a]	2.1[a]	111[a]	73[ab]
4mos-old	79.2[a]	27.2[b]	6.4[b]	2.3[a]	113[a]	70[a]
1.5yr-old	74.7[b]	15.5[c]	8.2[a]	4.2[b]	98.5[b]	76[b]

E: extractable PACNs; CP-B: crude protein-bound; NDF-B: neutral detergent fiber-bound; ADF-B: acid detergent fiber-bound; [a]Means with different superscripts in columns are significantly different ($P<0.05$).

Table 99.2. *In vitro* dry matter loss (mg/g) of *Calliandra* leaf samples incubated with mixed rumen contents in a consecutive batch culture system.

Leaves	\<br\>Consecutive cultures						
	1	2	3	4	5	6	7
Coppice	319[a]	218[a]	240[a]	260[a]	270[a]	298[a]	280[a]
4mos old	137[b]	195[b]	178[b]	208[b]	186[b]	219[b]	221[b]
1.5yrs old	155[b]	215[b]	216[b]	220[b]	270[c]	260[c]	267[c]

Means with different superscripts in columns are significantly different ($P<0.05$).

leaves (Table 99.3). Aqueous MeOH extracts generally reduced the CMCase activity more than did the water extracts, and the xylanase activity was more affected by these extracts than was the CMCase activity. Water extracts of *Sericea* that were also thought to contain polyphenols have been reported to inhibit rumen CMCase activities (Smart *et al.*, 1961). The observed enzyme inhibition may be due to binding of the tannins or other phenolics to the substrate as well as to the enzymes (Makkar *et al.*, 1990).

The rumen bacteria *P. bryantii* and *Selenomonas ruminantium* proved to be relatively insensitive to the presence of aqueous MeOH extracts of the leaves (Table 99.4). Jones *et al.* (1994) also reported that the condensed tannins of sainfoin (*Onobrychis viciifolia*) had little effect on the growth of *P. bryantii* strain B_14 in comparison with other bacteria. Odenyo and Osuji (1997) found a *Selenomonas* strain able to grow in a nutrient medium containing up to 8g/L condensed tannins, and able to hydrolyze gallic acid.

Table 99.3. The carboxymethyl cellulase (CMCase) and xylanase activities of cell free preparation of the rumen fungus *N. frontalis* (strain RE1) in the presence of extracts of *Calliandra* leaves.

Leaves	mg reducing sugar released/g/hr		% reduction in activity	
	CMCase	Xylanase	CMCase	Xylanase
Aqueous MeOH extracts				
1.5yr-old	6.23	9.84[a]	8.20[a]	18.9[a]
4mos-old	5.63	8.21[b]	17.0[b]	32.3[b]
Coppice	5.67	9.26[a]	16.4[b]	23.7[a]
Water extracts				
1.5yr-old	10.4[a]	13.7[a]	0.00[a]	30.5[a]
4 mos-old	8.64[b]	10.9[b]	10.2[b]	44.4[b]
Coppice	8.95[b]	16.1[c]	7.03[b]	18.3[c]

Means with different superscripts in columns are significantly different ($P<0.05$).

Table 99.4. Effect of aqueous methanol extracts of *Calliandra* leaf extracts on the growth of anaerobic gut bacteria.

	Absorbance (600nm) in 24hr - strain (extract volume)					
	17	16/4	D6/1	HD4	B₁4	BL2
Leaves	(50μl)	(100μl)	(100μl)	(200μl)	(200μl)	(100μl)
Coppice	0.09^a	0.16^a	0.62^a	0.40^a	0.88^a	0.14^a
4mos old	0.11^b	0.50^b	0.59^{ab}	0.43^b	0.91^{ab}	0.12^b
1.5yrs old	0.09^a	0.53^b	0.58^b	0.50^c	0.93^{bc}	0.09^c
Control	0.50^c	0.69^c	0.52^c	0.52^d	0.96^c	0.62^d

The volume of extract indicated was added to 9ml cultures; *Ruminococcus flavefaciens* (strain 17); *Butyrivibrio fibrisolvens* (strain 16/4); *B. fibrisolvens* (strain D6/1); *Selenomonas ruminantium* (strain HD4); *Prevotella bryantii* (strain B₁4); *Fibrobacter succinogenes* (strain BL2); ^{a}Means with different superscripts in columns are significantly different ($P<0.05$).

In the present study, the *Butyrivibrio* strains varied in their response, growth of the human fecal isolate (16/4) being inhibited in the presence of coppice leaf extracts. This may be an example of strain variation, but it is possible that human isolates may be more susceptible to some secondary compounds than similar strains from the rumen, which may be routinely exposed to such metabolites.

The sensitivity of *Fibrobacter succinogenes* to the *Calliandra* extracts is consistent with its known sensitivity to tannins from *Lotus corniculatus* (Bae *et al.*, 1993). The growth of *R. flavefaciens* strain 17 was reduced markedly by the leaf extracts, with the lowest level of extract (50μl) inhibiting growth.

Extracts of the 4mos-old leaves generally had more markedly inhibitory effects than the other extracts. Because young plants are particularly vulnerable to attack by insect pests and herbivores, these plants presumably synthesize biologically active compounds to deter herbivory. This may explain why their leaves had particularly low DM losses in CBC, and why their extracts inhibit rumen enzymes and rumen bacterial growth.

References

Bae, H.D., McAllister, T.A., Yanke, J., Cheng, K.J. and Muir, A.D. (1993) Effects of condensed tannins on endoglucanase activity and filter paper digestion by *Fibrobacter succinogenes* S85. *Applied Environmental Microbiology* 59, 2132-2138.

Caldwell, D.R. and Bryant, M.P. (1966) Medium without rumen fluid for non-selective enumeration and isolation of rumen bacteria. *Applied Microbiology* 14, 794-801.

Field, J.A. and Lettinga, G. (1992) Toxicity of tannic compounds to microorganisms. In: Hemingway, R.W. and Laks, P.E. (eds), *Plant Polyphenols*. Plenum Press, New York, NY, pp. 673-692.

Hungate, R.E. and Stack, R.J. (1982) Phenylpropanoic acid growth factor for *Ruminococcus albus*. *Applied Environmental Microbiology* 44, 79-83.

Jackson, F.S., Barry, T.N., Lascano, C. and Palmer, B. (1996) The extractable and bound condensed tannin content of leaves from tropical tree, shrub and forage legumes. *Journal of Science and Feed Agriculture* 71, 103-110.

Jones, G.A., McAllister, T.A., Muir, A.D. and Cheng, K.J. (1994) Effects of sainfoin (*Onobrychis viciifolia* Scop.) condensed tannins on growth and proteolysis by four strains of ruminal bacteria. *Applied Environmental Microbiology* 60, 1374-1378.

Makkar, H.P.S., Dawra, R.K. and Singh, B.(1990) *In vitro* effect of oak tannins on some hydrolytic and ammonia assimilating enzymes of the bovine rumen. *Journal of Animal Nutrition* 7, 207-210.

Marvin-Sikkema, F.D., Pedro-Gomes, T.M., Grivet, J.P., Gottschal, J.C. and Prins, R.A. (1993) Characterisation of hydrogenosomes and their role in glucose metabolism of *Neocallimastix sp.* L2. *Archives of Microbiology* 160, 388-396.

Odenyo, A.A. and Osuji, P.O. (1997) Tannin-resistant ruminal bacteria from East African ruminants. Reproductive Nutrition and Development Supplement. In: Chesson, A., Flint, H.J. and Stewart, C.S. (eds), *Evolution of the Rumen Microbial Ecosystem.* INRA, Paris, p. 78.

Rumney, C.J., Duncan, S.H., Henderson, C. and Stewart, C.S. (1995) Isolation and characteristics of a wheat-bran-degrading *Butyrivibrio* from human faeces. *Letters in Applied Microbiology* 20, 232-236.

Salawu, M.B., Acamovic, T., Stewart, C.S., Hovell, F.D.B. and McKay, I. (1997) Assessment of the nutritive value of *Calliandra calothyrsus*: *in sacco* degradation and *in vitro* gas production in the presence of *Quebracho* tannins with or without Browse Plus. *Animal and Feed Science Technology* 69, 219-232

Smart, W.W.G., Bell, T.A., Stanley, N.W. and Cope, W.A. (1961) Inhibition of rumen cellulase by an extract from *Sericeae* forage. Technical notes. *Journal of Dairy Science* 44, 1945-1946.

Steele, R.G.D. and Torrie, J.H. (1980) *Principles and Procedures of Statistical Analysis.* MacGraw-Hill, New York, NY.

Theodorou, M.K., Gascoyne, D.J., Akin, D.E. and Hartley, R.D. (1987) Effect of phenolic acids and phenolics from plant cell walls on rumen like fermentation in consecutive batch culture. *Applied Environmental Microbiology* 53, 1046-1050.

Williams, A.G. and Orpin, C.G. (1987) Polysaccharide degrading enzymes formed by three species of anaerobic rumen fungi grown on a range of carbohydrate substrates. *Canadian Journal of Microbiology* 33, 418-426.

Chapter 100

Chinaberry (*Melia azedarach*) Poisoning in Animals

W.R. Hare

National Animal Poison Control Center Department of Veterinary Biosciences, College of Veterinary Medicine, University of Illinois, Urbana, Illinois 61801, USA

Introduction

Chinaberry (*Melia azedarach*) poisoning has been reported in both man and animals, following ingestion of the plant or its fallen fruit. Pigs are the most frequently poisoned domestic animal. Adverse clinical signs are consistent with abnormalities of the gastrointestinal system and nervous system. Animals poisoned by ingestion of *M. azedarach* fruit generally survive no longer than 48hrs following the onset of clinical signs. Quick, aggressive efforts at decontamination, together with symptomatic and supportive treatment, are recommended.

Melia azedarach is a rapidly growing ornamental shade tree with worldwide distribution (Kingsbury, 1964). The tree belongs to the Mahogany family, and has become naturalized in both the North and South Temperate Zones. It is found in abundance throughout Africa, South America, Australia, New Zealand, Indonesia, Japan, the southern US, Hawaii, Bermuda and elsewhere. In the US, *M. azedarach* is commonly known as White Cedar, Texas Umbrella or Chinaberry tree. It grows to a height of 6-12m, with some rainforest varieties reaching 30-45m. The leaves are dark green and relatively large, arranged in an alternate, twice suboppositely (bipinnate) compound fashion. Leaflets are lanceolate, serrate or lobed, 2.5-7cm long and 1-4cm wide. Its purple flowers are approximately 2.5cm across, and are composed of five to six narrow petals. The tree bears a prolific production of fruits that ripen and drop to the ground over a short period of time. The fruits are approximately 1.5cm in diameter, initially green, smooth and round to ovoid, becoming pale-yellow, somewhat pitted and ribbed at maturity.

Poisonous Principle and Toxicity

The principle toxins responsible for *M. azedarach* poisoning, the cytotoxic limonoid tetranorterpenes meliatoxins A1, A2 and B1, are mostly found in the fruit (Oelrichs *et al.*, 1983; Oelrichs *et al.*, 1985). Numerous other potentially toxic limonoids have been isolated from *M. azedarach* (Ochi *et al.*, 1978; Srivastava, 1986; Ahn *et al.*, 1994; Wang *et al.*, 1994). The toxicity of the fruits varies with environmental factors and growth stage of the tree, making risk assessment difficult (Kingsbury, 1964).

Feeding experiments in pigs and sheep have established the toxic dose of *M. azederach* berries to be approximately 5g/kg (Kingsbury, 1964). Hothi *et al.* (1976) found that 0.6-0.7mg/kg was too low for oral dosing, but excessive for parenteral administration of fruit extract. Clinical reports indicate that ingestion of five or six fruits by young dogs (Hare *et al.*, 1997) or six to eight fruits by children (Everist, 1974) can be fatal. The duration of clinical signs in fatally poisoned individuals is no longer than 48hrs (Hothi *et al.*, 1976; Hare *et al.*, 1997).

Conditions of Poisoning

The leaves, bark, flowers and especially the fruits of *M. azederach* are poisonous (Vahrmeijer, 1981). Most cases of poisoning result from ingestion of fallen fruits, despite their bitterness. Poisoning has been reported in cattle, sheep, goats, pigs, dogs, rabbits, rats, guinea pigs and poultry (Steyn and Rindl, 1929; Watt and Breyer-Brandwick, 1962; Kingsbury, 1964; Everist, 1974; Hothi *et al.*, 1976).

Chinaberry poisoning manifests itself quickly, with clinical signs related to the gastrointestinal and/or nervous system appearing within 2-4hrs. The clinical signs include anorexia, vomiting, diarrhea, constipation, colic, excitement, convulsions, ataxia, depression, paresis and coma (Kingsbury, 1964; Kwatra *et al.*, 1974; Hothi *et al.*, 1976; Hare *et al.*, 1997), with circulatory collapse and respiratory failure.

Gross necropsy lesions tend to be non-specific. The most common gross finding is severe, diffuse renal and hepatic congestion (Hothi *et al.*, 1976), with secondary petechiation of the musculature due to seizure activity. In both natural and experimental intoxication of pigs, the most characteristic microscopic finding is a mild, diffuse cellular necrosis of the interglandular lamina propria throughout the gastrointestinal tract (Oelrichs *et al.*, 1985). There are also hepatic and renal fatty degeneration (Kingsbury, 1964), proteinaceous casts in the renal tubules and a moderate, diffuse, lymphoid necrosis of the mesenteric lymph nodes, spleen and intestinal lymphoid nodules when a day or two elapses before death (Kwatra *et al.*, 1974; Oelrichs *et al.*, 1985). Rarefaction of lymphoid nodules also occurs in poisoned rabbits, but fowl present with lymphoid hyperplasia (Hothi *et al.*, 1975). In rats, necrosis and fragmentation of skeletal muscle have been reported (Bahri *et al.*, 1992).

References

Ahn, J., Choi, S. and Lee, C. (1994) Cytotoxic limonoids from *Melia azedarach* var. *japonica*. *Phytochemistry* 36, 1493-1496.

Bahri, S., Sani, Y. and Hooper, P.T. (1992) Myodegeneration in rats fed *Melia azedarach*. *Australian Veterinary Journal* 69, 33.

Everist, S.L. (1974) *Poisonous Plants of Australia*. Angus and Robertson, Sydney, p. 369.

Hare, W.R., Schutzman, H., Lee, B.R. and Knight, M.W. (1997) Chinaberry (*Melia azedarach*) poisoning in two dogs. *Journal of the American Veterinary Medical Association* 210, 1638-1640.

Hothi, D.S., Singh, B., Kwatra, M.S. and Chawla, R.S. (1976) A note on the comparative toxicity of *Melia azedarach (DHREK)* berries to piglets, buffalo-calves and fowls. *Journal of Research* (Punjab Agricultural University) 13, 232-234.

Kingsbury, J.M. (1964) *Poisonous Plants of the United States and Canada*. Prentice-Hall, Englewood Cliffs, NJ, pp. 206-208.

Kwatra, M.S., Singh, B., Hothi, D.S. and Dhingra, P.N. (1974) Poisoning by *Melia azedarach* in pigs. *Veterinary Record* 94, 421.

Ochi, M., Kotsuki, H., Kataoka, T. and Tada, T. (1978) Limonoids from *Melia azedarach* L. var. *japonica* makino. III. The structure of ochinal and ochinin acetate. *Chemistry Letters* 331-334.

Oelrichs, P.B., Hill, M.W., Vallely, P.J., MacLeod, J.K. and Molinski, T.F. (1983) Toxic tetranortriterpenes of the fruit of *Melia azedarach*. *Phytochemistry* 22, 531-534.

Oelrichs, P.B., Hill, M.W. and Vallely, P.J. (1985) The chemistry and pathology of meliatoxins A and B constituents from the fruit of *Melia azedarach* L.var. *australasica*. In: Seawright, A.A., Hegarty, M.P., James, L.F. and Keeler R.F. (eds), *Plant Toxicology: Proceedings of the Australia-USA Poisonous Plants Symposium, Brisbane, Australia, May 14-18, 1984*. Queensland Poisonous Plants Committee, Yeerongpilly, Qld, pp. 387-394.

Srivastava, S.D. (1986) Limonoids from the seeds of *Melia azedarach*. *Journal of Natural Products* 49, 56-61.

Steyn, D.G. and Rindl, M. (1929) Preliminary report on the toxicity of the fruit of *Melia azedarach* (Syringa Berries). *Transcripts of the Royal Society of South Africa* 17, 295-308.

Vahrmeijer, J. (1981) *Poisonous Plants of Southern Africa that Cause Stock Losses*. Tafelberg Publishers Ltd, Cape Town, pp. 92-93.

Wang, W., Wang, Y. and Chiu, S. (1994) The toxic chemical factors in the fruits of *Melia azedarach* and the bio-activities toward *Pieris rapae*. *Acta Entomologica Sinica* 37, 20-24.

Watt, J.M. and Breyer-Brandwick, M.G. (1962) *The Medicinal and Poisonous Plants of Southern and Eastern Africa*, 2nd ed. E&S Livingstone, London, p. 745.

Chapter 101

The Purification and Isolation of Two Hepatotoxic Compounds from the Uruguay Sawfly *Perreyia flavipes*

P.B. Oelrichs[1], J.K. MacLeod[2], A.A. Seawright[1], J.C. Ng[1], F. Dutra[4], F. Riet-Correa[3] and M.C. Méndez[3]

[1]*The National Research Centre for Environmental Toxicology, 39 Kessels Road, Coopers Plains, Queensland 4108, Australia; [2]Research School of Chemistry, Australian National University, Canberra, Australia; [3]Laboratório Regional de Diagnóstico, Facutade de Brasil de Veterinária, Ufpel, 96010-900 Pelotas RS, Brazil; [4]Dilave Miguel C Rubino, Rincón 203, Treinta y Tres, Uruguay*

Introduction

The sawfly *Perreyia flavipes* (Pergidae) is widespread in Uruguay and has been recorded in Brazil, Argentina, Paraguay and Venezuela. Cattle and sheep that ingest the sawfly larvae are poisoned, and the reported mortalities are high. Outbreaks of the disorder occur from June to early October when the larvae emerge from the soil and are found on forage grass in large numbers (Dutra *et al.*, 1997).

Feeding experiments (Dutra *et al.*, 1997) showed conclusively that the larvae are responsible for the mortalities. Little is known about the feeding habits of the larvae, which have been seen in large numbers moving on the grass in orderly columns about 15cm long and 8cm wide. They have been seen feeding on green leaves of grasses and small shrubs, as well as on cattle feces. Pupation of the larvae occurs in the soil about 5mm below the surface, where the prepupae construct cocoons from the surrounding earth. After about 6mos, the adult fly emerges and is often seen flying over plants including *Senecio* spp., *Baccharis* spp. and *Eryngium* spp. (Dutra *et al.*, 1997).

There is a marked similarity between this disorder in cattle and sheep, sawfly larva poisoning of cattle in Queensland by *Lophyrotoma interrupta* (Callow, 1955, Oelrichs *et al.*, 1977; Williams *et al.*, 1983) and sheep poisoning in Denmark by *Arge pullata* (Thamsborg *et al.*, 1987; Kannan *et al.*, 1988). The latter two species of sawfly have a common active principle, namely the octapeptide lophyrotomin, a unique compound containing four D-amino acids (Fig. 101.1). To investigate the possibility that *P. flavipes* larvae have the same active principle, a dried *P. flavipes* sample obtained from Brazil was processed using a modification of the method previously reported for the isolation of lophyrotomin (Oelrichs *et al.*, 1977). Two hepatotoxic compounds were isolated and the structures of these compounds are being elucidated.

The Poisoning of Cattle, Sheep and Pigs in Uruguay and Brazil

From June to early October during 1993-95, at least 40 outbreaks of sawfly larva poisoning were reported in cattle and sheep in Uruguay (Dutra *et al.*, 1997). In 1995, the total loss of cattle probably exceeded 1,000 head, and mortalities as high as 11-28% occurred in some farms. Sheep were less frequently affected than cattle. In the Brazilian states of Santa Catarina and Rio Grande do Sul, the larvae of *P. flavipes* are known as *mata porco* (pig killer) where losses of these animals have occurred.

In field cases, cattle with clinical signs showed weakness, muscular tremors, depression, stupor and death. Some animals became highly agitated and aggressive. Most affected animals died within 2d, but jaundice and photosensitization were observed in animals that lived longer. The livers of all affected animals were mottled and showed extensive areas of fatty infiltration and necrosis. Gross and microscopic lesions were characterized by severe periacinar or massive hepatic necrosis with prominent edema of the gall bladder wall and its attachments (Dutra *et al.*, 1997).

Materials and Methods

The larvae of *P. flavipes* were collected from an area in Uruguay where losses of cattle had occurred. The larvae were dried at 100°C for 24hrs before being packed for transport to Queensland. On arrival, the dried larvae were stored at -80°C until being processed. Ground, dried larvae (50g) were extracted (3x) with boiling methanol (MeOH)/H_2O (1:1) and filtered, and the filtrate was concentrated under reduced pressure to a thick syrup. After adding water (500ml), the solution was made 0.1% NH_4OH and then extracted with ethyl ether (2x), followed by n-butanol (3x). Emulsions were broken by centrifugation. Butanol was removed from the aqueous layer, and the water solution was acidified to pH 1-2 with sulfuric acid and added to polyamide celite (1:1) (200g) held in a glass sinter funnel. The polyamide was washed with water until the pH became neutral. The toxin was displaced by washing

---

C_6H_5CO-(D)-Ala-(D)Phe-(L)-Val-(L)-Ile-(D)-Asp-(L)-Asp-(D)Glu(L)-Gln

Fig. 101.1. The structure for lophyrotomin.

with 0.1% NH$_4$OH. The last step was repeated with the solution that initially passed through the polyamide, and the two eluates containing the toxin were combined and dried under reduced pressure. The residue was dissolved in 0.1% NH$_4$OH, added to an XAD-2 reverse phase column (100g) (Serva, NY) with a particle size of 0.05-0.1mm, and the toxin eluted with a gradient of MeOH in this solvent. Fractions were monitored using TLC and biological testing with mice. The active fractions were combined and dried under reduced pressure. The toxins were further purified using a column of silica-gel (0.07-0.15mm) with a mobile phase of chloroform (CHCl$_3$)/MeOH/HOAc/H$_2$O (55:35:5:5).

Results and Discussion

Feeding experiments (Dutra *et al.*, 1997) showed conclusively that the larvae of *P. flavipes* were toxic to cattle and sheep and were the cause of the heavy mortalities. Single doses of 10, 20 and 40g/kg of whole larvae were lethal to sheep within 68, 43 and 14hrs, respectively. Feeding trials have shown that both live and dead larvae of *P. flavipes* had similar toxicity to sheep and cattle, and oven drying larvae had no effect on the activity (100°C/24 hrs). Recent feeding experiments have established that pigs are poisoned by dosing with the larvae (Chapter 58).

The clinical signs and pathological findings of the disorder are remarkably similar to the *L. interrupta* larvae poisoning of cattle in Australia (Callow, 1955) and that of sheep in Denmark by *Arge pullata* larvae (Thamsborg *et al.*, 1987). Sawfly larvae poisoning causes hepatic lesions ranging from mild periacinar necrosis to severe periacinar or massive necrosis (Dutra *et al.*, 1997). The neurologic signs were consistent with acute hepatic encephalopathy. Severe cattle losses occurred with *P. flavipes* poisoning in Uruguay during 1993-95, when total losses exceeded 1,000 animals. Similarly, in Australia serious losses occurred in 1972-81, when 5,284 cattle deaths occurred on 37 properties.

Despite the extensive occurrence of *P. flavipes* in Uruguay, the disorder has been seen only in the Durazno and Florida regions. Likewise toxic *L. interrupta* in Australia is found only in the Maranoa, Warrego and Leichardt pastoral districts of Queensland, although the sawfly *L. interrupta* is considered widespread in eastern Australia. Also, the host tree of *L. interrupta*, the silver leaf iron bark, occurs throughout eastern Australia. The silver birch tree sawfly, *A. pullata*, is apparently toxic wherever it is found, although the area involved in this case is probably smaller. No information is available on the toxicity of *A. pullata* in the Baltic states where it is supposed to have originated. A remarkable feature of this disorder in both Uruguay and Queensland is that animals eat the dead and living sawfly larvae, developing a considerable liking for this abnormal food, which in a short time causes them death. It has been suggested (Oelrichs, 1982) that outbreaks of sawfly larvae poisoning are confined to areas in which the sawfly develops a property that attracts cattle and sheep. It is known that the *L. interrupta* larvae are found in other areas of Australia but are not eaten by animals.

Both *L. interrupta* and *A. pullata* larvae have a common active principle, lophyrotomin (Fig. 101.1), found in a concentration of about 0.1% of dried sawfly larvae (Oelrichs *et al.*, 1977; Kannan *et al.*, 1988). The fact that no detectable lophyrotomin was found in the leaves of the host tree, *Eucalyptus melanophloia*, or in microorganisms present in or derived from the larvae suggests the toxicity is due to the larvae *per se* (Oelrichs and Williams, 1992). It is well known that other species of sawfly discharge an oily fluid if stressed and this serves primarily for protection. An oily fluid collected from *L. interrupta* contained a high concentration of lophyrotomin and this evidence points to a defensive role for the toxin.

Although the toxin in *P. flavipes* is not known, the similarity of the disorders caused by *L. interrupta* and *P. flavipes* suggest the latter may also contain lophyrotomin. The purification scheme described above yielded at least two compounds that are responsible for the liver damage observed in test mice. The two compounds, lophyrotomin and an unidentified minor component, had R_f values of 0.8 and 0.48, respectively, on Merck Kieselgel 60 developed using $CHCl_3/MeOH/HOAc/H_2O$ (65:25:5:5) as the solvent. A study to determine the structures of the two isolated liver toxins (yield: 80mg of lophyrotomin and 100mg of the unidentified minor) is ongoing. Preliminary work suggests that the first toxin (R_f=0.8) is lophyrotomin and the second is an unrelated peptide. Periacinar necrosis of hepatocytes was the main lesion observed in mice poisoned with the second peptide.

The mode of action of lophyrotomin is not known, but it has been suggested that the presence of both hydrophilic and hydrophobic groups in the peptide could disrupt cell membranes (Williams *et al.*, 1983). Attempts to cleave lophyrotomin enzymatically were unsuccessful due to the four D-amino acids in the octapeptide. Their abnormal stability to hydrolysis may account for its liver toxicity. It is not yet known whether other organs (such as the brain) are affected by long-term exposure to sublethal doses, or if lophyrotomin is eliminated from the body in the urine. Interestingly, there are two different peptide toxins in *P. flavipes*, both apparently causing the same liver lesions. Any similarity in the structures of the two toxins will need to be studied.

References

Callow, L.C. (1955) Sawfly poisoning in cattle. *Queensland Agricultural Journal* 81, 155-161.

Dutra, F., Riet-Correa, F., Méndez, M.C. and Paiva, N. (1997) Poisoning of Cattle and Sheep in Uruguay by Sawfly (*Perreyia flavipes*) Larvae. *Veterinary and Human Toxicology*)39, 281-286.

Kannan, R., Oelrichs, P.B., Thamsborg, S.M., and Williams, D.H. (1988) Identification of the octapeptide lophyrotomin in the European birch sawfly (*Arge pullata*). *Toxicon* 26, 224-226.

Oelrichs, P.B. (1982) Sawfly poisoning in cattle. *Queensland Agricultural Journal* 108, 110-112.

Oelrichs, P.B. and Williams, D.H. (1992) Two naturally occurring toxins causing stock losses. In: Watters, D., Lavin, M., Maguire, D. and Pearn, J. (eds), *Toxins and Targets.* Harwood Academic Publishers, Philadelphia, PA, pp. 119-130.

Oelrichs, P.B., Vallely, P.J., MacLeod, J.K., Cable, J., Kiely, D.E. and Summons, R.E. (1977) Lophyrotomin, a new toxic octapeptide from the larvae of sawfly *Lophyrotoma interrupta. Lloydia* 40, 209-214.

Thamsborg, S.M., Jorgensen, R.J. and Brummerstedt, E. (1987) Sawfly poisoning in sheep and goats. *The Veterinary Record* 12, 253-255.

Williams, D.H., Santikarn, S., De Angelis, F., Smith, R.J., Reid, D.G., Oelrichs, P.B. and MacLeod, J.K. (1983) The structure of at toxic octapeptide from the larvae of sawfly. *Journal of Chemical Society Perkin Translation* 1, 1869-1878.

Chapter 102

Toxicity of Monocrotaline (Pyrrolizidine Alkaloid) to the Liver of Chicken Embryos

T.N. Pereira[1, 2], A.A. Seawright[1], P.E.B. Reilly[2] and A.S. Prakash[1]

[1]National Research Centre for Environmental Toxicology (NRCET), 39 Kessels Road, Coopers Plains, Queensland 4108, Australia; [2]Department of Biochemistry, The University of Queensland, St Lucia, Queensland 4067, Australia

Introduction

Over 200 pyrrolizidine alkaloids (PAs) have been found in plants of many species (Furuya *et al.*, 1987) but only those of three unrelated families are toxic: Compositae, Boraginaceae and Leguminosae. Toxic plants are found in the genera *Senecio*, *Heliotropium* and *Crotalaria*, respectively. All of the alkaloids have two fused five-membered rings joined by a single nitrogen atom.

The parent alkaloids are not bioactive, but are metabolized to toxins in the liver. Miranda *et al.* (1991) showed that cytochrome P_{450}-3A4 is the primary enzyme catalyzing the bioactivation of senecionine in the human liver. Since the liver is the site of toxic pyrrole production, it is the main target organ. Although grazing animals do not normally forage on PA-containing plants, they do consume them in drought periods when other food is in short supply or when the feed is contaminated. Substantial differences in susceptibility exist between species. Poultry and pigs are most susceptible, while horses and cattle are less so, and sheep and goats are relatively resistant to PA toxicity.

Several studies have reported PA poisoning of pigs, poultry and ducks in Australia (Hooper and Scanlan, 1977; Pass *et al.*, 1979). Gaul *et al.* (1994) recorded a contamination of feed resulting in the poisoning of pigs and poultry in southern Australia when high summer rainfall aided the growth of heliotrope weed in wheat fields and delayed the harvest. A bioassay to monitor the presence of these toxins would help to prevent such losses to the poultry industry. The aim of this study was to evaluate the toxicity of monocrotaline in chicken embryos and to determine the suitability of this model as a bioassay for PA poisoning.

Materials and Methods

Eggs (supplied by Inghams, Murrarie, Qld) were incubated at 37°C, 70% humidity during the course of the experiment. On day five of development, an observation window was cut in the shell above the embryo and sealed with paraffin film to avoid contamination. Monocrotaline was dissolved in 500µl of saline and injected into the fluid surrounding the embryo on day ten of development, and the paraffin film seal was replaced. In the multiple dosing study, this was repeated on days 12, 14 and 16 of development. The controls received saline only. A range of doses from 10-60mg/kg egg weight was used. Windowing and dosing were carried out in a laminar flow hood. The experiment was stopped by decapitating the embryos on day 19 of development. The organs were removed and fixed in 10% neutral buffered formalin. For histological examination, the sections were cut and stained with hematoxylin and eosin (HE), and compared to controls (Fig. 102.1).

Results and Discussion

The PA monocrotaline elicits a toxic response in the liver of chicken embryos. A single *in ovo* dose of 60mg/kg resulted in hepatocyte enlargement (Fig. 102.2), metaphase arrest, necrosis (Fig. 102.3) and liver atrophy (Fig. 102.4). The multiple dosing schedule was a suitable model for long-term ingestion of the toxin.

The poisoning of poultry with feed contaminated with PA-containing seeds is well documented (Hooper and Scanlan, 1977; Pass *et al.*, 1979; Gaul *et al.*, 1994). Wet summers delayed the harvest and promoted growth of weeds in the wheat fields. Losses to the industry have been high due to acute death of animals and general ill health of those that survive. A bioassay to monitor the presence of these toxins will help prevent such losses to the poultry industry. This study has confirmed that the chicken embryo system is suitable for studying PA poisoning in poultry.

The minimum dose required to elicit a toxic response was 20mg/kg egg weight, suggesting that embryos are more sensitive to PAs than adult chickens. This may be due to the presence of fetal cytochrome P_{450} enzymes, which are more efficient at metabolizing PAs than are adult enzymes.

Megalocytosis characteristic of PA poisoning was seen throughout the liver, as expected in embryonic hepatocytes constantly stimulated for cell division. Larger amounts of liver needed for molecular analysis of PA-induced megalocytosis can be overcome by extending the experiment to allow the embryo to hatch and develop further; the extent and severity of megalocytosis will increase as the chicken grows.

Peterka *et al.* (1994) have studied the teratogenic effects of PAs in chicken embryos, but this is the first study assessing toxicity by histopathology. Based on these results, the chicken embryo system is a rapid, inexpensive and relevant model for studying PA poisoning *in vivo*, particularly in avian species.

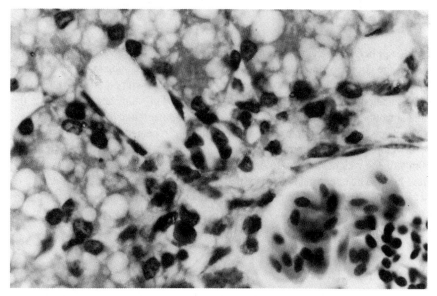

Fig. 102.1. Normal chicken embryo liver: 19d saline control (HE).

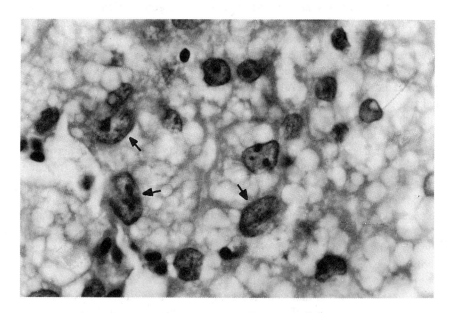

Fig. 102.2. Hepatic megalocytosis: 19d embryo, 60mg/kg monocrotaline (HE).

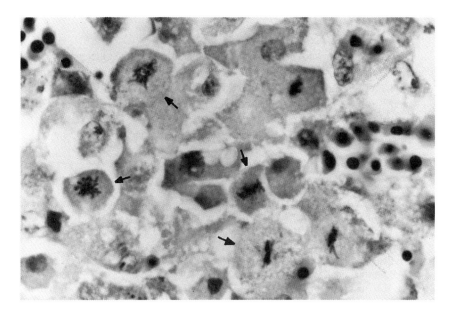

Fig. 102.3. Hepatic metaphase arrest and necrosis: 19d embryo, 60mg/kg monocrotaline (HE).

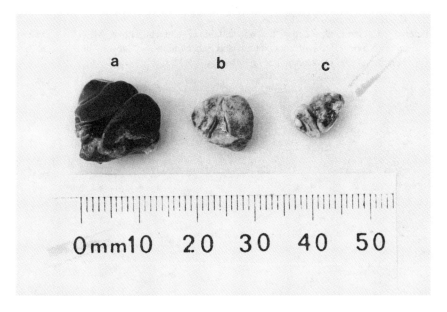

Fig. 102.4. Liver atrophy in 19d embryos treated with 60mg/kg of monocrotaline: **a**) untreated control embryo; **b**) chick 1; **c**) chick 2.

Acknowledgments

T. Pereira gratefully acknowledges a travel scholarship from the Queensland Cancer Fund to attend ISOPP 5. Chicken embryos *in ovo* were kindly supplied by Inghams, Murarrie, Queensland. NRCET is jointly funded by the National Health and Medical Research Council, Queensland Health, The University of Queensland and Griffith University.

References

Furuya, T., Asada, Y. and Mori, H. (1987) Pyrrolizidine alkaloids. In: Hirono, I. (ed), *Naturally Occurring Carcinogens of Plant Origin*. Elsevier Press, Tokyo, pp. 25-51.

Gaul, K.L., Gallagher, P.F., Reyes, D., Stasi, S. and Edgar, J. (1994) Poisoning of pigs and poultry by stock feed contaminated with heliotrope seeds. In: Colegate, S.M. and Dorling, P.R. (eds), *Plant-Associated Toxins: Agricultural, Phytochemical and Economic Aspects*. CAB International, Wallingford, Oxon, pp. 137-142.

Hooper, P.T. and Scanlan, W.A. (1977) *Crotalaria retusa* poisoning of pigs and poultry. *Australian Veterinary Journal* 53, 109-114.

Miranda, C.L., Reed, R.L., Guengerich, F.P. and Buhler, D.R. (1991) Role of cytochrome P450IIIA4 in the metabolism of the pyrrolizidine alkaloid senecionine in human liver. *Carcinogenesis* 12, 515-519.

Pass, D.A., Hogg, G.G., Russell, R.G., Edgar, J.A., Tence, I.M. and Rikard-Bell, L. (1979) Poisoning of chickens and ducks by pyrrolizidine alkaloids of *Heliotropium*. *Australian Veterinary Journal* 55, 283-288.

Peterka, M., Sarin, S., Roeder, E., Wiedenfeld, H. and Halaskova, M. (1994) Differing embryotoxic effects of senecionine and senecionine-*N*-oxide on the chick embryo. *Functional and Developmental Morphology* 4, 89-92.

Chapter 103

Experimental Poisoning in Rabbits Fed with *Senna occidentalis* Seeds

A.C. Tasaka[1], E.E. Calore[2], M.J. Cavalieri[2], M.L.Z. Dagli[1], M. Haraguchi[3] and S.L. Górniak[1]

[1]Department of Pathology of the Faculty of Veterinary Medicine and Zootechny of the University of São Paulo, São Paulo, Brazil; [2]Department of Pathology, Emilio Ribas Institute, São Paulo, Brazil; [3]Department of Pharmacology, Biological Institute of São Paulo, São Paulo, Brazil

Introduction

Senna occidentalis (*Cassia occidentalis*) from Caesalpinoideae family, is one of the most important toxic plants that contaminate animal feeds in the Brazilian territory (Joly, 1977). *Senna* spp. is also present in North America, Africa and Australia. This plant grows in corn, sorghum and soybean fields and contaminates the crops with its toxic seeds at harvest, leading to animal and human intoxication.

Signs of *Senna* poisoning include ataxia, diarrhea, myoglobinuria and sternal recumbency leading to death. Affected animals continue to eat and remain alert until shortly before death (Hebert *et al.*, 1983). Although the affected organs are the same, the severity of lesions varies in different species. In cattle and poultry, degenerative skeletal and myocardial myopathies prevail (Barros *et al.*, 1990; Calore *et al.*, 1997); in rabbits, myocardial lesions are frequently observed (O'Hara and Pierce, 1974a, b); in goats, alterations are seen in the small intestine, liver, kidneys, lungs and heart (Suliman *et al.*, 1982; El Sayed *et al.*, 1983). Intoxication is caused by the uncoupling of oxidative phosphorylation (Graziano *et al.*, 1983), which impairs mitochondrial function, leading to swelling, loss of mitochondrial matrix, fragmented cristae and glycogen depletion (Calore *et al.*, 1997).

Albimun, alkaloids, glycosides and anthroquinones contained in *S. occidentalis* seeds are said to be toxic principles. Hebert *et al.* (1983) observed that chickens responded to an aqueous extract of *Senna* seeds, suggesting that the toxin was polar in nature. Graziano *et al.* (1983) observed that when heated, the polar extract of *Senna* seeds continued to promote intoxication, but with less intensity. Others have shown that anthroquinones cause *Senna* intoxication (Kean, 1968). The present study was designed to investigate the levels of *S. occidentalis* toxic to rabbits.

Material and Methods

Senna occidentalis seeds were ground and mixed in a commercial ration at 2%. Twelve 35d-old rabbits were divided in two groups: eight experimental and four control animals. The experimental group received the commercial ration containing 2% *S. occidentalis* for 30d. The control group was fed the commercial ration containing no seed. Food and water were provided *ad libitum*. Body weight, water and feed consumption were measured weekly. Blood samples were obtained before the experiments and every 15d by cardiac puncture. Serum was analyzed for alkaline phosphatase, creatine kinase (CK), γ-glutamyl transferase, lactate dehydrogenase, aspartate aminotransferase (AST), alanine aminotransferase, creatinine and urea using a commercially available kit (Reactoclin-Celm®). After 30d of treatment, the rabbits were euthanized and samples of muscle, liver and kidney were collected for histopathological analyses.

Results

Body weight, water and feed consumption of treated animals were not significantly different from the controls. Sera obtained from rabbits fed the ration containing 2% *S. occidentalis* seeds for 30d had significantly elevated AST and CK on day 15. Creatine kinase was also significantly elevated in these animals on treatment day 30. There were no significant differences in the other biochemical parameters measured (Table 98.1). Microscopically, there was vacuolar degeneration of renal and hepatic samples from rabbits fed *S. occidentalis*. No microscopic lesions were observed in control animals.

Table 103.1. Serum aspartate aminotransferase (AST) and creatine kinase (CK) of rabbits consuming *Senna occidentalis* at 2% of the diet for 30d (mean±SD).

		Days of treatment		
		0	15	30
CK	c	180.7 ± 33.3	193.1 ± 9.8	212.6 ± 4.6
(U/L)	t	162.9 ± 27.9	335.4 ± 55.6*	297.7 ± 0.9*
AST	c	14.2 ± 1.5	12.2 ± 1.5	12.2 ± 2.2
(U/L)	t	11.7 ± 1.7*	13.5 ± 2.3	14.7 ± 1.3*

c= control; t= treatment; *P<0.05 (Student's *t* test)

Discussion

One of the most important effects of *S. occidentalis* intoxications is muscle degeneration, which is not well characterized. Dollahite and Henson (1965) reported a myodegenerative condition in Texas cattle whose pastures included large amounts of *S. occidentalis* and *S. obtusifolia*. After this report, several studies were performed in various species treated with seeds and/or plant extracts: cattle (Henson *et al.*, 1965; Henson and Dollahite, 1966; Mercer *et al.*, 1967; O'Hara *et al.*, 1969; Barros *et al.*, 1990), swine (Colvin *et al.*, 1986; Martins *et al.*, 1986; Rodrigues *et al.*, 1993), horses (Martins *et al.*, 1986; Irigoyen *et al.*, 1991), sheep (Dollahite and Henson, 1965), goats (Dollahite and Henson, 1965; Suliman *et al.*, 1982), rabbits (Dollahite and Henson, 1965; O'Hara and Pierce, 1974a,b) and chickens (Simpson *et al.*, 1971; Calore *et al.*, 1997).

Although there were no gross alterations in physical parameters (body weight, water and ration consumption) in the present study, biochemical analyses demonstrated elevations of AST and CK. Creatine kinase is related to muscle, and may be used as an indicator of myocardial lesions. Previous studies indicated that rabbits are more sensitive to cardiac lesions than other species (O'Hara and Pierce, 1974b). Alterations in AST may be related to damage of muscle or other tissues, and may reflect the vacuolar degeneration observed in the liver and kidney. At electronic microscopy, intoxication with *S. occidentalis* promoted alterations in mitochondrial morphology of different chicken tissues (Calore *et al.*, 1997). It is known that this intoxication is due to the blockage of cytochrome oxidase action in the tricarboxilic acid (TCA) cycle. The source of acetyl-CoA for the TCA cycle is fatty acids (Cardinett, 1989). The occurrence of vacuolar degeneration in hepatic and renal tissues is indicative of lipid accumulation as a result of this blockage. This study illustrates the early mild pathological process, verified only by biochemical alterations. A higher concentration of *S. occidentalis* in ration or a longer period of exposure will result in more accentuated pathologic lesions.

References

Barros, C.S.L., Pilati, C., Andujar, M.B., Graça, D.L., Irigoyen, L.F., Lopes, S.T. and Santos, C.F. (1990) Intoxicação por *Cassia occidentalis* (Leg. Caes.) em bovinos. *Pesquisa Veterinaria Brasiliera* 10, 47-58.

Calore, E.E., Cavaliere, M.J., Haraguchi, M., Górniak, S.L., Dagli, M.L.Z., Raspantini, P.C. and Calore, N.M.P. (1997) Experimental mitochondrial myopathy induced by chronic intoxication by *Senna occidentalis* seeds. *Journal of Neurological Sciences* 146, 1-6.

Cardinett, G.H. III (1989) Skeletal muscle function. In: Kaneko, J. (ed), *Clinical Biochemistry of Domestic Animals*, 4th ed. Academic Press, New York, NY, pp.462-495.

Colvin, B.M., Harrison, L.R., Sangster, L.T. and Gosser, H.S. (1986) *Cassia occidentalis* toxicosis in growing pigs. *Journal of the American Veterinary Medical Association* 189, 423-426.

Dollahite, J.W. and Henson, J.B. (1965) Toxic plants as the etiologic agent of myopathies in animals. *Journal of the American Veterinary Medical Association* 26, 749-752.

El Sayed, N.Y., Abdelbari, E.M., Mahmoud, O.M. and Adam, S.E.I. (1983) The toxicity of *Cassia senna* to Nubian goats. *The Veterinary Quarterly* 45, 80-85.

Graziano, J.M., Flory, W., Serger, C. and Hebert, C.D. (1983) Effects of a *Cassia occidentalis* extract in a domestic chicken (*Gallus domesticus*). *American Journal of Veterinary Research* 44, 1238-1244.

Hebert, C.D., Flory, W., Seger, C. and Blanchard, R.E. (1983) Preliminary isolation of a myodegeneration toxic principle from *Cassia occidentalis*. *American Journal of Veterinary Research* 44, 1370-1374.

Henson, J.B. and Dollahite, J.W. (1966) Toxic myodegeneration in calves produced by experimental *Cassia occidentalis* intoxication. *American Journal of Veterinary Research* 27, 947-949.

Henson, J.B., Dollahite, J.W., Bridges, C.H. and Rao, R.R. (1965) Myodegeneration in cattle grazing *Cassia* species. *Journal of the American Veterinary Medical Association* 147, 142-145.

Irigoyen, L.F., Graça, D.L. and Barros, C.S.L. (1991) Intoxicação experimental por *Cassia occidentalis* (Leg. Caes) em eqüinos. *Pesquisa Veterinaria Brasiliera* 11, 35-44.

Joly, A.B. (1977) *Botânica: Introdução à Taxonomia Vegetal*, 4th ed. Editora Nacional, São Paulo.

Kean, E.A. (1968) Rhein, an inhibitor of mitochondrial oxidations. *Archives of Biochemistry and Biophysics* 127, 528.

Martins, E., Martins, V.M.V., Riet-Correa, F., Soncini, R.A. and Paraboni, S.V. (1986) Intoxicação por *Cassia occidentalis* (Leguminosae) em suinos. *Pesquisa Veterinária Brasiliera* 6, 35-38.

Mercer, H.D., Neal, F.C., Himes, J.A. and Edds, G.T. (1967) *Cassia occidentalis* toxicosis in cattle. *Journal of the American Veterinary Medical Association* 151, 735-741.

O'Hara, P.J. and Pierce, K.R. (1974a) A toxic cardiomyopathy caused by *Cassia occidentalis*. I. Morphologic studies in poisoned rabbits. *Veterinary Pathology* 11, 97-109.

O'Hara, P.J. and Pierce, K.R. (1974b) A toxic cardiomyopathy caused by *Cassia occidentalis*. II. Biochemical studies in poisoned rabbits. *Veterinary Pathology* 11, 110-124.

O'Hara, P.J., Pierce, K.R. and Read, W.K. (1969) Degenerative myopathy associated with ingestion of *Cassia occidentalis* L.: Clinical and pathological features of the experimentally induced disease. *American Journal of Veterinary Research* 30, 2173-2180.

Rodrigues, U., Riet-Correa, F. and Mores, N. (1993) Intoxicação experimental em suinos com baixas concentrações de *Senna occidentalis* (Leg. Caes.) na ração. *Pesquisa Veterinária Brasiliera* 13, 57-66.

Simpson, C.F., Damron, B.L. and Harms, R.H. (1971) Toxic myopathy of chickens fed *Cassia occidentalis* seed. *Avian Diseases* 15, 284-90.

Suliman, H.B., Wasfi, I.A. and Adam, S.E.I. (1982) The toxicity of *Cassia occidentalis* to goats. *Veterinary and Human Toxicology* 24, 326-330.

Chapter 104

Toxicity and Molecular Shape of Pyrrolizidine Alkaloids

C.G. Logie, A.M. Craig, J.T. Hovermale, W.H. Johnston and L.L. Blythe
College of Veterinary Medicine, Oregon State University, Corvallis, Oregon 97331, USA

Introduction

There are over 6,000 plants containing more than 350 different pyrrolizidine alkaloids (PAs). The toxicity of PAs to liver and lung is well known, although the absolute toxicities of only a few have been investigated. Toxicity is dependent on the presence in the molecule of functional groups necessary to form reactive metabolites (pyrroles) and on its shape. The structures of isolated alkaloids can be determined by NMR spectroscopy supported by other techniques. Only alkaloids of the correct shape will bind to the biological receptor with the functional groups properly aligned. The structural requirements for toxicity have been determined (Mattocks, 1986).

Molecular modeling determines the preferred shape of molecules based on energy constraints. A few examples are given to show the envisaged route one could take to arrive at a suspected toxicity without actually testing the alkaloid.

Determining Structure and Stereochemistry

The structure of an isolated PA can be determined by ^{1}H and ^{13}C NMR spectroscopy (Roeder, 1990; Logie *et al.*, 1994). Many uses of NMR spectroscopy for determining PA structure have been published (Reina *et al.*, 1995). Once the structure is known, a prediction about the toxicity of the PA can be made. The shape of the PA now needs to be determined. A first step towards this is determining stereochemistry.

Relative stereochemistry can be determined by a simple NMR technique, nuclear Överhauser effect (NOE) difference spectroscopy. This method has been used to determine the stereochemistry of a PA in *Senecio chrysocoma* extracts (Logie and

Liddell, 1994). Absolute stereochemistry was determined by comparison of a synthesized isomer of known stereochemistry with the extracted natural product (Logie *et al.*, 1997). A mixture of synthetic and natural product and NMR chiral shift reagent were analyzed by NMR spectroscopy. Lack of doubled signals in the proton spectrum indicated that the PAs had the same stereochemistry. A mixture of synthetic and natural PAs was analyzed by chiral HPLC and GC. In both systems, only one alkaloid peak was seen, suggesting that both had the same *7R,8R* stereochemistry. Stereochemistry of the necine base portion may not be an important factor in determining toxicity as both *7S,8R* and *7R,8R* isomers produce similar lesions. The structure of the necic acid moiety is likely to influence toxicity due to steric hinderance (Mattocks and Bird, 1983).

Shape and Toxicity

Attempting to relate toxicity to shape of PAs is difficult because few extensive toxicity studies have been undertaken. Culvenor *et al.* (1976) tested 62 PAs for acute and chronic liver and lung toxicity in rats, and these results have been used for comparisons between shape and toxicity (Henry and Craig, 1989).

Theoretically, a single conformer of an active molecule binds productively to its receptor, and it is the relationship between this conformer and the response that is sought (Davies *et al.*, 1979). Rather than attempt to correlate shape to activity, a physical property of the compound can be compared to biological activity (Davies *et al.*, 1979). In this study the toxicities of 21 PAs (Fig. 104.1) were related to their octanol/water partition coefficient, and showed an apparent parabolic relationship (Hansch and Leo, 1979) (Fig. 104.2).

Conformation of a molecule is a component of shape and defines the location of the atoms in space; shape includes other properties of the atoms that create other sets of factors (Hopfinger and Burke, 1989). In solution, a molecule can exist in a number of conformations of equally low energy. This study attempts to find a link between molecular shape and toxicity. Pyrrolizidine alkaloids were modeled using MMP2 and CHEMLAB-II. A known set of low energy conformers were compared to the crystal structures for jacobine (Rohrer *et al.*, 1984) and retrorsine (Coleman *et al.*, 1980). Shape of the conformers was described by generating standard molecular shape descriptors. These included simple topographical descriptors (Kier's $^2\kappa$ index) (Kier, 1986a, b), complex geometric descriptors (Jurs shadow and length/breadth descriptors) (Rohrbaugh and Jurs, 1987), and Hopfinger's molecular shape analysis (MSA) volume descriptors (Hopfinger and Burke, 1989).

Soft independent modeling by class analogy (SIMCA) (Wold, 1976) was used to express direction as well as bulk and to allow for better comparison of structures. Reference compounds were used to compare structures and generate shape descriptors. Each compound was used as a reference structure. Every other structure was compared to the reference using the binding moment method, which requires superimposing the corresponding ends of the binding moment vector plus aligning

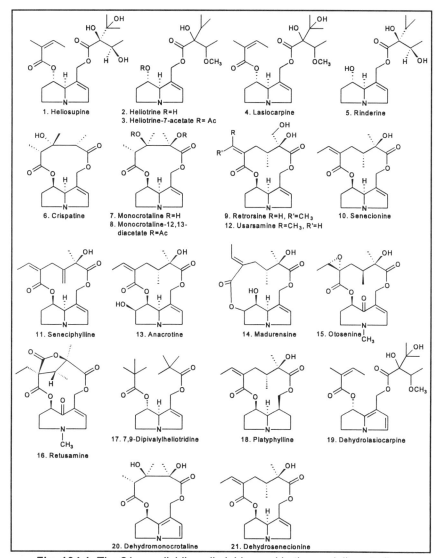

Fig. 104.1. The 21 pyrrolizidine alkaloids used in the modeling studies.

of the ring nitrogen atoms. Binding moments were calculated using Andrews' fragment binding constants (Andrews *et al*, 1984).

As it is a non-clustering technique, the atoms of each structure were initially clustered using Automated Data Analysis and Pattern Recognition Tool Kit (ADAPT). Three to five clusters were chosen for each structure and, as expected for bonded atoms, the clustering followed the bonding patterns, i.e. the atoms clustered together were those bonded to each other. The statistical variation of each structure

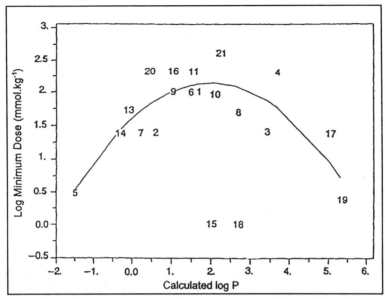

Fig. 104.2. Correlation between octanol/water coefficient (P) and acute morbidity of pyrrolizidine alkaloids (Source: Henry and Craig, 1989, reprinted with permission).

is included, therefore while each does not fit its models perfectly, it fits them better than any other structure. With senecionine (10) as the model, the standard deviations and F-values are as in Table 104.1.

The different shape descriptors were not intercorrelated (R≤0.7). The correlation of SIMCA F-value descriptors with biological activity gave poor results (correlation coefficients from <0.6 to 0.7). Results were similar for other descriptors. As both the

Table 104.1. Average and maximum standard deviation and F-value descriptors with senecionine as reference (maximum in parentheses).

	Fitted Compound		
Cluster	Senecionine 10	Seneciphylline 11	Dehydroseneionine 21
1 (SD)	0.218 (0.510)	0.643 (1.313)	0.873 (1.799)
1 (F)	0.469 (1.921)	3.590 (13.43)	5.020 (20.45)
2 (SD)	0.557 (1.390)	0.907 (1.610)	0.505 (1.322)
2 (F)	0.625 (2.350)	1.680 (4.030)	5.570 (26.24)
3 (SD)	0.230 (0.530)	0.663 (1.245)	0.958 (2.458)
3 (F)	0.547 (1.928)	1.500 (3.501)	4.436 (11.56)

Source: Henry and Craig, 1989, reprinted with permission.

biodata and the shape descriptors are multivariate, canonical correlation analysis was used to give a single measure of correlation between molecular shape and biological activity.

Table 104.2 gives the canonical correlation coefficients for relating the various shape descriptors to the five measures of bioactivity. The correlations shown are the highest obtained in the investigation. The biological response variables in the last column are the ones most highly associated with the first canonical variable of the biological data. Due partly to the larger number of descriptors, the Jurs shadow descriptors show the best correlation with biological activity. The same amount of computational work is required for MSA and SIMCA descriptors. The SIMCA descriptors do have directionality, which would allow the determination of the section of the molecule responsible for shape differences. There are more advanced computational methods for determining molecular shape and with newer *in vivo* assays, more precise relationships between shape, and toxicity are possible.

Table 104.2. Canonical correlation results.

Descriptor Set	Vars.	Canon. Corr.	Highest Corr. Act.
Kier $^2\kappa$	1	0.50	Liver (C)
Jur's Shadow	6	0.86	Death (P)
Jur's L/B	2	0.42	Death (P)
MSA			
Senecionine	3	0.72	Liver (A)
Usaramine	3	0.74	Liver (C)
Madurensine	3	0.64	Lung
Average		0.66	
SIMCA			
Retrorsine	3	0.80	Liver (A)
Rinderine	3	0.75	Liver (C)
Madurensine	4	0.78	Lung
Average		0.72	

Source: Henry and Craig, 1989, reprinted with permission.

References

Andrews, P.R., Craik, D.J. and Martin, J.L. (1984) Functional group contributions to drug receptor interactions. *Journal of Medicinal Chemistry* 27, 1648-1657.

Coleman, P.C., Coucourakis, E.D. and Pretorius, J.A. (1980) Crystal structure of retrorsine. *South African Journal of Chemistry* 33, 116-119.

Culvenor, C.C.J., Edgar, J.A., Jago, M.V., Outteridge, A., Peterson, J.E. and Smith, L.W. (1976) Hepato- and pneumotoxicity of pyrrolizidine alkaloids and derivatives in relation to molecular structure. *Chemico-Biological Interactions* 12, 299-324.

Davies, R.H., Sheard, B. and Taylor, P.J. (1979) Conformation, partition, and drug design. *Journal of Pharmaceutical Sciences* 68, 396-397.

Hansch, C. and Leo, A. (1979) *Substituent Constants for Correlation Analysis in Chemistry and Biology.* Wiley-Interscience, New York, NY.

Henry, D.R. and Craig, A.M. (1989) Statistical modeling of molecular shape, similarity and mechanism. In: Magee, P.S., Henry, D.R. and Block, J.H. (eds), *Probing Bioactive Mechanisms.* American Chemical Society, Washington, DC, pp. 70-81.

Hopfinger, A.J. and Burke, B.J. (1989) Molecular shape analysis of structure-activity tables. In: Fauchere, J.L. (ed), *QSAR: Quantitative Structure-Activity Relationships in Drug Design.* AR Liss, New York, NY, pp. 151-159.

Kier, L.B. (1986a) Shape indexes of orders one and three from molecular graphs. *Quantitative Structure-Activity Relationships* 5, 1-7.

Kier, L.B. (1986b) Distinguishing atom differences in a molecular graph shape index. *Quantitative Structure-Activity Relationships* 5, 7-12.

Logie, C.G. and Liddell, J.R. (1994) Novel structured pyrrolizidine alkaloids from *Senecio chrysocoma.* In: Colegate, S.M. and Dorling, P.R. (eds), *Plant-Associated Toxins: Agricultural, Phytochemical and Ecologial Aspects.* CAB International, Wallingford, Oxon, pp. 221-225.

Logie, C.G., Grue, M.G. and Liddell, J.R. (1994) Proton NMR spectroscopy of pyrrolizidine alkaloids. *Phytochemistry* 37, 43-109.

Logie, C.G., Liddell, J.R. and Kaye, P.T. (1997) Confirmation of the stereochemistry of 7β-angelyl-1-methylene-8 α-pyrrolizidine. *South African Journal of Chemistry* 50, 72-74.

Mattocks, A.R. (1986) *Chemistry and Toxicology of Pyrrolizidine Alkaloids.* Academic Press, London, p. 317.

Mattocks, A.R. and Bird, I. (1983) Pyrrolic and *N*-oxide metabolites formed from pyrrolizidine alkaloids by hepatic microsomes *in vitro*: relevance to *in vivo* hepatotoxicity. *Chemico-Biological Interactions* 43, 209-222.

Reina, M., Mericli, A.H., Cabrera, R. and González-Coloma, A. (1995) Pyrrolizidine alkaloids from *Heliotropium bovei. Phytochemistry* 38, 355-358.

Roeder, E. (1990) Carbon-13 NMR spectroscopy of pyrrolizidine alkaloids. *Phytochemistry* 29, 11-29.

Rohrbaugh, R.H. and Jurs, P.C. (1987) Molecular shape and the prediction of high-performance liquid chromatographic retention indexes of polycyclic aromatic hydrocarbons. *Analytical Chemistry* 59, 1048-1054.

Rohrer, D.C., Karchesy, J. and Deinzer, M. (1984) Structure of jacobine methanol solvate, $C_{18}H_{25}NO_6 \cdot CH_4O$. *Acta Crystallographica* 40, 1449-1452.

Wold, S. (1976) Pattern recognition by means of disjoint principal component models. *Pattern Recognition* 8, 127-139.

Chapter 105

Molecular Interactions of Pyrrolizidine Alkaloids with Critical Cellular Targets

G.L. Drew[1], F.R. Stermitz[2] and R.A. Coulombe, Jr[1]

[1]*Programs in Toxicology and Molecular Biology and Department of Animal, Dairy and Veterinary Sciences, Utah State University, Logan, Utah 84321, USA;* [2]*Department of Chemistry, Colorado State University, Fort Collins, Colorado 80523, USA*

Introduction

Pyrrolizidine alkaloids (PAs) are common plant toxins produced by several genera of flowering plants, including *Senecio*, *Crotalaria* and *Cynoglossum*. Pyrrolizidine alkaloid-containing plants pose significant health hazards to animals and to people who consume "natural" herbal teas and traditional folk remedies.

Pyrrolizidine alkaloids are activated by cytochromes P_{450} in liver and other tissues to reactive bifunctional pyrrolic electrophiles that bind to a variety of cellular nucleophiles, such as DNA and proteins. It is likely that DNA cross-links are a critical event in PA bioactivity. The cytotoxic, antimitotic, and megalocytic activities of PAs closely correspond with the formation of cross-links *in vitro* (Kim *et al.*, 1993). Pyrrolizidine alkaloids form both DNA interstrand and DNA/protein cross-links in equal amounts *in vitro* (Hincks *et al.*, 1991). Structure/activity studies have revealed that a continuous macrocyclic diester and α,β-unsaturation are critical structural determinants for DNA cross-link formation (Kim *et al.*, 1993). Due to the importance of DNA/protein cross-links in the toxicity of numerous agents and in anti-cancer activity of bifunctional alkylators such as mitomycin C and *cis*platin, characterization of the proteins involved in PA-induced cross-links was begun.

Initial experiments pointed to the presence of actin among proteins purified from DNA cross-links from PA-treated cells. A major protein purified from cross-links had a molecular weight of about 43kD, as assessed by polyacrylamide gel electrophoresis (SDS-PAGE), and an isoelectric point of approximately 5.0 (Coulombe *et al.*, 1994; Kim *et al.*, 1995). In addition, proteins purified from PA-treated cells had similar two-dimensional electrophoretic patterns to those seen in cells treated with *cis*platin, a benchmark bifunctional cross-linker known to cross-link DNA with actin. Confirmation of a possible role for actin was sought, and attempts were made to determine the possible pattern of characteristics of nucelophiles that may be involved in cross-links.

Identification of Actin in DNA/Protein Cross-links

Two representative pyrrolic PAs were used for this study: dehydrosenecionine (DHSN) and dehydromonocrotaline (DHMO). Two known DNA-protein cross-linkers, mitomycin C (MMC) and *cis*platin (*Cis*), were used as positive controls (Fig. 105.1). Actin and DNA are cross-linked by *Cis* (Miller *et al.*, 1991). Mitomycin C is reductively activated to a pyrrole that has structural similarities to PA pyrroles. Nuclei prepared from Madin-Darby bovine kidney (MDBK) and human breast carcinoma (MCF7) cells were exposed to 1mM DHSN, 1mM DHMO (each 4hrs), 400µM MMC or 6mM *Cis* (each 6hrs), in DMSO not exceeding 1% of the total volume, at 37°C. Pyrroles were prepared from their parent compounds by chemical oxidation (Mattocks *et al.*, 1989), and MMC was reductively activated with sodium dithionite (Borowy *et al.*, 1990). DNA/protein cross-links were purified, then the DNA was digested with DNAse I. The proteins were separated by SDS-PAGE and transferred to nitrocellulose membranes (Kim *et al.*, 1995). The blocked, phosphate-buffered saline (PBS)-rinsed membrane was incubated 2hr with a 1:500 dilution of mouse monoclonal anti-actin, then 1hr with affinity-purified goat anti-mouse IgG horse radish peroxidase conjugate at 1:40,000. It was washed in chemiluminescent solution (ECL, Amersham) and exposed to autoradiographic film.

Western immunoblot analysis confirmed the presence of an actin-immunoreactive protein in DNA/protein cross-links from normal (MDBK) and malignant (MCF7) cell nuclei treated with either DHSN or DHMO (Fig. 105.2). Actin was detected in much greater quantities in DNA/protein cross-links purified from nuclei exposed to the benchmark anti-cancer drugs *Cis* and MMC compared to those from DHSN and DHMO-treated nuclei. While actin involvement in *Cis*-induced DNA cross-links is documented, the discovery of actin cross-linking in MMC-treated cells or nuclei is novel.

Several chemical and physical carcinogens induce DNA/protein cross-links (Olenick *et al.*, 1987). A number of critical cellular targets, such as RNA polymerases (Bedinger *et al.*, 1983) are inactivated or otherwise altered when cross-linked. While other cellular proteins are almost certainly complexed by PAs, actin appears to be a major protein target. Because actin is abundant in the nuclear matrix and is involved in cell division, cell structure, and gene regulation, it is possible that the cytotoxic, anti-mitotic, and megalocytic action of PAs are mediated, at least in part, by their cross-linking with actin.

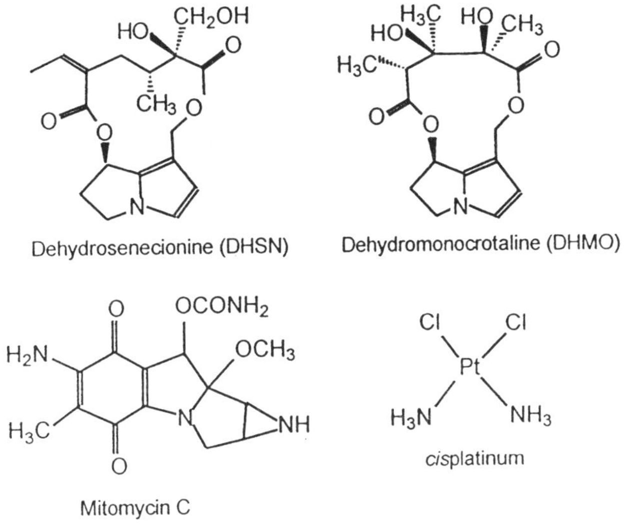

Dehydrosenecionine (DHSN)   Dehydromonocrotaline (DHMO)

Mitomycin C   *cis*platinum

Fig. 105.1. DNA-cross linkers used in this chapter.

Nucleophilic Competition for DNA/DNA Cross-links

Comparison of the competition of several cellular nucleophiles with DNA for reaction with pyrroles has relevance for detoxification and may aid determination of which proteins are involved in DNA/protein cross-links. Nucleophilic competition was studied in an *in vitro* cross-linking system using *Hin*d III-digested λ-phage DNA. Varying amounts of glutathione (GSH), cysteine (Cys), and methionine (Met) were added to determine their relative competition with DNA for reaction with the pyrrole. The DNA:nucleophile:pyrrole (added in that order) combinations were mixed and centrifuged, capped with N_2, held for 2hrs at 0°C, separated on 0.8% agarose gel, stained with ethidium bromide and photographed.

*Hin*d III has seven recognition sites within λ-phage, and digestion yields λ fragments of varying molecular weight; these were the targets for DNA cross-linking in this assay, and six of them appeared in the gels (Fig. 105.3, Lane 1). Nucleophiles alone had no effect on the electrophoretic mobility of control λ (Fig. 105.3, Lane 2), indicating that none interacted with λ DNA. However, addition of a bifunctional pyrrolic cross-linker (either DHSN or DHMO) caused an apparent increase in molecular weight, which is seen as a shift of these fragments toward the top of the gel, hence the term "gel shift" (Fig. 105.3, Lane 3).

Glutathione and Cys, but not Met, overcame DHSN- and DHMO-induced cross-

actin   control DHSN DHMO MMC *Cis* actin

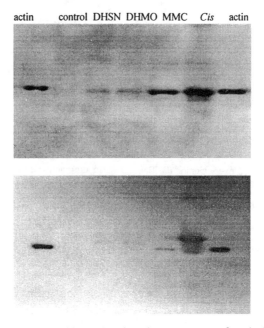

Fig. 105.2. Western immunoblots showing the presence of actin isolated from DNA/protein cross-links from MDBK (top) and MCF-7 nuclei treated with DMSO vehicle (control), dehydrosenecionine (DHSN), dehydromonocrotaline (DHMO), mitomycin C (MMC), or *cis*platin (*Cis*). Blots were probed with a monoclonal anti-actin antibody that recognizes an epitope conserved in all actin isoforms. α-skeletal actin was used as a standard.

links, indicating that GSH and Cys reacted with these pyrroles, preventing them from cross-linking λ-phage (Fig. 105.3, Lanes 4-7). Glutathione appeared to be a more potent nucleophile than Cys, because less GSH was needed to restore the electrophoretic pattern of cross-linked DNA. Dehydrosenecionine was a more potent cross-linker than DHMO, as indicated by comparative intensity of the gel shifts and the higher amounts of GSH and Cys needed to overcome DHSN-induced cross-links. These data indicate that free sulfhydryl groups are potential reactive nucleophiles in the DNA/protein cross-links. Free amino nitrogens were present in all nucleophiles examined here, but Met, which lacks a free sulfhydryl group, was not able to compete with DNA for the pyrrole at any ratio tested.

Glutathione is a universal detoxifying co-enzyme for reactive, electrophilic toxins and carcinogens, such as aflatoxin B_1, and this *in vitro* data support the findings that GSH is important in detoxification of PAs *in vivo*. Glutathione metabolites have been found in isolated rat liver perfused with various PAs (Yan *et al.*, 1995). Thus, it is likely that GSH may act to prevent the formation of DNA cross-links in PA-exposed animals.

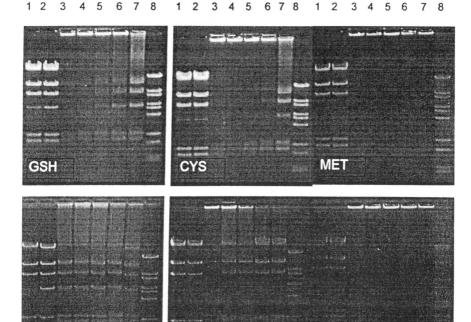

Fig. 105.3. Agarose gel-shifts showing the effect of added nucleophiles (glutathione: GSH; cysteine: CYS; and methionine: MET) on dehydrosenecionine (DHSN; top) and dehydromonocrotaline (DHMO; bottom) cross-links in *Hin*d III- digested λ DNA. DNA:pyrrole:nucleophile molar weight ratios of 1:2.5:0.5 (lane 4), 1:2.5:1 (lane 5), 1:2.5:2 (lane 6) and 1:2.5:4 (lane 7) were used. Controls are: 1:2.5µl DMSO:0 (lane 1); 1:2.5µl DMSO:4 (lane 2); DNA + DHSN or DHMO only 1:2.5:0 (lane 3); and *BstE* II-digested λ-phage DNA marker (lane 8).

Acknowledgments

The authors wish to acknowledge support from the Willard L. Eccles Charitable Foundation, Western Regional Research Project W-122, from the Colorado State University Experiment Station, and from the Utah Agricultural Experiment Station, where this paper is designated number 6004.

References

Bedinger, P., Hochstrasser, M., Jongeneel, C.V. and Alberts, B.M. (1983) Properties of the T4 bacteriophage DNA replication apparatus: the t4 dda DNA helicase is required to pass a bound RNA polymerase molecule. *Cell* 34, 115-123.

Borowy, H., Lipman, R., and Tomasz, M. (1990) Recognition between mitomycin C and specific DNA sequences for cross-link formation. *Biochemistry* 29, 2999-3006.

Coulombe, R.A. Jr, Kim, H.Y. and Stermitz, F.R. (1994) Structure-activity relationships of pyrrolizdine alkaloid DNA cross-links. In: Colegate, S.M. and Dorling, P.R. (eds), *Plant-Associated Toxins: Agricultural, Phytochemical and Ecological Aspects*. CAB International, Wallingford, Oxon, pp. 125-130.

Hincks, J.R., Kim, H.Y., Segall, H.J., Molyneaux, R.J., Stermitz, F.R., and Coulombe, R.A. Jr (1991) DNA cross-linking in mammalian cells by pyrrolizidine alkaloids: structure-activity relationships. *Toxicology and Applied Pharmacology* 111, 90-98.

Kim, H.Y., Stermitz, F.R., Wilson, D.W., Taylor, D. and Coulombe, R.A. Jr (1993) Structural influences on pyrrolizidine alkaloid induced cytopathology. *Toxicology and Applied Pharmacology* 122, 61-69.

Kim, H.Y., Stermitz, F.R. and Coulombe, R.A. Jr (1995) Characterization of pyrrolizidine alkaloid-induced DNA-protein cross-links. *Carcinogenesis* 16, 2691-2697.

Mattocks, A.R., Jakes, R. and Brown, J. (1989) Simple procedures for preparing putative toxic metabolites of pyrrolizidine alkaloids. *Toxicon* 27, 561-567.

Miller, C.A., Cohen, M.D. and Costa, M. (1991) Complexing of actin and other nuclear proteins to DNA by *cis*-diamminedichloroplatinum (II) and chromium compounds. *Carcinogenesis* 12, 269-276.

Olenick, N.L., Chiu, S., Ramakrishnan, N. and Xue, L. (1987) The formation, identification and significance of DNA-protein cross-links in mammalian cells. *British Journal of Cancer* 55, 135-140.

Yan, C.C., Cooper, R.A. and Huxtable, R.J. (1995) The comparative metabolism of four pyrrolizidine alkaloids, seneciphylline, retrorsine, monocrotaline, and trichodesmine in the isolated, perfused rat liver. *Toxicology and Applied Pharmacology* 133, 277-284.

Chapter 106

Factors Influencing Urinary Excretion of Immunoreactive Sporidesmin Metabolites in Sheep Dosed with Sporidesmin

B.L. Smith, L.R. Briggs, P.P. Embling and N.R. Towers
Toxicology and Food Safety Research Group, New Zealand Pastoral Agriculture Research Institute Ltd, Ruakura Research Centre, Private Bag 3123, Hamilton 2001, New Zealand

Introduction

Sporidesmin (Synge and White, 1959), the hepatotoxin produced by certain strains of *Pithomyces chartarum* which causes the hepatogenous photosensitization called facial eczema (FE), is a very cytotoxic compound. It causes cytopathic effects at concentrations as low as 0.4ng/ml in culture medium (Mortimer and Collins, 1968). Using cytotoxicity as a bioassay, sporidesmin administration to sheep results in "cytotoxicity" in the approximate ratios of 1:10:100 for plasma, urine and bile, respectively (Mortimer and Stanbridge, 1968). The concentration of sporidesmin in bile may be the main reason for the typical pericholangitis and obstructive biliary hepatopathy of this mycotoxicosis (Mortimer and Stanbridge, 1969). The urinary bladder cytotoxicity sporidesmin causes is probably the reason for the hemorrhagic cystitis seen in sheep with severe lesions of FE (Done *et al.*, 1960).

The detection of sporidesmin, or its metabolites, in tissues and fluids by chromatography has not been possible due to insufficient sensitivity. This work used radiolabeled sporidesmin (Towers, 1972; Fairclough and Smith, 1983) but the specificity of this method is limited. Two ELISAs (enzyme-linked immunosorbent assays) have been developed recently, each specific for different regions of the sporidesmin molecule (Briggs *et al.*, 1994) and optimized for use in bile and urine.

Progeny and performance testing has been used to select sheep both resistant and susceptible to the hepatic injury caused by sporidesmin A (Morris *et al.*, 1995). Research is focusing on determining the pathophysiological basis of resistance and susceptibility to sporidesmin toxicity. This paper reports on the detection of epitopes of sporidesmin by ELISA in urine after administration of sporidesmin to sheep, unselected and selected for resistance and susceptibility to sporidesmin. Both sporidesmin A (Ronaldson *et al.*, 1963) (SPDM-A), the recognized toxin of FE, and sporidesmin D (Jamieson *et al.*, 1969) (SPDM-D) a nontoxic analogue of SPDM-A, were used throughout these experiments. SPDM-D was used to examine the repeatability of results without the results being influenced by injury caused by SPDM-A. The possible influence of potentiating effects of SPDM-A on the toxicokinetics of subsequent sporidesmin doses was also examined.

Materials and Methods

Female Romney sheep, non-selected or selected either for resistance (R) or susceptibility (S) to sporidesmin toxicity, were put in metabolism cages and fitted with a 12 gauge Foley catheter for urinary collection. The sheep were given SPDM-A or SPDM-D by a single intraruminal intubation. Urine was collected, measured and frozen at -20°C. Blood samples were taken and analyzed for serum glutamate dehydrogenase (GDH) and gamma glutamyltransferase (GGT).

In Experiment 1, three sheep, unselected for R to FE, were dosed on three separate occasions with SPDM-D, with 42 and 137d intervals, at 0.2mg/kg, followed 21wks later by a final dose of SPDM-A at 0.2mg/kg to two of the sheep. Experiment 2 used eight R and eight S sheep, each given two doses of SPDM-D at 0.2mg/kg. Each group was subdivided into two groups, with 13 and 105d between doses for each group. Experiment 3 used ten randomly bred sheep divided into two equal groups, one of which was given a smaller potentiating dose of SPDM-A at 0.03mg/kg 10d before both groups received 0.2mg/kg. In Experiment 4, seven S and seven R sheep were potentiated with SPDM-A at 0.03mg/kg and 10d later dosed with 0.2mg/kg.

Sporidesmin metabolite was detected by competitive ELISA using a monoclonal antibody that binds to sporidesmin in a region distal to the disulfide bridge and includes the chlorine grouping. The sporidesmin and metabolite(s) are expressed as µg of immunoreactive equivalents to sporidesmin A ("SPDM-A") or sporidesmin D ("SPDM-D"), depending on whether SPDM-A or SPDM-D was used to generate the standard curves. All samples were diluted 1:50 to reduce matrix effects, and further diluted up to 1:500 to give concentrations within the working range of the ELISA.

Results

The cumulative excretion of SPDM-A following three repetitive dosings of SPDM-D followed by one dose of SPDM-A showed similar relativity in each of three sheep. Excretion of immunoreactive material continued for 50hrs in the case of SPDM-D but up to 75hrs in the case of SPDM-A. The total SPDM-A detected by this method (approximately 2-4.5mg) appeared to represent only a small proportion of that dosed to the sheep (about 15-20mg). The total SPDM-A excreted by the two sheep dosed SPDM-A also appeared to be less than that recorded for the SPDM-D dosed sheep for each occasion. The maximum rates of SPDM-A urinary excretion occur at approximately 2-8hrs after SPDM-D dosing and at 15-30hrs for SPDM-A dosing. The liver injury response, as measured by serum GGT and GDH was greatest for the sheep with the highest cumulative outputs of SPDM-A.

The urinary output of SPDM-D from R and S sheep after two repetitive doses of SPDM-D is shown in Fig. 106.1. For the first dose of SPDM-D to both R and S sheep there was much variability in response. The responses in both groups of sheep after the second dose had much less variability, with a reduced urinary output of SPDM-D from both R and S sheep. The mean cumulative urinary SPDMD excretion after both dosings was greater for S than R sheep.

Potentiation caused a more severe liver injury to SPDM-A as seen in the serum GGT changes (Fig. 106.2). Urinary output of SPDM-A (Fig. 106.3) was less in the potentiated group of sheep. The difference in output arose late in the post-dosing period and was due to urinary flows in the potentiated sheep ceasing at different times after dosing. Peak concentrations of SPDM-A in urine ranged from 2,688-1,561ng/ml for control animals and from 1,200-1,944ng/ml for potentiated animals. The cumulative excretion rates of SPDM-A in R and S sheep dosed with SPDM-A after the small challenge with SPDM-A (potentiation) demonstrated no difference between R and S sheep. Urinary excretion continued for at least 80hrs with maximum rates occurring at 10-30hrs. The urine production of the S sheep was lower and less variable, and their urinary concentrations of SPDM-A were consequently higher. The peak urinary SPDM-A concentrations for the R sheep were 2.72µg/ml (range 1.40-4.32) and the S sheep, 3.61µg/ml (range 1.99-6.52).

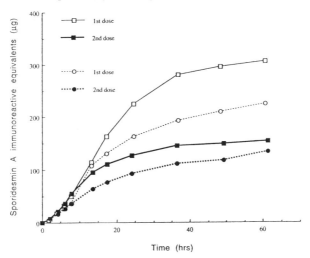

Fig. 106.1. Urinary output of sporidesmin metabolite immunoequivalents after two repetitive doses of sporidesmin D in sheep resistant (O) and susceptible (□) to facial eczema.

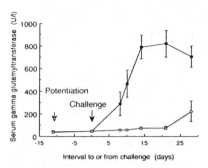

Fig. 106.2. Effect of potentiation (■) with a small dose of sporidesmin on the subsequent liver injury caused by a later dose of sporidesmin A (0.2mg/kg).

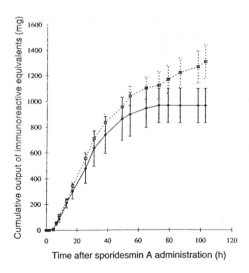

Fig. 106.3. Effect of potentiation with a small dose of sporidesmin on the subsequent cumulative urinary excretion of a later dose of sporidesmin A (0.2mg/kg). Susceptible (-----) and resistant (- - - -) sheep.

Discussion

Although the immunological method is specific for molecular epitopes, these may be common to a number of metabolites of the molecule under investigation. Unless the configuration of the epitope is changed and its immunological recognition altered, different metabolic products will be detected by the method. Changes or differences in the molecule outside the epitopic region may also alter the affinity of the molecule for the specific antibody. For these reasons the immunoreactive products detected in urine were all described by the generic term "immunoreactive sporidesmin equivalents" (SPDM-A or SPDM-D), which may include both the metabolites and the unmetabolized toxin. Preliminary HPLC work showed that urine from sheep dosed with SPDM-A contained three different peaks with ELISA immunoreactivity, suggesting at least three different metabolites.

The three consecutive doses of SPDM-D showed that the three sheep maintained a similar relationship to one another in the efficiency with which they eliminated SPDM-D *via* urine. All eliminated sporidesmin over 40-50hrs with maximum rates of elimination occurring 1-10hrs after dosing. The maximum rate of elimination of SPDM-A occurred later than for SPDM-D, and maximum cumulative plateau for SPDM-A elimination was lower. This may be due to either differences between the immunoreactivity of the SPDM-D and SPDM-A metabolites or to the toxic effect of sporidesmin on metabolism.

When SPDM-D was dosed to R and S sheep, similar patterns of urinary excretion were seen. A notable reduction in variability in SPDM-D excretion was observed after the second dose of SPDM-D, as were reduced rates of excretion and maxima of cumulative excretion for both the R and S sheep. At this point the difference between R and S sheep just reached significance ($P<0.05$), with the S sheep excreting more rapidly and to higher cumulative totals. It is possible that SPDM-D might stimulate the hepatic microsomal drug metabolizing system (HMDMS) or associated mechanisms. In animals with induced HMDMS, a more complete metabolism of the SPDM-D might reduce the immunoreactivity of the metabolites and hence explain the results. Any induction of the HMDMS might also explain the reduced variability after the second dose of SPDM-D.

There was greater liver injury for the potentiated sheep as measured by serum GGT and GDH concentrations. The cumulative output of SPDM-D was lower for the potentiated and more severely affected sheep, and the urine outputs ceased for the potentiated sheep after 75hrs. This is consistent with the results for the second dose of SPDM-D in Experiment 2.

The potentiation of R and S sheep prior to the main challenge was introduced to explore the hypothesis that the difference between R and S sheep might rest in their abilities to metabolize the toxin; that is, to depend upon the ability of the R sheep to either resist sporidesmin destruction of this ability or be stimulated by the toxin itself with the respective inverses operating in the S sheep. No differences between R and S sheep were demonstrated in the cumulative excretion of the SPDM-A, but there was a difference between the R and S sheep in the volume of urine voided. The S sheep voided less urine - probably due to injury to the urinary tract, which occurs in more severely affected potentiated sheep. Higher-peak SPDM-A concentrations in the S sheep support this contention.

References

Briggs, L.R., Towers, N.R. and Molan, P.C. (1994) Development of an enzyme-linked immunosorbent assay for analysis of sporidesmin A and its metabolites in ovine urine and bile. *Journal of Agricultural and Food Chemistry* 42, 2769-2777.

Done, J., Mortimer, P.H. and Taylor, A. (1960) Some observations on field cases of facial eczema: liver pathology and determination of serum bilirubin, transaminase and alkaline phosphatase. *Research in Veterinary Science* 1, 76-83.

Fairclough, R.J. and Smith, B.L. (1983) Sporidesmin concentrations in the bile of sheep resistant or susceptible to sporidesmin dosing. *Proceedings of the New Zealand Society of Animal Production* 43, 213-215.

Jamieson, W.D., Rahman, R. and Taylor, A. (1969) Sporidesmins. Part VIII. Isolation and structure of sporidesmin D and sporidesmin F. *Journal of the Chemistry Society* C, 1564-1567.

Morris, C.A., Towers, N.R., Wheeler, M. and Wesselink, C. (1995) Selection for and against facial eczema susceptibility in Romney sheep, as monitored by serum concentrations of a liver enzyme. *New Zealand Journal of Agricultural Research* 38, 211-219.

Mortimer, P.H. and Collins, B.S. (1968) The *in vitro* toxicity of the sporidesmins and related compounds to tissue-culture cells. *Research in Veterinary Science* 9, 136-142.

Mortimer, P.H. and Stanbridge, T.W. (1968) Excretion of sporidesmin given by mouth to sheep. *Journal of Comparative Pathology* 78, 505-512.

Mortimer, P.H. and Stanbridge, T.W. (1969) Changes in biliary secretion following sporidesmin poisoning in sheep. *Journal of Comparative Pathology* 79, 267-75.

Ronaldson, J.W., Taylor, A., White, E.P. and Abraham, R.J. (1963) Sporidesmins. Part 1. Isolation and characterization of sporidesmin and sporidesmin B. *Journal of the Chemistry Society* 3172-3180.

Synge, R.L.M. and White, E.P. (1959) Sporidesmin: A substance from *Sporidesmium bakeri* causing lesions characteristic of facial eczema. *Chemistry and Industry* 1959, 1546-1547.

Towers, N.R. (1972) Absorption and excretion of 35-sulphur by biliary fistulated rats and guinea pigs following administration of 35 S-labeled sporidesmin. *Life Sciences* 11, 691-698.

Chapter 107

Disease in Cattle Dosed Orally with Oak or Tannic Acid

K.H. Plumlee, B. Johnson and F.D. Galey
California Veterinary Diagnostic Laboratory System, University of California, Davis, California 95617, USA

Introduction

Oak trees and shrubs can be found worldwide, as can the occurrence of oak toxicosis. In April, 1985, an estimated 2,700 cattle died from oak toxicosis in northern California alone (Spier *et al.*, 1987). Most incidents of poisoning occur when cattle ingest immature leaves or freshly fallen acorns, which contain the highest levels of tannic acid (Harper *et al.*, 1988), a hydrolyzable gallotannin that undergoes microbial and acid hydrolysis to release phenolics such as gallic acid (Zhu *et al.*, 1992). Gallic acid is then believed to be metabolized to pyrogallol and resorcinol (Murdiati *et al.*, 1992).

Range cattle with oak toxicosis often develop constipation, followed by bloody diarrhea. Discolored urine, roughened hair coats, anorexia and a "tucked up" posture have also been reported (Harper *et al.*, 1988). A 1919 USDA Bulletin reported that oak toxicosis could not be reproduced in cattle using tannic acid (Dollahite *et al.*, 1962), but tannic acid, gallic acid and pyrogallol given orally to rabbits all produced lesions similar to those seen with oak toxicosis (Dollahite *et al.*, 1962; Pigeon *et al.*, 1962). Preliminary work indicated that a calf dosed with tannic acid had brownish blood with high levels of methemoglobin (metHb), while a calf dosed with oak had normal blood. Sheep dosed with tannic acid also developed methemoglobinemia, while sheep dosed with oak developed renal disease (Zhu *et al.*, 1992; Zhu and Filippich, 1995). The following experiments were designed to compare patterns of disease in cattle that were dosed with either oak leaves or commercial tannic acid.

Materials and Methods

Eleven female calves (73-164kg) were maintained on alfalfa (tested negative for tannic acid) prior to and during the experiments, except as noted. All animals had fully functional rumens and appeared healthy upon physical examination.

Immature blue oak (*Quercus douglasii*) leaves were collected during the spring and frozen for about 6mos prior to the experiment. Analysis by GC/MS (Tor *et al.*, 1996) determined a gallic acid content of 1,542ppm. Commercial tannic acid (Mallinckrodt Chemical Company, St Louis, MO) was found to contain 45% gallic acid. Gallic acid is the precursor of pyrogallol, the metabolite to be studied.

Two calves (A, B) matched in size and weight were fed thawed oak leaves *ad libitum*. The weight of leaves consumed was recorded daily, and the daily amount of gallic acid ingested was calculated. Blood was collected for CBC and metHb analysis on days zero through four, seven and eight. Serum was analyzed by standard biochemistry panel on days zero through nine and 11. Urinalysis was performed on free-catch urine collected on the first 9d. Calves A and B were euthanized and necropsied on day 11.

Two calves (C, D) were dosed daily for 7d with commercial tannic acid in amounts equal to the oak eaten by calves A and B, based on gallic acid content. Aqueous tannic acid was dosed *via* stomach tube. Serum biochemistry, CBC and urinalysis were monitored on days zero, two, four, seven and nine. Serum and urine were collected from animals A-D at variable intervals for pyrogallol analysis (Tor *et al.*, 1996).

Seven heifers (E-K) were dosed once *via* stomach tube with aqueous commercial tannic acid such that the gallic acid content would be 2.0, 1.0, 1.5, 2.0, 2.5, 2.0 and 2.5g/kg, respectively. An up-and-down dosing procedure (Bruce, 1985) was followed. Methemoglobin was measured at 0, 3, 6, 24 and 48hrs, and pyrogallol was monitored in the serum and urine.

Results

Calves A and B each consumed about 1.9, 2.7, 0.9, 1.4, 1.6, 0.7 and 0.2kg of oak leaves on days one through seven, respectively, and refused to eat any more oak after 7d. Calves C and D were dosed with 67, 94, 31, 49, 55, 24 and 7mg/kg of tannic acid on days one through seven, respectively. Therefore, each of the four calves received about 30, 42, 14, 22, 25, 11 and 3mg/kg of gallic acid on days one through seven, respectively.

Oak-fed calves (A, B) developed moderate constipation with a small amount of mucus on day two, became depressed on day six and anorectic after 7d. The hair coats appeared roughened and the abdomens tucked-up. Clinical signs were more pronounced beginning at 7d. Urine from both calves briefly changed to a slightly brown color. The tannic-acid-fed calves (C, D) developed no clinical signs throughout the experiment.

The serum biochemistry parameters for calves A and B remained normal except for blood urea nitrogen (BUN), creatinine, sorbitol dehydrogenase (SDH), and aspartate aminotransferase (AST). The BUN became elevated at 6d and continued to increase through day 11 in both animals. Both had slightly elevated creatinine levels at 0-5d, with a dramatic increase occurring on day six. The SDH became elevated on day five for calf A and day six for calf B, and the values returned to normal on day 11. The AST became elevated in calf A at 24hrs and returned to normal by day 11. Calf B had elevated AST from days six to nine.

Urinalysis remained normal in calves A and B except for protein, blood, and glucose. Urine from both calves had detectable amounts of blood beginning at 4d and glucose at 5d. Calf A was proteinuric on day two, and calf B on day three.

Calves A and B had small amounts of pyrogallol in their serum only at 3hrs and 6hrs after beginning to ingest oak leaves; calf A had pyrogallol in the urine 3-60hrs and calf B 3-48hrs after the first oak ingestion (Table 107.1).

At necropsy, calves A and B had massive perirenal edema, extensive retroperitoneal edema, marked ascites, and hydrothorax. Microscopically, there was a severe nephrosis with secondary nephritis in both animals. Calf B also had a multifocal rumenitis.

Table 107.1. Serum and urine pyrogallol levels (ppm) from calves dosed *per os*. with oak leaves for 7d (A,B) or a single dose of gallic acid (GA) (E-K).

Calf	GA g/kg	Time (hrs) after first dosing								
		0	3	6	9	12	24	36	48	60-216
Serum										
A	oak	ND*	1.7	2.0	ND	ND	ND	ND	ND	ND
B	oak	ND	1.7	1.7	ND	ND	ND	ND	ND	ND
E	1.0	ND	17.0	15.0	1.8	0.9	ND	ND	ND	-
F	1.5	ND	19.0	27.0	5.6	1.3	ND	ND	ND	-
G	2.0	ND	18.0	28.0	20.0	4.6	ND	ND	ND	-
H	2.0	ND	20.0	25.0	26.0	17.0	0.5	ND	ND	-
I	2.0	ND	15.0	22.0	20.0	10.0	0.7	ND	ND	-
J	2.5	ND	15.0	21.0	22.0	4.1	ND	ND	ND	-
K	2.5	ND	14.0	13.0	14.0	17.0	1.2	ND	ND	-
Urine										
A	oak	ND	48.0	550	500	85.0	8.6	150	48	2.4
B	oak	ND	52.0	360	370	91.0	7.5	1.7	1.2	ND
E	1.0	ND	1180	1230	220	22.0	0.7	ND	ND	-
F	1.5	ND	840	400	210	28.0	2.7	ND	ND	-
G	2.0	ND	1260	960	1340	240	2.5	0.6	ND	-
H	2.0	ND	210	1590	1500	1110	10.0	1.4	ND	-
I	2.0	ND	1660	740	640	490	53.0	2.5	ND	-
J	2.5	ND	1340	1150	1280	880	9.9	1.4	ND	-
K	2.5	ND	1550	2420	990	610	130	1.8	ND	-

*ND = not detected, with a detection limit of 0.5ppm.

Calves C and D had no abnormalities in serum biochemistry or urinalysis and no pyrogallol in serum or urine during the experiment. They were not euthanized.

Calves E-K, which received single doses of commercial tannic acid, had normal CBC, urinalysis and serum biochemistry values throughout the experiment. Pyrogallol was detected in serum and urine from calves E-K beginning at 3hrs after dosing, in serum until 12-24hrs and in urine until 24-36hrs (Table 107.1). Both calves dosed with 2.5g/kg of gallic acid and one of three dosed with 2.0g/kg developed significant methemoglobinemia 24hrs after dosing (Table 107.2).

Discussion

The oak-fed calves in this study developed gross and histopathologic lesions consistent with those previously reported for oak poisoning. Oak toxicosis occurs most often in range cattle (Harper *et al.*, 1988), and the clinical signs observed in these study animals provide insight into the vigilance required for early diagnosis of oak poisoning. Constipation would easily be missed by the casual observer, as would the transient discoloration of the urine. Cattle suffering from oak toxicosis often have significant renal disease before any clinical signs are noted.

In the oak-fed calves, urinalysis revealed sequential abnormalities - proteinuria, hematuria, then glucosuria - preceding the sharp rise in serum BUN and creatinine on day 6. Increased AST and SDH, with normal alkaline phosphatase, GGT and bilirubin, indicated that hepatocellular necrosis occurred without cholestasis. The cause for this transient increase in liver enzymes is unclear.

Tannic acid does not appear to have the same effect as oak in ruminants. Sheep dosed orally with tannic acid did not develop renal disease (Zhu *et al.*, 1992). Calves dosed orally with tannic acid (whether with multiple low doses or single high doses) did not have any of the laboratory abnormalities or clinical signs seen with oak-fed calves. Therefore, commercial tannic acid given orally cannot be used as a substitute for oak in studies of oak toxicosis in cattle.

Table 107.2. Percent of methemoglobin in blood from calves dosed orally with oak for 7d (A,B) or with single doses of gallic acid (GA) (E-K).

| Calf | GA g/kg | Time (hrs) after first dosing | | | | | | | | |
		0	3	6	24	48	72	96	168	192
A	oak	1.0	-	-	2.0	2.5	0.7	1.3	1.7	1.0
B	oak	2.0	-	-	1.9	2.1	1.0	1.7	1.5	1.6
E	1.0	1.0	3.0	2.0	1.1	0.7	-	-	-	-
F	1.5	1.2	0.9	2.6	0.3	1.9	-	-	-	-
G	2.0	1.0	1.6	5.0	2.6	1.7	-	-	-	-
H	2.0	6.0	0	1.5	6.1	5.2	-	-	-	-
I	2.0	1.1	1.0	6.8	22.0	28.0	-	-	-	-
J	2.5	0	1.2	3.3	30.4	40.7	-	-	-	-
K	2.5	0.8	0.5	1.9	24.5	29.2	-	-	-	-

The oak-fed calves had detectable levels of pyrogallol in their serum only at 3hrs and 6hrs after the first ingestion of oak leaves, but in the urine 48hrs and 60hrs. After this, pyrogallol could no longer be detected in either body fluid, even though the animals continued to eat the leaves for several days. Because it had disappeared from these samples before the appearance of most clinical signs or abnormal clinical pathology findings, pyrogallol in serum or urine was not a useful indicator of oak toxicosis.

References

Bruce, R.D. (1985) An up-and-down procedure for acute toxicity testing. *Fundamental and Applied Toxicology* 5, 151-157.

Dollahite, J.W., Pigeon, R.F. and Camp, B.J. (1962) The toxicity of gallic acid, pyrogallol, tannic acid, and *Quercus havardi* in the rabbit. *American Journal of Veterinary Research* 23, 1264-1266.

Harper, K.T., Ruyle, G.B. and Rittenhouse, L.R. (1988) Toxicity problems associated with the grazing of oak in intermountain and southwestern USA. In: James, L.F., Ralphs, M.H. and Neilson, D.B. (eds), *The Ecology and Economic Impact of Poisonous Plants on Livestock Production*. Westview Press, Boulder, CO, pp. 197-206.

Murdiati, T.B., McSweeney, C.S. and Lowry, J.B. (1992) Metabolism in sheep of gallic acid, tannic acid and hydrolysable tannin from *Terminalia oblongata*. *Australian Journal of Agriculture Research* 43, 1307-1319.

Pigeon, R.F., Camp, B.J. and Dollahite, J.W. (1962) Oral toxicity and polyhydroxyphenol moiety of tannin isolated from *Quercus havardi* (shin oak). *American Journal of Veterinary Research* 23, 1268-1270.

Spier, S.J., Smith, B.P., Seawright, A.A., Norman, B.B. Ostrowski, S.R. and Oliver, M.N. (1987) Oak toxicosis in cattle in northern California: clinical and pathological findings. *Journal American Veterinary Medical Association* 191, 958-964.

Tor, E.R., Francis, T.M., Holstege, D.M. and Galey, F.D. (1996) GC/MS determination of pyrogallol and gallic acid in biological matrices as diagnostic indicators of oak exposure. *Journal Agriculture and Food Chemistry* 44, 1275-1279.

Zhu, J. and Filippich, L.J. (1995) Acute intra-abomasal toxicity of tannic acid in sheep. *Veterinary and Human Toxicology* 37, 50-54.

Zhu, J., Filippich, L.J. and Alsalami, M.T. (1992) Tannic acid intoxication in sheep and mice. *Research Veterinary Science* 53, 280-292.

Chapter 108

Jatropha curcas Toxicity: Identification of Toxic Principle(s)

H.P.S. Makkar and K. Becker
Institute for Animal Production in the Tropics and Subtropics (480), University of Hohenheim, D-70593 Stuttgart, Germany

Introduction

The genus *Jatropha* belongs to the family Euphorbiaceae and contains approximately 170 species. *Jatropha curcas* (physic nut) is native to Central America. It is a small, drought-resistant tree that can reach a height up to 8m. The plant survives on poor soils and can easily be propagated. The oil, which comprises about 60% of seed weight, is strongly purgative and is also used topically for skin diseases and to alleviate pain such as that caused by rheumatism. The seed extract has molluscicidal activity (Rug *et al.*, 1997), and the seed meal is used as fertilizer.

Mampane *et al.* (1987) described giddiness, vomition and diarrhea in humans accidentally poisoned by jatropha seeds. Adam (1974) observed dose-related mortality rates, pathological changes observed and times to death after feeding mice ground seeds. Adverse effects and mortality have been reported in chicks fed diets containing jatropha seeds (El Badwi *et al.*, 1992). When an aqueous slurry of ground jatropha seeds was given to male goats, mortality occurred within 6-25d (Ahmed and Adam, 1979a). Postmortem findings included large areas of diffuse hemorrhage on rumen and reticulum mucosae, hemorrhagic or catarrhal abomasitis and enteritis and small ulcers in the small intestine. Ahmed and Adam (1979b) reported toxicity to calves at single doses of 0.025g/kg/d and above.

The kernels from *J. curcas* are consumed by humans in certain regions of Mexico, suggesting the presence of a nontoxic variety of *J. curcas* (Cano *et al.*, 1989). The nontoxic and toxic nature of the seeds, obtained from Mexico and Nicaragua respectively, was verified, using rats and fish. Toxic and antinutritional factors in seeds of the two varieties were compared, and the toxic principle was confirmed by reproducing the appropriate clinical signs in fish fed a diet containing isolated pure toxin. Effects of detoxified seed meal were studied in fish and rats.

Confirmation of Toxic and Nontoxic Nature of Seeds

The protein efficiency ratio (PER) for rats on a diet containing Mexican jatropha seed meal (defatted using diethyl ether) was 1.29 ± 0.28, while it was 3.52 ± 0.48 for the control group (casein-containing diet). Heat treatment (66% moisture, 121°C, 30min) increased the PER to 3.02 ± 0.31 (86% of control PER). Rats fed diets with raw and heated jatropha meal weighed 23 and 7% less than controls on a casein diet. Feed intakes by the heated jatropha meal group did not differ significantly from those of the casein group, but intakes by the raw meal group were 21% below the control. Heat-treatment allowed higher protein degradability and inactivation of trypsin inhibitors and lectins, and thus better performance from the animals.

Diets containing either raw or heated Mexican jatropha meal were fed for 14d to fish (carp, *Cyprinus carpio*) weighing 3-5g. Half of the crude protein (CP) of fishmeal in the standard diet was replaced with jatropha meal (32% jatropha meal), so that all diets contained 40% CP. Each diet was fed at 5x the maintenance rate, split into seven portions per day. Body weight increases were (mean±SD, %): control 83.6 ± 8.7, jatropha raw 63.4 ± 6.3 and jatropha heated 62.9 ± 5.9. Both groups fed jatropha meal had some mucus in the feces. This indicated the presence of a minute amount of some heat-stable toxin in the non-toxic Mexican jatropha meal. Fish on both raw and heated jatropha diets gained weight, while the rats on heated jatropha gained significantly more than those on the raw diet. This suggests either that fish are less sensitive to trypsin inhibitors and lectins than rats or that heat treatment does not aid the digestibility of jatropha proteins for fish.

Rats fed a diet with 10% CP contributed entirely by Nicaraguan jatropha seed meal (16% in the diet) experienced severe weight loss (33% *vs.* controls) over 4d, with 50% mortality. Feed intake was negligible (1.9g/d/rat). Diets containing 32% of either raw or heated Nicaraugan jatropha meal fed to fish weighing 3-5g led to a 7.0 ± 0.21% decrease in body mass in 4d. The fish refused the diet on the second day of feeding. An abundance of mucus was seen in the aquarium, but no fish died.

The seed meal from the Mexican variety, although not entirely free from antinutrients, is apparently edible and non-lethal, as suggested above. However, seeds from the Cape Verde variety cultivated in Nicaragua are toxic.

Nutrients and Antinutrients in the Two Varieties

Jatropha kernels consist mainly of lipid (58%) and protein, with very little moisture (5-6%) and ash (4%). The CP content of the kernels was 22.2 and 27.2% for the toxic and the nontoxic varieties, respectively, giving CP of 56.4% and 63.8% in defatted meal. The amino acid composition of the two varieties was similar. The high PER in rats and the rapid growth observed in fish fed nontoxic jatropha meal suggested that the protein quality of jatropha seeds is very high.

Amylase inhibitor activity, tannins, cyanogens and glucosinolates were not detected in any meal samples. Trypsin inhibitor activity (mg trypsin inhibited/g of seed) was 21.3 and 26.5mg/g in the toxic and the nontoxic varieties, respectively. Lectin activity was higher in meal from the toxic variety (102 *vs.* 51) based on hemagglutination (Gordon and Marquardt, 1974). Toxicity of *J. curcas* seeds has been attributed to their lectin content (Stirpe *et al.*, 1976), but similar lectin values in both varieties implied that this was not the major toxic principle. Phytate content was 9.4% for the toxic and 8.9% for the nontoxic variety.

High concentrations of phorbol esters were present in the kernel of the toxic seeds, while there were very low concentrations in the nontoxic variety (2.70 *vs.* 0.11mg/g; as phorbol-12-myristate-13-acetate equivalent). Jatropha seeds from two trees in Quintana Roo were consumed by both humans and chickens, but ingestion of seeds from a tree in Laguna Guerreo, Mexico, by humans caused diarrhea, giddiness and vomiting. Phorbol esters in the Quintana Roo and Laguna Guerreo seeds were 0.09, 0.03 and 2.49mg/g kernel, respectively. The levels of lectin, trypsin inhibitors, saponins and phytate were similar in all three samples. Phorbol esters have caused purgation, skin-irritant effects and tumor promotion (Adolf *et al.*, 1984). Plants from the Euphorbiaceae and Thymelaeaceae that biosynthesize diterpene esters of the phorbol type cause severe toxic symptoms in livestock (Kingsbury, 1964). Phorbol esters are heat stable at temperatures as high as 160°C for 30min. These results suggested that the toxicity of *J. curcas* seeds could be attributed to phorbol esters present in higher amounts in the toxic variety.

Phorbol Esters as the Toxic Principle

Oil of the toxic *J. curcas* variety was methanol (MeOH) extracted 5x, evaporated, and mixed in standard fish feed for a phorbol ester content of 2.5mg/g. Five fish, starved for 24hrs, were fed an average of 8g (5x maintenance) split into seven portions per day. The fish ate all feed offered on the first day, and 1hr after initial consumption fine tubings of mucus filled with feed were observed, suggesting membrane irritation and purgative effects. After this, the fish refused the feed. The chamber was cleaned daily, and production of mucus tubes continued for 5d. Fish were sluggish and grouped near the bottom of the tank. After 7d, the mean weights for control and treated groups were 10.8g (+35%) and 7.3g (-9%), respectively.

Phorbol esters purified from toxic *J. curcas* seed by HPLC (Adolf *et al.*,1984) were mixed with standard fish feed (2mg/g) and fed to 5g fish as above. Symptoms of jatropha toxicity included mucus production, feed refusal, and weight loss (6.2%), confirming phorbol esters as the major agent of jatropha toxicity.

Water-Soluble Toxin(s) in Jatropha Seeds

A 5% (w/v) aqueous extract of toxic jatropha meal (residual oil 1%) was prepared, and 280ml was added to 20L aerated water. Five fish (average weight 3.5g) appeared normal at 1hr, but all were found dead at 18hrs, with hemorrhagic lesions near the anus and the lower gills. Jatropha extract from the nontoxic variety, from heat-treated toxic meal or with heat-treated extract of the toxic meal caused no mortality. These results indicated the presence of water-soluble heat-labile toxin(s) in jatropha meal along with the phorbol esters. Hemorrhagic lesions and inactivation by heat suggested that the toxins were lectins, and further studies will focus on their identification. Adding *J. curcas* meal extract to tanks of fish (carp; weight <5g) could be used as a bioassay for identifying toxic *Jatropha* varieties, and for identifying the water-soluble toxin(s). Although fish exposed to the extract from the nontoxic variety for 18hrs did not die, their feed intake was lower for 2d, suggesting that the non-toxic variety has low levels of water-soluble toxin(s).

Detoxification of Seed Meal from the Toxic Variety

Inactivation and extraction of toxic factors from jatropha meal was attempted. Heat treatment (121°C, 30min, 66% moisture) inactivated lectins and trypsin inhibitors (Aderibigbe *et al.*, 1997). Ethanol (EtOH) (80%) or MeOH (92%) [1:5 w/v] reduced both saponins and phorbol esters by 95% after four extractions.

A diet with 10% CP entirely from "detoxified" jatropha meal (16% in diet) was fed to rats for 10d. The mean feed consumption and weight gain were 13.6±0.29g/d and 2.39±0.21g/d, respectively, while rats on the toxic meal containing diet ate 9.0g/d and gained 1.95g/d. Based on this, the detoxification scheme would allow the use of jatropha meal as a feed source.

Conclusions

Phorbol esters are the primary heat-stable toxic agents in jatropha. The trypsin inhibitor, lectin and phytate might aggravate adverse effects, but do not seem responsible for short-term toxicity. The protein and amino acid compositions of meal from the nontoxic and toxic varieties are similar. The meal from both varieties contained high levels of heat labile trypsin inhibitor and lectin. The presence of high levels of heat-stable phytate can decrease the bioavailability of minerals, but the high PER (85% of that for casein) observed in rats and rapid growth of fish on a diet containing 32% jatropha meal suggest that phytate does not adversely affect animal performance. The nontoxic variety of jatropha could be a source of oil for human consumption, and the seed meal could be a protein source for humans and livestock. Comparative evaluation of seed yield and resistance to diseases for the nontoxic and toxic varieties are in progress in Nicaragua, Zimbabwe, Mexico and India. Extraction with 92% aqueous MeOH or 80% EtOH may effectively detoxify jatropha meal obtained after extraction of oil.

References

Adam, S.E.I. (1974) Toxic effects of *Jatropha curcas* in mice. *Toxicology* 2, 67-76.

Aderibigbe, A.O., Johnson, C.O.L.E., Makkar, H.P.S., Becker, K. and Foidl, N. (1997) Chemical composition and effect of heat on organic matter and nitrogen-degradability and some antinutritional components of Jatropha meal. *Animal Feed Science and Technology* 67, 223-243.

Adolf, W., Opferkuch, H.J. and Hecker, E. (1984) Irritant phorbol derivatives from four *Jatropha* species. *Phytochemistry* 23, 129-132.

Ahmed, O.M.M. and Adam, S.E.I. (1979a) Toxicity of *Jatropha curcas* in sheep and goats. *Research in Veterinary Science* 27, 89-96.

Ahmed, O.M.M. and Adam, S.E.I. (1979b) Effects of *Jatropha curcas* on calves. *Veterinary Pathology* 16, 476-482.

Cano, A.L.M., Plumbly, R.A. and Hylands, P.J. (1989) Purification and partial characterization of the hemagglutinin from seeds *of Jatropha curcas. Journal of Food Biochemistry* 13, 1-20.

El Badwi, S.M.A., Mousa, H.M., Adam, S.E.I. and Hapke, H.J. (1992) Response of Browne Hissex chicks to low levels of *Jatropha curcas, Ricinus communis* or their mixture. *Veterinary and Human Toxicology* 34, 304-306.

Gordon, J.A. and Marquardt, M.D. (1974) Factors affecting haemagglutination by concanavalin A and soybean agglutinin. *Biochimica et Biophysica Acta* 36, 144.

Kingsbury, J.M. (1964) *Poisonous Plants of the United States and Canada.* Prentice-Hall, Englewood Cliffs, NJ, pp. 190-191.

Mampane, K.J., Joubert, P.H. and Hay, I.T. (1987) *Jatropha curcas*: use as a traditional Tswana medicine and its role as a cause of acute poisoning. *Phytotherapy Research* 1, 50-51.

Rug, M., Sporer, F., Wink, M., Jourdane, J., Henning, R. and Ruppel, A. (1997) Investigation of molluscidal properties of *Jatropha curcas* against snails transmitting human schistosomes. In: *Proceedings of Jatropha 97: International Symposium on Biofuel and Industrial Products from* Jatropha curcas *and other tropical oil seed plants.* Managua, Nicaragua, pp. 23-27.

Stirpe, F., Pession-Brizzi, A., Lorenzoni, E., Strochi, P., Montanaro, L. and Sperti, S. (1976) Studies on the proteins from the seeds of *Croton tiglium* and *Jatropha curcas. Biochemical Journal* 156, 1-6.

Chapter 109

Species Differences in Bioactivation and Detoxification of Pyrrolizidine Alkaloids

P.R. Cheeke and J. Huan

Department of Animal Sciences, Oregon State University, Corvallis, Oregon 97331, USA

Introduction

Animal species and individuals within species vary in their susceptibility to plant toxins, largely because of differences in hepatic enzyme activities (Cheeke, 1994a, b; 1997). Pyrrolizidine alkaloids (PAs) are metabolized in the liver to produce pyrroles or dihydroxypyrrolizine (DHP) derivatives, the active metabolites of PAs. Some of the enzymes involved in bioactivation of PAs include cytochrome P_{450} enzymes (CYP) and flavin-containing monooxygenases (FMO); detoxification mechanisms include hepatic esterases and glutathione (GSH) conjugation.

In this study, species differences in hepatic microsomal enzymes involved in metabolism of PAs, including CYP, FMO and hepatic esterases, and the roles of the CYP isozymes CYP3A and CYP2B in bioactivation and detoxification of the PA senecionine (SN) were investigated. Sheep and hamsters were used as ruminant and non-ruminant animals resistant to PA toxicity (Cheeke, 1997).

Experiment 1

Species differences in DHP and SN N-oxide formation among eight species varying in susceptibility to PA toxicity were measured by *in vitro* liver microsomal incubation. Four adult male animals of each species (hamster, rat, gerbil, rabbit, chicken and Japanese quail) were terminated by cervical dislocation, the livers removed immediately and microsomes prepared by standard procedures. Livers of sheep and cattle were obtained from an abattoir. Microsomal incubations were conducted as described by Kedzierski and Buhler (1986a, b).

There was no strong correlation between DHP (pyrrole) formation and species susceptibility among the tested species (Table 109.1). In hamsters, the formation of DHP greatly exceeded the rate of SN N-oxide formation. In contrast, SN N-oxide was the major metabolite in sheep. In general, susceptible species tended to have a lower rate of DHP formation than resistant animals, except for Japanese quail, which had very little SN metabolism. Almost 92% of the SN was unchanged for this species, in agreement with earlier studies (Buckmaster *et al.*, 1977).

Based on the results in Table 109.1, hamsters and sheep were selected as examples of resistant animals whose mechanisms of resistance appeared to differ. Several chemical inhibitors and anti-NADPH cytochrome P_{450} reductase IgG were used. The inhibitor SKF 525A almost completely inhibited DHP formation in both species, while N-oxidation was reduced by 7.6% in sheep and 34% in hamsters, indicating that CYP is involved in formation of DHP but not N-oxide from SN, which was confirmed by the use of the more specific inhibitor, anti-P_{450} reductase IgG (Table 109.2). Two inhibitors of FMO, methimazole and thiourea, reduced N-oxide formation in both species (Table 109.2), indicating that FMO activity is a major route of N-oxidation. Phenylmethylsulfonyl fluoride (PMSF) and tri-*ortho*-cresyl phosphate (TOCP), inhibitors of hepatic esterases, were used to assess the involvement of esterases in SN metabolism. The results (Table 109.2) suggest a higher esterase activity in sheep than in hamsters, using TOCP as the inhibitor. Caboxylesterase is strongly inhibited by PMSF (Dueker *et al.*, 1992). Sheep have a higher rate of esterase hydrolysis than hamsters (Table109.2), which may reduce the amount of PA substrate available for conversion to toxic metabolites, reducing the toxicity of dietary PA.

Experiment 2

The roles of CYP3A and CYP2B in bioactivation and detoxification of PA in sheep and hamsters were evaluated. Antibodies for sheep hepatic CYP3A and CYP2B

Table 109.1. *In vitro* metabolism of senecionine (SN) by liver microsomes to form DHP[1] and N-oxide in different animal species.

Species	DHP	N-oxide	DHP/ N-oxide	% SN unchanged
PA-susceptible				
Bovine	0.23	0.59	0.38	61.1
Chicken	0.22	0.52	0.44	84.5
Rat	0.85	0.70	1.25	69.7
PA-resistant				
Sheep	0.45	1.76	0.26	27.5
Hamster	3.55	1.55	2.29	23.1
Rabbit	1.71	0.81	2.11	30.1
Japanese quail	0.08	0.28	0.30	91.5
Gerbil	1.34	0.97	1.39	42.2

[1] (±) 6,7-Dihydro-7-hydroxy-1-hydroxymethyl-5H-pyrrolizine

Table 109.2. Inhibition of DHP and N-oxide formation from SN in sheep and hamster liver microsomes by chemicals and specific antibody.

Inhibitors	DHP (as % of control)		N-oxide (as % of control)	
	sheep	hamster	sheep	hamster
SKF-525A (0.5mM)	ND	1.8	92.4	65.9
Methimazole (0.25mM)	6.1	ND	62.3	20.1
Thiourea (0.25mM)	30.8	14.8	44.2	29.0
PMSF (1.0mM)	69.2	89.6	74.0	105.7
TOCP (0.1mM)	9.2	76.9	72.9	84.4
Anti-P_{450} reductase IgG	32.7	17.8	91.9	45.3

Results are expressed as percentage of control (no inhibitor) and are given as the mean of four animal liver microsomal incubations containing 0.5mg microsomal protein. ND means the values are non-detectable. Each incubation was done in duplicate at a concentration of 30mg/nmol P_{450} with preimmune IgG as control.

were prepared as previously described (Pineau *et al.*, 1990). Microsomal preparations containing the antibodies were incubated with SN as described by Huan (1995). The rates of DHP and SN N-oxide formation from sheep and hamster microsomes (Table 109.3) indicated that the two species differ markedly in their metabolism of SN. Hamsters produced much more DHP than N-oxide, while the reverse was true with sheep. In both sheep and hamsters, incubation with anti-CYP3A IgG markedly reduced DHP formation, but had less effect on N-oxide (Table 109.4). Thus CYP3A is the major enzyme involved in PA bioactivation, as well as being partially involved in PA detoxification. CYP2B has only a minor contribution in SN biotransformation in both species.

Discussion

These results help to elucidate species differences in PA metabolism. Sheep and hamsters are both very resistant to PA toxicity (Cheeke and Pierson-Goeger, 1983). Sheep have a low rate of pyrrole (DHP) formation while hamsters have a high rate (Table 109.1). Susceptibility to PA intoxication thus is not solely determined by rate of pyrrole formation, but also by its rate of disposition. The N-oxide is the major metabolite from SN in sheep, while in hamsters the rate of DHP formation greatly exceeded that of SN N-oxidation. The resistance of sheep to PA toxicosis appears to be due to a low pyrrole production rate, a high esterase activity, and a high rate of N-oxide formation due to a high FMO activity. Hamsters have a high rate of pyrrole formation, and of conjugation of DHP with GSH (Dueker *et al.*, 1994; Reed *et al.*, 1992), but the complete explanation of their resistance is unclear.

Immunoinhibition data indicated that CYP3A is the major enzyme involved in PA metabolism in both sheep and hamsters, as is also true in rats and humans (Williams *et al.*, 1989). Steroids, macrolide antibiotics and phenobarbital induce CYP3A (Gonzalez, 1988), but phenobarbital pretreatment of sheep did not influence

Table 109.3. *In vitro* metabolism of SN by hepatic microsomes to form DHP and N-oxide in sheep and hamsters.

Species four, male	DHP nM/min/mg	N-oxide nM/min/mg	Ratio of DHP/N-oxide
Sheep	0.45 ± 0.07[a]	1.76 ± 0.08[a]	0.26 ± 0.05[a]
Hamster	3.55 ± 0.08[b]	1.55 ± 0.01[a]	2.29 ± 0.07[b]

Mean±SE in the same column followed by different superscripts are different (*P*<0.050). Hepatic microsomal protein was 0.5mg in the incubation mixtures.

Table 109.4. Effect of cytochrome P_{450} 3A (CYP3A) and 2B(CYP2B) inhibition on metabolism of SN by sheep and hamster microsomes.

	% Inhibition of rate of formation			
	sheep		hamsters	
Inhibitor	DHP	N-oxide	DHP	N-oxide
AntiCYP3A IgG	96.5	38.8	69.5	41.3
AntiCYP2B IgG	47.4	24.6	32.5	35.5

PA toxicity (Swick *et al.*, 1983). The CYP2B also had affected PA metabolism, but was less active than CYP3A in both species. Low rates of pyrrole production coupled with efficient GSH conjugation may explain the resistance of sheep to SN, which high rates of GSH conjugation of DHP may be a major factor responsible for hamsters'resistance to SN intoxication.

The major metabolic routes for PA in animals are dehydrogenation to pyrrolic derivatives, conversion to N-oxides and hydrolysis (Mattocks, 1986). Resistance or susceptibility of animals to PA intoxication rests on the balance of these pathways with other secondary detoxification pathways such as GSH conjugation. The specific involvement of distinct enzyme systems together with unique substrate specificities accounts for species differences in PA metabolism in animals.

References

Buckmaster, G.W., Cheeke, P.R., Arscott, G.R., Dickinson, E.O., Pierson, M.L. and Shull, L.R. (1977) The response of Japanese quail to dietary and injected pyrrolizidine (*Senecio*) alkaloid. *Journal of Animal Science* 45, 1322-1325.

Cheeke, P.R. (1994a) A review of the functional and evolutionary roles of the liver in the detoxification of poisonous plants, with special reference to pyrrolizidine alkaloids. *Veterinary and Human Toxicology* 36, 240-247.

Cheeke, P.R. (1994b) The role of the liver in the detoxification of poisonous plants. In: Colegate, S.M. and Dorling, P.R. (eds), *Plant-Associated Toxins: Agricultural, Phytochemical and Ecological Aspects.* CAB International, Wallingford, Oxon, pp. 281-286.

Cheeke, P.R. (1997) *Natural Toxicants in Feeds, Forages, and Poisonous Plants.* Interstate Publishers, Danville, IL.

Cheeke, P.R. and Pierson-Goeger, M.L. (1983) Toxicity of *Senecio jacobaea* and pyrrolizidine alkaloids in various laboratory animals and avian species. *Toxicology Letters* 18, 343-349.

Dueker, S.R., Lame, M.W. and Segall, H.J. (1992) Hydrolysis of pyrrolizidine alkaloids by guinea pig hepatic carboxylesterases. *Toxicology and Applied Pharmacology* 117, 116-121.

Dueker, S.R., Lame, M.W., Jones, D., Morin, D. and Segall, H.J. (1994) Glutathione conjugation with the pyrrolizidine alkaloid, jacobine. *Biochemical and Biophysical Research Communications* 198, 516-522.

Gonzalez, F.J. (1988) The molecular biology of cytochrome P450s. *Pharmacology Review* 40, 243-288.

Huan, J. (1995) Species Differences in Bioactivation and Detoxification of Pyrrolizidine (Senecionine) Alkaloids. MS Thesis. Oregon State University.

Kedzierski, B. and Buhler, D.R. (1986a) Method for determination of pyrrolizidine alkaloids and their metabolites by high performance liquid chromatography. *Analytical Biochemistry* 152, 59-65.

Kedzierski, B. and Buhler, D.R. (1986b) The formation of 6,7-dihydro-7-hydroxyl-1-hydroxymethy-5H-pyrrolizine, a metabolite of pyrrolizidine alkaloids. *Chemico-Biological Interactions* 57, 217-222.

Mattocks, A.R. (1986) *Chemistry and Toxicology of Pyrrolizidine Alkaloids*. Academic Press, Orlando, FL.

Pineau, T., Galtier, P., Bonfils, C., Derancourt, J. and Maurel, P. (1990) Purification of a sheep liver cytochrome P450 from the P450IIIA gene subfamily. *Biochemical Pharmacology* 39, 901-909.

Reed, R.L., Miranda, C.L., Kedzierski, B., Henderson, M.C. and Buhler, D.R. (1992) Microsomal formation of a pyrrolic alcohol glutathione conjugate of the pyrrolizidine alkaloid senecionine. *Xenobiotica* 22, 1321-1327.

Swick, R.A., Miranda, C.L., Cheeke, P.R. and Buhler, D.R. (1983) Effect of phenobarbital on toxicity of pyrrolizidine (*Senecio*) alkaloids in sheep. *Journal of Animal Science* 56, 887-894.

Williams, D.E., Reed, R.L., Kedzierski, B., Ziegler, D.M. and Buhler, D.R. (1989) The role of flavin-containing monooxygenase in the N-oxidation of the pyrrolizidine alkaloid senecionine. *Drug Metabolism and Disposition* 17, 380-386.

Chapter 110

Bog Asphodel (*Narthecium ossifragum*) Poisoning in Cattle

F.E. Malone[1], S. Kennedy[2], G.A.C. Reilly[2] and F.M. Woods[3]

[1]*Veterinary Sciences Division, Department of Agriculture for Northern Ireland, 43 Beltany Road, Omagh, County Tyrone BT78 5NF, UK;* [2]*Veterinary Sciences Division, Department of Agriculture for Northern Ireland, Stormont, Belfast BT9 5PX, UK;* [3]*Erne Veterinary Group, Church Road, Lisnaskea, County Fermanagh, UK*

Introduction

Bog asphodel (*Narthecium ossifragum*) is a perennial plant of the family Liliaceae that is common on wet heaths, bogs and moorland throughout Britain and Ireland (Butcher, 1961; Webb, 1963). It has two rows of grass-like leaves, flattened into a sheath at ground level and a stem 15-30cm long, which during summer bears a slender raceme of bright yellow flowers approximately 1cm in diameter. Bog asphodel poisoning in animals has previously been reported only in sheep and lambs (Cooper and Johnston, 1988), where it is associated with hepatogenous photosensitization colloquially known as "alveld" in Norway (Ender, 1955), "yellowses," "head greet" or "plochteach" in Scotland, "saut" in Cumbria (Ford, 1964) and "hard lug" in Antrim, Northern Ireland (Lamont, 1952).

In summer of 1989, a farm in County Fermanagh, Northern Ireland, experienced an outbreak of disease in 25 suckler cows and their calves on pasture that had been limed at the end of July, 1989. By mid-August, 15 of the cows were in poor body condition, becoming anorexic over the next 2wks.

Blood samples collected from affected animals on three occasions during the disease outbreak were analyzed for calcium, magnesium, phosphate, copper, γ-glutamyl transpeptidase, urea, creatinine and plasma proteins, were subjected to standard hematological examination and were cultured for bovine viral diarrhea (BVD) virus. The mean plasma urea and creatinine concentrations were high and plasma albumin levels were low in the affected cows (Table 110.1), but there were no consistent changes in the other parameters. No sera yielded BVD virus.

Table 110.1. Clinical pathology of 13 affected cows sampled on 15 August (a), 21 August (b) and 1 September (c), 1989.

Normal:	Urea (mM/L) (3.3-8.3)			Creatinine (µM/L) (0-130)		
Cow	a	b	c	a	b	c
1	36.3	-	-	629	-	-
2	42.8	-	-	1,028	-	-
3	31.1	-	-	461	-	-
4*	11.8	-	-	105	-	-
5	40.5	99.8	-	830	1,276	-
6	35.9	42.0	-	496	579	-
7	-	156.0	-	-	1,332	-
8	-	74.5	-	-	1,338	-
9	-	148.0	-	-	1,431	-
10	-	144.4	-	-	1,678	-
11*	-	-	15.2	-	-	174
12*	-	-	14.1	-	-	165
13*	-	-	20.6	-	-	280
Mean	33.1	110.8	16.6	592	1,272	206
±SD	11.2	46.3	3.5	292	368	64

* These cows subsequently recovered.

Eleven cows became recumbent and died or were euthanized; the other four recovered. Necropsy of one euthanized cow (A) and of two that died (B, C) found all three animals emaciated, with fusiform ulcers approximately 0.5-1.0cm long in the mucosae of the esophagus, ruminal pillars and abomasum. Intestinal contents were extremely fluid, indicating an enteritis, but Peyer's patches appeared normal. There was extensive consolidation of the cranial lobes of the lungs in cow B. The urine of cow A had a specific gravity of 1.015 with proteinuria and glycosuria. Tissues collected were processed and evaluated by standard histopathological methods, and samples of lung, liver, kidney and small and large intestines were cultured for bacteria. The spleen, mesenteric lymph nodes and esophageal and ruminal ulcers were tested for BVD virus. Rumen and abomasal contents were subjected to Reinsch tests for heavy metals, and kidneys were assayed for mercury.

No significant bacteria were isolated, nor heavy metals detected. A mesenteric lymph node of Cow B, but not spleen or esophagus, yielded BVD virus.

Diffuse renal tubular necrosis, characterized by epithelial cells with swollen, hypereosinophilic cytoplasm and pyknotic nuclei, mild focal mononuclear cell infiltration and basophilic mineral particles, was present in all three animals. Collapsing tubular lamina were filled with necrotic debris and proteinaceous material with occasional oxalate crystals, and there were a few small foci of peritubular fibrosis. Other histopathological changes included mild, hydropic hepatocellular degeneration with increased sinusoidal leukocytes in cows A and B, and lesions of necrotizing bronchopneumonia in cow C.

Investigation

The farm comprised a main pasture of 11.5ha, an abandoned orchard of 0.1ha and a 2ha area of peat bog used for turf production. Cattle had access to all three areas. A survey of the farm for poisonous substances was undertaken. Metals detected in limestone, paint, putty and "clay pigeon" fragments collected were present at non-toxic levels. Small quantities of marsh ragwort (*Senecio aquaticus*) grew in the main pasture, and a profuse growth of *N. ossifragum* covered a 2,500m^2 area of the peat bog, which had been grazed due to scarcity of grass in the main pasture. The density of bog asphodel, calculated by averaging 20 randomly thrown 0.5x0.5m quadrats (Moore and Chapman, 1986), was 5.9±7.8 plants/quadrat.

Samples of bog asphodel were collected from the farm in early August (sample A, in flower) and early October (sample B, in seed) and stored at -20°C until use. Blood samples were drawn for baseline clinical evaluation from two 14wk Friesian steers maintained on hay and commercial concentrate. The calves were each fed 1.6kg bog asphodel mixed with concentrate, one receiving sample A and the other sample B. Clinical examination and blood sampling were performed daily for 7d.

The calf fed sample A was alert with a normal appetite throughout the trial. However, its plasma urea and creatinine levels significantly increased at 2d, and the phosphorus level also increased somewhat (Fig. 110.1). At 3d, its ruminations decreased from two or three normal to one weak contraction/min. The calf fed sample B remained normal at all times.

The kidneys of the calf fed sample A were pale and severely edematous, especially at the hilar regions. There was also mesenteric edema and approximately 1L of peritoneal fluid. The glucosuric urine had a specific gravity of 1.011. Microscopically, the kidneys were diffusely necrotic, with marked regeneration of the tubular epithelium characterized by irregular accumulations of enlarged epithelial cells and common mitoses. Many tubular lumina contained necrotic debris and proteinaceous material, and occasional small foci of mononuclear cell infiltration were present in the cortex.

Discussion

Renal tubular damage in cattle is related to a variety of nephrotoxic agents (Maxie, 1993). No medications had been administered to the cows before this disease outbreak, and a careful search of the farm for known nephrotoxic substances and plants was negative. The experimental reproduction of renal tubular necrosis by feeding flowering bog asphodel to a healthy calf provides strong evidence that the outbreak was due to this plant. The failure to induce renal disease in another calf by feeding bog asphodel bearing seed heads, suggests that the flowering plant is more toxic than the mature plant. Therefore, bog asphodel should be considered as a differential diagnosis of renal disease in cattle which have access to this plant.

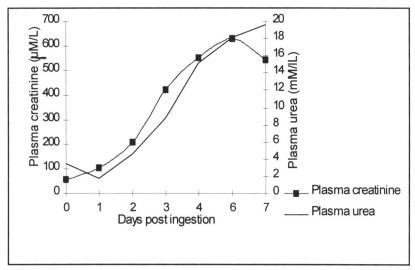

Fig. 110.1. Plasma urea and creatinine concentrations in calf fed bog asphodel.

Several plants produce renal lesions in cattle. Oak (*Quercus* spp.) poisoning is causes acute renal tubular necrosis and gastroenteritis, most frequently in cattle ingesting acorns (Divers *et al.*, 1982). Similar renal lesions, with perirenal edema, are characteristic of pigweed (*Amaranthus retroflexus*) poisoning in cattle and pigs in North America (Buck *et al.*, 1966; Brown, 1974). Renal damage due to the ingestion of *Isotropis* species has been reported in cattle and sheep in Western Australia by Gardiner and Royce (1967). In this study, perirenal edema developed in the calf fed flowering bog asphodel, but not in field cases of the disease. This lesion is thought to be due to tubular back-leakage with subsequent impairment of lymphatic drainage and leakage into the perirenal connective tissue (Maxie, 1993).

Narthecium asiaticum, a species related to bog asphodel, has poisoned grazing cattle in Japan (Suzuki *et al.*, 1985). Clinically, affected animals developed ascites and were depressed, anorexic and oliguric. At necropsy, swelling of the kidneys was the predominant finding, with ascites, perirenal edema and hemorrhage of various organs also present in most cases. Microscopically, there was severe renal tubular degeneration and necrosis, with intraluminal proteinaceous material and oxalate crystals. Hypoplasia of the erythroblast series in the bone marrow, lymphatic atrophy and degeneration and focal hepatic necrosis were also observed.

Raws (1983) stated that *N. ossifragum* was controlled by sheep and that when protected from grazing, it spread and dominated local vegetation. Drier areas of the farm were grazed preferentially by the cattle, except when grass became scarce. Despite its name, bog asphodel does not thrive in water-logged soils (Daniels, 1975). The dry summer and aeration of the peat bog by turf cutting operations may also have contributed to the proliferation of bog asphodel on the site.

Acknowledgments

The authors thank Dr A.D. Courtney, Department of Agriculture for Northern Ireland for the identification of the plants; Mr A.W. Johnston, Commonwealth Agricultural Bureaux for assistance with library material and Mr P.J. McParland and Mr N.S. Bell, Department of Agriculture for Northern Ireland for professional and technical assistance.

References

Brown, C.M. (1974) Chronic *Amaranthus* (pigweed) toxicity in 5 of 100 400lb Hereford heifers. *Veterinary Medicine Small Animal Clinician* 69, 1551-3.

Buck, W.S., Preston, K.S., Abel, M. and Marshall, V.L. (1966) Perirenal edema in swine: A disease caused by common weeds. *Journal of the American Veterinary Medical Association* 148, 1525-1531.

Butcher, R.W. (1961) *A New Illustrated British Flora*. Leonard Hill, London, p. 632.

Cooper, M.R. and Johnston, A.W. (1988) *Poisonous Plants and Fungi, An Illustrated Guide*. Her Majesty's Stationery Office, London, pp. 59-60.

Daniels, R.E. (1975) Observations on the performance of *Narthecium ossifragum* (L.) Huds. and *Phragmites communis* Trin. *Journal of Ecology* 63, 965-977.

Divers, T.J., Crowell, W.A., Duncan, J.R. and Whitlock, R.H. (1982) Acute renal disorders in cattle: A retrospective study of 22 cases. *Journal of the American Veterinary Medical Association* 181, 694-699.

Ender, F. (1955) Etiological studies on "alveld" - a disease involving photosensitization and icterus in lambs. *Nordisk Veterinaraermedicin* 7, 329-377.

Ford, E.J.H. (1964) A preliminary investigation of photosensitization in Scottish sheep. *Journal of Comparative Pathology* 74, 37-44.

Gardiner, M.R. and Royce, R.D. (1967) Poisoning of sheep and cattle in Western Australia due to species of *Isotropis* (Papilionaceae). *Australian Journal of Agricultural Research* 18, 505-513.

Lamont, H.G. (1952) Urticaria. Photosensitization. Blue nose in horses. Oedema of the bowel in pigs. *Reports of the Proceedings of the Conference on Metabolic Diseases*. British Veterinary Association, pp. 85-100.

Maxie, M.G. (1993) Nephrotoxic acute tubular necrosis. In: Jubb, J.V.F., Kennedy, P.C. and Palmer, N. (eds), *Pathology of Domestic Animals, Vol 2*, 4th ed. Academic Press, London, pp. 489-495.

Moore, P.D. and Chapman, S.B. (1986) *Methods in Plant Ecology*. Blackwell, London, pp. 448-460.

Raws, M. (1983) Changes in two high altitude blanket bogs after the cessation of sheep grazing. *Journal of Ecology* 71, 219-235.

Suzuki, K., Kobayashi, M., Ito, A. and Nakgawa, M. (1985) *Narthecium asiaticum* Maxim. poisoning of grazing cattle: observations on spontaneous and experimental cases. *Cornell Veterinarian* 75, 348-365.

Webb, D.A. (1963) *An Irish Flora*. Dundalgan Press, Dundalk, Ireland, p. 175.

Chapter 111

Livestock Poisoning by *Baccharis coridifolia*

C.S.L. de Barros

Departamento de Patologia, Universidade Federal de Santa Maria, Santa Maria, RS, Brazil

Introduction

Baccharis coridifolia (Compositae) is one of the most recognized toxic plants in southern Brazil, where it grows in the states of Rio Grande do Sul, Santa Catarina and Paraná. The plant is also found of the Brazilian southeastern state of São Paulo, in large areas of Uruguay and in northern Argentina and Paraguay (Barros, 1993). In Brazil the plant is commonly known as "mio-mio" and, in the Spanish-speaking countries, as "romerillo." The natural toxicosis affects mainly cattle, less frequently sheep and only rarely horses. It has been reproduced experimentally in various species including cattle (Tokarnia and Döbereiner, 1975; Varaschin *et al.*, 1997), sheep (Tokarnia and Döbereiner, 1976), horses (Costa *et al.*, 1995) and rabbits (Döbereiner *et al.*, 1976; Rodrigues and Tokarnia, 1995). Lesions induced by the ingestion of *B. coridifolia* are mainly confined to the gastrointestinal tract.

Baccharis coridifolia is a dioecious plant that sprouts in spring (October through November) and flowering in late summer and early fall (end of February to April). Although the plant is more toxic when it blooms, the toxicosis can occur at any time of year, and heavy cattle losses have occurred in the spring (Barros, 1993). Doses as low as 0.25-0.50g/kg of the green flowering plant cause deaths in cattle. In the sprouting period, 2g/kg are required for the same effect (Tokarnia and Döbereiner, 1975). Based on this, the concentration of the toxic principle is about 4x higher in plants during the flowering period (Jarvis *et al.*, 1991). Sheep are comparatively resistant, and the toxic dose for this species is about 2x that for cattle (Tokarnia and Döbcreiner, 1976). Horses are particularly susceptible, with administration of a single dose of 0.06g/kg causing severe disease and doses from 0.125-0.5g/kg being lethal (Costa *et al.*, 1995). When dried, *B. coridifolia* retains about 50% of its potency for at least 17mos (Tokarnia and Döbereiner, 1975).

Several factors influence the appearance of the toxicosis. Cattle that are raised in pastures where *B. coridifolia* exists will graze it very rarely, if ever (Barros, 1993). Typically, the toxicosis occurs when naive animals raised in areas free of *B. coridifolia* are transferred to pastures infested by the plant. The risks of toxicosis increases considerably if, while being transported, the animals are subjected to such stress factors as fatigue, hunger or thirst. (Tokarnia and Döbereiner, 1975; Barros, 1993). Toxicosis has also been reported in suckling lambs in the first month of life, when they begin to graze (Tokarnia and Döbereiner, 1976).

Although the livestock losses by *B. coridifolia* poisoning are no doubt important, it is difficult to assess accurately the incidence of the disease. Most cases of the toxicosis are not reported to the veterinary diagnostic laboratories, largely due to the fact that farmers and practicing veterinarians recognize the toxicosis and implement measures to control it. The mortality is very high and may approach 80% in naive cattle transported to mio-mio infested areas and released without having being previously rested, fed and their thirst quenched.

Clinical Signs

Typically, ingestion of *B. coridifolia* induces an acute disease. In fatal cases in cattle and sheep, clinical signs start 5-29hrs and 2-23hrs, respectively, after ingestion of the plant, and death occurs 3-23 and 2-42hrs after the onset of clinical signs (Tokarnia and Döbereiner, 1975, 1976; Barros, 1993; Varaschin *et al.*, 1997). In cattle there is partial anorexia, mild to moderate bloat, instability of hindlimbs, muscle tremors, dry muzzle, serous ocular discharge, dry feces or diarrhea, excessive salivation, thirst, vocalization, rapid and labored breathing, tachycardia and restlessness. Animals eventually assume sternal recumbency, followed by lateral recumbency and death (Tokarnia and Döbereiner, 1975; Barros, 1993; Varaschin *et al.*, 1997). Sheep present similar signs (Tokarnia and Döbereiner, 1976). In horses that die from the intoxication, the clinical signs start 5-13hrs after ingestion of the plant and death occurs after a clinical course varying from 12-26hrs, consisting mostly of gastrointestinal disturbances. There is anorexia, colic, diarrhea, fever, dehydration, polydipsia, tachycardia, tachyphgmia, tachypnea, dyspnea, restlessness, gait instability and death (Costa *et al.*, 1995).

Pathology

Necropsy findings are mainly related to the gastrointestinal tract and are similar for cattle and sheep (Tokarnia and Döbereiner, 1975, 1976; Barros, 1993; Varaschin *et al.*, 1997). There are variable degrees of reddening, edema and erosions of the mucosa of the forestomachs. The mucosae of abomasum and intestines may be hyperemic, with petechiae. Intestinal contents are usually fluid and may be blood-tinged. In cattle, there is marked transmural edema and mucosal reddening of the colon and cecum (Barros, 1993). Lymphnodes may be swollen, moist and reddened (Varaschin *et al.*, 1997) and the liver is usually pale, with an enhanced lobular pattern (Tokarnia and Döbereiner, 1975). Epi- and endocardial hemorrhages, although nonspecific lesions, are present in most cases. In some experimental and spontaneous cases, gross lesions may be absent or difficult to detect. In horses there is edema of the gastric wall, hyperemia of the mucosae of the glandular part of the stomach, duodenum, jejunum, cecum and large colon and distention of the distal third of jejunum, cecum and large colon by large amounts of fluid. The mucosae of the large colon and cecum are spotted by petechiae and covered by pseudomembranous fibrinous exudate (Costa *et al.*, 1995).

Microscopic lesions consist of degenerative and necrotic changes in the epithelial lining of the forestomachs (mainly rumen and reticulum) and in the lymphoid tissue. The early lesion in the stratified epithelim of the forestomachs is acute cellular swelling that progresses to ballooning degeneration, vesicular change and necrosis. Transepithelial migration of leukocytes is frequent. The degenerative and necrotic epithelial changes induce detachment of the epithelial lining from the lamina propria with cleft formation (Tokarnia and Döbereiner, 1975; Barros 1993). In horses the main lesions are mucosal necrosis and submucosal edema in the cecum, large colon and, to a lesser degree, in the stomach (Costa *et al.*, 1995). In spontaneous and experimentally induced poisoning in cattle, necrosis of the lymphoid tissue has been observed (Barros, 1993; Varaschin *et al.*, 1997), affecting both lymph nodes and lymphoid aggregates. In lymph nodes the necrotic changes are located in lymph follicles, suggesting that B-lymphocytes are involved (Varaschin *et al.*, 1997).

Dilatation of the space of Disse in the liver and splenic congestion are frequent findings. Tokarnia and Döbereiner (1975) saw occasional neutrophilic infiltrates and hemorrhages in the intestinal lamina propria. Perivascular hemorrhages inconsistently seen in the brain are probably either agonal or artifactual.

Toxic Principle

All parts of *B. coridifolia* are toxic. Flowers and seeds contain the highest concentrations of the toxic principles (Jarvis *et al.*, 1988), which are macrocyclic trichothecenes: roridin A and E, miotoxin A, B, C and D, miophitocen A and B and verrucarol. These are produced by soil fungi of the genus *Myrothecium*, mainly *M. roridum* and *M. verrucaria,* growing in the soil near the roots of *B. coridifolia*, and are later absorbed and translocated by the plant (Busam and Habermehl, 1982; Habermehl *et al.*, 1985). Experiments with *B. coridifolia* in rabbits (Rodrigues and Tokarnia, 1995) and in cattle (Varaschin *et al.*, 1997) suggested that female specimens of the plant were by far the more toxic, and this has been confirmed by monthly chemical analysis of female and male plants (Jarvis *et al.*, 1996).

References

Barros, C.S.L. (1993) Intoxicações por plantas que afetam o tubo digestivo. Intoxicação por *Baccharis coridifolia*. In: Riet-Correa, F., Méndez, M.C. and Schild, A.L. (eds), *Intoxicações Por Plantas e Micotoxicoses em Animais Domésticos*. Editorial Hemisferio Sur, Montevideo, pp. 159-169.

Busam, L. and Habermehl, G.G. (1982) Accumulation of mycotoxins by *Baccharis coridifolia*: A reason for livestock poisoning. *Natturwissenschaften* 69, 391-393.

Costa, E.R., Costa, J.N., Armién, A.G., Barbosa, G.D. and Peixoto, P.V. (1995) Intoxicação experimental por *Baccharis coridifolia* (Compositae) em equinos. *Pesquisa Veterinária Brasileira* 15, 19-26.

Döbereiner, J., Resende, A.M.L. and Tokarnia, C.H. (1976) Intoxicação experimental por *Baccharis coridifolia* em coelhos. *Pesquisa Agropecuária Brasileira Série Veterinária* 11, 27-35.

Habermehl, G.G., Busam, L., Heydel, P., Mebs, D., Tokarnia, C.H., Döbereiner, J. and Spaul, M. (1985) Macrocyclic trichothecenes: causes of livestock poisoning by the Brazilian plant *Baccharis coridifolia*. *Toxicon* 23, 731-745.

Jarvis, B.B., Midiwo, J.O., Bean, G.A., Abdoul Nasr, M.B. and Barros, C.S. (1988) The mystery of trichothecene antibiotics in *Baccharis* species. *Journal of Natural Products* 4, 736-744.

Jarvis, B.B., Mokhari-Rejali, N., Schenkel, E.P., Barros, C.S. and Matzembacher, N.I. (1991) Trichothecenes mycotoxins from Brazilian *Baccharis* species. *Phytochemistry* 30, 789-797.

Jarvis, B.B., Wang, S., Cox, C., Varaschin, M.S. and Barros, C.S. (1996) Brazilian *Baccharis* toxins: livestock poisoning and isolation of macrocyclic trichothecenes glucosides. *Natural Toxins* 4, 58-61.

Rodrigues, R.L. and Tokarnia, C.H. (1995) Fatores que influenciam a toxidez de *Baccharis coridifolia* (Compositae): um estudo experimental em coelhos. *Pesquisa Veterinária Brasiliera* 15, 51-69.

Tokarnia, C.H. and Döbereiner, J. (1975) Intoxicação experimental em bovinos por "mio-mio," *Baccharis coridifolia*. *Pesquisa Veterinaria Brasiliera Série Veterinária* 10, 79-97.

Tokarnia, C.H. and Döbereiner, J. (1976) Intoxicação experimental em ovinos por "mio-mio," *Baccharis coridifolia*. *Pesquisa Veterinária Brasiliera Série Veterinária* 11, 19-26.

Varaschin, M.S., Barros, C.S.L. and Jarvis, B.B. (1998) Intoxicação experimental por *Baccharis coridifolia* (Compositae) em bovinos. *Pesquisa Veterinária Brasiliera* (in press).

Chapter 112

Narthecium ossifragum Associated Nephrotoxicity in Ruminants

A. Flåøyen

National Veterinary Institute, PO Box 8156, Dep., N-0033 Oslo, Norway /
Norwegian College of Veterinary Medicine, PO Box 8146 Dep., Oslo, Norway

Introduction

During the summer of 1992, renal failure was diagnosed in 232 grazing cattle in 85 herds on the West Coast of Norway (Flåøyen *et al.*, 1995a). The salient clinical signs were depression, anorexia and melena or fresh blood in the feces. Diarrhea was also commonly observed. Serum concentrations of creatinine, urea, magnesium (Mg) and phosphorus were above normal. The activities of glutamate dehydrogenase (GLDH), aspartate aminotransferase (AST) and creatine kinase were also high. Post mortem examinations consistently revealed renal tubular necrosis. In some cases, liver necrosis was present, as were erosions at the base of the tongue, in the esophagus and in the jejunum and colon. A similar disease caused by ingestion of the plant *Narthecium ossifragum* was reported in Northern Ireland in 1989 (Malone *et al.*, 1992). In the 1992 outbreak in Norway, *N. ossifragum* was found in all pastures in which it was specifically sought (Flåøyen *et al.*, 1995a).

Narthecium asiaticum, a plant closely related to *N. ossifragum*, reportedly causes a similar disease in cattle in Japan (Suzuki *et al.*, 1985). Steroidal saponins have been suggested to be the nephrotoxic principle of the plant (Kobayashi *et al.*,1993; Inoue *et al.*, 1995), but the disease had not been reproduced by dosing saponins isolated from *N. asiaticum* to cattle. Steroidal saponins of the same type isolated from *N. ossifragum* and eight other plants are implicated in hepatogenous photosensitization of sheep (Flåøyen, 1996), but none of them has been reported to cause kidney lesions of the type seen in *N. asiaticum* poisoning.

Narthecium ossifragum is a loosely to densely clonal, perennial herb, 5-40cm tall, with a creeping rhizome. It belongs to the Lileacea, and it occurs on oligotrophic, mesotrophic and eutrophic peat deposits in Scandinavia, the British Isles, the Netherlands, Spain and eastern Portugal (Summerfield, 1974). In Norway the plant is found on the western coast as far north as Tromas County, and in Eastern Norway as far east as Ringerike. It is common on bogs and heathery moors which are low in calcium (Ca) (Fægri, 1960; Høeg, 1974).

Feeding Experiments

Calves and lambs were fed a mixture of bog plants containing 15g (wwt) *N. ossifragum*/kg/d (Flåøyen *et al.*, 1995b; Flåøyen *et al.*, 1995c). The lambs ate the *Narthecium*-containing feed for 10d, but the calves had reduced appetite for it on the second day of feeding and refused it the following days (Flåøyen *et al.*, 1995b).

Both calves and lambs were susceptible to the nephrotoxin in *N. ossifragum*, but the calves were more so. In the calves, serum creatinine concentrations were above normal as early as 1d after feeding started and reached their maximum 8d later. The serum creatinine concentrations in the lambs were significantly increased on days 3, 4 and 5 and reached maximum on day 4. In the same period, serum Mg concentrations were increased both in the calves and the lambs, while increased serum urea and decreased serum Ca concentrations occurred only in the calves.

Histopathological examination of the kidneys of animals with increased serum creatinine and urea concentrations revealed tubular epithelial cell degeneration and necrosis, with changes found mainly in the proximal convoluted tubules, in the renal medulla and, to a lesser degree, in the proximal convoluted tubules. All the calves, but none of the lambs, had increased activities of AST, GLDH and γ-glutamyl transferase, indicating liver dysfunction, but signs of liver damage were not seen on histopathological examination.

Interestingly, both the calves and the lambs developed strategies to avoid further poisoning after being fed the first doses of *N. ossifragum*-containing feed. The calves developed aversion to the feed, whereas the lambs developed resistance to the toxin. The increased creatinine concentrations in serum from 1-4d after dosing commenced suggests that, although the first doses ingested were toxic, later doses became non-toxic. Lambs grazing on *N. ossifragum* are known to develop hepatogenous photosensitization, but natural cases of nephrotoxicity due to grazing on the plant have not been reported. The development of resistance to the nephrotoxin seen in the lambs fed *N. ossifragum* may explain why natural cases of the diseases have not been diagnosed.

The concentration of the nephrotoxin is probably much higher in the flower stems of *N. ossifragum* than in the leaves (Flåøyen *et al.*, 1997a). One calf that was dosed intraruminally with *N. ossifragum*-flower stems at 25g (wwt)/kg was terminated for humane reasons 7d after dosing. At that time, the calf was severely depressed and anorectic. Additionally, rumination had ceased and melena was observed. After dosing, serum creatinine increased from 100μM/L to 1,400μM/L, serum urea from <5mM/L to 75mM/L and serum Ca dropped from >2.3mM/L to 0.8mM/L. Histopathological examination of the kidneys revealed severe tubular necrosis and degeneration. Minimal reaction was seen in another calf that was dosed with the same amount of *N. ossifragum* leaves.

Results from other dosing experiments have shown that the nephrotoxin is water soluble (Flåøyen *et al.*, 1997a, Flåøyen *et al.*, 1997b). Kidney damage was observed in a calf dosed with an aqueous extract of *N. ossifragum* at 30g/kg. Signs of liver damage but no signs of kidney damage were observed in another calf that was dosed with the insoluble plant residue. The results indicate that the nephrotoxin is water-soluble, while the hepatotoxin is water insoluble.

Goats have experimentally been found to be susceptible to the nephrotoxin (Flåøyen *et al.*, 1997b). They are apparently almost as susceptible to the nephrotoxin as calves, but less susceptible to the liver toxin.

Moose, red deer and reindeer are susceptible to the *N. ossifragum* nephrotoxin. Moose and red deer seem to be as susceptible as goats are, whereas reindeer are less susceptible (Flåøyen *et al.*, unpublished). The histopathological changes in the kidneys of moose, red deer and reindeer dosed with the nephrotoxin are similar to the changes of the calves, goats and lambs. Fallow deer seem not to be susceptible to the toxin. The species difference in susceptibility to the toxin is probably due to differences in microsomal or cytosolic enzyme activities in the liver/kidney tissues or to differences in the microbial metabolism in the rumen.

Conclusions

Feeding of cattle with subclinical doses of *N. ossifragum* before they are put on pasture may result in aversion to the plant and prevent further intake. Calves and lambs probably have a similar metabolism of the toxin. Induction of the microsomal or cytosolic enzymes involved in the metabolism of the toxin or manipulation of the rumen microorganisms may also prevent the disease.

Acknowledgments

Thanks to Bjørn Bratberg, Arne Frøslie, Hallstein Grønstøl, Kjell Handeland, Wenche Langseth, Peter G. Mantle, Thorbjørn Refsum, Kathrine A. Ryeng and Anita von Krogh, who all have been important contributors to the works referred to in this chapter.

References

Fægri, K. (1960) *Narthecium ossifragum.* In: Fægri, K., Gjærevoll, O. and Lid, J. (eds), *Maps of Distribution of Norwegian Plants.* Oslo University Press, Oslo, p. 95.
Flåøyen, A. (1996) Do steroidal saponins have a role in hepatogenous photosensitization diseases of sheep. *Advances in Experimental Medicine and Biology* 405, 395-404.

Flåøyen, A., Binde, M., Bratberg, B., Djønne, B., Fjølstad, M., Grønstøl, H., Hassan, H., Mantle, P.G., Landsverk, T., Schönheit, J. and Tønnesen, M.H. (1995a) Nephrotoxicity of *Narthecium ossifragum* in cattle in Norway. *Veterinary Record* 137, 259-263.

Flåøyen, A., Bratberg, B., Frøslie, A. and Grønstøl, H. (1995b) Nephrotoxicity and hepatotoxicity in calves apparently caused by experimental feeding with *Narthecium ossifragum*. *Veterinary Research Communications* 19, 63-73.

Flåøyen, A., Bratberg, B. and Grønstøl, H. (1995c) Nephrotoxicity in lambs apparently caused by experimental feeding with *Narthecium ossifragum*. *Veterinary Research Communications* 19, 75-79.

Flåøyen, A., Bratberg, B., Frøslie, A., Grønstøl, H., Langseth, W., Mantle, P.G. and von Krogh, A. (1997a) Further studies on the presence, qualities and effects of the toxic principles from *Narthecium ossifragum* plants. *Veterinary Research Communications* 21, 137-148.

Flåøyen, A., Bratberg, B., Frøslie, A., Grønstøl, H., Langseth, W., Mantle, P.G. and von Krogh, A. (1997b) Nephrotoxicity in goats by dosing with water extract from *Narthecium ossifragum* plants. *Veterinary Research Communications* 21, 499-506.

Høeg, O.A. (1974) *Narthecium ossifragum* (L.) Huds., rome. In: Høeg, O.A. (ed), *Planter og tradisjon. Floraen i levende tale og tradisjon i Norge 1925-1973*. Universitetsforlaget, Oslo, pp. 464-466.

Inoue, T., Mimaki, Y., Sashida, Y. and Kobayashi, M. (1995) Structures of toxic steroidal saponins from *Narthecium asiaticum* Maxim. *Chemical Pharmaceutical Bulletin* 43, 1162-1166.

Kobayashi, M., Suzuki, K., Nagasawa, S. and Mimaki, Y. (1993) Purification of toxic saponins from *Narthecium asiaticum* Maxim. *Journal of Veterinary Medical Science* 55, 401-407.

Malone, F.E., Kennedy, S., Reilly, G.A.C. and Woods, F.M. (1992) Bog asphodel (*Narthecium ossifragum*) poisoning in cattle. *Veterinary Record* 131, 100-103.

Summerfield, R.J. (1974) *Narthecium ossifragum* (L.) Huds. *Journal of Ecology* 62, 325-339.

Suzuki, K., Kobayashi, M., Ito, A. and Nakgawa, M. (1985) *Narthecium asiaticum* Maxim. poisoning of grazing cattle: Observations on spontaneous and experimental cases. *Cornell Veterinarian* 75, 348-365.

Index

α-tumor necrosis factor, 330
β-carboline, 413
β-glucosidase, 283
γ-coniceine, 221, 304, 345, 347
γ-glutamyl-β-cyanoalanine, 369
2-n-pentyl-3,4,5,6-tetra-hydropyri-
           dine, 419
4-methoxypyridoxine, 453
5-hydroxy-2-n-pentyl-piperidine, 419

Abies spp., 343
    A. concolor, 343
    A. grandis, 343
    A. lasiocarpa, 343
Abrus precatorius, 14
Acacia spp., 351
Acetylanabasine, 224
Acetylanatabine, 224
Acetylcholine, 205, 400, 403
Acetylisocupressic acid, 340
Acetylnornicotine, 224
Achillea millefolium, 116
Achromotrichia, 115
Acremonium spp., 69, 465, 475
Actaee rubra, 15
Acyl transferase, 475
Aeschynomene spp., 2
Agropyron trichophorum, 116
Agrostis avenacea, 49, 165
Agrostis spp., 116
Albizia spp., 453
    A. tanganyicensis, 453
    A. versicolor, 453
Alfalfa, 116
Alkali disease, 380, 389
Allomatrine, 297
Alternaria spp., 465, 475
Alveld, 564
Amaranthus retroflexus, 567

American yew, 78
Aminopropanol, 240
Aminopyrine, 369
Ammi majus, 10
Ammodendrine, 304, 345, 347
Ammodendron spp., 345
Anabasine, 235, 304, 345, 347
Anacardium occidentale, 356
Anaerovibrio lipolytica, 74
Anagallis arvensis, 3
Anagyrine, 297, 304
Anguina spp., 165
Annual beard grass, 49
Annual ryegrass, 49, 201
Anthroquinone, 527
Antimethanogens, 213
Antinutrients, 106
Araucaria spp., 342
Arge pullata, 517
Arginine, 267
Arisaema triphyllum, 15
Aristida spp., 489
Aristolochia fangchi, 369
Arnica montana, 15
Artemisia absinthium, 363
Arthrinium, 154
Asclepias spp., 435
    A. fascicularis, 435
    A. incarnata, 435
    A. mexicana, 435
    A. pumila, 435
    A. subverticillata, 435
    A. verticillata, 435
Aspergillus spp., 154, 174, 397, 481
    A. flavus, 465
    A. parasiticus, 465
Astragalus spp., 19, 154, 243, 297,
    303, 428
    A. bisculcatus, 385
    A. canadensis, 243

A. cibarius, 243
A. lentiginosus, 20, 286
A. miser, 154, 239, 243
A. mollissimus, 20
A. pterocarpus, 243
A. pubentissimus, 20
Astrebla spp., 464, 483, 487, 497
A. elymoides, 489
A. pectinata, 489
A. squarrosa, 489
Atropa belladonna, 277
Avena sativa, 15
Avocado, 131

Baccharis spp., 292, 517, 569
Baptifoline, 297
Baptisia australis, 297
Beard grass, 165
Bentgrass, 116
Binns, 6
Black pearl, 369
Black soil blindness, 464, 483, 487, 497
Blackbrush, 351
Blanche fleur, 369
Blastogenesis, 260
Blind staggers, 380, 389
Blowing disease, 271
Blown grass, 49, 165
Blue grass, 483
Blue wild indigo, 297
Bluebell, 279
Bog asphodel, 564, 566
Bovine spongiform encephalopathy, 282
Bowenia serrulata, 447
Brachiaria brizantha, 175
Bracken fern, 255, 329
Bromegrass, 116
Bromus inermis, 116
Broom snakeweeds, 307
Bufadienolide, 407
Bufotenine, 414, 415
Burning bush, 504

Butyrivibrio fibrisolvens, 213, 509

Caffeine, 368
Calamagrostis rubescens, 240
Calliandra calothyrsus, 509
Calystegia sylvatica, 277
Calystegine glycosidase, 276, 429
Campanula rotundifolia, 280
Canada Bluegrass, 116
Canaline, 267
Canavalia ensiliformis, 267
Canavanine, 267
Carcinogenicity, 126, 249
Cardamine hirsuta, 42
Cardenolide, 407, 437
Cardiac glycosides, 124
Carica spp., 345
Cashew, 356
Cassia spp., 345, 527
Castanospermum australe, 276
Ceramide synthase, 475
Chaparral, 369
Chenopodiaceae, 504
Chinaberry, 514
Chinese yew, 78
Cholangiohepatitis, 138
Chrysopogon fallax, 489
Chui fong, 369
Cicuta maculata, 368
Cinnamoyl, 437
Clasmadendrocytes, 481
Clavibacter spp., 49, 165, 179, 201, 276
Claviceps spp., 397, 465, 464, 493
C. paspali, 465
C. purpurea, 464, 493
Clavines, 493
Colchicum autumnale, 15
Collidium spp., 345
Comfrey, 368, 369
Conhydrine, 221, 420
Conidia, 493, 499
Conidiomata, 497
Conidiophores, 499

Coniine, 221, 304, 345, 347, 420
Conium spp., 10, 15, 223, 233, 267, 303, 345, 419
    C. maculatum, 10, 15, 223, 233, 367, 419
Convicine, 283, 369
Convolvulus avensis, 279
Corallocytostroma spp., 464, 483, 489, 497
    C. ornicopreoides, 483, 489, 497
Corals, 483, 497
Coronilla spp., 154
Corticosteroids, 369
Corydalis aurea, 221
Corynetoxins, 49, 165, 179, 201
Cottonseed, 97, 149
Cotyledon spp., 407
Cotyledoside, 409
Coyotillo, 440
Crazy cow syndrome, 276
Crofton weed, 271
Crotalaria spp., 55, 305, 312, 369, 522, 537
    C. retusa, 305
    C. spectabilis, 312, 305, 315
Cupressus spp., 342, 343
    C. macrocarpa, 343
    C. sempervirens, 342, 343
Cyanide, 416
Cyanogenic glycosides, 369
Cycas spp., 4, 447
    C. revoluta, 4, 447
Cyclodextrin, 180
Cyclopamine, 303
Cyclopiazonic acid, 466
Cycloposine, 303
Cymarose, 437
Cynanchum africanum, 436
Cynoglossum spp., 305, 537
    C. officinale, 305, 369
Cyprinus carpio, 555

Dactylis glomerata, 116, 282
Dallisgrass, 465
Datura spp., 10, 233

D. metaloides, 233
D. sanguinea, 10
D. wrightii, 233
Deacetylnudicauline, 402
Death camas, 224
Dehydromonocrotaline, 538
Dehydroretronecine, 249
Dehydrosenecionine, 538
Delphinium spp., 14, 23, 205, 243, 402
    D. barbeyi, 227, 233, 243, 402
    D. glaucescens, 243, 402
    D. glaucum, 243, 402
    D. occidentale, 23, 243, 402
Deltaline, 235, 402
Deoxynivalenol, 465
Derris, 282
Desulfotomaculum ruminis, 75
Desulfovibrio spp., 74, 75
Diaporthe toxica, 62, 191, 196, 459
Dichanthium spp., 483, 489
Dichroa spp., 345
Digitalis spp., 368
Dihydroxypyrrolizine, 559
Dimethyltyramine, 414
Diplodia maydis, 464
Diterpene, 205, 309, 339, 397, 556
Dollahite, 17
Dopaminergic, 493
Douglas Fir, 239
Duboisia spp., 345

Ehrharta calycina, 413
ELEM, 474, 479
Eleutherococcus senticosus, 369
Elymus junceus, 116
English yew, 233
Enteque seco, 323
Ephedra spp., 368
Epilobium spp., 116
Equine leukoencephalomalacia, 474
Equisetum arvense, 15
Eragrostis xerophila, 120
Erechtites hieracifolia, 4
Ergotamine, 493

Ergotism, 464, 493
Ergotoxine, 493
Ergovaline, 45, 465
Eryngium spp., 291, 517
Espichamento, 323
Ethnobotany, 378
Eucalyptus melanophloia, 520
Eucalyptus oil, 368
Eupatorium spp., 220, 271
        E. adenophorum, 271
        E. riparium, 271
        E. rugosum, 220
Euphorbia pulcherrima, 368
European yew, 78

Facial eczema, 465, 469, 543
Favism, 369
Fennel, 159
Ferula communis, 159
Ferulosis, 159
Fescue foot, 464
Festuca spp., 15, 116. 239, 282
        F. arundinacea,15, 282
        F. rubra, 116
        F. scabrella, 239
Fibrobacter succinogenes, 512
Flinders grass, 483
Floodplain staggers, 165
Fumonisin, 466, 474, 482
Fusarium spp., 71, 465, 466 ,474,
           476, 479, 482
        F. babinda, 475
        F. beomiforme, 476
        F. graminearum, 465, 475
        F. moniliforme, 466, 474,
           479
        F. napiforme, 476
        F. nygamai, 475
        F. proliferatum, 475, 482
        F. solani, 71
        F. subglutinans, 475, 482

Gallic acid, 549
Gallotannin, 549
Genista spp., 345

Ginseng, 368
Glycosidase, 276
Glycosyltransferase, 276
Gossypol, 97, 149
Gramine, 413-415
Grewia flava, 139
Guajillo, 351
Gutierrezia spp., 309

Halimium brasiliense, 1
Halogeton glomeratus, 169
Hard lug, 564
Harmane, 417
Head greet, 564
Hedeoma pulegoides, 367
Helenium hoopseii, 15
Heliotropium spp., 55, 369, 522
        Heliotropium europeaum,
           55
Hemlock, 419
Henbane, 10
Heracleum mantegazianum, 12
Herbal remedies, 367
Hogweed, 12
Homeria spp., 407
Hoof dystrophy, 389
Hoplomachus affiguratus, 23
Hordenine, 351, 414, 415
Horsfieldia superba, 416
Horsfiline, 416
Houndstongue, 305
Hyacinthoides non-scripta, 279
Hydrangea spp., 279, 345
        H. orientalis, 279
Hydrochlorothiazide, 369
Hyoscyamine, 235, 368
Hyoscyamus niger, 15
Hypothyroidism, 335
Hystrine, 347

Ilex paraguariensis, 368
Imbricatoloic, 340
Indigofera spp., 154
Indole diterpenoid, 397
Indolizidine, 428

Indomethacin, 369
Interleukins, 289
Ipomoea spp., 276, 429
   I. carnea, 429
   I. crassicaulis, 430
   I. fistulosa, 430
   I. polpha, 432
Iseilema spp., 483, 489
Isocupressic acid, 308, 339, 340
Isolinamarin, 369
Isotropis spp., 567
Iva spp., 297, 298
   I. angustifolia, 298
   I. annua, 297
   I. ciliata, 297
Ivermectin, 423

Jack bean, 267
Jacobine, 185
Japanese yew, 78
Jatropha curcas, 554
Jimsonweed, 233
Juniperus spp., 342, 343
   J. communis, 342
   J. monosperma, 343
   J. osteosperma, 343
   J. scommunis, 343
   J. scopulorum, 342, 343

Kalanchoe spp., 407
   K. lanceolata, 410
Kalmia latifolia, 223
Karwinol A, 440
Karwinskia humboldtiana, 440
Kentucky bluegrass, 116
Kochia scoparia, 504
Krimpsiekte, 408

Labdane resin acids, 307, 308, 340
Lameness, 115, 382
Lanceotoxin B, 410
Lantana camara, 4
Larkspur, 29, 205, 243, 402
Larrea tridentata, 369

Lavendula stoechas, 91
Lens culinaris, 369
Leucaena leucocephala, 213
Leukocytosis, 66
Leukoencephalomalacia, 465, 479
Limberleg, 351
Limonoid, 515
Linum usitatissimum, 15
Liparia spp., 345
Liseola, 474
Lobelia spp., 345
Locoism, 276, 285, 304
Locoweed, 285, 297, 428
Lolitrem B, 45, 69, 397
Lolium spp. 49, 165
   L. perenne, 45, 69, 282, 465
   L. rigidum, 49, 165
Lonchocarpus, 282
Lophyrotoma interrupta, 517
Lophyrotomin, 518, 520
Lotus spp. 512
   Lotus corniculatus, 512
Lupin, 143
Lupinosis, 62, 191, 196, 459
Lupins, 459
Lupinus spp., 144, 191, 303, 345, 349
   L. angustifolius, 144, 191
   L. arbustus, 345
   L. cosentinii, 459
   L. formosus, 345
Lycoctonine, 235
Lysosomal storage diseases, 428

Má huáng, 368
Macrozamia spp., 447
Magnolia officinalis, 369
Maldronksiekte, 276
Manduca sexa, 268
Manganese, 115
Mannosidase, 19, 285, 428
Marsh-elder, 297
Mata porco, 291, 518
Matrine, 297
Medicago sativa, 116
Mefenamic, 369
Megasphaera elsdenii, 76

Melaleuca alternifolia, 367
Melia spp., 514
    M. azedarach, 514
Meliatoxins, 515
Menispermum canadense, 15
Mentha puleguim, 367
Mercurialis annua, 15
Methylanabasine, 224
Methyllycaconitine, 205, 402
Methyltyramine, 414
Metsulfuron, 310
Mexican fireweed, 504
Micropolysporum faeni, 71
Milkvetches, 154, 243
Mimosa, 112
Mimosine, 213
Mio-mio, 569
Miophitocen, 571
Miotoxin, 571
Miserotoxin, 239
Mistletoe, 368
Mitchell grass, 464, 483, 487, 497
Monensin, 81
Monocrotaline, 249, 312, 522
Monterey cypress, 343
Moraea, 407
Morgania floribunda, 124
Mucor spp., 174
Mycotoxicosis, 464, 487, 497
Myrothecium spp., 571
    M. verrucaria, 571
    M. roridum, 571

N-acetylhystrine, 304
N-methyl coniine, 304
N-methyl-β-phenethylamine, 351
N-methyl-tyramine, 351, 415
Narthecium spp., 564, 567, 573
    N. asiaticum, 567, 573
    N. ossifragum, 564, 573
Necine, 185
Nenta, 408
Neocallimastix frontalis, 509
Neotyphodium spp., 45, 397
Neriine, 215
Nerium oleander, 5, 131, 215, 368
Nicotiana spp., 233, 345

    N. glauca, 233, 345
    N. tabacum, 15, 223
Nicotine, 224, 234
Nicotinic acid, 403
Nierembergia hippomanica, 2
Nitrate, 74, 154, 335
Nitropropanol, 74, 154, 239, 243
Nitropropionic acid, 74, 239, 243
Nordihydroguaretic acid, 369
Norditerpenoid alkaloids, 243, 402
Norharmane, 417
Nortropane, 276
Nortropine, 235
Numinbah horse sickness, 271

Oak, 549, 567
Ochratoxin A, 466
Oleander, 131
Oleandrin, 215
Oleandrigen, 235
Onobrychis viciifolia, 116, 511
Orchardgrass, 116
Organophosphate, 416
Oxalate, 169, 174, 504
Oxalis cernua, 169
Oxalobacter formigenes, 169
Oxymatrine, 297
Oxytropis spp., 19, 304, 297, 428
    O. sericea, 20

Panicum spp., 470
    P. dichotomiflorum, 470
Papaver somniferum, 11
Paraguay tea, 368
Paraperreyia dorsuaria, 291
Paspalum spp., 465
    P. dilatatum, 465
Patulin, 365, 466
Paxilline, 397
Peastruck, 276
Penicillic acid, 466
Penicillium spp., 154, 174, 397, 466, 481
    P. crustosum, 466
    P. oxalicum, 174
Penitrem, 397, 466

Pennistum merckerii, 176
Peptococcus heliotrinereducans, 213
Peptostreptococcus, 213
Perennial ryegrass, 45, 69
Periploca sepium, 369
Perreyia spp., 291, 517
    P. flavipes, 291, 517
    P. lepida, 291
Persea americana, 86, 131
Persin, 86, 132
Phalaris spp., 354, 413, 416
    P. aquatica, 413, 416
    P. cerulescens, 413
Phalaris staggers, 354
Phleum pratense, 116
Phomopsis spp., 62, 191,196, 465,
    459
    P. leptostromiformis, 465
Phoradendron flavescens, 368
Phorbol esters, 556
Photosensitization, 465, 504, 543,
    573
Phylloerythrin, 470
Physic nut, 554
Phytomenadione, 160
Picea engelmannii, 343
Pig killer, 291
Pinegrass, 240
Pinus spp., 342, 345
    P. aristata, 343
    P. arizonica, 343
    P. contorta, 116, 342, 343
    P. echinata, 343
    P. elliotii, 343
    P. flexilis, 343
    P. jeffreyi, 343
    P. monophylla, 343
    P. montezumae, 343
    P. palustris, 343
    P. ponderosa, 307, 339, 343
    P. taeda, 343
Piperidine alkaloids, 304, 345, 419
Pithomyces chartarum, 465, 469, 543
Pithomycotoxicosis, 465
Plakkies, 407
Plochteach, 564
Plumeria rubra, 317
Plumieride, 319

Poa spp., 116
    P. compressa, 116
    P. pratensis, 116
Poinsettia, 368
Poison hemlock, 223, 233, 367, 419
Poison ivy, 357
Poison pea, 428
Pokeweed, 285
Polioencephalomalacia, 385
Polyethylene glycol, 101, 113
Polypogon monspeliensis, 49, 165
Ponderosa pine, 307, 339
Precocenes, 220
Pregnane, 437
Prevotella bryantii, 510
Proanthocyanidins, 112, 509
Prosopis spp., 345
Prototheca spp., 373
Protothecosis, 373
Pseudotsuga menziesii, 239, 343
Ptaquiloside, 255, 329
Pteridum spp., 255, 329
    P. aquilirium, 14
Punica spp., 345
Purkinje cells, 71, 432
Purple mint, 71
Putrescine, 55
Pyridoxine, 453
Pyrogallol, 549
Pyrroles, 531, 559
Pyrrolizidine alkaloid, 55, 185, 249,
    304, 312, 369,
    552, 531, 537,
    559

Quercus spp., 550, 567
    Q. douglasii, 550
Quinolizidine alkaloids, 297, 304

Ramguard®, 471
Ranunculus spp., 15, 42
    R. abortivus, 15
Red fescue, 116
Red grass, 139
Red lentils, 369
Resorcinol, 549

Retrorsine, 234
Rhamnose, 437
Rhizopus spp., 174
Riddell's groundsel, 305
Roebourne plains grass, 120
Romerillo, 569
Roquefortine, 466
Roridin, 571
Rosyntjiebos, 139
Rumex spp., 169
Russian wild rye, 116
Ryegrass, 165, 179, 470, 493
Ryegrass staggers, 45, 69

Sainfoin, 116
Salmonella typhimurium, 126
Sandplain lupins, 459
Saponins, 309, 505
Saut, 564
Sawfly, 517
Scilla, 279
Sclerotia, 499
Sclerotium, 493
Scopolamine, 235, 368
Sedum spp., 345
Selenium, 74, 260, 389
Selenocystine, 261, 381
Selenomethionine, 260, 380, 381, 390
Selenomonas ruminantium, 74, 511
Selenosis, 260, 380, 389
Semecarpus australiensis, 356
Sendaverine, 221
Senecio spp., 40, 55, 185, 292, 305,
            312, 517, 522,
            537, 531, 566
    S. aquaticus, 40, 566
    S. crysocoma, 531
    S. jacobaea, 40, 185, 305,
        312
    S. lautus, 55
    S. longilobus, 305
    S. riddellii, 305
Senecionine, 57, 534
Seneciphylline, 185
Senna spp., 527, 529
    S. obtusifolia, 529
    S. occidentalis, 527, 529

Serotonergic, 493
Shore pine, 116
Shrubby morning glory, 430
Silky sophora, 297
Silver grass, 413
Silverleaf nightshade, 424
Slangkop, 407
Solanidine, 234, 424
Solanine, 424
Solanum spp., 276, 279, 303, 323,
            424
    S. dimidiatum, 276
    S. dulcamara, 279
    S. eleagnifolium, 424
    S. kwebense, 276
    S. malacoxylon, 323
Sophocarpine, 297
Sophora spp., 297
    S. arizonicus, 297
    S. nuttallianna, 297
    S. sericea, 297
    S. stenophylla, 297
Sophoridine, 297
Sorghum spp., 15, 55
    S. halepense, 15
    S. vulgaris, 55
Spermidine, 55
Sphinganine, 475
Sphingosine, 475
Spinifex, 487
Sporidesmin, 465, 471, 543
Stemodia florulenta, 124
Stemodia kingii, 120
Stephania tetranda, 368
Steroidal glycosides, 505
Steroidal saponins, 573
Streptomyces spp., 179, 201, 276, 282
    S. lysosuperificus, 179, 201
Strychnine, 234, 416
Succinylisocupressic acid, 340
Swainsona spp., 276, 304, 428
Swainsonine, 19, 276, 285, 304, 428
Sweet potatoes, 71
Symphytum spp., 369

Tall larkspur, 23, 227, 233, 402
Tallebudgera horse disease, 271

Tannic acid, 101, 106, 549
Tannins, 101, 106, 111
Tansy ragwort, 185, 305
Taxine, 79
Taxus spp., 78, 131, 223, 224, 233
    T. baccata, 78, 131, 223, 233
    T. brevifolia, 78
    T. canadensis, 78
    T. chinensis, 78
    T. cuspidata, 78
Tea tree, 367
Themeda triandra, 139
Thevetia peruviana, 368
Thevetose, 437
Thiaminase, 255
Threadleaf groundsel, 305
Thyroxine, 334
Timothy, 116
Tobacco, 224, 367
Toxicodendron radicans, 357
Tree tobacco, 233
Tremetone, 220
Tribulus terrestris, 417
Trichothecenes, 465, 571
Triodia spp., 487
Tullidora, 440
Tulp poisoning, 407
Tunicaminyluracil, 51, 179, 201, 276
Tunicamycin, 50, 179, 201
Tussilago farfara, 369
Tussock grasses, 120
Tylecodon spp., 407
    T. grandiflorus, 409
    T. ventricosus, 408, 409
    T. wallichii, 408
Tyledoside D, 409
Tyramine, 351, 413, 415, 416

Urginea spp., 407
Ursolic acid, 91
Urushiol, 357

Veillonella alcalescens, 74
Veldt grass, 413
Venoocclusive, 249
Veratrum spp., 303
Verrucarol, 571
Vetch, 369
Vicia spp., 283, 369
    V. cracca, 283
    V. sativa, 369
Vicianin, 369
Vicine, 283, 369
Vitamin D, 323
Vitamin K, 159
Vomitoxin, 465
Vryburg hepatosis, 138
Vulpia spp., 413

Western yew, 78
Wheatgrass, 116
White cedar, 514
White snakeroot, 220
Wild rye, 116
Willowherb, 116
Wimmera ryegrass, 49
Withania spp., 345
Wolinella succinogenes, 74
Wormwood, 363

Xanthium strumarium, 3

Yarrow, 116
Yellowses, 564
Yew, 78, 131

Zamia spp., 447
Zamia staggers, 447
Zygadenine, 224
Zygadenus venenosus, 223